Biotechnology of Blood

BIOTECHNOLOGY SERIES

1. R. Saliwanchik — *Legal Protection for Microbiological and Genetic Engineering Inventions*

2. L. Vining (editor) — *Biochemistry and Genetic Regulation of Commercially Important Antibiotics*

3. K. Herrmann and R. Somerville (editors) — *Amino Acids: Biosynthesis and Genetic Regulation*

4. D. Wise (editor) — *Organic Chemicals from Biomass*

5. A. Laskin (editor) — *Enzymes and Immobilized Cells in Biotechnology*

6. A. Demain and N. Solomon (editors) — *Biology of Industrial Microorganisms*

7. Z. Vaněk and Z. Hošťálek (editors) — *Overproduction of Microbial Metabolites: Strain Improvement and Process Control Strategies*

8. W. Reznikoff and L. Gold (editors) — *Maximizing Gene Expression*

9. W. Thilly (editor) — *Mammalian Cell Technology*

10. R. Rodriguez and D. Denhardt (editors) — *Vectors: A Survey of Molecular Cloning Vectors and Their Uses*

11. S.-D. Kung and C. Arntzen (editors) — *Plant Biotechnology*

12. D. Wise (editor) — *Applied Biosensors*

13. P. Barr, A. Brake, and P. Valenzuela (editors) — *Yeast Genetic Engineering*

Biotechnology of Blood

Edited by

Jack Goldstein
Cell Biochemistry Laboratory
The New York Blood Center
New York, New York

Butterworth–Heinemann
Boston   London   Oxford   Singapore   Sydney   Toronto   Wellington

Editorial and production supervision by Science Tech Publishers, Madison, WI 53705.

Library of Congress Cataloging-in-Publication Data
Biotechnology of blood / edited by Jack Goldstein
   p. cm. — (Biotechnology series : 19)
Includes bibliographical references and index.
ISBN 0-7506-9120-4 (casebound : alk. paper)
1. Blood products—Biotechnology. I. Goldstein, Jack, 1930–
II. Series.   III. Series: Biotechnology (Reading, Mass.); 19.
   [DNLM: 1. Biotechnology. 2. Blood Preservation. 3. Blood
Substitutes. 4. Blood Transfusion.   W1 B1918M no. 19 / WH 460
B616]
TP248.65.B56B56 1991
615′.39—dc20
DNLM/DLC
for Library of Congress
                      91-4543
                      CIP

British Library Cataloguing in Publication Data
Goldstein, Jack
Biotechnology of blood.
   1. Humans. Blood. Use of Biotechnology.
I. Title II. Series
615.39

   ISBN 0-7506-9120-4

Butterworth–Heinemann
80 Montvale Avenue
Stoneham, MA 02180

10 9 8 7 6 5 4 3 2 1

Printed in the United States of America

John W. Adamson
Hematopoietic Growth Factors
    Laboratory
The New York Blood Center
New York, New York

William F. Bennett
Cardiovascular Research
    Department
Genentech, Inc.
South San Francisco, California

Alessandra Bini
Coagulation and Fibrinolysis
    Research Unit
Centro di Ricerche
    Farmacologiche e Biomediche
Consorzio Mario Negri Sud.
S. Maria Imbaro (Chieti), Italy

Birger Blombäck
Blood Coagulation Biochemistry
    Laboratory
The New York Blood Center
New York, New York, and
Karolinska Institutet
Stockholm, Sweden

Christopher B. Brown
Division of Hematology
University of Washington
Seattle, Washington

Thelma H. Carter
Hemazyme, Inc.
New York, New York

David Ciavarella
Hudson Valley Blood Services
Valhalla, New York

Désiré Collen
Center for Thrombosis and
    Vascular Research
University of Leuven
B-3000 Leuven, Belgium

Beth Goins
Center for Bio/Molecular Science
    and Engineering
Naval Research Laboratory
Washington, D.C.

Jack Goldstein
Cell Biochemistry Laboratory
The New York Blood Center
New York, New York

Bernard Horowitz
Blood Protein Biochemistry
    Laboratory
The New York Blood Center
New York, New York

Robert J. Kaufman
HemaGen/PFC
St. Louis, Missouri

Kenneth Kaushansky
Division of Hematology
University of Washington
Seattle, Washington

Bohdan J. Kudryk
Blood Coagulation Biochemistry
    Laboratory
The New York Blood Center
New York, New York

Kotoku Kurachi
Human Genetics Department
University of Michigan Medical
    School
Ann Arbor, Michigan

Leslie L. Lenny
Cell Biochemistry Laboratory
The New York Blood Center
New York, New York

Frances S. Ligler
Center for Bio/Molecular Science
    and Engineering
Naval Research Laboratory
Washington, D.C.

Gerald L. Moore
Blood Research Division
Letterman Army Institute of
    Research
Presidio of San Francisco,
    California

Stephen Petersdorf
Division of Hematology
Department of Medicine
University of Washington
Seattle, Washington

Scott F. Rosebrough
Radiology Department
S.U.N.Y. Health Services Center
Syracuse, New York

Alan S. Rudolph
Center for Bio/Molecular Science
    and Engineering
Naval Research Laboratory
Washington, D.C.

August J. Salvado
Division of Medicine
Walter Reed Army Institute of
    Research
Washington, D.C.

Thomas F. Schaible
Clinical Research
Centocor, Inc.
Malvern, Pennsylvania

Steven R. Snyder
Biochemistry Department
University of Iowa
Iowa City, Iowa

Linda Stehling
1032 Tramway Lane NE
Albuquerque, New Mexico

Jay Valinsky
Special Diagnostics Department
The New York Blood Center
New York, New York

Johan Vandersande
Hyland Division
Baxter Healthcare Corporation
Los Angeles, California

Joseph A. Walder
Biochemistry Department
University of Iowa
Iowa City, Iowa

Howard L. Zauder
Department of Anesthesiology
School of Medicine
University of New Mexico
Albuquerque, New Mexico

CONTENTS

ix

PART
I

Oxygen Delivery Systems

Biotechnology, Economics, and the Business of Blood

Thelma H. Carter

Transfusion of blood and blood constituents is an essential element in the practice of modern medicine and surgery. Providing blood for therapy is a complicated process and relies on the successful integration of the efforts of both not-for-profit and for-profit organizations. Worldwide, it has become accepted practice to obtain all red cells and components from volunteer donors. The collection and distribution of donated blood is handled by large, not-for-profit organizations like the Red Cross societies. Even though donors receive no payment for their blood, there are charges to patients for transfusions which reflect the costs of operating these eleemosynary agencies, and their costs for collecting and distributing blood. Plasma and plasma products, on the other hand, are derived from blood from paid donors, and the supply and distribution of plasma products is largely within the for-profit business sector. The supply and distribution of blood and plasma products is big business. In the United States alone, the cost of blood and blood products represents about 1% of all healthcare expenditures (OTA 1985). The value of all components at the supplier level has been estimated at between $900 million and $2 billion. In 1986, costs to patients of all

The author wishes to thank Adam Carter, who, like Gabriel García Márquez's Colonel Aureliano Buendia, is a model for tenacity, perseverance, and intellectual achievement.

3

components transfused was estimated to be between $4.5 and $7.2 billion (Zuck 1988). The Federal Register (May 30, 1989) lists revenue/budgets for blood, plasma, and tissue centers at $1.42 billion.

The dollar value of blood and blood products in the USA alone could account for the interest of businesses, consumers, and scientists in these products; however, in addition to the commercial value of blood, there is also a vast potential for the application of new technology. Because human donor blood is now the primary resource for blood components and products, it is well understood, by both the for-profit and not-for-profit sectors, that this human resource is not unlimited in quantity, nor uniformly of pristine quality. It has been commonplace for blood banks, physicians, and patients to experience shortages of red cells or of plasma components like antihemophilic clotting factor and other important blood products. Those of us who live in big cities have heard urgent appeals for blood on the radio, or seen newspaper ads seeking donors, particularly at holiday times, or when there are local catastrophes and emergencies. Further, it is well known that there can be significant health risks to recipients of human-derived blood and blood factors, AIDS and hepatitis being the most well-known examples of transfusion risk. Both the public and industry will look to the new capabilities of biotechnology to solve supply and safety problems.

It is very likely that the need for transfusion blood will increase as a consequence of improvements in medical care, increasing bravery on the part of surgeons (and their patients), and the way we live our longer and more adventuresome lives. In the USA, we have a growing and aging population; between 1960 and 1986 our population increased from 179,323,000 to 241,078,000, and the percentage of individuals 65 years of age or older went from 9.2 to 12.1% (U.S. Bureau of the Census 1988). The number of cardiovascular operations and procedures performed have increased from 1,740,000 in 1982 to 3,116,000 in 1987 (Division of Health Care Statistics 1988); between 1983 and 1987, liver transplant operations increased from 164 to 1,182, heart transplant operations from 172 to 1,512, and kidney transplant operations from 6,112 to 8,967 (Office of Organ Transplantation 1988). These are just a few dramatic examples of the increasingly complex surgical procedures now undertaken and now possible, because of advances in surgical techniques and pharmaceutical sciences. In the USA, severe injury is the leading cause of death up to age 44, and up to age 34, trauma kills more people than all other diseases combined (American College of Surgeons 1989). New modalities of treatment for cancer, AIDS, and many other diseases may have the side effect of producing severe anemia. Patients suffering trauma, undergoing complex surgical procedures, dialysis, or taking marrow-depressing medications are all potential transfusion recipients. Surveys conducted by the American Red Cross (ARC) reported that 17% of American adults have been transfused, and that approximately 2% of Americans were transfused in the year preceding the survey (1985/1986). The ARC has calculated that in every minute of every day, 23 units of whole blood or red cells (38 units of all components) are transfused (Cumming et al. 1987). The solution to the problems of shortages and to meeting the increased demands made on our blood supply will not rest only with better public relations campaigns to increase the numbers of donors, but also will require the more efficient utilization and amplification of available resources using innovations provided by biotechnology.

Many new products are in development to offer new approaches for solving the problems of insufficient supply and provide safety assurance, including: typeless red cells, supplemental oxygen transport materials, cell formation stimulants, and sensitive diagnostic screening tests. In addition, new technologies can now provide substances which, in the past, were so scarce and difficult to produce that they remained research curiosities. The ability to manufacture therapeutic quantities of these rare or unavailable serum components, such as hormones or growth and clotting factors, or equivalent products, better by design in bioengineered cells, than the natural components, can initiate completely new approaches to the amelioration of illness. The following overview of the blood business is presented in order to provide background of the industry and to point up problems and perhaps some solutions, actual or contemplated, that rely on the scientific advances made possible by biotechnology.

1.1 CELLULAR COMPONENTS AND BIOTECHNOLOGY

While the history of transfusion medicine extends over 300 years, the practicality of transfusion has been achieved only in the very recent past. In 1657, a man better known for his skills as an architect, Sir Christopher Wren, invented a prototype for the syringe and needle, a device that made it possible to inject fluid into veins. Experimentation with the injection of blood ranged from the use of lamb's blood to cure the insane to human transfusion to cure the effects of blood loss. The first successful human transfusion was performed in 1829, at Guy's Hospital, by James Blundell, an obstetrician, to save the life of a hemorrhaging postpartum patient. Unperceived incompatibilities made the transfusion procedure unreliable at best until after 1901, when Karl Landsteiner discovered the blood groups. The difficulties of preventing coagulation of blood, necessitating the use of "warm donors" in the bed next to the transfusion recipient, made the process unwieldy. The later invention of anticoagulants made transfusion practical, and, soon after World War II, increasing numbers of blood banks became established in order to supply blood (Greenwalt 1989). The transfusion of the cellular elements has been implemented not only by the understanding of biological compatibility but also by the refinement of devices and instrumentation that ranges from plastic blood bags to automatic machinery for testing donor/recipient compatability and the absence of infectious agents, to computer inventory management programs. New packaging and handling techniques, sterility safeguards, separation techniques, and storage technologies are outgrowths of advances in the biological sciences that impact on the ability to transfuse blood cells and components. These technologies also impact on the cost of blood transfusion. Anticipated advances in transfusion medicine, based on biotechnology, will transform the field in much less time than it took to establish transfusion of the cellular elements of blood as safe and practical.

1.1.1 Blood Collection and Distribution

It has been estimated that worldwide, more than 75 million units of blood are collected annually (Leikola 1988). In the United States, Japan, and other developed nations, blood donation is a praiseworthy civic responsibility, and virtually all red cell and component resources are obtained on an entirely voluntary (no fee to donors) basis. In the USA, most donated blood is collected by regional and community blood centers. Organizations collecting blood are organized under the auspices of the ARC, accounting for over half of the collections, the American Association of Blood Banks (AABB), collecting somewhat less than half the donated units, and the Council of Community Blood Centers (CCBC), individual, independent, regional centers, and smaller blood bank facilities at hospitals collect the rest. Many of the collecting organizations in the USA have overlapping memberships which make precise annual collection data difficult to obtain. The ARC does not make annual collection and transfusion data available; however, data from the annual reports of the AABB listing collections and transfusions at member facilities reflect the demands for transfusion blood. Table 1–1 shows this data for the years 1978–1988. Collections by AABB affiliates increased from 1978 through 1984 (from 3,288,000 to 7,550,389 units) and decreased somewhat in 1985, 1986, and 1987, concurrently with public awareness and misperceptions about the relationship between AIDS and donation, and the initiation of diagnostic testing for the disease. In 1988, actual donations drawn were the highest ever attained (8,464,267 units).

Blood collection occurs at large collecting facilities, substations of large collecting facilities, or in mobile vans that participate in blood drives, on-site at participating companies, educational institutions, and other places where donors are likely to be found. In the USA, the AABB distinguishes three types of facilities involved in the collection and transfusion of human blood: type A facilities, which collect and distribute blood; type B facilities, hospitals with large blood banking operations which collect more than 100 units of blood each year; and type C facilities, which collect less than 100 units per year (Table 1–2).

In Europe and Japan, similar and related organizations collect blood from unpaid donors. Japanese collection data are shown in Table 1–3 and are similar to U.S. data in that the numbers of units collected peaked about mid-decade (1985 in Japan, 1984 in the USA, but U.S. transfusions peaked

TABLE 1-1 Annual U.S. Blood Collection and Red Cell Consumption Data[1]

			Number of Units			
Year Drawn	Drawn	Transfused[2]	Difference Trans.-Col.	Outdated[2]	Outdated vs Drawn (%)[3]	Deferred
1978	3,288,000	4,396,000	1,108,000	428,000	13.0	
1979	4,351,000	5,194,000	843,000	460,000	10.6	
1980	5,404,689	6,330,000	935,311	513,000	9.5	
1981	5,660,345	7,258,044	1,597,690	363,794	6.7	
1982	6,253,802	8,012,147	1,758,345	389,916	6.2	
1983	6,328,927	7,483,843	1,154,921	400,835	6.3	
1984	7,550,389	8,355,240	804,851	443,109	5.9	
1985	7,222,840	8,560,087	1,337,238	403,447	5.6	722,669
1986	6,996,130	8,155,547	1,159,417	443,445	6.3	728,553
1987	6,562,000	8,065,576	1,503,576	1,012,313	15.4	784,310
1988	7,738,400	7,333,500	−404,900	571,344	7.4	725,867

[1] Data shown in the table are from AABB Annual Reports, 1979–1989. The AABB collects somewhat less than half of all the red cells collected in the USA. Starting with 1985, the number of units given in the second column designated as "Drawn" by the AABB does not include those units that were "Drawn and Deferred." Numbers of units actually collected were: 1985, 7,945,509; 1986, 7,724,682; 1987, 7,346,310, and 1988, 8,464,267. American Red Cross data from *Annual Blood Facts*, 1986/1987, (data collected for the period of one year, mostly running through the 1986 calendar year but including some data through June 1987) state that in the USA, of the 14.8 million donors who presented at collection facilities, 9% were deferred (AABB Annual Reports show 9.1% deferred in 1985, 9.4% deferred in 1986, 11.2% in 1987, and 8.6% in 1988) and 13.2 million units were successfully collected. ARC data (1986/1987) state total losses due to testing for infectious disease markers was 5.4% or about 700,000 units, AABB data were similar in 1985 and 1986 but increased in 1987.

[2] Only includes red cell units transfused; platelet and plasma products are not included.

[3] Number of units of red cells outdated divided by units collected × 100.

in 1985). The Japanese collections comprise units of 200 ml and 400 ml, and the numbers shown in Table 1-3 represent the total for donations of either size. The Japanese Red Cross collects blood at many centers and subfacilities across the nation and uses blood mobiles to recruit donors.* Many European countries collect more blood than can be used, and excess blood is sold to nations that do not have sufficient blood. The USA is a blood importer, and recent information indicates that almost 300,000 units are imported annually (Altman 1989).

Most donated blood units are divided into component parts for transfusion, and ARC data for 1986/87 state that, on average, 2.4 units of com-

---

* In 1984–1985, the Japanese Red Cross Society included 64 Blood Centers, 107 Sub-Blood Centers, 132 Collecting Centers/Rooms, and 359 Blood Mobiles operated by the society for the collection of blood. About 1,300 vehicles were used for transporting donors, for public relations, and for delivering blood and blood products to medical institutions. At the end of 1985 there were 82 doctors, 443 pharmacists, 594 laboratory technicians, 803 nurses, 482 practical nurses, 2,346 clerical staff, and 290 other workers, for a total of 5,040 workers in the Japanese Blood Program.

TABLE 1-2   Data on Facilities Collecting and Transfusing Red Blood Cells[1]

Year	Type A Facility			Type B Facility			Type C Facility	
	Number of Facilites Col/Tra[2]	% Collected	% Transfused	Number of Facilities Col/Tra[2]	% Collected	% Transfused	Number of Facilities Trans	% Transfused
1978	80/15	76.0	1.3	455/438	23.9	43.3	957	55.4
1979	100/17	79.6	2.0	488/484	20.2	39.3	1149	58.7
1980	116/17	83.0	1.5	478/480	16.9	37.4	1305	61.1
1981	113/17	81.6	2.1	456/457	18.3	35.0	1414	62.9
1982	126/13	84.3	1.6	455/465	15.6	33.4	1540	65.0
1983	111/10	84.5	1.6	333/394	15.4	33.3	1387	65.0
1984	121/18	77.6	1.4	439/449	22.3	34.2	1598	64.4
1985	108/15	85.9	1.7	436/439	14.0	33.3	1609	65.0
1986	111/15	86.4	5.6	379/389	13.5	20.9	1518	70.9
1987	108/104	85.0	5.7	374/393	15.0	22.1	1449	72.1
1988	115/120	87.8	1.7	333/449	11.3	37.0	1336	60.0

[1] Data are taken from AABB Annual Reports 1978–1989. There is no column for the number of type C hospitals which report collections because these hospitals collect less than 0.1% of the total amount collected annually (except for 1988 when collections rose to 0.9%). The actual number of type C facilities collecting blood declined from 171 hospitals in 1978, collecting 4,000 units, to 76 in 1987, collecting 9,915 units and deferring use of 3,634, for a total of 6,281 transfusable units. In 1988, 91 type C facilities collected 78,411 units and deferred use of 5,444, for a total of 72,967 transfusable units. These data show that A facilities are the major collectors and C facilities are the major users of red blood cell transfusions. Data from the ARC for one year during the period 1986–1987 state that 90% of the blood donated is collected by regional blood centers and 10% is collected by hospital blood banks. Note: Published AABB totals for 1986 and 1987 reported an increase in the number of A facilities transfusing blood or components.

[2] Number of facilities collecting/number of facilities transfusing.

TABLE 1–3 Data on Blood Collection and Distribution in Japan[1]

| | | | Number of Units Distributed | | | |
| | Number of Units Collected | | Whole Blood | | Red Cells | |
Year	200 ml	400 ml	200 ml	400 ml	200 ml	400 ml
1980	6,178,741					
1981	6,866,833					
1982	7,149,803					
1983	7,680,029					
1984	8,307,975		1,723,361		4,241,190	
1985	8,696,105		1,503,543		4,570,631	
1986	7,962,322	616,595	1,287,846	54,902	4,117,626	260,141
1987	7,112,514	1,048,917	1,205,502	105,926	3,550,964	496,083
1988	6,621,888	1,250,661	1,085,246	125,497	3,446,498	655,702

[1] Data from the Japanese Red Cross Society, *Review of Activities, 1984–1985*, and Dr. Takemitsu Hosoi, Japan Society of Blood Transfusions, personal communication. Between 1986 and December 1988, 31 of 19,330,000 donors were confirmed positive for anti-HIV-1 (0.00016% or 1.6 per million). ARC experience for the period between April, 1985 and December, 1987 was that 2,497 units of 12,864,511 donated were Western-blot positive (Cumming et al. 1989) (0.0194% or 194 per million for the entire period). In 1987 the number of positives found by ARC was 831 for about 6.5 million units tested (approximately 133 per million). In Japan in 1988, 229,886 liters of plasma were recovered from voluntary blood donations to Red Cross Blood Centers; 44,533 liters were collected from paid donors by firms; and 264,933 were imported for fractionation.

ponents were prepared per unit of collected blood (Cumming et al. 1987). Other recent data (Cumming et al. 1989) state that the average donated unit of whole blood is converted into 1.54 component units, exclusive of plasma components (Cumming et al. 1989). Cell component transfusions on which the AABB collects data include packed red cell units; frozen and thawed red cell units (for special antigen configurations); pediatric-size red cell units; washed, leucocyte-poor red cell units; platelets; granulocytes; and whole blood units.

Collections must provide enough blood so that transfusion units are available when needed, despite fluctuations in donation rates based on vagaries in the availability of the donating public, decreases in available units based on risk-reducing diagnostic tests, and the biological effects on red cells of aging, which result in outdating. The data in Table 1–1 show that until 1988, AABB collections did not provide enough blood to meet members' needs for red cell transfusions, and collections were augmented by non-AABB collectors. This discrepancy demonstrates that blood resources must be shared among collecting agencies, and those with collections in excess of their needs would ship blood to institutions with shortages. These arrangements may be coordinated by Red Cross or AABB national blood exchange offices, or by blood bankers contractual arrangements with suppliers, or ad hoc arrangements between facilities. Most blood moves "off record" and some individual units of blood move several times to avoid outdating and to meet needs (Gaul 1989). A specifically unknown but considerable portion of blood moves around the USA and internationally every year (OTA 1985).* Blood must be shipped by couriers that will rapidly move it to its destination and requires special handling to keep it cool and the packaging intact. Some shippers classify blood as a "dangerous material" and charge additional fees for special handling, which add to the considerable expense of shipping blood. Monitoring blood requirements at user sites and moving blood from site to site to match locally variable usage adds significantly to the cost and complexity of providing transfusion blood.

Even with the support of altruistic donors there are considerable costs associated with collecting and distributing this "free" resource. The collection process costs include the direct costs for maintaining the collection services, as well as advertising and promotional expenses for donor recruitment. For example, data made available by the Canadian Transfusion Service shows that for each dollar budgeted, $0.33 was spent for collection, $0.11 for donor recruitment, $0.07 for distribution, $0.24 for processing, $0.14 for administration of centers, $0.07 for the national Transfusion Service office, and $0.04 for a National Reference Laboratory (OTA 1985). Older data for the distribution of costs for collection and distribution for four medium and large blood service centers are remarkably similar: 33% of total costs are for collection (74% of that for labor and 14% for travel and vehicles), 10% for recruiting, 10% for distribution, 33% for processing and testing, and 14% for administration (Cumming et al. 1974). Across the USA, costs for collection vary, based on local variations in operating costs. A Pulitzer-Prize winning series of articles about the blood supply which appeared in the Philadelphia Inquirer (Gaul 1989) stated that, for ARC facilities, the national average ratio of unit collections to number of employees is 821, but some blood banks collect as many as 3,000 units per employee, while others collet 377 units per employee. (Similar data for the Japanese Red Cross indicate that in 1985 collections were 1,687 units per worker, including all employees.) It could be calculated, if one were to assume an average salary of $20,000 per blood collection employee, according to the collection data cited in the Inquirer article, that the national average for labor costs would be $24.36 per unit ($20,000 divided by 821) but in some areas it would be $6.66 per unit, and others $53.05. These figures do not consider that, on average, 1.54 units of components are made from a single collected unit (if all products were considered, labor costs would be lower for collection, but higher if production costs for components

---

* Available data from the AABB and the ARC indicate that in 1980 their national blood exchanges promoted the shipment of 650,696 units, 976,772 in 1983; the AABB alone in 1986 shipped 402,618 units. Estimates indicate that in 1980, national exchange shipments represented 3.8% of all shipments and 4.9% of all shipments in 1985. (From AABB Annual Report 1987 and OTA 1985.)

and products were included). Data from the Inquirer article indicate that blood units from the collection agency in the city in which the labor cost was $6.66 per unit, sold blood to hospitals for $33 per unit, and the range of prices for red cells around the country was between $33 to just under $80. It is of historic interest (and perhaps also depressing) to note that in fiscal 1972, the average price charged by the 56 ARC blood centers was $13.01 for whole blood ($11.92 for packed red cells, $10.99 for fresh frozen plasma, $11.62 for platelets, and $10.44 for AHF cryoprecipitate) with a range of prices for whole blood between $8.50 and $22.00 (Cumming et al. 1974). It has been estimated that the cost of collection to corporate sponsors of blood drives in terms of lost productivity and recruitment activities could be $20 per unit. Over a quarter of a billion dollars (assuming a $20 per unit cost) is conservatively estimated to be the cost to blood drive sponsors in business and industry annually (Zuck 1988).

Blood bank managers have found that in most circumstances the unit cost of shortages significantly exceeds the cost of outdated units of blood (Cohen and Pierskalla 1979; Cohen et al. 1983), and therefore they will attempt to minimize the cost of inventory shortages by overcollecting whenever possible. Excess collections above predicted needs are sold and shipped to areas experiencing shortages (as described above). Outdating can be the consequence of overcollecting or of the inability or inefficiency of matching available donor types to recipient needs. Development of new approaches to the storage of blood, like freeze-drying, could provide new ways to substantially ease the problems of shortages based on donor availability and outdating. To help achieve ideal inventory levels, in addition to overcollecting, blood bank managers have adopted management policies like "type and hold" systems (Huang et al. 1980) and maximum surgical blood-ordering schedules (Kuriyan and Kim 1989) to make the most efficient use of inventories. Such procedures determine the numbers of required transfusion units based on past experience with specific surgical procedures and reserve blood in accord with these predicted needs. Blood bank management practice attempts to influence the utilization of blood (without affecting patient safety or good medical practice) and to increase the availability of units of blood by reducing crossmatches and holding times for each transfusion unit (Kuriyan and Kim 1989). Units are not reserved for specific patients, so when a doctor calls for a transfusion unit, the blood bank crossmatches and rushes the unit to the patient quickly. While such blood bank management practices have been shown to reduce costs and outdating, the hospital blood bank needs to closely monitor all aspects of blood utilization. Disruption of the ideal operation of any one inventory control procedure can negate good effects realized by strenuous efforts to positively influence all the other utilization parameters (Cohen et al. 1983).

Materials and methodology for increasing the shelf life of blood have proven useful in decreasing outdating and thereby reducing shortage rates. The following equation developed to estimate optimal inventory levels (S)

by Cohen et al. (1983) relates transfusion to crossmatch ratio, shelf life, cross match release time, and mean daily demand for a blood type:

$$S = \frac{4.755 \times (d_M)^{0.6964} \times (p)^{0.1146} \times (L)^{0.1332}}{(D)^{0.0453}}$$

where (d_M) is the mean daily demand for a given blood type, p is the average ratio of transfusion to crossmatch, L is the maximum shelf life in days, and D is the crossmatch release time in days. Taken singly over the respective ranges of each variable, with the others held constant, the effect of p (crossmatch to transfusion ratio), D (crossmatch release time), or L (shelf life) on S is, at most, 6–8% (Cohen et al. 1983). Calculations based on such mathematical relationships show that additives, such as citrate-phosphate-dextrose or citrate-phosphate-dextrose-adenine, could permit reduction of shortage rates from 18–20% per year to 6–8% per year (Brodheim and Hirsch 1979) by increasing shelf life (L) from 21 days to 35 to 42 days. Indeed, these additives have been effective, but have not completely solved the problems of outdating (see Table 1–1).

Type-specific daily demand is the major determinant of the number of units of blood that must be available in the blood bank since, as shown in the equation above, the small values of 0.1146 for the power of p, 0.1332 for power of L, and 0.0453 for the power of D, cannot influence S as much as d_M (raised to the power of 0.6964 does) (Cohen et al. 1983). If all the other conditions which rely for change on modification of human behavior remain the same in the blood bank, i.e., the transfusion/crossmatch ratio; the length of time a unit is allowed to remain on the shelf; and the crossmatch release time, it is the need to supply patients with type-matched blood or with type O (universal donor) blood that drives collection or purchasing and shipping decisions in the blood bank.

Patients who make autologous blood donations prior to surgery provide most of their own units of appropriate type-specific blood and decrease the need for transfusion units from the general pool. New emphasis by blood collecting agencies and medical practitioners on autologous donation may reduce some of the inventory blood banks need in order to provide adequate amounts of appropriately typed blood. Autologous donation reduces or eliminates a patient's need for bank blood; one estimate was that predeposited blood could replace as much as 72% of an average patient's needed units and as much as 10% of all transfusions could be supplied by autologous units (Toy et al. 1987). If autologous donations can decrease the amount of blood needed by blood banks by only 1 or 2% they would be an important resource for the national blood supply (Surgenor 1987). Intraoperative salvage instrumentation can maximize the patient's ability for autologous donation (see Chapter 3). Hospitals must purchase the instruments required, and costs for plastic software (tubing, centrifuge cups, etc.) are about $150 per patient. These techniques for autologous donation may be useful for relieving some of the problems of blood shortage (Toy et al. 1987), even though they are not entirely based on new developments in biotechnology.

1.1.2 Blood Cells, Components, Supply, and Biotechnology

Biotechnology can positively influence blood availability and address inventory problems based on donor fluctuation or shortages of blood to meet type-specific daily demands by several new developments. Examples of new technologies that can impact on the amounts of blood that the blood bank manager will need to have on hand to meet daily demands, and which are discussed in this book, include enzymatic treatment of cells to eliminate ABO type specificity (Chapter 4), the development of blood substitutes (Chapters 5–7), and the bioengineering of cell production regulators such as erythropoietin (EPO) (Chapter 14), thrombopoietin, and the various colony-stimulating factors (Chapter 15). Other new technologies for safety testing, and the purification (Chapters 17 and 18) and preservation of blood (Chapter 2) will influence the availability of blood by decreasing the wastage of collected units. New technologies which influence inventory levels may be expected to have a more reliable, significant influence on blood bank management practice than attempts to influence the blood ordering habits of physicians and surgeons by management practices and ordering schedules.

The stimulation of a patient's blood-forming cells for autologous donation by EPO can relieve some of the demand on collections, and it can provide most, if not all, of the blood of the right type for specific patients. The use of EPO in patients planning surgery can allow for presurgical donation units to be increased by at least one unit over the amounts that can be given by unstimulated autologous donation. EPO stimulation of red cell production can actually provide 50% more red cells, using hematocrit as a measure, and may influence the need for transfusion (Goodnough et al. 1989). This procedure is not yet FDA approved, and criteria to identify those patients that will have the most benefit from the procedure need to be defined. EPO can have the effect of decreasing the need for bank blood if patients are able to plan elective surgical procedures and are well enough to travel to blood collection centers for presurgical donation. It can be expected to be an important choice in preoperative planning for well-motivated individuals. At this time, EPO is approved for use in patients with anemia associated with kidney failure and dialysis (Eschbach et al. 1987). It has been effective in reducing or eliminating blood transfusion in such patients. Other applications for EPO are under study. The use of EPO will be measured against the likelihood of patient compliance and the costs of providing blood collected by classical means compared with costs for these engineered factors. The effect of EPO on the blood supply overall remains to be demonstrated, and it may not globally resolve the inventory problems of quantity or type and outdating.

Where pretransfusion planning is not possible or does not provide adequate amounts of blood, the ability to convert non-O red cells enzymatically to type O (Goldstein 1989) can allow for the universal usage of any donated cells on a first-in, first-out basis. Most of the blood that outdates is either type A or B and not the more useful type O (Axelrod et al. 1987). Increased availability of universal donor type O blood would be of value in emergency situations where typing and crossmatching are impossible or difficult and blood bank supplies of the appropriate type are inadequate. Such converted blood can be frozen and stored and provide resources for the major suppliers as well as the small, noncollecting facilities (type C facilities described above) that are the major consumers of collected units. This technology could have a major overall effect on blood supply by reducing or even eliminating the need for overcollection, and reducing outdating based on mismatches between available blood type and needs. It could be particularly valuable for multiply transfused individuals with reactivities to lesser blood groups by eliminating the need for crossmatching for ABO compatibility in addition to the rare blood group (Issitt 1985).

At least 12 commercial organizations (Andrews 1990) and many research laboratories are working on the development of blood substitutes to address problems of adequate supply, as well as safety from infectious agents. Research on oxygenating solutions for use as blood substitutes has taken two paths: the development of fully synthetic chemicals such as perfluorocarbons and the adaptation of biological molecules such as hemoglobin. While perfluorochemicals have been shown to take up useful quantities of oxygen, the history of in vivo studies with these chemicals has been disappointing. Recent results addressing problems of concentration and stability hold promise (Kaufman 1988), and one such product has FDA approval for special application during balloon angioplasty procedures. Perfluorocarbons may be useful in other biotechnology applications, such as providing oxygen in cell culture systems (King et al. 1989).

Providing hemoglobin-based oxygen transport substances centers on finding and producing stable, nontoxic forms of the hemoglobin molecule and on having an adequate supply of hemoglobin. Free tetrameric hemoglobin will dissociate into dimers that are excreted rapidly and may produce toxicity and kidney injury. In addition, extracellular hemoglobin without its normal regulatory, 2,3-bisphosphoglycerate, will not unload its oxygen. New methods of chemical modification may provide a means to convert human hemoglobin (from blood or genetically engineered microorganisms) into useful, stable, and nontoxic molecular configurations, and the development of cross-linking methods (based on chemical changes) may provide commercially useful compounds (Snyder et al. 1987; Kiepert et al. 1989). The possibility of combining all the required modifications and producing properly configured molecules in microorganisms may seem a daunting challenge, but biotechnology may be able to provide such a useful molecule by fermentation. Ideally, this genetically engineered molecule will not require additional chemical modification, and the production of these molecules will be cost effective. Genetically engineered yeast have been shown to produce intracytoplasmic human hemoglobin correctly configured to have useful oxygen binding properties (Wagenbach et al. 1991).

In some studies, hemoglobin for blood substitute research has been derived from outdated human cells, but this resource is limited and not likely to be able to supply enough hemoglobin to meet a significant portion of the demand that can be anticipated for such a product. The challenge to genetic engineers to produce sufficient quantities of hemoglobin is formidable; if a single unit of blood contains 45 g of hemoglobin and costs $120, the hemoglobin in a unit of blood is purchased at less than $3 per g. To produce the amount of hemoglobin in 10 million units of blood, 450,000 kg of hemoglobin would have to be manufactured by engineered organisms. At a competitive cost of $3 or less per g the total cost for the equivalent amount of hemoglobin that would be found in 10 million units of blood would be $1,350,000,000 (Walder 1988). It may be that the major role of genetic engineering will not be to genetically engineer microorganisms, but rather, to engineer large animals for the production of human hemoglobin. Herds of cows milked for human hemoglobin sounds like Epcot Center technology, but even today, there are examples of the potential of such genetic engineering feats (Simons et al. 1988). When hemoglobin-based blood substitutes become available, it seems most likely that they will be important adjuncts to red cell therapy, but will not completely replace blood in transfusion medicine because of their biological effects, their short useful half-life (Schmitz 1988), and the costs of production.

Progress in the area of blood substitutes has been slow (the attempt to use hemoglobin as a resuscitation fluid began over 50 years ago; Kiepert et al. 1989), and at times disappointing, but recent advances in hemoglobin research may provide new materials for clinical trials. Oxygen-transporting compounds may prove to be better than blood in delivering oxygen to tissues where blood vessels are occluded or otherwise compromised, and the relatively large size of red cells restricts their passage. The blood substitute products presently under study, unlike human blood cells, have useful half-lives of only hours to a few days, and thus, function principally as short-term oxygen providers. Hemoglobin-based or perfluorocarbon substitutes, which can be used in emergencies without need for preliminary typing, could be useful to provide interim, crisis blood units for patients and could prove of great value to trauma victims or on the battlefield. Presently there is inadequate knowledge about patient response to massive amounts of blood substitutes or whether patients can be maintained on alternate oxygen delivery systems. Clinical trials with modified hemoglobins have been disappointing, but a number of commercial entities are investigating their use and have invested considerable research funds into their development (Pool 1990). The problems with the use of hemoglobin as a blood substitute in patients include the ability of such substitutes to be tolerated in clinically useful doses, oxygen affinity and release, half-life, shelf life, antigenic potential, capability of inducing renal failure, oncotic pressure, cost, and availability (Winslow 1989). The genetic engineering of substitutes based on biological molecules will need to solve significant problems of molecular configuration, scale of production and toxicity, before human blood will no longer be required for emergency transfusions.

Biotechnology has provided the capability for large-scale production of antibodies and antigens for diagnostic testing, which make increased amounts of reagents available for safety assurance testing. While the blood supply is probably safer now than it has ever been (Cumming et al. 1989), recent data have shown that a significant portion of collected blood is not used because diagnostic tests indicate contamination with infectious agents (see deferral data in Table 1–1). Since 1985, the AABB has been reporting on donated units deferred or not used for transfusion because of the results of screening tests: in 1985, 9.1% of donations were deferred, in 1986, 9.4%, in 1987, 11.2%, and in 1988, 8.6%, before diagnostic testing for hepatitis C (non-A non-B) virus became readily available. Type A facilities deferred less of their collections than type B or type C facilities, perhaps reflecting better targeted recruitment efforts for potential donors. We can expect increasing numbers of diagnostic tests (the new test for parenterally transmitted non-A, non-B (C) hepatitis is one example) which will increase safety but decrease the numbers of useful donated units. Estimates of the size of the market for a new test for type C hepatitis are about $115 million in the USA alone, $67 million for screening tests and $48 million for clinical testing (Cowen and Co. 1990). It is reasonable to expect biotechnology companies with special techniques and capabilities in design and implementation of diagnostic tests will compete for worldwide markets with improved, rapid, more sensitive, and reliable blood tests.

The rapid development of tests for the AIDS virus (HIV-1) and the use of these tests are examples of how biological innovation can be mobilized to solve a problem. Many biotechnology companies have developed AIDS tests, and later-generation tests using DNA or RNA probes or other diagnostic systems are under development (Van Brunt 1988). When the public became aware of AIDS and its association with blood transfusion, many potential donors erroneously confused the possibility of getting AIDS by donating blood with transmission of the virus by transfusion and donations dropped. Blood collections increased again after testing procedures were adopted and efforts to educate the public were begun. While the Centers for Disease Control (CDC) and the Public Health Service have estimated that there are between 1 and 1.5 million Americans infected with the AIDS virus (Booth 1988), the testing procedure is very effective in identifying contaminated units; the remaining risk to transfusion recipients comes primarily from donors recently infected who are not yet HIV-seropositive. For 1987, testing coupled with education, donor recruitment, and self-screening was 99.83 to 99.95% effective in eliminating HIV-infected units from transfusion units. Increasing the numbers of donors not likely to be infected, such as repeat female donors who have an estimated undetected infection rate of 0.48 per million, could make the blood supply even safer (Cumming et al. 1989).

Improvements in safety need to be made in conjunction with maintaining the adequacy of the blood supply, and careful donor selection can work with new technologies to continually decrease transfusion risk. Application of new, extremely sensitive tests, such as those for gene amplification using the polymerase chain reaction (PCR) or related techniques (Van Brunt 1990) have just begun to be implemented for important usage in the blood bank (Ehrlich et al. 1986). Treatment of components and products to inactivate infectious agents will increase the safety of transfusion and will also impact on the availability of collected units by maximizing the ability to use collected units. Such technologies have already had an impact (not all positive) on the cost and availability of plasma fractions. Inactivation technologies for the cellular components have not yet been achieved, but are under active study (Prodouz and Fratantoni 1988). Until actual costs for inactivation processing are known, it will not be clear whether it will be less expensive to develop diagnostic tests and discard positive units or less expensive and safer to use techniques for inactivation of all infectious agents.

In the past, availability, not cost, has been the principle governing blood supply operations. The advent of prospective payment systems, medical insurers' constraints, and escalating need will intensify scrutiny of blood supply procedures and costs (OTA 1985). Increased costs based on scientific innovation can reduce the acceptability of technological changes or may influence providers to reduce or cease production altogether of components and products which become too expensive to make. The price of an individual component or product is linked to the price of all the components and products made from the same unit of blood. Advances in biotechnology may eliminate price supports for red cells by reducing the sale of plasma for plasma product manufacture. Other technologies may alter the financial relationships between the cellular components. New technology may present very significant safety and convenience advantages to patients and providers, but can also disrupt pricing strategies and present new challenges to presently accepted industry practice. Those technologies that can influence the supply parameters may have a profound effect on inventory practices, and hence on shortages, outdating, the cost of collecting blood for transfusion, and ultimately, the acceptability of new technology.

1.1.3 Transfusion Usage

While one study (Surgenor et al. 1988) has described a leveling off in the use of red cells for transfusion between 1980–1985 in certain hospital settings, statistics from the AABB (somewhat less than half of the national data) for the same period show increased usage annually, except for 1983 (see Table 1-1). Over the decade 1978–1987, the number of transfusions of red cells reported by organizations belonging to the AABB increased from almost 4,400,000 to over 8,000,000, peaking in 1985 at 8,560,078 units. Decreases observed in utilization data could reflect decreased availability of units to transfuse, rather than a decrease in need.

The AABB data confirm the fact that facilities collecting less than 100 units annually are the major transfusion users. The numbers of transfusions given at such facilities has gone from 55% in 1978 to as much as 72.1% in 1987 (see Table 1-2). The number of transfusions given at larger hospital centers (type B facilities) declined (along with collections) from 43.3% (23.9% collected) to 22.1% (15.0% collected) of annual AABB collections in 1987. A marked increase in transfusions at type B facilities was recorded in 1988 (37.0%) with an accompanying decline in collections (to 11.3% of all collections), perhaps reflecting renewed confidence in the safety of blood collected at major collecting centers. The statistics over the decade preceding 1988 may reflect changes in healthcare delivery practices across the nation: technological advances may allow for more complicated procedures to be performed at small hospitals, ambulatory surgical centers, renal dialysis centers, other special treatment facilities, as well as the availability of more sophisticated home care. As an example of the shift in healthcare delivery practices, it is interesting to note that between 1980 and 1987 the number of ambulatory surgical procedures performed in hospital-affiliated ambulatory care centers increased by 204.3%, and represented 44.2% (9,758,000 procedures) of all surgical procedures performed in hospitals in 1987. Increasing usage of blood in such facilities can be anticipated and new technologies will find applications at nontraditional sites. Problems and, hence, opportunities for making blood available for transfusion at sites distant from traditional hospital blood banks, such as surgical ambulatory care centers, have not yet been fully appreciated (Moore et al. 1987) but certainly will be targeted by new developments in transfusion technology. While biotechnological advances need sophisticated laboratories for discovery and development, these developed technologies will have to furnish useful information or provide capabilities for blood utilization in the absence of sophisticated laboratories. The observed changes in the geography of healthcare delivery will emphasize the difference between the markets for collecting and processing technology and transfusion technology. This difference has become more distinct over the past decade and could be of importance to developers of typing and screening diagnostic tests, processing and storage instrumentation, autologous blood management, and blood substitutes.

Over the past decade the transfusion of platelets has increased dramatically as shown in data from the Annual Reports of the AABB (Table 1-4). Red blood cell transfusion has declined from 69.9% of all component transfusions in 1978 to 54.5% in 1987 (the actual number of transfusions,

TABLE 1-4 Annual U.S. Blood Components Transfusion Data[1]

Year	Number of Units		Red Cells— % of Total	No. of Platelets	Platelets— % of Total	No. of Platelets (Pheresis)[2]	% of Total
	All Components	Red Cells					
1978	6,287,000	4,396,000	69.9	1,081,000	17.2	44,000	0.7
1979	7,602,000	5,194,000	68.3	1,330,350	17.5	68,400	0.9
1980	9,615,000	6,330,000	66.8	1,730,700	18.0	57,700	0.6
1981	11,282,010	7,258,044	64.3	2,110,000	18.7	79,000	0.7
1982	12,661,882	8,012,147	63.3	2,645,100	20.9	82,300	0.7
1983	12,307,803	7,483,843	60.8	2,798,000	22.7	80,000	0.7
1984	14,106,619	8,355,240	59.2	3,316,000	23.5	129,800	0.9
1985	14,522,503	8,560,078	58.9	3,534,800	24.3	124,900	0.9
1986	14,081,484	8,155,547	57.9	3,535,900	25.1	166,200	1.2
1987	13,820,000	8,065,576	54.5	3,767,300	27.3	217,000	1.6
1988	12,810,355	7,333,500	57.7	3,162,900	24.7	205,000	1.6

[1] Data are from AABB Annual Reports 1979–1989.
[2] Platelet units obtained from plasmapheresis.

however, has gone from 4,396,000 to 8,065,576, down from the high for the decade of 8,560,078 in 1985). The transfusion of platelets, as a portion of all components transfused, has increased, both in number and percentage, in the decade from 1978 to 1987 (from 17.9% or 1,125,373 units to 28.9% or 3,984,300 units). Table 1–4 shows a trend towards increasing use of platelets derived from plasmapheresis procedures, which has decreased the need for donated whole blood units. Whole blood collections are now divided into an average of 1.54 components (Cumming et al. 1989) and total costs for collecting units are shared by several products made from whole blood units. Increasing platelet availability by means of pheresis could have the side effect of decreasing the availability or increasing the cost of red cells. The use of platelets has found increased therapeutic use, even though significant technical problems in providing the component remain. The inability to store platelets beyond 5 days results in significant outdating and the need to collect platelets frequently in response to demand (Ledman and Groh 1984). New containers have been helpful in keeping platelets viable (Holme et al. 1989) but better understanding of the physiology of platelets is required to provide modalities for maintenance or storage of these fragile components. Completely synthetic means of producing platelets in tissue culture seem a distant future achievement, but there is active study of growth and differentiation factors in vitro to learn their mechanisms of action (Sachs 1987) and potential for application to in vitro production. Clinical trials are underway to learn how to use some of these proteins to stimulate a patient's own platelet production to reduce transfusion needs. Platelet-stimulating agents may follow the model of EPO for the enhanced production of red cells and may provide alternative means of supply.

It is interesting to note that while biotechnologists are seeking new ways of providing cells for transfusion or blood equivalents, a recent review of the current problems in transfusion medicine awaiting solution discussed the need for definition of the transfusion trigger (Chernoff et al. 1989). At this time, there is no consensus about the minimal laboratory value for a patient's hemoglobin which would indicate the necessity for treatment by transfusion. There are different standards in different places, and in some circumstances patients are undertransfused and in others, overuse of blood may occur. Can biotechnology contribute to providing the measuring devices for parameters that will help the physician or the surgeon decide if the patient needs blood? Sensors for measuring oxygen consumption, cardiac output, organ function, and blood volume would be useful in developing acceptable baseline standards and assessing patient values to determine when conditions indicate transfusion is appropriate therapy (Chernoff et al. 1989). It is the need for transfusion blood which is the essential question, and if the need is proven, then decisions about providing blood by autologous donation, hormonal stimulation, or from safe donors can be made.

1.2 PLASMA PRODUCTS

In 1984, the plasma products business represented about $1.7 billion in sales (IFPMA 1987). According to plasma industry sources, the USA supplies 60% of the world's fractionated plasma with paid donors providing 75% of this plasma (ABRA 1988). In contrast to red cell and component supplies, the collection and processing of plasma is largely within the for-profit sector. However, not-for-profit organizations harvest plasma from donated whole blood, and most U.S. blood centers separate and fractionate fresh plasma immediately after collection. In other countries, plasma is harvested from outdated blood, and this technique provides about 1 million liters (Leikola 1989). In the USA, the use of whole blood for transfusion of red cells has declined from 27.8% of red cell type transfusions in 1978 to 3% in 1988, increasing the available amount of plasma from donor collections. Plasma separated from whole blood is either sold to commercial plasma producers or fractionated into plasma products at facilities owned by not-for-profit collecting agencies. The not-for-profit sector realizes significant revenue from selling their plasma or from their own fractionating activities, and these funds are important in providing the financial resources which support their organizations. Thus, the not-for-profit sector, which is responsible for the provision of red cells and components, is financially tied to the plasma products industry, and the cost and availability of red cells can be influenced by sales of plasma or plasma products within the plasma industry.

1.2.1 Plasma Collection
Most plasma is collected by means of plasmapheresis (taking the plasma and returning the red cells to the donor) at stations around the USA, which in total, provides 60% of the world's supply (Leikola 1989). The plasma is pooled and transported to plasma fractionation plants to produce various plasma products. In 1984 there were 95 plasma fractionation plants worldwide, 57 operated by noncommercial organizations and 38 operated by industry. These facilities are able to process 15 million liters of plasma annually, with 75% of the capacity in the control of the for-profit sector. In 1984 over 12 million liters of plasma were fractionated, with 9.1 million liters processed in the USA and 3.3 million liters processed in Western Europe. Of the 9.1 million liters processed in the USA, 67% came from plasmapheresis collection programs (IFPMA 1987). Many commercial organizations have their own plasmapheresis collection programs or they have arrangements with independent commercial plasmapheresis centers. Leikola (1989) reports that in 1987, worldwide, 15 million liters of plasma were fractionated. He further states that 8 million liters (54%) came from plasmapheresis, 6 million liters from whole blood donations, and 1 million liters from placental material. In the not-for-profit sector, only 11% of the 2.9

million liters of plasma processed in 1984 were collected by plasmapheresis (IFPMA 1987), the remainder collected by separation from whole blood donations. In 1987, similarly, 90% of the plasmapheresis source plasma came from commercial efforts (Leikola 1989).

Statistics about paid plasma donors show that most often they are repeating donors; 99% of a study group of almost 10,000 donors appeared at collection centers more than once in a three-month survey period. Further, in the three-month survey, a significant number of donors (31%) appeared more than 18 times. Donors were mostly young (80% under 40), mostly male (84%), and mostly white (64%) (Rodell 1987). For comparison, unpaid donors of red cells, are, on average, young (in their early 30s), educated (some college or technical training), with a median household income of $30,000, mostly male, but with a much greater proportion of female donors (40%), and 85% white (Cumming et al. 1987). In the USA, the FDA maintains regulatory standards to insure plasma donor health. Plasma donors are tested, at the time of collection, for normal plasma protein levels, and other factors. First-time donors have a physical examination, urinalysis, and tests for syphilis, liver enzymes, AIDS, and hepatitis. Repeat donors have a brief physical and laboratory tests are repeated (IFPMA 1987). In the USA, FDA regulations allow a maximum donation of 50 or 60 liters per year per donor; other countries have lower allowable donation limits. For example, Mexico allows 15 liters per year, per donor; Canada allows 30 liters; France allows 10 liters; and Japan allows 12 liters. A recent study (Cumming et al. 1989) of unpaid blood donors indicates that the estimated relative risk of the likelihood of AIDS virus infection in donated blood decreases in groups that are repeat donors, and previously tested for AIDS. Data from ARC sources shows that repeating and previously tested female donors are the least likely to have positive AIDS tests. This group is nine times safer than the group of males donating for the first time, and blood from the riskiest female donor group is safer than blood from the safest male group. Further complicating the safety assurance of donated blood is the fact that in the very early stage of HIV infection, the virus may be present but is undetected by tests that look for the presence of antibodies in the serum. The estimated rate for seronegative, but HIV-positive blood units, for repeating, previously tested, female donors is less than one per 2 million donations (Cumming et al. 1989). In contrast to paid donors, the average nonpaid donor gives blood 1.5 times each year (Cumming et al. 1987). Obviously, the paid donor sector should look at methods for increasing the number of female donors, and the nonpaid donor sector should attempt to bring in tested female donors more frequently to even further decrease the risk of exposure to AIDS or other infectious agents.

1.2.2 Plasma Utilization

Plasma products are used for both diagnostic and therapeutic purposes. Therapeutic products that are currently approved for use in the USA include factor VIII, albumin, plasma protein fraction, factor IX complex, intravenous gamma globulin, intramuscular gamma globulin, and alpha-1 protease inhibitor. In Europe or Japan there are six additional products available, and clinical trials are in progress in the USA for other plasma-derived proteins including fibronectin, single factor IX, transferrin, Von Willebrand factor, factor XIII, factor X, and proteins C and S (Stagnaro 1989). Albumin products represent the majority (49%) of all plasma products sales, and in the USA, 2.5 million patients receive albumin therapeutically every year. Coagulation products represent 16% of the worldwide market for plasma products, and immune and hyperimmune globulins represent 35% of plasma products sold worldwide annually. About 10% of total plasma collections are used for blood grouping sera, hematology and chemistry controls, coagulation test reagents, and disease identification controls (IFPMA 1987).

The plasma industry can expect to see significant changes occurring as genetically engineered products become available. The safety from infectious disease organisms and the specificity of genetically engineered proteins can make engineered products a very desirable alternative to human-plasma-derived products. While the plasma products industry has been careful to screen for infectious agents (where tests exist) there is now increasing industry emphasis on methods of virus inactivation for plasma fractions. These inactivation techniques, while making products safer, can negatively effect yields and may markedly increase the cost of producing plasma products.

Recent problems precipitated by the contamination of antihemophilic factors by AIDS virus have made both the public and the industry even more aware of the need for safe products. It is estimated that, worldwide, there are about 225,000 people with hemophilia and about 15-20,000 hemophiliacs live in the USA (ABRA 1988). Many of these patients were able to lead an almost normal life because of treatment with clotting factors derived from human plasma. Individuals with hemophilia used products made from pooled human plasma, and as many as 10,000 donors could make up the pool. Patients taking treatment more than once each week were subject to exposure to blood from between 800,000 and 1,000,000 individual donations per year. While the industry exerts vigorous efforts to keep this vast plasma pool disease-free, before AIDS virus testing was available, many hemophiliacs became infected. It has been estimated that as many as 12,000 hemophiliacs over the age of five may now be infected, and many have transmitted the disease to mates and offspring (Lyon 1989).

Technologies for the inactivation of all viruses in the plasma pools used for fractionation into the various plasma products have decreased coagulation factor yields and added to the costs of production. The decreased yields of virus-inactivated factor VIII occur in the face of a decrease in demand and in the price that can be charged for albumin. Producing sufficient factor VIII by processing more pooled plasma, to make up for viral inactivation losses, without the potential for profit from sales of albumin, makes the production of clotting factor even more expensive. In the past, the demand for albumin was the driving factor in fractionating plasma, but now factor VIII is the primary demand stimulant for the plasma-fractionating industry (ABRA 1988; Leikola 1989). The plasma industry will need to seek additional markets for its products or to identify new products that can be made from processed plasma in order to keep fractionation a reasonable business activity.

The costs to patients will reflect the changes in the plasma processing industry. The cost for clotting factors for an individual patient have increased about sevenfold, and patients may have annual bills of $60,000-$300,000 per year (Lyon 1989). Where bioengineered therapeutic factors can be effectively used in relatively small quantities, production costs may be anticipated to be less than the cost of producing factors from human plasma, and engineered factors will provide strong competition. Production costs for factors that patients need to take on a continuing basis in significant amounts may become competitive as the costs rise for producing plasma products. Costs to consumers, both for plasma-derived or bioengineered products will probably rise. However, disease-free clotting factors and therapeutically improved factors produced by biotechnology will have a receptive audience when they become available.

1.2.3 Effects of Biotechnology on Plasma Products

Efforts are underway to produce proteins, in addition to factor VIII, of significance for patients with clotting disorders. Some factors that relate to the clotting mechanisms of the blood have already had more publicity than many genetically engineered products can anticipate over their useful lifetime. Studies on tissue plasminogen activator (tPA) are followed by the popular press like baseball games. Hopefully, the pubic interest in tPA is a forecast of their ongoing interest in the development and use of important new products made possible by advances in biotechnology. The high cost of tPA and the lower cost of competitive products emphasize the influence of production costs and marketplace factors on acceptance. New products will not succeed just because they are new and may represent a scientific tour-de-force. It will be necessary to convince users, and especially third-party payers of medical bills, of the superiority of, and necessity for, the biotechnology-derived product over less costly alternative products. While tPA and enzymes derived from bacteria are not strictly comparable to bioengineered plasma factors and blood-derived factors, the experience of introducing tPA into the marketplace may serve as a model for plasma products that have alternative methods of production. Although the introduction of the product was anxiously awaited by the medical community, its utilization still faces strong competition from less expensive products. Needless to say, the last chapter on the use of this product or of its second- and later-stage equivalents has not yet been written.

One of the oldest production capabilities of the new biology is the ability to make large quantities of specific monoclonal antibodies. Such capabilities have found application in the production of new sensitive diagnostic tests and can be expected to be instrumental in the development of targeted therapeutic molecules. Products based on antibodies to clotting factors will soon be available and will be important in providing diagnostic imaging information or new modalities of treatment, and several biotechnology companies have products close to approval for use by the FDA. The use of protective hyperimmune gamma globulin from plasma will face strong competition from products which are designed and manufactured to have specific antibody activities. Products for the treatment of Gram-negative sepsis are near approval by the FDA and will provide modalities of treatment not available with conventional antisera.

Monoclonal antibodies may be replaced in new applications for diagnostic and therapeutic medicine by completely engineered molecules that have desired antibody characteristics without foreign antigenicity. It even seems likely that antibody constructs can be produced inexpensively by fermentation techniques, cloning the genes for antibodies into microorganisms. Such genetically engineered organisms could generate molecules with much higher antibody affinity than those induced in animals, or could produce antibodies that catalyze enzyme reactions, function as biosensors, or be used in other applications to facilitate clinical and basic research (Guyer and Koshland 1989). These molecules may also be useful for the preparation and purification of plasma products or for diagnostic tests used in blood banks.

There has been much emphasis on research to find inexpensive methods of production in order to compete with plasma-derived proteins. Despite the fact that a product like albumin is relatively inexpensive when made from human sources and that there are virus-inactivating processes that leave the albumin intact, bioengineers have been seeking methods to efficiently produce albumin in many organisms. For example, it has been shown that albumin genes can be inserted and expressed in potato plants. The plants expressed albumin in leaf tissue and in suspension cultures. The protein secreted was indistinguishable from the authentic human protein (Sijmons et al. 1990) and provides an example of how levels of production might be increased to produce large quantities in a potentially economic manner. Another approach being taken by some biotechnology firms is the genetic engineering of large animals to produce plasma proteins like albumin (Klausner 1988). There are other, rare products that the plasma industry provides that are particularly expensive. These high-cost, human-derived products are likely targets for biotech entries into the competitive plasma business. One plasma-processing company makes alpha-1 antitrypsin, which is used to treat congenital emphysema. The drug costs an average of $22,000 per year per patient and is in limited supply. Several biotechnology companies are developing engineered molecules that could replace this product and they anticipate that patient costs will be lower with these bioengineered proteins (Webber 1990).

Commercial-scale production of factors that stimulate the production of blood cells was discussed in Section 1.1.1. EPO has been approved for use in anemic patients on dialysis (Eschbach et al. 1987) and approval for use in AIDS patients on anemia-producing drugs is expected. The cost effectiveness and convenience of the use of EPO for preoperative patients wishing to make autologous donations prior to undergoing elective surgery (Goodnough et al. 1989) is not yet established, nor is the use of colony-stimulating factors or thrombopoietin. Therapeutic applications for most of these molecules is likely to be economically significant for the firms providing these factors, and this will spur research, development, and the in vivo studies that will expand clinical applications. While it now seems almost routine to be able to make the genes for these therapeutic compounds, expression and economical production of these proteins in biologically active form and with reasonable in vivo half-lives is not currently routine. Increasingly, scientists have become aware of the role of the carbohydrate moieties in biological activity of cloned proteins, clearance from the circulating blood, or potential antigenicity. Both EPO (Goto et al. 1988) and tPA (Wilhelm et al. 1990) have been shown to require appropriate carbohydrate configuration, which may be controlled by expression systems, for modulating the biological activity of proteins and perhaps for clearing them from the circulation. Similar studies of the production of biologically active colony-stimulating factors have been carried out to identify useful hosts for production (Ernst et al. 1987; Weaver et al. 1988).

As described in Section 1.2.2, the production economics of the plasma industry have been based on producing multiple products from the available plasma resources. The plasma products industry is very competitive and has a past history of relatively low pricing for most products; coupled with high research and development costs, these factors make the industry particularly vulnerable to changes in the marketplace (ABRA 1988). The plasma products industry is presently doing research on expanded applications for the proteins now available and on new products that can spread the cost of preparing other plasma products (Stagnaro 1989). The plasma industry is trying to provide purer, safer, and more effective products and, simultaneously, to keep costs down to remain competitive with engineered equivalent products. For the biotechnology industry, particularly in the areas where biotechnology will be providing comparable, not innovative products, there is the daunting challenge of producing the amounts of products needed as inexpensively as possible. The cost of engineered products, perhaps more than their freedom from viral contamination (particularly where the equivalent plasma products have been treated to inactivate viruses), will be a major factor in marketplace decisions as to which product to use, plasma-derived or the equivalent engineered product, and the competition may provide some advantage to consumers. Engineered molecules like albumin or hemoglobin, now relatively inexpensive when derived from human blood, may significantly change the plasma business, and they may have the overall effect of increasing prices and decreasing choices.

Because genetically engineered products will compete with and in some cases may replace products now marketed by fractionators, the choice of any single engineered product over the presently marketed, plasma-derived product will effect the overall profitability of processing human plasma. Where engineered products are not or will not be available, because of technical difficulties or production costs, human plasma may need to remain the basic resource. However, as individual products made from human plasma are replaced by bioengineered products, the present pricing structure of products, which depends on the production of many saleable units from a single plasma resource, may have to be redesigned. It is unlikely that, within a few years, biotechnology could be economically able to replace all the products now made from human plasma and supplied by the plasma industry. The availability of engineered products will increase gradually and they probably will never completely replace all human-derived blood products. Increased production costs in the plasma industry, shared over fewer final products or smaller markets, coupled with the availability of only a few competitive, bioengineered products could cause blood industry providers to decrease or terminate production of needed products even though they are not available by bioengineering. From the point of view of the patient or end user, the timing of the integration of engineered products in a way that does not adversely affect the supply of needed proteins may be complex and difficult to orchestrate. However, where genetically engineered products are able to increase supplies of rare molecules, like engineered alpha-1 antitrypsin (or its equivalents), market effects when such products are introduced may be less complicated, and end users will enjoy the benefits of new technology.

The effect of the entry of engineered products into the marketplace may be to raise the prices not only of existing plasma products, but of red cells and components as well. The not-for-profit red cell and component supply sector of the blood industry has a financial relationship to the plasma industry, through its sales of plasma products and as a plasma source for industry fractionators, and it depends on these revenues to finance operations. Balances will need to be struck between new and classically prepared, inactivated plasma products, both in U.S. markets and abroad. The plasma industry now sees a difficult future and looks to expanding its markets into countries that are not presently consumers of plasma products. Many questions remain about the role of biotechnology worldwide, and the access to new developments by patients in poorer countries. Introduction of bioengineered products in less wealthy nations may be slow, and such nations may remain primarily the users of not-so-high tech products and dependent on U.S. plasma resources. The biotechnology industry and the classical plasma industry will have to work together in order to provide the benefits to patients of all the products that the new science can design.

REFERENCES

Altman, L.K. (1989) *New York Times* 5 Sept. 1989, A-1.

American Association of Blood Banks (AABB) Annual Reports 1979–1989.

American Blood Resources Association (ABRA) (1988) 15 Oct. 1988.

American College of Surgeons (1989) Socioeconomic Affairs Department, Chicago, IL.

Andrews, E.L. (1990) *New York Times* 19 Feb. 1990, (Business Section) 9.

Axelrod, F.B., Grindon, A.J., and Vroon, D.H. (1987) *Transfusion* 27, 219–221.

Booth, W. (1988) *Science* 239, 253.

Brodheim, E., and Hirsch, R. (1979) *Transfusion* 19, 105–107.

Chernoff, A.I., Klein, H.G., and Sherman, L.A. (1989) *Transfusion* 29, 711–742.

Cohen, M.A., and Pierskalla, W.P. (1979) *Transfusion* 19, 444–454.

Cohen, M.A., Peirskalla, W.P., and Sansetti, R.J. (1983) *Transfusion* 23, 54–58.

Cowen and Co. (1990) *Perspectives* 11 Jan. 1990, 1–3.

Cumming, P.D., Wallace, E.L., Surgenor, D.M., et al. (1974) *Medical Care* 12, 743–753.

Cumming, P.D., Schorr, J.B., and Wallace, E.L. (1987) *Annual Blood Facts United States Totals—1986/87.* American Red Cross. 16 Sept. 1987.

Cumming, P.D., Wallace, E.L., Schorr, J.B., and Dodd, R.Y. (1989) *New England J. Medicine* 321, 941–946.

Division of Health Care Statistics, National Center for Health Statistics (1988) *Data from the National Hospital Discharge Survey for 1982–1987*, U.S. Department of Health and Human Services, U.S. Government Printing Office, Washington, DC.

Ehrlich, H., Sheldon, E.L, and Horn, G. (1986) *Bio/technology* 4, 975.

Ernst, J.F., Mermod, J.-J., DeLamarter, J.F., Mattaliano, R.J., and Moonen, P. (1987) *Bio/technology* 5, 831–834.

Eschbach, J.W., Egrie, J.C., Downing, M.R., Browne, J.K., and Adamson, J.W. (1987) *New England J. Medicine* 316, 73–78.

Federal Register (1989) Part II Department of Labor, Occupational Safety and Health Administration 29 CFR Part 1910, 30 May 1989.

Gaul, G.M. (1989) *Philadelphia Inquirer* 24 Sept. 1989, 13A.

Goldstein, J. (1989) *Transfusion Medicine Reviews* 3, 206–212.

Goodnough, L.T., Rudnick, S., Price, T.H., et al. (1989) *New England J. Medicine* 321, 1163.

Goto, M., Akai, K., Murakami, A., et al. (1988) *Bio/technology* 6, 67–71.

Greenwalt, T.J. (1989) *Transfusion* 29, 248–258.

Guyer, R.L., and Koshland, D.E. (1989) *Science* 246, 1543.

Holme, S., Heaton, A., and Momoda, G. (1989) *Transfusion* 29, 159–164.

Hosoi, T. (1989) in *Meeting of International Group of Red Cross Blood Transfusion Experts,* 12-13 June 1989, Montreal, Canada, Japanese Red Cross Society, Tokyo, Japan.

Huang, S.T., Lair, J., Floyd, D.M., and Cole, G.W. (1980) *Transfusion* 20, 725–728.

International Federation of Pharmaceutical Manufacturers Association (IFPMA) (1987) *A Study of Commercial and Non-Commercial Plasma Procurement and Plasma Fractionation: Revised—1987*, Working Group on Human Blood Products and Related Substances, December.

Issitt, P. (1985) in *Applied Blood Group Serology* 3rd ed. p. 159, Montgomery Scientific Publications, Miami, FL.

Japanese Red Cross Society (1986) in *Review of Activities, 1984–1985* p. 22.

Kaufman, R.J. (1988) in *Biotech USA, Proceedings*, 14–16 Nov. 1988, San Francisco, pp. 371–380, Conference Management, Norwalk, CT.

Kiepert, P.E., Adeniran, A.J., Kwong, S., and Benesch, R.E. (1989) *Transfusion* 29, 768–773.

King, A.T., Mulligan, B.J., and Lowe, K.C. (1989) *Bio/technology* 7, 1037–1041.

Klausner, A. (1988) *Bio/technology* 6, 663–670.

Kuriyan, M., and Kim, D.U. (1989) *Vox Sang* 57, 152–154.

Ledman, R.E., and Groh, N. (1984) *Transfusion* 24, 532–533.

Leikola, J. (1988) *Vox Sang.* 54, 1–5.

Leikola, J. (1989) *Beitr. Infusionsther.* 24, 69.

Lyon, J. (1989) *The Chicago Tribune Magazine* 23 April 1989, 12.

Moore, S.B., Reisner, R.K., Losasso, T.J., and Brockman, S.K. (1987) *Transfusion* 27, 359–361.

Office of Organ Transplantation (1988) *Organ Transplantation Q & A Pamphlet*, September, U.S. Department of Health and Human Services, Washington, DC.

Office of Technology Assessment (OTA) (1985) *Blood Policy and Technology*, OTA-H-260, January, 1985, U.S. Congress, Washington, DC.

Pool, R. (1990) *Science* 250, 1655–1656.

Prodouz, K., and Fratantoni, J.C. (1988) *Transfusion* 28, 2–3.

Rodell, M.B. (1987) *Plasmapheresis* September, pp. 53–55.

Sachs, L. (1987) *Science* 238, 1374–1378.

Schmitz, T.H. (1988) in *Biotech USA Proceedings*, 14–16 Nov. 1988, San Francisco, p. 371, Conference Management, Norwalk, CT.

Sijmons, P.C., Dekker, B.M.M., Schrammeijer, B., et al. (1990) *Bio/technology* 8, 217–221.

Simons, J.P., Wilmut, I., Clark, A.J., Archibald, A.L., and Bishop, J.O. (1988) *Bio/technology* 6, 179–183.

Snyder, S.R., Welty, R.Y., Walder, R.Y., Williams, L.A., and Walder, J.A. (1987) *Proc. Natl. Acad. Sci.* 84, 7280–7284.

Stagnaro, T.P. (1989) *Plasmapheresis* September, pp. 206–209.

Surgenor, D.M. (1987) *New England J. Medicine* 316, 542–544.

Surgenor, D.M., Wallace, E.L., Hale, S.G., and Gilpatrick, M.W. (1988) *Transfusion* 28, 513.

Toy, P.T.C.Y., Strauss, R.G., Stehling, L., et al. (1987) *New England J. Medicine* 316, 517–520.

U.S. Bureau of the Census (1988) *Statistical Abstract of the U.S.* U.S. Government Printing Office, Washington, DC.

Van Brunt, J. (1988) *Bio/technology* 6, 259–264.

Van Brunt, J. (1990) *Bio/technology* 8, 291–294.

Wagenbach, M., O'Rourke, R., Vitez, L., et al. (1991) *Bio/technology* 9, 57–61.

Walder, J.A. (1988) *Biotech USA, Proceedings*, 14–16 Nov. 1988, San Francisco, pp. 357–362, Conference Management, Norwalk, CT.

Weaver, J.F., McCormick, F., and Manos, M.M. (1988) *Bio/technology* 6, 287–290.

Webber, D.S. (1990) *Research* 18 April 1990, p. 21, Alexander Brown and Sons, Baltimore, MD.

Wilhelm, J., Lee, S.G., Kalyan, N., et al. (1990) *Bio/technology* 8, 321–325.

Winslow, R.M. (1989) *Transfusion* 29, 753–754.

Zuck, T.F. (1988) *Transfusion Medicine Reviews* 2, 245.

Long-Term Storage and Preservation of Red Blood Cells

Gerald L. Moore

Humans have experimented with blood transfusions for over 300 years and have attempted to preserve human blood since the early 1900s. The first modern attempts to store blood were stimulated by World War I when blood was stored in citrate-glucose solutions (Robertson 1918; Rous and Turner 1916). During World War II, the increased need for blood plasma and whole blood resulted in the development of a solution called acid-citrate-dextrose (ACD) for 21-day refrigerated storage of blood. A slight variation of ACD, called CPD, was introduced in the late 1950s. For CPD, phosphate was added to the citrate and dextrose, which slightly improved the viability of stored red cells, although the dating period was held to 21 days. Blood preservation solutions remained unchanged until the late 1970s when adenine was first added to CPD to produce CPDA-1, which extended the shelf life of blood to 35 days (Peck et al. 1981). CPDA-1 appears to be the industry's final attempt to modify the anticoagulant solution for better blood preservation. However, the success of U.S. and European blood banks with CPDA-1 has encouraged the development of modern additive solutions for component-specific preservation.

Red cell preservation research has traditionally centered around three issues, maximizing viability and function, while minimizing cell lysis, and assuring that the storage systems maintain sterility. Obviously, any preservatives must be nontoxic. Percent viability is defined as the percentage of the stored red cells remaining in circulation for 24 hours after infusion. For many years the U.S. Food and Drug Administration (FDA) set this mean percentage at 70%, but in 1985 this percentage was raised to 75%. Viability must be measured by in vivo red cell survival, but adenosine triphosphate (ATP) has been traditionally used in developmental experiments as an indicator assay for viability since some correlation exists between ATP level and viability (Peck et al. 1981). This correlation is best defined as a threshold, since it is known that if ATP levels drop below about 30% of normal, the cells will have low viability, although a high ATP level will not necessarily insure good viability. Red cell function, i.e., oxygen delivery, is closely associated with the level of cellular 2,3-diphosphoglycerate (2,3-DPG). During storage in any commercially available system, the 2,3-DPG level falls to near zero in about 2 weeks. This results in a 50% reduction in the cells' ability to deliver oxygen to tissues, all other things being equal (Moore 1983). Red cell lysis during storage is limited to 1% by FDA policy. Red cells stored with white cells or without any plasma are most likely to lyse, but the degree varies dramatically among donors.

This chapter covers the use of modern additive solutions for red cell storage, the use of freezing to preserve red cells, some methods being developed to make freezing of blood more practical, and new methods of processing stored red cells. The storage of platelets, white cells, and plasma components will not be discussed.

2.1 CURRENT STATE OF LIQUID PRESERVATION AT 4°C

In 1983 the USA and parts of Europe shifted from preserving red cells by anticoagulant fortification to using CPD coupled with separate additive solutions. Employing this approach, blood was drawn into a basic anticoagulant and processed into components. The red cell component was then

TABLE 2-1   Composition of Commercial Additive Solutions for Red Cells

Component	CPD	ADSOL	AS-3[1]	SAG	SAGM	Cir/Pk[1]
Adenine (mg)	—	27	17	17	17	7
Glucose (g)	1.61	2.2	0.4	0.9	0.9	0.4
NaH$_2$PO$_4$ (mg)	140	—	285	—	—	285
Na Citrate (g)	1.66	—	588	—	—	588
Citric acid (mg)	206	—	42	—	—	42
Mannitol (g)	—	0.75	—	—	0.52	—
NaCl (mg)	—	900	718	877	877	718
Water (ml)	63	100	100	100	100	100

[1] AS-3 and Cir/Pk also use double glucose in their CPD.

mixed with an isotonic solution containing a nutrient mixture designed for 42-day red cell preservation. The four principle solutions are ADSOL (or AS-1), AS-3 (or Nutricel), and in Europe, SAG and SAGM. Table 2–1 gives recipes for these solutions.

The first additive solution was developed in Sweden in the late 1970s and contained saline, adenine, and glucose (and thus was named SAG). SAG was later modified by the addition of mannitol (SAGM) to retard lysis (Hogman et al. 1978a, 1987b, 1981). Hogman was the first to show that white cell enzymes contaminating red cell suspension will increase red cell lysis rates (Hogman et al. 1978b) and that lysis can be reduced to manageable levels by adding mannitol to the storage solution (Hogman et al. 1981). Buffy-coat-poor red cells can be stored in SAGM for 35 days.

In 1983 Fenwal Laboratories (Deerfield, IL) introduced ADSOL (AS-1) solution for 49-day red cell storage (Heaton et al. 1984). In 1985 a controversy over the viability of red cell in this product (Heaton et al. 1985; Page 1985; Beutler 1985; and Valeri 1985) resulted in a reduction of the storage time in AS-1 to 42 days.

Also in 1983, Cutter Laboratories (Berkeley, CA) introduced Nutricel additive solution, and in 1984 a modified version (AS-3) was introduced for the 42-day storage of red cells (Figure 2–1) Both AS-1 and AS-3 in 100-ml volumes are added to packed red cells after removal of platelet-rich or platelet-poor plasma. Similar additive solutions have recently been introduced by Turumo (OPTISOL), and Tuta Corp. (Australia) makes Circle Pack (Cir/Pk).

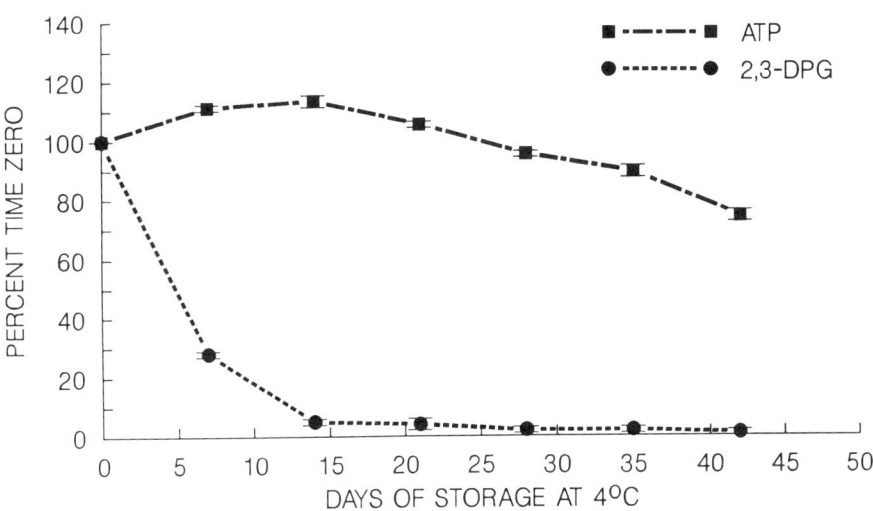

FIGURE 2–1 The time course for the retention of ATP and 2,3-DPG by packed red cells stored in AS-3 solution for 42 days at 4°C. Results in ADSOL are similar.

The switch to additive solutions for red cell preservation resulted in an increase of storage time from 35 days (in CPDA-1) to 42 days. Other advantages of the additive systems include lower viscosity, elimination of excessive nutrients in platelets, and better control of optimal ratios of red cells to nutrients. The current 42-day storage limit is felt by many to be all that is needed in modern blood banking. Further extension in storage time would require dramatically different approaches since the drop in pH and membrane changes become critical after 42 days. None of these additive solutions preserve 2,3-DPG beyond 7–14 days. Expanded reviews of additive solutions are presented elsewhere (Heaton 1986; Moore 1987).

2.2 CURRENT RESEARCH IN NONFROZEN SYSTEMS

2.2.1 Long-Term Liquid Storage

In 1986, Meryman investigated the use of osmotic swelling to extend red cell storage time in an attempt to retard lysis (Meryman et al. 1986). He used ammonium salts of low concentration and found that they were effective in maintaining red cell ATP for long periods (50% left after 12–16 weeks). In vivo red cell survivals were measured on these cells after storage at 4°C for 84–131 days. Percent survival varied from 46–86%. Lysis varied from 0.5–7%. The in vitro portion of these studies were repeated in our laboratory, with 50% ATP remaining at 8–10 weeks with a rate of lysis under 1%. The system does have limitations, however, since the 2,3-DPG drops to near zero by week 2, the pH (at 37°C) drops to 6.0 by 7 weeks, and the cells must be washed extensively prior to use.

2.2.2 Maintaining 2,3-DPG for Preservation of Function

The concept of maintaining 2,3-DPG during storage has been investigated for two decades. Many metabolites have been tested as elevators of 2,3-DPG, including dihydroxyacetone, inosine, ascorbate (active component is oxalate), and methylene blue. To date these compounds have been either of only marginal benefit, or toxic (Moore 1983, 1987).

Several studies have been published using a modified xanthone, 2-hydroxyethoxy-6-(5-tetrazoyl) xanthone, which was named BW A440C by its developers (Hyde et al. 1984). Our data (Moore, unpublished observations) supports the findings of Hyde et al. (1984) and Paterson et al. (1988) that BW A440C (440C) elevates 2,3-DPG and P_{50}, while not affecting ATP, pH, the use of glucose, or the production of lactose. We tested the 440C both as a supplement to ADSOL storage for 42 days and as a supplement to an adenine/glucose/mannitol solution used in post-thaw preservation (Figure 2–2). We showed that the xanthone could bind to pure A0 hemoglobin (the predominant genetic form) and raise the P_{50} of hemoglobin in a manner similar to the addition of 2,3-DPG (Table 2–2). The P_{50} effect on red cells

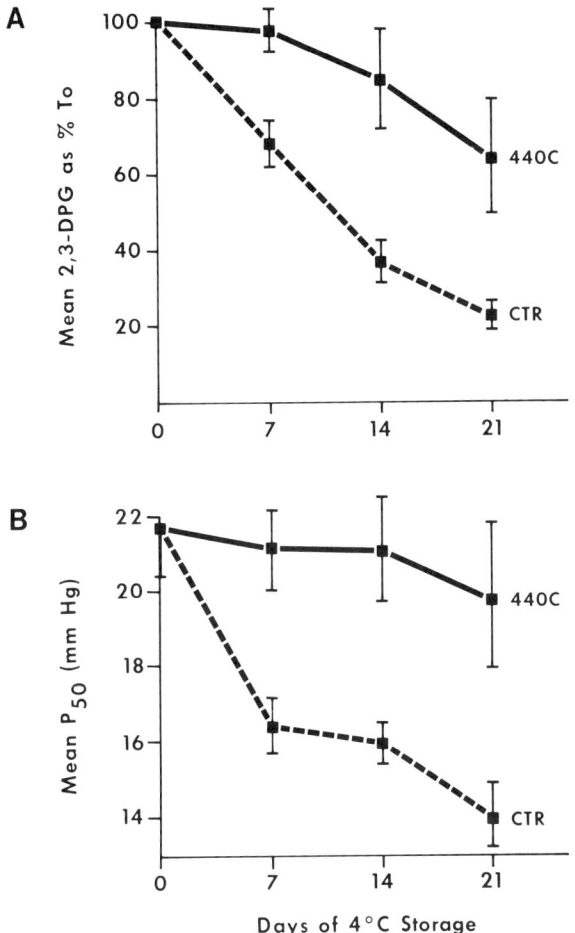

FIGURE 2–2 Post-thaw maintenance of (A) 2,3-DPG and (B) P_{50} levels of red cells stored in an additive solution of adenine, glucose, and monosodium phosphate with and without 440C.

is a combined response to 2,3-DPG maintenance and a direct binding effect with the hemoglobin. A detailed analysis of the effect of 440C on red cell enzymes was done by Beutler et al. (1988). They found that the compound inhibited several enzymes, including pyruvate kinase. This inhibition is known to cause elevation of 2,3-DPG levels. Beutler et al. also showed that the xanthone greatly reduced the viability of rabbit red cells, perhaps by shutting down their metabolism. Our data and those of the Hyde-Paterson group argue against significant reduction in metabolism since 440C did not retard conversion of glucose to lactate and acid. The compound may also

TABLE 2–2   Mean P_{50} Values as Measured on Hemoscan at 37°C and pH 7.4

Sample	Mean P_{50} (N = 3)
Red cells in buffer	19.0
Red cells plus 440C	19.0[1]
Hemoglobin A0 in buffer	17.5
Hemoglobin plus 440C	27.5
Hemoglobin plus 2,3-DPG	32.0
Hemoglobin plus 440C and 2,3-DPG	31.5

[1] 440C and 2,3-DPG were added at time of P_{50} assay.

have some hypotensive activity, which could preclude its general use as a blood additive (unpublished data, Burroughs Wellcome Co.).

Meryman discussed a washing procedure which used nonionic buffers such as citrate to remove plasma components from red cells (Meryman 1989). Subsequent storage of red cells in this buffer appears effective in maintaining 2,3-DPG, perhaps by producing hydroxyl ions, which enter the red cells in order to replace the lost chloride ions. Other efforts will undoubtedly be made to maintain 2,3-DPG during storage, if a nontoxic additive can be found.

2.2.3 Concern Over Phthalate Toxicity

Polyvinylchloride (PVC) bags have been used for 30 years to store blood. PVC is made pliable by addition of up to 40% of di(2-ethylhexyl)phthalate (DEHP). The DEHP is mechanically trapped in the vinyl matrix and is insoluble in water, but it will leach into hydrophobic materials such as plasma lipoproteins or cell membranes. The acute toxicity of DEHP is very low, but it does bind to red cells, platelets, and plasma during blood storage, and is converted into the monoester MEHP (Rock et al. 1978). There has been a long-standing concern that DEHP may have toxic effects. An excellent and current review of this subject raises new concern that DEHP may promote, if not induce, cancer in some test animals (Rubin and Ness 1989). This effect cannot be reproduced in human hepatocytes (Turnbull and Rodricks 1985). The primary component of concern in vivo is the water soluble monoester MEHP formed by lipase activity on DEHP (Rubin and Ness 1989).

The concern over DEHP/MEHP has influenced the search for alternate nonleachable plastics for blood components. Another concern, which will not be discussed here, is the need for greater oxygen permeability to allow 5-day platelet storage. Platelet storage bags of polyolefins or of PVC with the plasticizer tri-(2-ethylhexyl)trimellitate have been available for several years. Shimizu et al. (1989) recently reported good results with platelet storage using a PVC bag containing di-*n*-decyl phthalate. There has however been some reluctance to modify the storage bag used with red cells since DEHP actually stabilizes red cell membranes during storage (Estep et al. 1984). Fenwal Laboratories recently has developed an entirely new bag system based on a citrate-plasticized PVC. This new bag is made of PVC containing butyl-trihexylcitrate and is effective for the storage of both red cells and platelets (Buchholz et al. 1989a, 1989b). Other bag companies will probably follow Fenwal with a non-DEHP plasticized PVC bag.

2.2.4 Methods of Measuring Red Cell Survival

The efficacy of red cell preservation can only be measured by tagging the cells with a radioactive label, reinjecting them into the donor and evaluating their survival. The two common methods for this procedure are defined as the single and double label chromium methods. With the single-label method, red cells are tagged with ^{51}Cr, injected, and blood volume is calculated by back-extrapolation of the 5–15 min dilutions of tagged red cells. In the double-label method, red cells are tagged with ^{51}Cr, but blood volume is measured by a separate isotope, usually ^{125}I bound to albumin. A standardized method for each of these approaches has been published (Moroff et al. 1984), with the recognition that each method contains assumptions which can reduce its accuracy.

The controversy concerning relative accuracy and problems with single versus double-label methods has been discussed (Moore 1987). To evaluate the single label method independently, Beutler proposed an alternate double label method using ^{99m}Tc and fresh red cells to measure blood volumes (Beutler and West 1984). He showed that if the viability was above 80% the two methods gave identical results, but below 80% viability, the single label technique overestimated blood volumes. However, he showed that the error in absolute percentage of viable red cells remains small because the large (10 to 20) percentage error is applied to a small percentage of remaining viable red cells, making the largest overestimation of viability only 4% (Beutler and West 1984). Marcus et al. (1987) studied red cell survival measurements using isotopes of chromium, technetium, and indium, and showed that chromium produced higher and more accurate viabilities, due to lower isotope elution rates. Heaton developed a modified technetium procedure which minimized the elution of the technetium label, and confirmed the 3-4% higher values found with single versus double label methods (Heaton et al. 1989a). AuBuchon and Brightman studied five methods of indium-labeling red cells, and they found that four of them are effective as measures of blood volume (AuBuchon and Brightman 1989). One problem with the use of indium is its overlap with the energy window with chromium, so that the two cannot be used together in most gamma counters (AuBuchon and Brightman 1989).

The double label methods, while providing slightly superior viability data, subject the donor to additional radiation and are technically much more difficult. To overcome this, a double label method has been developed using nonradioactive Cr for blood volume measurements and ^{51}Cr for stored cell recovery (Heaton et al. 1989b, 1989c). This technique has a high correlation with older double-label methods ($66 \pm 5\%$ vs. $69 \pm 8\%$) but requires a Zeeman electrothermal atomic absorption spectrophotometer to measure ^{52}Cr. Another nonradioactive, red cell tagging method is being developed using rabbit red cells. In this technique, the cells are tagged with biotin by reaction with N-hydroxysuccinimidobiotin (Suzuki and Dale 1987).

2.3 FROZEN RED CELLS

The technology currently employed for freezing red cells was developed in the 1960s and early 1970s by the American Red Cross Research Lab (Washington, D.C.) and the U.S. Naval Blood Research Lab (Boston, MA). This method, known as the "high glycerol" procedure, has been extensively reviewed (Valeri 1970, 1976, 1988; Meryman and Hornblower 1972; Meryman 1979). The high glycerol method consists of mixing packed red cells with 6 M glycerol and freezing in a special polyolefin freezing bag at $-80°C$. Cells stored in this manner can be kept for at least 21 years (Valeri 1988; Valeri et al. 1989). Upon demand, the cells are thawed in a 37°C water bath and deglycerolized by centrifugal washing with 2–3 liters of sterile saline solution (Widmann 1985). While fresh red cells are the usual starting material for this process, Valeri et al. (1979) and Valeri (1988) have shown that outdated red cells may be rejuvenated with a solution of pyruvate, inosine, phosphate, and adenine (PIPA) for 1 h at 37°C and then frozen. PIPA restores the levels of the red cell ATP and 2,3-DPG to fresh blood levels, but must be removed by washing prior to infusion since inosine promotes hypotension. The deglycerolizing step removes the residual PIPA after freezing and thawing. Several washing machines developed by IBM (now COBE Labs, Denver, CO), and Haemonetics (Braintree, MA) have been approved by the FDA for washing thawed red cells.

This frozen red cell technology has been available for over two decades, but has not gained popularity, except for very rare blood types, for several reasons. First and foremost, the procedure is labor intensive and expensive, costing two to three times as much as nonfrozen red cells. Early hopes that the washing step might remove viruses from red cells was shown to be unfounded (Haugen 1979). An additional difficulty is the FDA requirement that a thawed-washed unit be used within 24 hours due to both the possible compromise of sterility, and the lack of adequate nutrient support for the cells in the final wash solution. Another problem was identified when attempts to ship the frozen units in their special bags resulted in an unacceptably high rate (10–15%) of breakage (Valeri 1988).

In 1981, the U.S. Naval Blood Research Lab developed a minor, but important, change in the freezing/thawing/washing procedure. This change allowed the red cells to be frozen in the primary bag in which the cells had been drawn (Valeri et al. 1981). The size of the primary bag for this procedure was increased from 600 to 800 ml. This change eliminated the need for transferring red cells to a freezing bag and reduced the breakage rate of red cells shipped in these bags from 15% to 1% (Valeri 1988). The oversized primary bag containing CPDA-1 and attached to the usual two or three satellite bags can then be used for component preparation and either frozen or nonfrozen red cell storage. If 4°C-stored cells are not used in one week, they may be frozen in the same bag; if not used within 35 days, they may be rejuvenated with PIPA and frozen in this same bag. In 1989, the U.S. Military Blood Program adopted this bag system, with the intention of developing a stockpile of several hundred thousand units of frozen type O red cells for emergency use. The frozen cells will also be used routinely to maintain turnover and familiarity with frozen blood manipulation. Stockpiled units will be maintained at various depots worldwide and kept for up to 21 years.

The explosive development of the biotechnology industry in recent years has stimulated many improvements in membrane technology designed to facilitate the separation of cells from supernatant solutions. Thus, it was natural that researchers consider using membranes to deglycerolize frozen-thawed red cells. Membrane technology offers several potential advantages over the current centrifugal washing to remove glycerol. First, the membrane and its integral connecting tubing harness could be sterilized. This would remove the FDA objection to the nonclosed nature of the centrifugal bowl, potentially allowing for extended storage of the washed red cells. Other potential advantages include smaller, less expensive hardware, faster wash times, less operator interface, and the ability to control hematocrit in the finished product.

To deglycerolize red cells, the Millipore Corp., Sterimatics Division (Boston, MA), has developed a prototype device which is a modification of their plasmapheresis machine. One prototype machine was extensively evaluated in the Naval Blood Research Lab and another was examined in our lab (Moore, unpublished observations). These prototypes are microprocessor-controlled and use three pumps and pressure transducers to regulate the rate of flow of saline into the wash, the rate of flow across the membrane, and the rate of permeate removal. The membrane is a 10-stack, tangential flow cartridge, 11.5 × 5.7 cm in size. When red cells are washed with the flow program established by C.R. Valeri, the red cell lysis, potassium leak, morphology, ATP, and 2,3-DPG levels are similar to those obtained with the centrifugal washer (Moore, unpublished observations). Unfortunately, we found that the time required to wash a red cell unit was 1.2 to 1.5 hours, which is much slower than the 0.6 hours required using the Haemonetics model 115 washer. However, development of a larger, 20-stack membrane could make the wash times competitive with the centrifugal methods. The Millipore Sterimatics washing tube set (with membrane and wash bag) includes sterilizing microfilters on all input lines, and sterile splicing tabs to attach the thawed red cells to the wash bag, thereby assuring sterility throughout the process.

Feasibility studies have also been done to test the ability of hollow fiber membrane systems to deglycerolize red cells. Numerous hemofilter and plasmapheresis devices were tested, and the best were superior to centrifugal washing, producing washed red cell units in 19 ± 4 min using $1,750\pm130$ ml of saline (Radovich 1989). At least eight red cell units per cartridge could be run with no decrement in membrane performance. Further development of hollow-fiber washing devices appears desirable as a quick, low-cost alternative to washing glycerol from red cells.

The issue of sterility of thawed-washed red cells has been one of the main reasons the technique has received limited use in blood banking. The FDA has limited dating of the final product to 1 day at 4°C, since sterility may have been compromised during processing. This could occur with the several additions of solutions: glycerol when freezing, then 12% and 0.9% saline when washing. Also, most centrifugal wash systems have a spinning bowl with a stationary center hub which is "closed to air," but not technically sealed. This step could, in theory, allow the introduction of bacteria. However, after processing "thousands" of units by this method, the Naval Blood Research reports no problems with bacterial contamination, even on units kept at 4°C for up to one week (Valeri 1988).

In November 1987, the FDA approved the Haemonetics platelet preparation system for use and agreed that the system was a "technically closed" unit. The platelet preparation unit is a spinning centrifugal bowl similar to the unit used to wash red cells. Haemonetics is currently gathering data for submission to the FDA to show that their red cell washing bowl is also "technically closed," with the hope that the question of compromised sterility can be overcome (D. Mareci, personal communication). This may give another washing system the potential for use in extended 4°C storage of thawed washed red cells.

2.4 POST-THAW PRESERVATION OF RED CELLS

The 24-hour shelf life of post-thawed red cells is a factor which seriously limits their practical use. For example, blood thawed for the elective surgery of a particular patient will usually be discarded if the surgery is postponed at the last minute, or if excess units are thawed. Such units are not recycled through the blood bank inventory. Should these units be autologous or rare, the loss is even greater. Another example of the limitations inherent in a 24-hour shelf life is the large fluctuation in combat casualties seen in military surgical hospitals during wartime. Maintaining the balance between frozen-thawed red cells and casualty admission rates would be very difficult, at best, with 24-hour dating. The use of post-thaw, 4°C, dating period for red cells would eliminate these types of problems and make frozen blood a more practical product of both homologous and autologous transfusions.

The requirements for extended post-thaw storage of blood are similar to regular blood bank storage: 1) sterility must be maintained; 2) red cell post-transfusion viability must be adequate (>75% ?); 3) hemolysis must be low (<1% ?); and 4) any additive must be nontoxic. A post-thaw storage time of 2–3 weeks at 4°C appears to be optimal for most needs. The sterility issues for post-thaw preservation have been discussed above, and will probably resolve themselves as a combination of closed washing systems, sterile splicing, and in-line filtration of wash solutions. Additive solutions are needed to maintain viability and to retard lysis.

We have evaluated in vitro several solutions as potential post-thaw additive solutions. ADSOL preserves ATP adequately to infer good viability on 21-day stored cells; however, ADSOL does not work as well with post-thaw cells as with fresh blood (Moore et al. 1987a). In our studies the red cells were washed on a Haemonetics model 115 centrifugal washer and restored in ADSOL (Figure 2–3). The lysis produced in this system was unacceptably high, exceeding 2% of the cells by day 21 (Figure 2–4). Ross et al. (1989), working with cells that were washed on the IBM washer and restored in ADSOL, showed 1% lysis over 21 days. This study also reported in vivo red cell survivals of 90±1% after 10 days of storage in ADSOL. The difference in lysis rates in the two studies may reflect the more gentle washing technique in the IBM cell washer, which resulted in less cell-membrane damage.

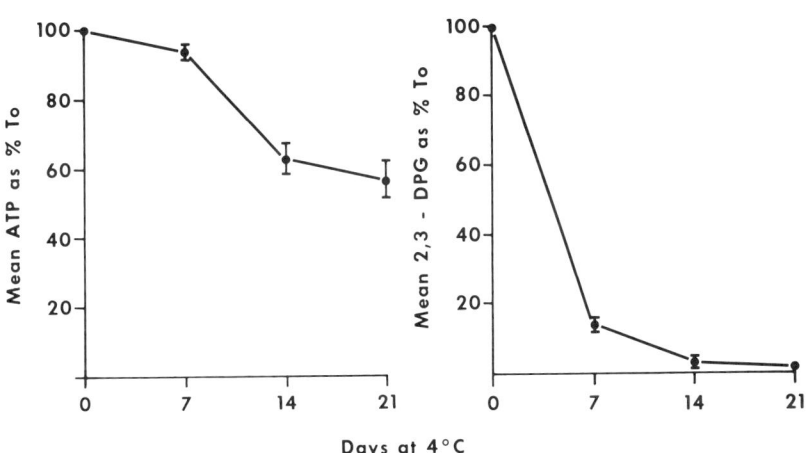

FIGURE 2–3 Post-thaw storage of red cells in ADSOL at 4°C. ATP and 2,3-DPG levels are shown as a percent of time 0, which was equivalent to fresh cells.

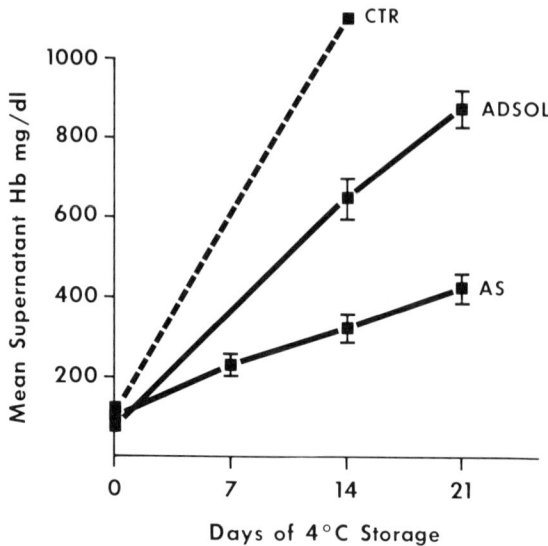

FIGURE 2-4 Supernatant hemoglobin levels of post-thaw red cells stored in saline/ glucose (CTR), ADSOL, or adenine/glucose/PO$_4$/mannitol (AS). In this study 1% lysis was about 300–350 mg/dl Hb.

Since thawed-washed red cells have been stripped of all plasma, which provided them with some buffering capacity and protection from lysis, and have also been depleted of intracellular metabolites due to the washing process, their metabolic needs are somewhat different from those of fresh cells. Therefore, we have examined modified additive solutions (AS) for post-thaw preservation (Moore et al. 1987a, 1987b). A solution containing adenine, glucose, mannitol, and trisodium phosphate showed greater potential than ADSOL. This AS produced lower lysis (Figure 2–4) and higher 2,3-DPG than ADSOL. By using D-optimality experimental design techniques we were able to optimize the formulation of this mixture to produce the maximal levels of both ATP and 2,3-DPG; however, the lysis of red cells in this mixture still exceeded 1% in some units (Figure 2–5). In an effort to lower lysis rates, we studied many antilysis/antioxidant reagents, including plasma, as components to the additive system. Mannitol helps retard lysis, but it alone is not sufficient. The mechanism of action of mannitol is not known, in spite of extensive studies of its actions (Beutler and Kuhl 1988). We have looked at the changes in membrane oxidation due to mannitol, but have seen no differences. Plasma and specifically a low-molecular-weight fraction of plasma is very effective in retarding lysis (Figure 2–6). Further investigation revealed that the active component was citrate, and substitution of citrate for mannitol in the AS solution was effective in reducing lysis by 50%, while only reducing the 2,3-DPG by 20% (Moore,

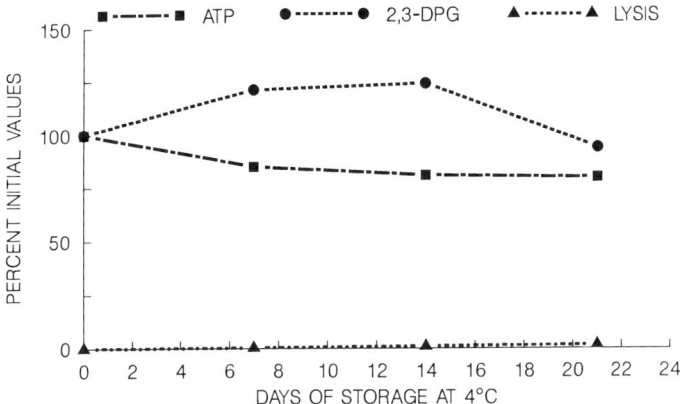

FIGURE 2–5 Post-thaw storage of red cells in an optimized additive solution (AS) containing adenine, glucose, sodium phosphate, and mannitol. Mean values for ATP, 2,3-DPG, and cell lysis are shown.

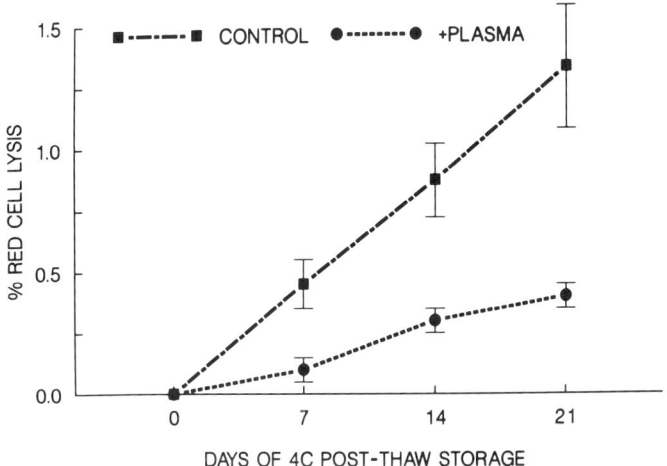

FIGURE 2–6 Lysis during post-thaw preservation of red cells in AS solution containing mannitol with or without 50 ml of added autologous plasma.

unpublished observations). Clinical trials on this additive solution are in progress.

In a preliminary report from Australia, a new method of freezing at $-20°C$ and re-storing red cells at $4°C$ was described (Lovric and Klarkowski 1989). This procedure is done in a five-bag circle pack and uses manual mixing and centrifugation to wash the thawed red cells. The cells are frozen at $-20°C$ in a mixture of glycerol and glucose, and they are viable for 6 months. Manual washing was done with an attached saline/glucose solution. The final wash was with an additive solution containing adenine, glucose, citrate, and disodium phosphate. The cells could be restored at 4°C for up to 35 days with "good in vitro parameters." Red cell survivals on such cells after storage at 4°C for 14–21 days were 70–81%. In its present configuration this system is bulky, slow, and labor intensive. However, it does show a potential for storing cells frozen in a commercial, −20°C freezer in a closed bag system. If such a system could be automated and optimized, it could have the potential of offering low-cost frozen storage. However, the maximum length of frozen storage time must be carefully explored.

2.5 CONCLUSIONS

Research in the 1960s and 1970s provided blood banks with quality red cell products having a 42-day, 4°C dating or frozen cells with a 3-year storage time. However, current medical and social challenges have created the need for new and different approaches to banked blood. With the advent of AIDS and increased awareness of other blood-borne viral diseases, autologous blood programs are growing at exponential rates. The need for autologous, long-term-stored, red cells are receiving new attention. In addition, the capability of frozen red cells to be stored for 10–20 years, and to be post-thaw-preserved for 2–3 weeks adds to the attractiveness, safety, and flexibility of frozen blood programs. The U.S. Department of Defense plan to stockpile frozen type O cells may lead to a reevaluation of stockpiling frozen cells for civilian emergencies. An additional impetus for developing new strategies for storing and processing red cells comes from the recent advances in the biotechnology field, where new innovations in membrane filtration, sterile transferring, and product manipulation are constantly appearing. Many of these technical advances will be incorporated into future blood products.

REFERENCES

Aubuchon, J.P., and Brightman, A. (1989) *Transfusion* 29, 143–147.
Beutler, E. (1985) *New Eng. J. Med.* 312, 1392.
Beutler, E., and Kuhl, W. (1988) *Transfusion* 28, 353–357.
Beutler, E., and West, C. (1984) *Transfusion* 24, 100–104.
Beutler, E., Forman, L., West, C., and Gelbart, T. (1988) *Biochem. Pharm.* 37, 1057–1060.
Buchholz, D., Aster, R., Menitone, J., et al. (1989a) *Transfusion* 29, 8s (Abstract).
Buchholz, D., Aster, R., Menitone, J., et al. (1989b) *Transfusion* 29, 51s (Abstract).
Estep, T.N., Pederson, R.A., Miller, T.J., and Stupar, K.R. (1984) *Blood* 64, 1270–1274.
Haugen, R.K. (1979) *New Eng. J. Med.* 301, 393–395.

Heaton, A., Miripol, J., Aster, R., et al. (1984) *Brit. J. Haem.* 57, 467–478.

Heaton, A., Aster, R.A., and Button, L. (1985) *New Eng. J. Med.* 312, 1391.

Heaton, W.A. (1986) in *New Frontiers in Blood Banking* (Wallas, C.H., and McCarthy, L.J., eds.), pp. 89–125. American Association of Blood Banks, Arlington, VA.

Heaton, W.A., Keegan, T., Holme, S., and Momoda, G. (1989a) *Vox Sang.* 57, 37–42.

Heaton, W.A.L., Hambury, C.M., Keegan, T.E., Pleban, P., and Holme, S. (1989b) *Transfusion* 29, 696–702.

Heaton, W.A.L., Keegan, T., Hambury, C.M., Holme, S., and Pleban, P. (1989c) *Transfusion* 29, 703–710.

Hogman, C.F., Hedland, K., Ackerblom, O., and Venge, P. (1978a) *New Eng. J. Med.* 299, 1377–1382.

Hogman, C.F., Hedland, K., Ackerblom, O., and Venge, P. (1978b) *Transfusion* 18, 233–241.

Hogman, C.F., Hedland, K., and Sahlestrom, Y. (1981) *Vox Sang.* 41, 274–281.

Hyde, R.M., Paterson, R.A., Livingstone, D.J., Batchelor, J.F., and King, W.R. (1984) *Lancet* 2, 15–16.

Lovric, V.A., and Klarkowski, D.B. (1989) *Lancet* 1, 71–73.

Marcus, C.S., Myhre, B.A., Angulo, M.L., et al. (1987) *Transfusion* 27, 415–419.

Meryman, H.T. (1979) *Prog. Hemat.* 11, 193–227.

Meryman, H.T. (1989) Public Communication, American Assoc. Blood Banks 42nd Annual Meeting, New Orleans, LA.

Meryman, H.T., and Hornblower, M. (1972) *Transfusion* 12, 145–156.

Meryman, H.T., Hornblower, M., and Syring, R.L. (1986) *Transfusion* 26, 500–505.

Moore, G.L. (1983) *Diag. Med.* 6(Sept.), 33–43.

Moore, G.L. (1987) *CRC Crit. Revs. in Clin. Lab. Sci.* 25, 211–229.

Moore, G.L., Ledford, M.E., Mathewson, P.J., and Hankins, D.J. (1987a) *Transfusion* 27, 496–498.

Moore, G.L., Ledford, M.E., Mathewson, P.J., Hankins, D.J., and Shah, S.B. (1987b) *Vox Sang.* 53, 15–18.

Moroff, G., Sohmer, P.R, Button, L.N., et al. (1984) *Transfusion* 24, 109–114.

Page, P.L. (1985) *New Eng. J. Med.* 312, 1391–1392.

Paterson, R.A., Dawson, J., Hyde, R.M., et al. (1988) *Transfusion* 28, 34–37.

Peck, C.C., Moore, G.L., and Bolin, R.B. (1981) *CRC Crit. Revs. in Clin. Lab. Sci.* 13, 173–212.

Radovich, J.M. (1989) U.S. Army Medical Research and Development Command Contract DAMD-17-86-E-6142, Final Report, Fort Dietrick, MD.

Robertson, O.H. (1918) *Br. Med. J.* 1, 691–695.

Rock, G., Secours, V.E., Franklin, C.A., Chu, I., and Villeneve, D.C. (1978) *Transfusion* 18, 553–558.

Ross, D.G., Heaton, W.A.L., and Holmes, S. (1989) *Vox Sang.* 56, 75–79.

Rous, P., and Turner, J.R. (1916) *J. Expl. Med.* 23, 219–237.

Rubin, R.J., and Ness, P.M. (1989) *Transfusion* 29, 358–361.

Shimizu, T., Kouketsu, K., Morishima, Y., et al. (1989) *Transfusion* 29, 292–297.

Suzuki, T., and Dale, G.L. (1987) *Blood* 70, 791–795.

Turnbull, D., and Rodricks, J.V. (1985) *J. Am. Coll. Toxicol.* 21, 111.

Valeri, C.R. (1970) *CRC Crit. Revs. in Clin. Lab. Sci.* 1, 381–425.

Valeri, C.R. (1976) *Blood Banking and the Use of Frozen Blood Products*, CRC Press, Boca Raton, FL.

Valeri, C.R. (1985) *New Eng. J. Med.* 312, 1392–1393.

Valeri, C.R. (1988) *Methods in Hemat.* 17, 277–304.

Valeri, C.R., Valeri, D.A., Dennis, R.C., Vecchione, J.J., and Emerson, C.P. (1979) *Crit. Care Med.* 7, 439–447.

Valeri, C.R., Valeri, D.A., Anastasi, J., et al. (1981) *Transfusion* 21, 138–149.

Valeri, C.R., Pivacek, L.E., and Gray, A.D. (1989) *Transfusion* 29, 429–437.

Widmann, F.K. (1985) in *Technical Manual*, 9th ed., pp. 59–69, AABB Press, Arlington, VA.

3

Autologous Blood Salvage Procedures

Linda Stehling
Howard L. Zauder

Salvage and reinfusion of shed blood is employed intraoperatively and postoperatively. In addition, patients sustaining chest trauma may benefit from the technique whether or not surgical exploration is required. The concept is not new: Blundell (1819) autotransfused 10 women with postpartum hemorrhage; five of the patients survived. Highmore (1874) is credited with bringing the procedure to the attention of the medical profession in a communication to the journal *Lancet* in 1874. The first case report in the American literature was published by Lockwood (1917) who reinfused 750 ml of blood obtained from the spleen of a patient with Banti's disease. Davis and Cushing (1925) reported use of autotransfusion in 22 patients undergoing intracranial procedures. Tiber (1934) described experience with the technique in 123 patients with ruptured ectopic pregnancies. Griswold and Ortner (1943) reported 100 cases of trauma to the abdomen or thorax during which autotransfusion was employed.

Sporadic reports of reinfusion of blood retrieved from patients sustaining traumatic hemothoraces appeared in the literature from 1931 (Brown

This research was supported in part by NHLBI Transfusion Medicine Academic Award number 1KO7 HLO1257.

and Debenham 1931) through the early 1970s. The procedure was usually utilized only as a life-saving measure and little was known about the quality of the product reinfused. In 1966, Symbas (1978) initiated laboratory and clinical studies which provided information about the technique, which he and his coworkers employed in 400 patients with acute traumatic hemothoraces.

The use of instrumentation developed specifically for intraoperative salvage and reinfusion was first reported by Dyer (1966) who designed a two-chambered pyrex unit attached to a suction apparatus. Either systemic heparinization or chamber anticoagulation was employed, and three sets of filters were used to remove debris from the blood prior to reinfusion. Two years later, Wilson and Taswell (1968) published a preliminary report on the use of a continuous-flow centrifuge bowl capable of separating and concentrating erythrocytes in the irrigating fluid returned during transurethral resection of the prostate. They reported clinical experience with the technique the following year (Wilson et al. 1969).

The first commercially available apparatus for intraoperative salvage and reinfusion was developed by Klebanoff and marketed by Bentley Laboratories. Initial clinical trials were performed in Vietnam and reported in 1970 (Klebanoff 1970). The device consisted of a disposable suction apparatus, cardiotomy reservoir and reinfusion tubing, and a DeBakey-type roller pump. Systemic anticoagulation was not employed, but the reservoir was primed with a heparinized crystalloid solution. The rate of reinfusion was regulated by controlling the air pressure in the reservoir. Although the device was equipped with an air-detecting sensor and a safety alarm, several cases of air embolism did occur, leading to withdrawal of the unit from the market. However, clinical experience with the technique in patients sustaining blunt and penetrating trauma to the chest and abdomen, as well as those undergoing a variety of elective and emergency intraabdominal procedures, demonstrated the merits of the technique and paved the way for development of more sophisticated devices (Stehling et al. 1975).

In 1974, the Haemonetics Corp. introduced a system incorporating washing and concentrating of salvaged red blood cells prior to reinfusion. At the present time, four companies market systems which employ cell washing. In addition, there are a variety of devices for the collection of shed blood which do not incorporate a cell-washing process. Blood salvaged with these devices may be washed prior to infusion or administered without washing. While the advantages of autologous salvage procedures are the same for all types of apparatus, the applications, contraindications, and potential complications differ. In addition, the characteristics of the product reinfused vary with the collection and processing technique.

3.1 ADVANTAGES OF AUTOLOGOUS TRANSFUSION

Avoidance of the potential complications of homologous transfusion is the primary motivation for employing autologous transfusion. Aside from the infectious diseases which may be transmitted, febrile, allergic, and acute and delayed hemolytic transfusion reactions can occur following infusion of bank blood. Technical and clerical errors associated with blood sample procurement and crossmatching are avoided when autologous blood is utilized. In addition, immunization to red blood cell, white blood cell, and platelet antigens, which can limit the availability, safety, and efficacy of subsequent transfusions, is eliminated. The potential immunomodulatory effects of homologous transfusion such as impaired response to infection and earlier tumor recurrence may not be associated with autologous transfusion.

Occasionally, autologous transfusion is the only option. Compatible homologous blood may not be readily obtained for patients with rare blood phenotypes. Blood loss can be so rapid that blood salvage is the only means of providing sufficient quantities of blood. While volunteer donors supply sufficient quantities of blood to meet the needs of virtually all patients, blood shortages do occur. Utilization of autologous transfusion conserves the homologous blood supply for patients who are not candidates for autologous transfusion. Finally, the technique is acceptable to some patients who refuse homologous transfusion for religious reasons.

3.2 CLINICAL APPLICATIONS

Blood salvage should be considered whenever it is anticipated that significant blood loss will occur from a clean wound, permitting retrieval without excessive hemolysis. A favorable cost:benefit ratio often exists when one or more of the following criteria are present: the anticipated blood loss is greater than 20% of the patient's blood volume; blood would ordinarily be crossmatched for the procedure; the mean transfusion for the proposed operation exceeds one unit; more than 10% of patients undergoing the procedure require transfusion.

Intraoperative salvage is employed extensively in cardiac (Ottesen and Froysaker 1982; Breyer et al. 1987; Giordano et al. 1988a; McCarthy et al. 1988; Davies et al. 1988; Dietrich et al. 1989) and vascular surgery (Tawes et al. 1986; Hallett et al. 1987; Stanton et al. 1987) and in orthopaedics (Bovill et al. 1986; Lennon et al. 1987; Goulet et al. 1989; Wilson 1989). Centers with active liver transplantation programs also use the technique (Dzik and Jenkins 1985; Williamson et al. 1989). There is less experience in gynecology, urology, and neurosurgery (Santrach et al. 1989; Keeling et al. 1983). Patients with blunt and penetrating injuries of the chest and abdomen are often appropriate candidates (Boudreaux et al. 1983; Timberlake and McSwain 1988).

It is generally considered that intraoperative autotransfusion is cost effective if more than two units of blood are salvaged and reinfused (Popovsky and Devine 1985). However, a recent study suggests that cost equivalence with homologous blood may not be achieved unless greater volumes of blood are salvaged (Solomon et al. 1988). Such financial calculations do not include the potential benefits of avoidance of transfusion-transmitted disease.

3.3 CONTRAINDICATIONS

Most contraindications are relative. Salvage of blood shed during cancer surgery has been considered contraindicated because of concerns about dissemination of tumor cells (Yaw et al. 1975). Laboratory investigations have demonstrated that when blood containing tumor cells is processed with a cell-washing device, the tumor cells are resuspended with the red blood cells and would therefore be reinfused (Dale et al. 1988). However, there are reports of intraoperative salvage in several patients undergoing surgery for genitourinary cancer without apparent ill effects (Klimberg et al. 1986; Hart et al. 1989).

Contamination of the wound with bowel contents or retrieval of blood from an infected site is usually a contraindication to intraoperative blood salvage. There are, however, reports of patients who received enteric-contaminated shed blood. The postoperative course of the patients, all of whom received parenteral broad-spectrum antibiotics, was not different from other patients sustaining similar injuries (Timberlake and McSwain 1988). In vitro studies have demonstrated that the bacterial count in the retrieved, contaminated blood decreases after the blood is processed. However, high concentrations of some bacteria, especially anaerobes, remain (Rumisek and Weddle 1981). In the rare circumstance when the alternative is exsanguination, autotransfusion should be considered.

A recent case report in which blood was aspirated during hepatic transplantation in a patient with sickle cell trait indicates that caution should be used in employing the technique in such patients (Brajtbord et al. 1989). While blood samples drawn from the patient and from the reservoir prior to processing showed no sickling, the blood drawn after processing exhibited a 50% incidence of sickling.

Use of the technique is not recommended when topical hemostatic agents such as thrombin or Avitene® are employed because they may initiate coagulation (Robicsek et al. 1986). Blood containing wound irrigants such as Betadine® or antibiotics not intended for parenteral use should not be salvaged. Blood should not be aspirated during application of methylmethacrylate, the cement used to stabilize joint prostheses. It is theorized that the exothermic reaction associated with "curing" of the cement could produce hemolysis. Aspiration of blood should cease prior to introduction of any of the aforementioned substances into the wound and the surgical field should be well irrigated before blood salvage is reinstituted. In obstetrics, the technique should be restricted to aspiration of blood not containing significant amounts of amniotic fluid. Intraoperative salvage is, however, acceptable in patients with ruptured ectopic pregnancies (Merrill et al. 1980).

3.4 INTRAOPERATIVE BLOOD SALVAGE DEVICES

There are three basic techniques of blood salvage, which are described below. Cell-processing devices utilize semicontinuous flow centrifugation to wash and concentrate red blood cells for reinfusion. Blood may also be aspirated into sterile, single-use, plastic reservoirs.

3.4.1 Canister Devices

The canister collection device was one of the earliest systems introduced for blood salvage. A sterile plastic liner is inserted into a rigid plastic canister (Figure 3–1). Appropriate connections are made to a vacuum source and a controlled volume of anticoagulant is delivered to the suction tip through the smaller lumen of a two-lumen tube. The larger of the lumens returns the salvaged, anticoagulated blood to the reservoir. When the liner is full or transfusion is indicated, an interface port is connected to the cell salvaging

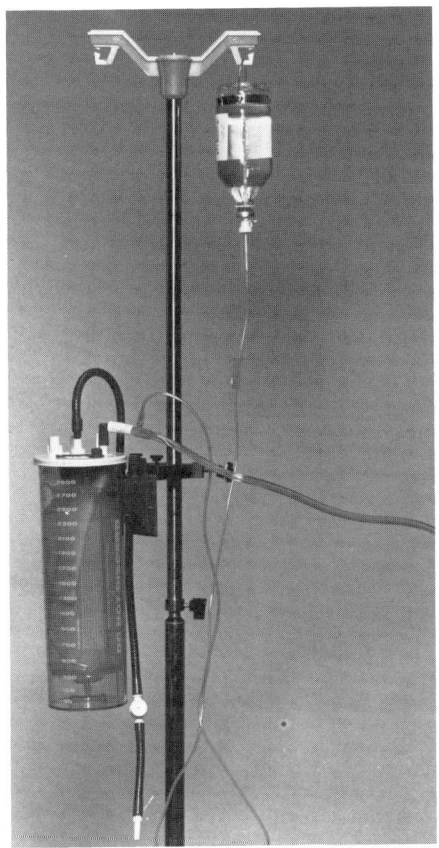

FIGURE 3–1 Rigid canister containing sterile disposable liner for blood collection. Anticoagulant solution and suction apparatus are attached to canister.

unit for processing. Alternately, the plastic liner may be taken to a remote location for processing of the blood. Least desirable is the direct administration of unprocessed blood. While primarily designed for operating room use, these units have been employed in emergency rooms and for chest drainage in post-thoracotomy patients.

Although canister devices have been used successfully in numerous types of operative procedures, obstruction of the filters with fat, bone fragments, and other debris has complicated their use in orthopaedics. For this reason, cardiotomy reservoir collection is employed during orthopaedic surgery in some institutions (Willis et al. 1989). The manufacturers of all devices recommend that a microaggregate filter be used when administering salvaged blood.

3.4.2 Cell-Processing Units

Semicontinuous flow devices filter, wash, pack, and return to the patient shed red blood cells collected from the surgical field. Four vendors, Cobe Laboratories (Lakewood, CO), Electromedics (Englewood, CO), Haemonetics (Braintree, MA), and Shiley (Irvine, CA) supply a variety of machines (Figure 3-2, 3-3, and 3-4) that process blood which would otherwise be discarded. Though they differ somewhat in the degree of sophistication (Figure 3-5, and 3-6), depending upon the microprocessor technology employed, they share schematic similarities (Figure 3-7). Basic to all are three pneumatic or electric solenoid valves, controlled manually or by the system's microprocessor, which determine the pathway of the salvaged blood at each stage of processing. The choices of apparatus depends upon the degree of automation desired, initial cost, and the availability of service. The units weigh about 250 pounds, require 3 square feet of floor space, and cost between $20,000 and $35,000. An exception is the Haemonetics' HaemoLite® (Figure 3-8), a portable, somewhat slower machine designed for intraoperative use as well as for transport to recovery rooms and intensive care units for postoperative processing.

Whole blood is salvaged from the surgical field through a two-lumen tube and small mixing chamber, which is attached to any conventional suction tip. The larger of the two lumens is attached to an appropriate reservoir, which in turn is connected to either wall suction or an integral suction pump. The vacuum in the collection reservoir should be maintained as far as possible at less than 100 mm Hg. Under no circumstances should it be permitted to exceed 150 mm Hg. Higher negative pressure, small orifice suction tips, "skimming" (i.e., aspirating air with the blood), and inadequate wash volume in orthopaedic procedures have been associated with gross hemolysis and renal dysfunction (Emergency Care Research Institute 1990). The smaller of the tubes terminates in a drip chamber and spike, which is inserted into a bag of anticoagulant solution. The anticoagulant merges with the blood suction tubing near the suction tip. Either heparin 30,000 units

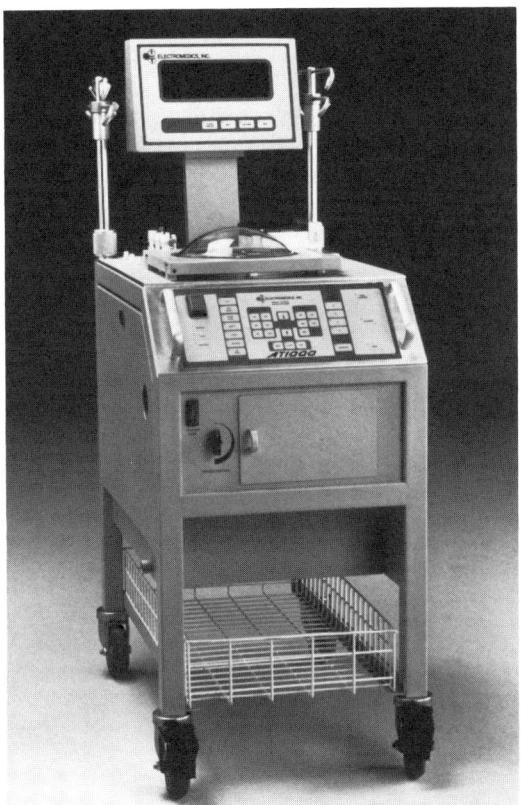

FIGURE 3-2 Electromedics AT 1000 Autotransfusion System. Courtesy of Electromedics, Inc.

in 1,000 ml of normal saline or citrate-phosphate-dextrose (CPD) may be used. Heparin is most often utilized for those procedures where the patient is heparinized (or is to be heparinized) while CPD is more frequently used for other patients. More important than the choice of anticoagulant is the need for meticulous attention to detail, by the operator, to insure that anticoagulant is introduced at a rate proportional to the rate of blood collection. As a general guideline, the ratio of citrate solution to blood should range between 1:5 and 1:10, or approximately 70 ml of citrate to 500 ml of salvaged blood. When heparin is used the flow is regulated so that approximately 15 ml of the solution is added to each 100 ml of blood, maintaining a heparin to blood ratio of approximately 1:7.

Salvaged blood is usually collected in a 3,000 ml cardiotomy reservoir fitted with a 120-, 40-, or 20-μm filter. Because blood collection is independent of the remainder of the processing cycle, a basic pack containing suction tubing and a collection bag may be substituted for the complete kit (Figure

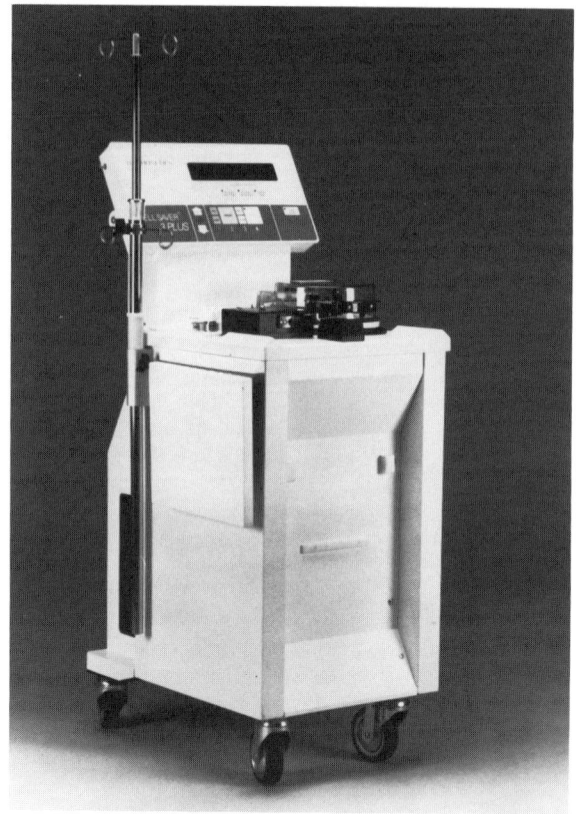

FIGURE 3-3 Haemonetics Cell Saver™ 3 Plus. Courtesy of Haemonetics Corp.

3-9) if there is uncertainty regarding the volume of blood that may be collected or whether the surgical wound may be contaminated. The option to process is retained while minimizing cost. When the standard 225-250 ml centrifuge bowl is used, at least 500 ml of aspirate should accumulate in the reservoir prior to the manual or automatic initiation of the fill cycle. This volume should be appropriately adjusted when larger (375-ml) or smaller (125-, 165-, or 175-ml) bowls are employed. The roller pump is activated, introducing blood into the centrifuge bowl spinning at 5,000 rpm (Figure 3-10). Because of their greater density, the red blood cells are packed against the outer wall of the bowl as it is filled (Figure 3-11). The supernatant flows through the outlet port to the collection bag. The level of red blood cells in the spinning bowl is sensed by an optical detector or the operator. The fill valve is closed, occluding the path from the reservoir to the bowl. At the same time the wash valve is opened, pumping normal saline into the spinning centrifuge bowl. Substances less dense than red blood cells,

FIGURE 3–4 Cobe's Baylor Rapid Autotransfuser: BRAT®. Courtesy of Cobe Laboratories, Inc.

e.g., anticoagulant, plasma-free hemoglobin, red blood cell stroma, and clotting factors, are washed out. While the majority of manufacturers use a Latham centrifuge bowl (Figure 3–12) the configuration of the Cobe (or Baylor) bowl (Figure 3–13) differs somewhat. The variation in design is said to allow for more rapid processing. However, while no problems have been identified clinically, an evaluation by the Emergency Care Research Institute (1988) indicates that this bowl is somewhat inefficient in washing out heparin added to an in vitro system. Washing continues until a predetermined volume of saline, usually 1,000 ml, is pumped through the bowl. At this point the machine is switched, either automatically or manually, to the empty cycle. The centrifuge is stopped, the roller pump reversed, and the empty valve opened while the fill and wash valves remain closed. The washed, packed, red blood cells are pumped into the reinfusion bag for administration. A number of devices are equipped with an air and foam detector that automatically shuts down the roller pump, thus minimizing

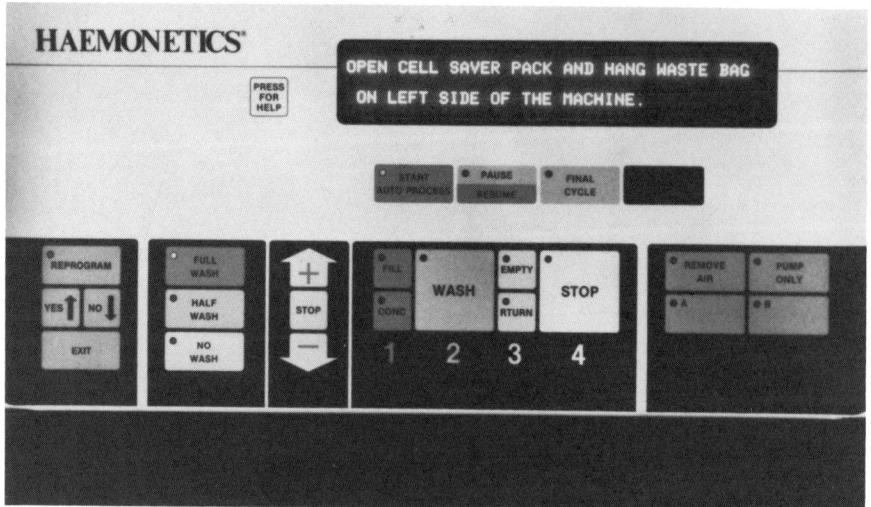

FIGURE 3–5 Operator panel of Haemonetics Cell Saver™. Courtesy of Haemonetics Corp.

FIGURE 3–6 Schematic of Cobe Brat™ control panel. Courtesy of Cobe Laboratories.

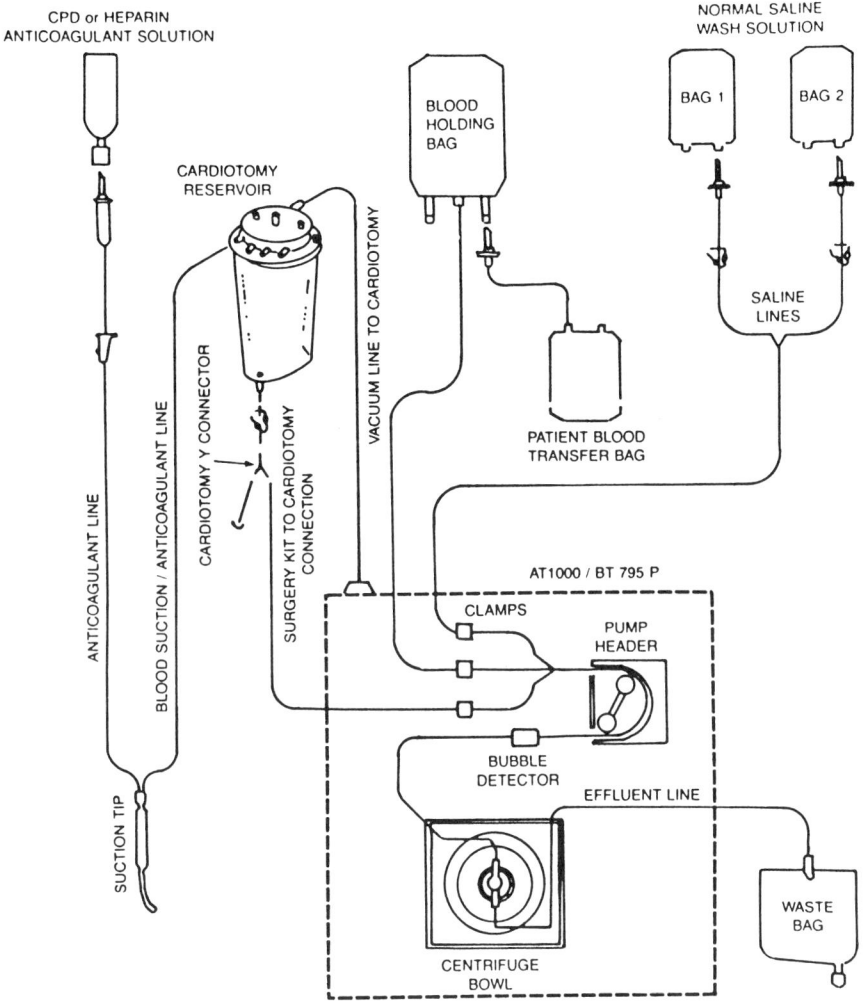

FIGURE 3–7 Schematic of Electromedics AT 1000 Autotransfusion System. Courtesy of Electromedics, Inc.

the danger of air embolism. From 3 to 10 minutes are required for a processing cycle depending upon the equipment and the processing parameters selected.

In a further attempt to decrease the use of homologous blood components, some investigators have advocated the use of autologous platelets (Giordano et al. 1988b). The platelets can be obtained by plateletpheresis two or three days prior to surgery. However, the relatively short shelf life of platelets and the uncertainty that surgery will proceed at the scheduled time makes it more desirable to collect platelet-rich plasma in the immediate

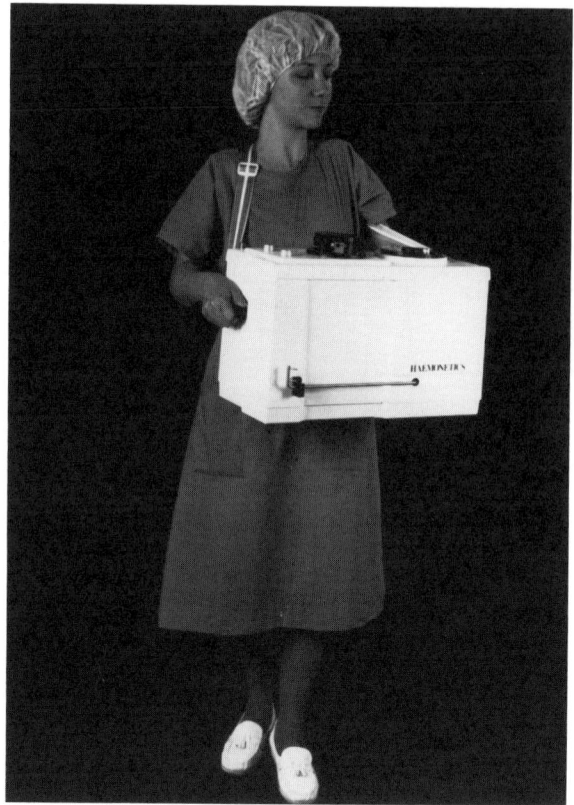

FIGURE 3–8 Haemonetics HaemoLite™ portable unit. Courtesy of Haemonetics Corp.

preoperative period. Haemonetics manufactures a modification of their Cell Saver® called the Plasma Saver®, which makes it possible to collect and store, from a single access site, within a 30-min period, 500 ml of plasma with a total platelet count of $1.0–2.0 \times 10^{11}$ (Giordano et al. 1988b). The apparatus requires a significant financial outlay and clutters the already overcrowded operating room with yet another piece of equipment. Some vendors have circumvented these objections with an additional disposable pack, the Plasma Sequestration Kit, that interfaces with their cell salvage systems. The tubing normally connected to the outlet of the cardiotomy reservoir is attached to a patient drawing/anticoagulant line. The pump and centrifuge speeds are slowed and blood is withdrawn from the patient at the rate of 25 ml/min. As with blood salvage, the blood is drawn into the spinning bowl. The supernatant, platelet-rich plasma, is collected in an appropriate reservoir while the residual red blood cells are pumped into a second bag for reinfusion. When the process of platelet sequestration is

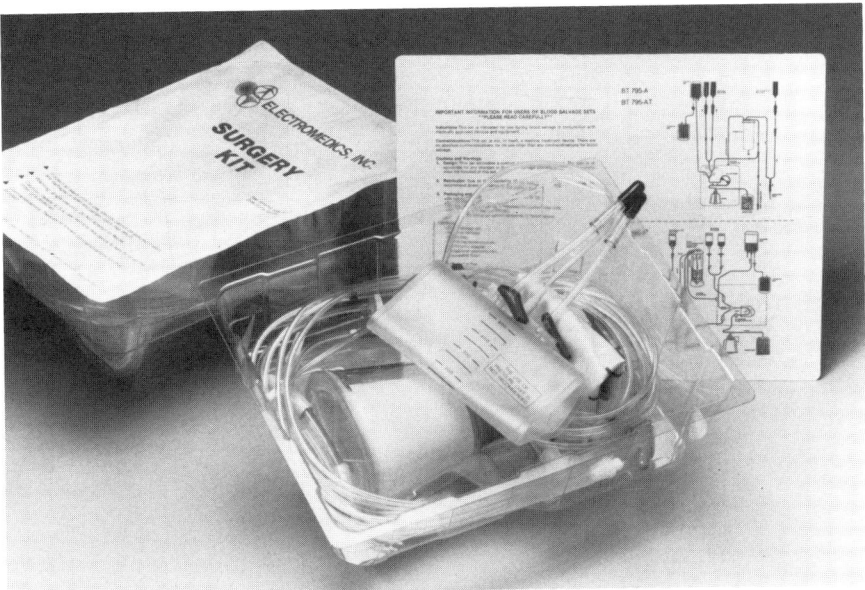

FIGURE 3-9 Electromedics Surgery Kit containing centrifuge bowl, harness, and reinfusion and waste bags. Courtesy of Electromedics, Inc.

FIGURE 3-10 Electromedics centrifuge well with bowl in place. Courtesy of Electromedics, Inc.

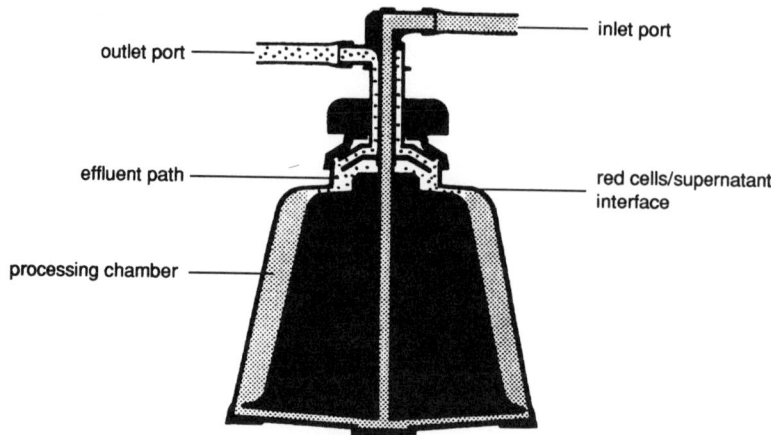

outlet port

inlet port

effluent path

red cells/supernatant interface

processing chamber

FIGURE 3–11 Schematic of Latham centrifuge bowl.

FIGURE 3–12 Latham centrifuge bowl.

complete, the patient is disconnected from the machine. At this time the sequestration disposables are discarded, the cell-processing equipment is configured in the conventional manner, and the equipment is ready to use for intraoperative cell salvage. The only additional expense is that of the disposables used to sequester the platelet-rich plasma.

3.4.3 Immediate Return Devices

The desire to eliminate the need for expensive systems, which require a dedicated operator, has led several manufacturers to develop and market devices that return unwashed blood suctioned from the surgical field. Since

FIGURE 3-13 Baylor centrifuge bowl. Courtesy of Cobe Laboratories, Inc.

elimination of plasma-free hemoglobin and debris is not possible, these units are best suited for use during major vascular procedures.

The O.R. Blood Banker® is a disposable device marketed by International Technidyne Corporation (Edison, NJ) supplied in a sterile package ready for use. The apparatus is positioned on an IV pole, attached to unregulated wall suction, flushed with saline, and primed with anticoagulant. Vacuum in the range of 50–80 mm Hg is automatically set by a regulator built into the blood reservoir. Anticoagulation is accomplished by mixing CPD solution with the blood as it is aspirated. An orifice in the suction tip allows the operator to control suction and thus the flow of anticoagulant. Occlusion of the orifice increases the negative pressure within the suction wand entraining anticoagulant. The vacuum regulator, valved delivery lines, and orifice diameters are designed to maintain a constant blood:citrate ratio, under variable conditions, that approximates that used in a standard unit of banked whole blood. A hard-shell, plastic, 2-liter, cardiotomy reservoir fitted with a 150-μm polyurethane filter coated with an antifoam agent collects the blood. Following filtration and defoaming the blood drains from the reservoir, by gravity, into an integral blood reinfusion bag that is placed below the reservoir. Any air that has accumulated in the bag can be purged through a closed circuit line that runs from the bag back into the reservoir. The reinfusion bag is elevated and after passage through a built in 20-μm micropore filter blood is returned to the patient. The maximum amount that can be processed is about 18 liters, the antifoaming capacity of the reservoir being the limiting factor.

A somewhat similar unit, the Intra-Op® (Figure 3–14) by Davol (Cranston, RI), is a self-contained, disposable device packaged ready for use. The conventional anticoagulant/suction double-lumen tube terminates in a trigger-controlled suction wand that allows the surgeon to control anticoagulant delivery. The system contains, in addition to the suction device and its tubing, an integral dial control to adjust the level of suction, a container with a large-surface-area, 265-μm filter, and a 1-liter collection/reinfusion bottle. A replacement kit facilitates continued collection while blood is being reinfused. The system, then, is not completely closed. Thus, there is danger of contamination when the urgent need for rapid collection and reinfusion may lead to less than meticulous attention to detail.

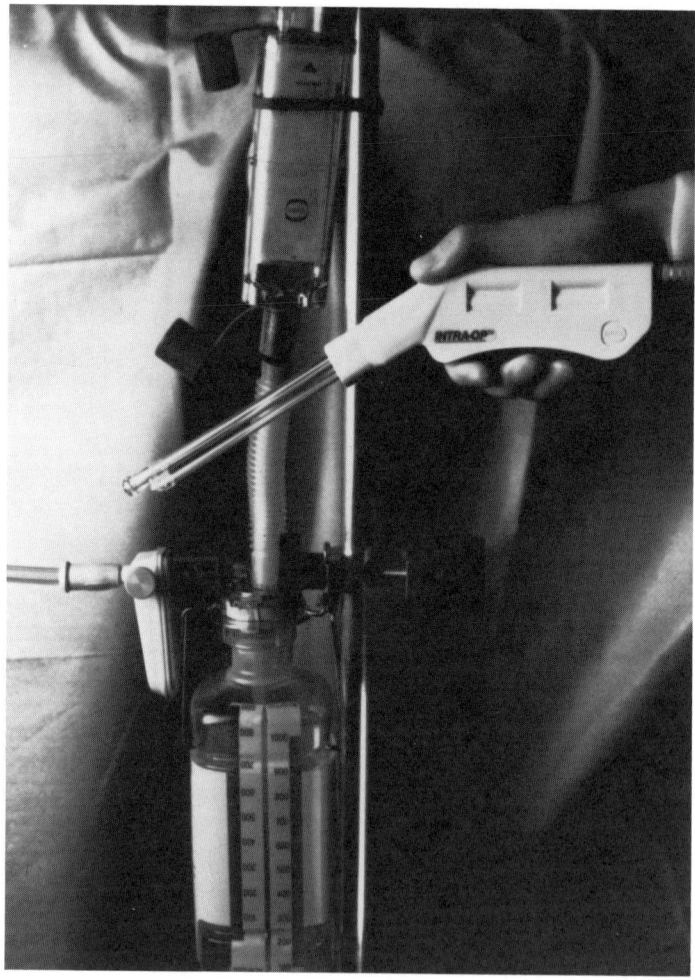

FIGURE 3–14 Davol Intra-Op® autotransfusion system. Courtesy of Davol, Inc.

A simple, disposable, self-contained unit, the Solcotrans®, marketed by Sulco Basle (Hingham, MA), differs from the other immediate return devices in that anticoagulant is added to the device prior to use. The need for a double-lumen anticoagulant/suction tube is thus eliminated. The system consists of a 600-ml PVC blood bag in a rigid container. The unit—together with its suction tubing, prefilter, and aspirator—is passed to the scrub nurse who adds 40 ml of acid-citrate-dextrose-adenine (ACD-A) solution to the blood bag and connects a source of vacuum. Blood aspirated from the surgical field passes through a prefilter located just proximal to the suction tip. The aspirate traverses the suction tubing to enter the blood bag where it mixes with the anticoagulant. The full bag in its rigid container is handed off the surgical field to the anesthesiologist for administration.

3.5 POSTOPERATIVE BLOOD SALVAGE

Blood salvage from the mediastinum and thoracic cavity is frequently employed following cardiac surgery. Similarly, reinfusion of blood collected from wound drainage following total joint replacement has become popular. In either case the need to anticoagulate the shed blood is controversial. A number of studies indicate that blood drained after it has come into contact with serosal surfaces becomes defibrinated and thus will not clot. There is some uncertainty that the same may not be true of blood drained from other sites.

Mediastinal blood may be collected following cardiac surgery in a reservoir attached to a conventional or modified chest drainage system. The Pleur-evac® Autotransfusion System (Figure 3–15), a modification of Deknatel's (Fall River, MA) standard underwater drainage set, and similar devices utilize a 1,000-ml plastic bag that hooks to the side of the water seal chest drainage unit. A 200-μm filter removes particulate matter as it enters the bag. When the unit is full or if transfusion is indicated earlier, the bag is removed and attached to an infusion set and microaggregate filter prior to administration. Changing reinfusion bags with this system requires interruption of the negative pressure applied to the chest tubes. A system provided by Atrium Medical Corp. (Valencia, CA) utilizes a smaller reinfusion bag that may be filled and transferred without disconnecting the chest tubes from the underwater seal (Figure 3–16). The Davol intraoperative autotransfusion system easily converts to their Thora-Klex® chest drainage system. These devices have also been used to salvage blood from patients who have sustained thoracic trauma.

Blood salvaged from the mediastinum contains virtually no fibrinogen because it coagulates upon contact with serosal surfaces and then undergoes fibrinolysis. Consequently, anticoagulation is not usually employed during collection of shed mediastinal blood. Since the blood does contain fibrin degradation products, its administration may confuse the laboratory diag-

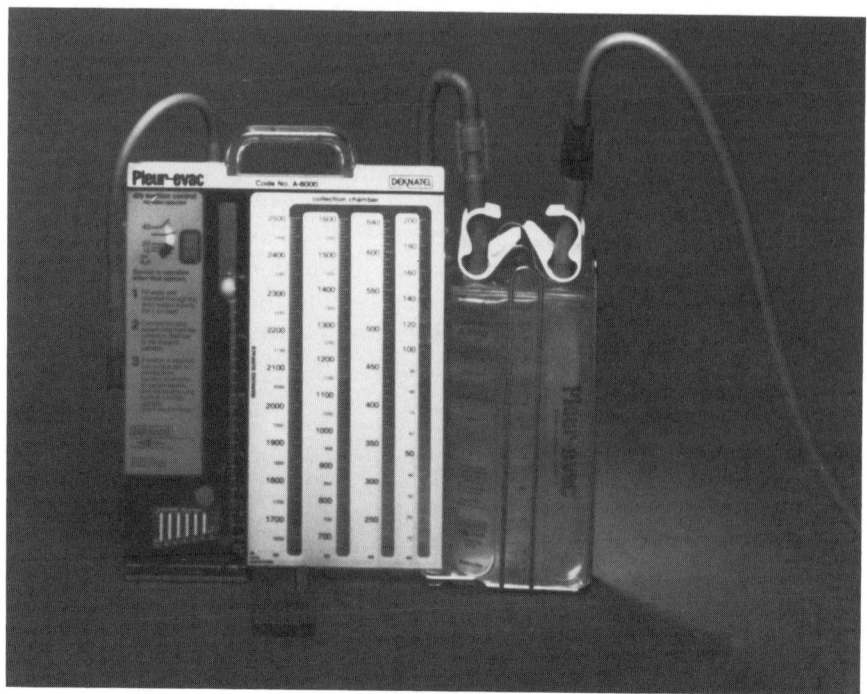

FIGURE 3-15 Deknatal Pleur-evac® Autotransfusion System. Courtesy of Deknatal, Inc.

nosis of disseminated intravascular coagulation or fibrinolysis and lead to unnecessary blood component administration (Griffith et al. 1989). Washing the infusate eliminates this problem. A number of cardiac surgeons utilize the cardiotomy reservoir taken from the heart-lung bypass machine at the termination of surgery as both a collection chamber and reinfusion receptacle for mediastinal drainage. In this case volume is measured hourly and the blood is returned by an infusion pump over a similar time period (Cosgrove et al. 1985).

The hematocrit of the blood salvaged from the mediastinum ranges from 20 to 25% (Thurer et al. 1979; Carter et al. 1981; Solem et al. 1987; Hartz et al. 1988). For this reason, many clinicians prefer to process the blood prior to transfusion.

Blood salvage following orthopaedic procedures is most frequently accomplished with a modified Solcotrans® apparatus (Figure 3-17) available from Richards Medical Co. (Memphis, TN). When this apparatus is used for postoperative blood salvage, a Y connector is inserted in place of the suction wand. A somewhat more complex device, the ConstaVac® distributed by Stryker Surgical (Kalamazoo, MI), is a closed system with integral

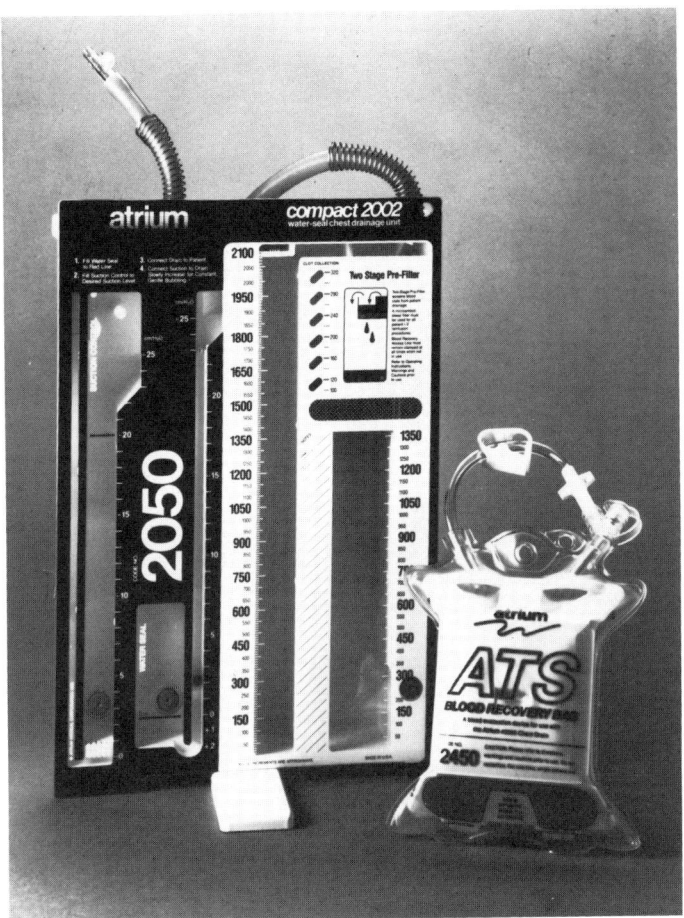

FIGURE 3–16 Atrium® blood recovery system. Courtesy of Atrium Medical Corp.

vacuum pump that prefilters and collects blood in an 800-ml reservoir. Blood is then transferred to a connected reinfusion bag for administration. A special feature of this device is the retention of the top 100 ml of fluid in the collection reservoir. Blood salvaged following total joint replacements often contains fat. Since the fat rises to the top of the chamber, reinfusion of this material is automatically precluded. Alternatively, a cell-washing unit may be moved to the recovery room to process blood drained from the operative site or the salvaged blood may be taken to the blood bank for processing.

The Standards of the American Association of Blood Banks require that the time from initiation of blood collection to completion of reinfusion not exceed 6 hours (Holland 1989).

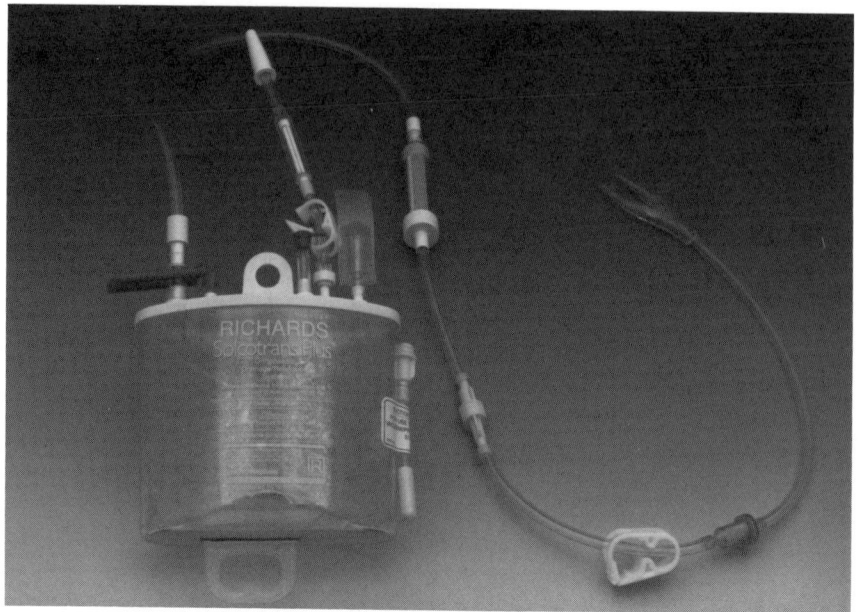

FIGURE 3–17 Richards-Sulcotrans® orthopaedic drainage and reinfusion system. Courtesy of Richards Medical Co.

3.6 COMPLICATIONS

Hemolysis occurs with high suction levels or if the technique of aspiration entails skimming of blood from the wound with significant air-blood interfacing. Inadequate washing results in infusion of blood containing significant free hemoglobin, which can produce renal damage. However, the amount necessary to induce nephrotoxicity is unknown. There is a recent report of two patients who required hemodialysis for renal dysfunction following infusion of shed, washed blood. Both were orthopedic patients. It was determined that excessive vacuum levels were used and insufficient normal saline wash volumes were employed (Emergency Care Research Institute 1990). In order to avoid this complication, it is recommended that suction regulators have factory-set maximum levels of 150 to 200 mm Hg and that the vacuum level be maintained no higher than 100 mm Hg whenever possible. Suction regulators should be inspected periodically for accuracy.

Coagulopathy can result from infusion of residual anticoagulant or dilution of coagulation factors. Hypocalcemia is a theoretical complication when citrate is used and very large volumes of salvaged blood are infused, especially in patients with severe liver disease. As with homologous transfusion, monitoring of coagulation parameters is recommended when large volumes of autologous blood are administered. The initiation of disseminated intravascular coagulation is a potential problem, especially in patients with poor perfusion secondary to prolonged hypotension. However, it has not been conclusively demonstrated to occur during intraoperative blood salvage, even with reinfusion of large quantities of processed blood (National Blood Resource Education Program Expert Panel 1990).

Salvaged blood contains variable amounts of leukocytes and platelets; however, the clinical significance of reinfusing these cellular elements is unknown. While limited animal studies suggest a detrimental effect on pulmonary function (Bull et al. 1988), the canine model probably does not accurately reflect the human clinical situation. A significant number of patients who sustain massive blood loss develop pulmonary dysfunction. However, extensive clinical experience appears to indicate that the incidence is no greater when autologous blood is reinfused intraoperatively. There is virtually no published data on the postoperative administration of blood shed following nonthoracic procedures. Many questions remain unanswered regarding the maximum amount of blood that can be safely reinfused and the potential for infection and pulmonary dysfunction.

Air embolism should not occur if the devices are used according to the manufacturers' directions. Salvaged and processed blood, especially that obtained during orthopedic procedures, does appear to contain residual fat. However, there are no documented cases of fat embolism.

3.7 CHARACTERISTICS OF SALVAGED BLOOD

Considerable data attest to the safety and efficacy of salvaged blood which has been processed prior to administration. While blood salvaged both intraoperatively and postoperatively has been reinfused without washing, data regarding the characteristics of the blood are noticeably absent from the literature. Thus, the following discussion relates to processed blood only.

Survival of the red blood cells is at least comparable to that of transfused homologous red blood cells. One group of investigators (Ray et al. 1986) using ^{51}Cr-tagged cells demonstrated that cells salvaged during spine surgery and processed with a Haemonetics Cell Saver® survived at least 24 days. The survival of reinfused autologous red blood cells obtained by preoperative phlebotomy was 45% and of homologous cells 43%. Using a similar technique, another group (O'Hara et al. 1983) demonstrated that survival of red blood cells salvaged during abdominal aortic surgery was 65%, measured 4-7 days following reinfusion. When a dual-isotope-labeling technique was employed in patients undergoing cardiovascular procedures, no significant difference was found in immediate or long-term (up to 288 hours) survival between processed reinfused red blood cells and those obtained by venipuncture following cardiopulmonary bypass (Ansell et al. 1982). Another study in cardiac patients also failed to demonstrate a difference between survival of salvaged red blood cells and those of normal individuals not undergoing surgery (Cordell and Lavender 1981). While morphologic changes in red blood cells are evident on microscopic examination, the significance of this finding is not known (Yawn and Bull 1988).

The 2,3-diphosphoglycerate (2,3-DPG) level of salvaged blood has been measured as a marker of red blood cell function. Processed blood consistently has higher 2,3-DPG levels than homologous blood (O'Hara et al. 1983; McShane et al. 1987; Yawn and Bull 1988).

Plasma-free hemoglobin levels in the range of 200 to 500 mg/ml are common in salvaged blood (O'Hara et al. 1983). However, 50–70% is removed during the washing process. Leukocytes, mostly neutrophils, are present in the salvaged blood as are low concentrations of coagulation proteins and fibrin degradation products (Yawn and Bull 1988). Complement activation during processing of salvaged blood has been demonstrated, but the washed product was complement-free (Siriwardhana et al. 1989).

Residual anticoagulant in the reinfused blood does not appear to be a frequent problem. One group of investigators, who assumed that the allowable level of heparin should be no greater than 0.1–0.2 units/ml, demonstrated in a simulated clinical situation that washing with a minimum of 500 ml of saline should be adequate, but concluded that 700 ml was advisable (Umlas and O'Neill 1981). Reported residual heparin levels with the various devices range from 0.019 to 0.064 units/ml (Ottesen et al. 1982; McShane et al. 1987).

In contrast to bank blood, the pH of salvaged, processed blood is usually alkaline. Sodium and potassium levels are in the physiological range (Yawn and Bull 1988). The hematocrit depends upon the processing parameters, but is usually 50–60%.

The fate of endogenous ligands in salvaged blood has not been extensively investigated. However, blood salvaged during resection of pheochromocytoma has been demonstrated to contain high levels of catecholamines after processing (Rice et al. 1987). Reinfusion of the blood was associated with significant hypertension (Smith et al. 1983).

There has been concern about the washout of therapeutically active drugs. However, studies involving the muscle relaxant d-tubocurarine (Shanks et al. 1985) and the narcotic fentanyl (Hanowell et al. 1989) have demonstrated that the amount of drug removed is not clinically significant.

3.8   ADMINISTRATIVE CONSIDERATIONS

Several questions should be addressed prior to establishing a blood salvage program. The types of operative procedures and the frequency with which they are performed determine which devices should be purchased and the personnel requirements. An individual, usually an anesthesiologist, surgeon, or transfusion medicine specialist, must be willing to assume responsibility for the program. The hospital transfusion committee must also be involved in reviewing protocols for the handling of salvaged blood and evaluating utilization of the service for safety and appropriateness.

When cell-processing devices are used, the equipment may be purchased by the hospital and hospital personnel utilized to provide the service. Alternatively, in some areas of the country it is possible to enter into a contractual arrangement with a blood center or private company for the service. Whatever the arrangement, the service should be available 24 hours a day, 7 days a week. If autologous transfusion represents good patient care during the day, it should also be a standard of care at night and on weekends.

Only a trained, dedicated operator should operate cell saving devices (National Blood Resource Education Program Expert Panel 1990). Duties of the operator include set-up of the machine, selection of processing parameters, regulation of the anticoagulant, monitoring processing cycles, and changing wash solutions and waste bags. While perfusionists are ideally suited to operate the devices, nurses or technicians can also be trained to perform these functions.

There should be written protocols defining responsibilities for the service, scheduling procedures, and the qualifications required of those operating the devices. Procedures for handling salvaged blood should conform to the Standards of the American Association of Blood Banks (Holland 1989). A procedure note documenting the type of apparatus used, the amount of blood salvaged and reinfused, the name of the operator, and any complications should be placed in the patient's chart. It is also desirable to include the manufacturer, lot number, and expiration date of all disposables used. Figure 3–18 is an example of a useful summary sheet.

Intraoperative and postoperative autologous transfusion employing methods of blood salvage other than cell processing must also be documented. Review of the autologous transfusion program should be a part of the hospital's quality assurance program. Audit methods include both patient-specific chart reviews and system-wide review of the program.

3.9 SUMMARY

Autologous transfusion is not a luxury. Some states have passed legislation mandating that surgical patients be informed of the alternatives to homologous blood administration as well as the risks and benefits of transfusion. Technological advances have made autologous blood salvage in surgery, the postoperative period, and certain trauma situations a safe and relatively inexpensive procedure. While there is minimal data documenting the quality of blood which is administered without processing, extensive clinical experience attests to the safety of the procedure. Ample clinical and laboratory data support the safety of reinfusing processed blood.

Intraoperative Autologous Transfusion Summary Sheet

Date __ / __ / __

Surgeon _____

Anesthesiologist _____

Operator _____

Diagnosis _____

Surgical procedure _____

Surgery start time _____ Surgery finish time _____

Total shed blood collected _____ ml

Total RBCs reinfused _____ ml

Washed/unwashed
Wash volume (total) _____ ml

Disposables	Manufacturer	Lot no.	Exp. Date
Reservoir			__ / __ / __
Surgical kit			__ / __ / __
Suction			__ / __ / __
Heparin/citrate			__ / __ / __

Comments: _____
_____
_____

Operator's signature

Name

Address

City   State   Zip

Hospital no.

Room no.

FIGURE 3–18 An example of an intraoperative autologous transfusion summary sheet.

REFERENCES

Ansell, J., Parrilla, N., King, M., et al. (1982) *J. Thorac. Cardiovasc. Surg.* 84, 387–391.

Blundell, J. (1819) *Med. Chir. Trans.* 10, 296–298.

Boudreaux, P., Bornside, G.H., and Gohn, L. (1983) *J. Trauma* 23, 31–35.

Bovill, D.F., Moulton, C.W., Jackson, W.S., et al. (1986) *Orthopaedics* 9, 1403–1407.

Brajtbord, D., Johnson, D., Ramsay, M., et al. (1989) *Anesthesiology* 70, 878–879.

Breyer, R.H., Engelman, R.M., Rousou, J.A., and Lemeshow, S. (1987) *J. Thorac. Cardiovasc. Sur.* 93, 512–522.

Brown, A.L., and Debenham, M.W. (1931) *JAMA* 96, 1223–1225.

Bull, M.H., Bull, B.S., Van Arsdell, G.S., and Smith, L.L. (1988) *Arch. Surg.* 123, 1073–1078.

Carter, R.F., McArdle, B., and Morritt, G.M. (1981) *Anaesthesia* 36, 54–59.

Cordell, A.R., and Lavender, S.U. (1981) *Ann. Thorac. Surg.* 31, 421–425.

Cosgrove, D.M., Amiot, D.M., and Meserko, I.J. (1985) *Ann. Thorac. Surg.* 40, 519–520.

Dale, R.F., Kipling, R.M., Smith, M.F., et al. (1988) *Br. J. Surg.* 75, 581.

Davies, M.J, Picken, J., Buston, B.F., and Fuller, J.A. (1988) *Med. J. Aust.* 149, 517–519.

Davis, L.E., and Cushing, H. (1925) *Surg. Gynecol. Obstet.* 40, 310–322.

Dietrich, W., Barankay, A., Dilthey, G., Mitto, H.P., and Richter, J.A. (1989) *J. Thorac. Cardiovasc. Surg.* 97, 213–219.

Dyer, R.H. (1966) *Am. J. Surg.* 112, 874–878.

Dzik, W.H., and Jenkins, R. (1985) *Arch. Surg.* 120, 946–948.

Emergency Care Research Institute. (1988) *Health Devices* 17, 219–242. Plymouth Meeting, PA, ECRI.

Emergency Care Research Institute. (1990) *Health Devices* 19, 25–26. Plymouth Meeting, PA, ECRI.

Giordano, G.F., Goldman, D.S., Mammana, R.B., et al. (1988a) *J. Thorac. Cardiovasc. Surg.* 96, 382–386.

Giordano, G.F., Rivers, S.L., Chung, C.K.T., et al. (1988b) *Ann. Thorac. Surg.* 46, 416–419.

Goulet, J.A., Bray, T.J., Timmerman, L.A., et al. (1989) *J. Bone Joint Surg.* 71A, 3–7.

Griffith, L.D., Billman, G.F., Daily, P.O., and Lane, T.A. (1989) *Ann. Thorac. Surg.* 47, 400–406.

Griswold, R.A., and Ortner, A.B. (1943) *Surg. Gynecol Obstet.* 77, 167–177.

Hallett, J.W., Popovsky, M.A., and Ilstrup, D. (1987) *J. Vasc. Surg.* 5, 601–701.

Hanowell, L.H., Eisele, J.H., and Erskine, E.V. (1989) *Anesth. Analg.* 69, 239–241.

Hart, O.J., Klimberg, I.W., Wajsman, Z., and Baker, J. (1989) *Surg. Gynecol. Obstet.* 169, 302–306.

Hartz, R.S., Smith, J.A., and Green, D. (1988) *J. Thorac. Cardiovasc. Surg.* 96, 178–182.

Highmore, W. (1874) *Lancet.* 1, 89.

Holland, P.V. (1989) *Standards for Blood Banks and Transfusion Services* (13th ed.), American Association of Blood Banks, Arlington, VA.

Keeling, M.M., Gray, L.A., Brink, M.A., et al. (1983) *Ann. Surg.* 197, 536–541.

Klebanoff, G. (1970) *Am. J. Surg.* 120, 718–722.

Klimberg, I., Sirois, R., Wajsman, Z., and Baker, J. (1986) *Arch. Surg.* 121, 1326–1329.

Lennon, R.L., Hosking, M.P., Gray, J.R., et al. (1987) *Mayo Clinic Proc.* 62, 1090–1094.

Lockwood, C.D. (1917) *Surg. Gynecol. Obstet.* 25, 88–191.

McCarthy, P.M., Popovsky, M.A., Schoaff, H.V., et al. (1988) *Mayo Clinic Proc.* 63, 225–229.

McShane, A.J., Power, C., Jackson, F.J., et al. (1987) *Br. J. Anaesth.* 59, 1035–1039.

Merrill, B.S., Mitts, D.L., Rodgers, W., and Weinberg, P.C. (1980) *J. Reprod. Med.* 24, 14–16.

National Blood Resource Education Program Expert Panel. (1990) *JAMA* 263, 414–417.

O'Hara, P.J., Hertzer, N.R., Santilli, P.H., and Bevan, E.G. (1983) *Am. J. Surg.* 145, 215–226.

Ottesen, S., and Froysaker, T. (1982) *Scand. J. Thorac. Cardiovasc. Surg.* 16, 263–268.

Popovsky, M.A., and Devine, P.A. (1985) *Mayo Clinic Proc.* 60, 125–134.

Ray, J.M., Flynn, J.C., and Bierman, A.H. (1986) *Spine* 11, 879–882.

Rice, J.M., Violante, E.V., and Kreul, J.F. (1987) *Anesthesiology* 67, 1017.

Robicseck, F., Duncan, G.D., Born, G.V.R., et al. (1986) *J. Thorac. Cardiovasc. Surg.* 92, 766–770.

Rumisek, J.D., and Weddle, R.L. (1981) in *Autotransfusion* (Hauer, J.M., Thurer, R.L., and Dawson, R.B., eds.), pp. 105–113, Elsevier, New York.

Santrach, P.J., Williamson, K.R., Taswell, H.F., Cucchiara, R.F., and Piepgras, D.G. (1989) *Transfusion* 29, 23S.

Shanks, C.A., Avram, M.J., Ronai, A.K., and Boswher, D.J. (1985) *Anesthesiology* 62, 161–165.

Siriwardhana, S.A, Kawas, A., Lipton, J.L., et al. (1989) *Can. J. Anaesth.* 36, S120.

Smith, D.F., Mihm, F.G., and Mefford, I. (1983) *Anesthesiology* 58, 182–184.

Solem, J.O., Steen, S., Tengborn, L., Lindgren, S., and Olin, C. (1987) *Scand. J. Thorac. Cardiovasc. Surg.* 21, 149–152.

Solomon, M.D., Rutledge, M.L., Kane, L.E., and Yawn, D.H. (1988) *Transfusion* 28, 379–382.

Stanton, P.E., Shannon, J., Rosenthal, D., et al. (1987) *South. Med. J.* 80, 315–319.

Stehling, L.C., Zauder, H.L., and Rogers, W. (1975) *Anesthesiology* 43, 337–345.

Symbas, P.N. (1978) *Surgery* 84, 722–728.

Tawes, R.L., Schribner, R.G., Duval, T.B., et al. (1986) *Am. J. Sur.* 152, 105–109.

Thurer, R.L., Lytle, B.W., Cosgrave, D.M., and Loop, F.D. (1979) *Ann. Thorac. Surg.* 27, 500–507.

Tiber, L.J. (1934) *Calif. West. Med.* 41, 16–19.

Timberlake, C.A., and McSwain, N.E. (1988) *J. Trauma* 28, 855–857.

Umlas, J., and O'Neill, T.P. (1981) *Transfusion* 21, 70–73.

Williamson, K.R., Taswell, H.F., Rettke, S.R., and Krom, R.A. (1989) *Mayo Clinic Proc.* 64, 340–345.

Willis, E.A., Williamson, K.R., Taswell, H.F., et al. (1989) *Transfusion* 29, 23S.

Wilson, J.D., and Taswell, H.F. (1968) *Mayo Clinic Proc.* 43, 26–35.

Wilson, J.D., Utz, D.C., and Taswell, H.F. (1969) *Mayo Clinic Proc.* 44, 374–386.

Wilson, W.W. (1989) *J. Bone Joint Surg.* 71A, 8–14.

Yaw, P.B., Sentany, M., Link, W.J., Wahle, W.M., and Glover, J.L. (1975) *JAMA* 231, 490–491.

Yawn, D.H., and Bull, B.H. (1988) in *Autologous Blood Transfusion: Current Issues* (Maffei, L.M., and Thurer, R.M., eds.), pp. 43–55, American Association of Blood Banks, Arlington, VA.

The Production of Group O Cells

Leslie L. Lenny
Jack Goldstein

The ability to maintain a constant supply of "universal blood type," or group O red blood cells (RBC)—cells that can be transfused to recipients of any blood group—would have major impact on both the management of blood inventories and the efficiency of transfusion therapy. Moreover, an inventory composed *solely* of group O RBC would eliminate the need for maintaining groups A, B, and AB blood (which sometimes are outdated and discarded) and make ABO typing of each unit unnecessary. An all-group O inventory would have benefit not only in specialized areas (e.g., combat casualty treatment; cases involving rare blood groups in which the rare type may be found but the cells remain incompatible with the recipient because of differing ABO groups) but also in transfusion medicine in general in the day-to-day management of every patient.

The goal of the studies in our laboratory is threefold: 1) to develop procedures to provide a constant supply of blood group O RBC; 2) to do so by enzymatic methods which convert groups A, B, and AB into O; and 3) to demonstrate that these enzymatically converted O RBC (ECO RBC)

The authors gratefully acknowledge the support of the United States Naval Medical Research and Development Command (N00014-84-C-0242, N00014-84-C-0307) and the Office of Naval Research (84-C-0543) in these studies.

survive and function normally in vivo and can be used in transfusion therapy in the same manner as native group O cells.

4.1 BIOCHEMISTRY, GENETICS, AND FORMATION OF THE ABO BLOOD GROUP ANTIGENS

Blood groups B, A, and O have similar carbohydrate-defined specificities but it is the terminal sugar residue at the nonreducing end of the carbohydrate chain that confers A, B or O antigenicity and distinguishes one group from another (Lloyd and Kabat 1968; Kabat 1970; Rovis et al. 1973; Ginsburg et al. 1971). The glycomoiety is then attached at its reducing end to either protein or lipid, which, in turn, is inserted into the RBC membrane. Hence, while the protein or lipoprotein specificities of other RBC antigens are integral to membrane structure (e.g., Rh), the immunodominant sugar residues of the ABO carbohydrate moieties are not. This convenient location external to the surface of the membrane suggests that their manipulation without compromising membrane integrity might be possible. Also fortuitous is the manner in which the ABO antigens are formed in vivo, since the terminal sugar residue carrying A or B specificity will be attached to the already-assembled group O carbohydrate chain: in the case of A antigen, the immunodominant sugar is N-acetyl-D-galactosamine (GalNAc) and for B, it is D-galactose (Gal), each in alpha linkage (Figure 4–1).

Assembly of the ABO carbohydrate chains is under the control of the ABH genes, which code for specific enzymes called transferases that attach the immunodominant sugar to the core polysaccharide (Watkins 1980). These transferases require that the specific sugar be attached to a nucleoside diphosphate: uridine diphosphate for A and B sugars and guanosine diphosphate for H (for group O specificity) sugar. The sequence of assembly is precise. In the presence of an H gene, a fucosyltransferase attaches L-fucose in α-1,2 linkage to the core polysaccharide. This H structure then acts as substrate for the further action of the A and B genes which, when present, code for N-acetylgalactosaminyltransferase or galactosyltransferase, respectively, attaching either GalNAc or Gal to the H structure. In the presence of both the A and B genes, some H structures will be converted to A and some to B, thereby conferring group AB specificity to the cells. Since this transfer of sugar to H-active structures is not entirely efficient, all A, B, and AB cells contain small amounts of unconverted H. The O gene is believed to be silent and no further modification of H substrate occurs. Group O cells thus are phenotypically H and contain L-fucose as their immunodominant sugar, and the designations ABO and ABH are sometimes used synonymously. As mentioned, this sequence of events is most precise since, in the absence of an H gene, no modification of the core carbohydrate structure occurs, even in the presence of A or B genes (Levine et al. 1955).

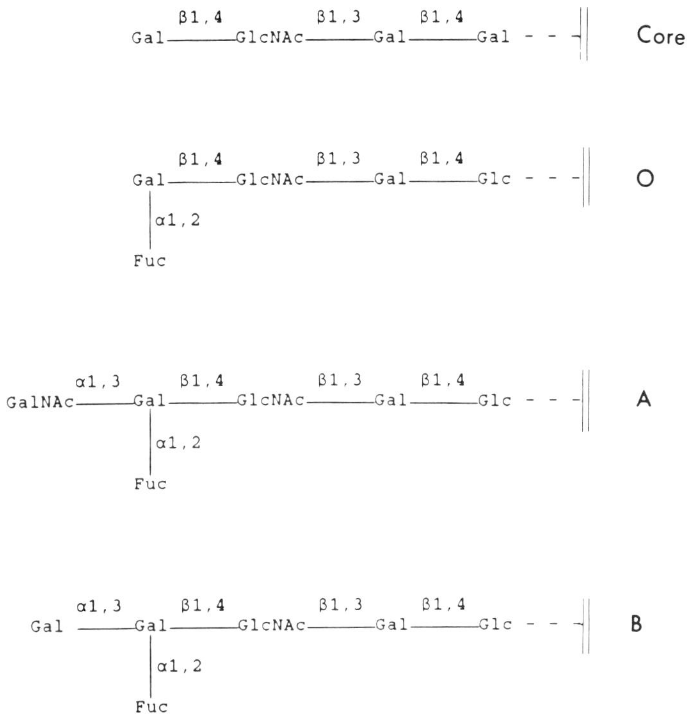

```
        β1,4          β1,3      β1,4
Gal————————GlcNAc————————Gal————————Gal - - -‖   Core
```

```
        β1,4          β1,3      β1,4
Gal————————GlcNAc————————Gal————————Glc - - -‖   O
  |α1,2
  Fuc
```

```
  α1,3      β1,4          β1,3      β1,4
GalNAc————Gal————————GlcNAc————————Gal————————Glc - - -‖   A
           |α1,2
           Fuc
```

```
  α1,3      β1,4          β1,3      β1,4
Gal ————Gal————————GlcNAc————————Gal————————Glc - - -‖   B
         |α1,2
         Fuc
```

FIGURE 4–1 Simplified carbohydrate structures of the polysaccharides of the red cell core and of blood groups A, B, and O. Abbreviations: Fuc, L-fucose; Gal, D-galactose; GalNAc, N-acetyl-D-galactosamine; Glc, D-glucose; GlcNAc, N-acetyl-D-glucosamine; - - - , glycoprotein or glycolipid; ‖, RBC membrane.

Cells lacking such structures are termed "Bombay" type and are extremely rare.

The actual number of A and B antigenic sites present on adult red cells has been estimated using several different techniques. It is generally believed that B cells contain 610,000–830,000 sites/cell and A cells, 120,000–1,170,000 sites/cell (Economidou et al. 1967). The wide range in the number of A sites is due to the existence of several subgroups of blood group A, each of which may contain a different amount of antigen (but see also Section 4.4.2). These investigators have also shown that AB red cells generally contain somewhat greater amounts of A antigen than of B.

The carbohydrate structures depicted in Figure 4–1 represent the most simplified forms of the ABO blood group specificities; in fact, these chains usually are longer and more complex, and they may exist as linear or branched structures (Koscielak et al. 1976; Dejter-Juszynski et al. 1978; Gardas 1978; Finne et al. 1978). Despite this complexity, which will be discussed in further detail in Section 4.4.2, it is clear that A and B specificities are built upon group O (H). It should be possible, then, to reverse this process in vitro by treatment with specific exoglycosidases that cleave the terminal immunodominant sugar residue from A and B glycoprotein or glycolipid structures, thereby producing group O.

4.1.1 History of the Enzymatic Degradation of Blood Group ABH Determinants from Membranes or Intact Red Cells

Several reports describing enzymes that degrade membrane blood group activity have appeared in the literature. Among the first were studies by Morgan (1947) who found that an enzyme preparation from *Clostridium welchii* serologically inactivated stroma from group O RBC, and by Watkins and Morgan (1954) who prepared a fucosidase from *Trichomonas foetus* and showed that it could produce a similar result with intact group O cells. Furukawa and Aminoff (1970), working with various *Clostridia* species, isolated and partially purified three enzymes, an α-galactosaminidase, an α-galactosidase, and an α-fucosidase, each capable of destroying A, B, and H activity, respectively. After reacting both crude and purified enzyme preparations with red cells having the appropriate blood group specificity, they found that cells treated with crude enzymes showed a certain amount of hemolysis and became polyagglutinated (i.e., agglutinable by all normal human sera regardless of blood type). On the other hand, those cells treated with purified enzymes showed no significant hemolysis and did not develop any polyagglutinability. Kinetic studies revealed that the specific agglutinability of A cells was lost after 4 hours, and that of B cells after 8 hours of treatment with their respective purified glycosidases. Following this, Aminoff and his colleagues studied the in vivo effect of converted type A cells in chicken and monkeys, but they were plagued by contamination of their enzyme preparations with appreciable amounts of sialidase, resulting in a premature destruction of these cells in vivo (Levy and Aminoff 1978). They have subsequently shown that enzyme preparations some 50 times more purified still contain appreciable sialidase activity, i.e., 0.1% of the activity of α-N-acetylgalactosaminidase (Levy and Aminoff 1980).

With regard to α-galactosidases, two groups have isolated and studied their effects upon B antigenicity. Flowers and collaborators, using enzyme from coffee bean and soy bean were able to remove B antigenicity from both ghosts and intact cells (Harpaz et al. 1975; 1977). This was done, however, under conditions that damaged the cells and made them unfit for transfusion. Oisha and Aida (1976) have isolated an α-galactosidase from a species of *Streptomyces* and showed that it could remove most but not all B antigen from the surface of red cells as measured by hemagglutination. Kubo (1989) recently obtained similar results with preparations from different strains of fungi which contained both α-N-acetylgalactosaminidases and α-galactosidases.

4.2 TREATMENT CONDITIONS COMPATIBLE WITH RBC VIABILITY

While the carbohydrate moieties of the ABO blood group antigens may not be integral to RBC membrane structure, the conditions under which the enzymatic treatment of intact red cells is carried out should provide for maintenance of RBC integrity and viability as well as optimal enzyme activity. We initially began these studies working with group B cells since α-galactosidases (B-zymes) were more readily available than α-N-acetyl-galactosaminidases (A-zymes). Using an α-galactosidase derived from green (unroasted) coffee beans, we first defined conditions by which group B red cells lost their reactivity with anti-B sera, yet maintained their structural and functional integrity.

The ability of α-galactosidase to cleave galactose residues from the red cell surface is pH dependent; peak enzyme activity occurs at pH 4.5–6.0, with a sharp decline between pH 6.0 and 6.5 and virtual absence of activity at pH 7.0 (Harpaz et al. 1977). Thus, to achieve maximal kinetics within a practical amount of time, cell treatment is performed at a pH below 6.0, and this requires that cell pH be reduced prior to and during enzymatic treatment. These conditions have been refined and optimized over the course of our studies. Briefly, group B cells are converted to group O in the following manner (Lenny et al. 1991): Packed red cells are washed in saline to remove residual plasma, platelets, and leukocytes, and then in pH 5.5–5.6 buffer to reduce their pH. The appropriate concentration of α-galactosidase is added and the mixture of cells and enzyme is incubated at 26°C until loss of reactivity with various test antisera is achieved. The enzyme is removed from the cells and the cell pH is restored to physiological pH by washing in phosphate-buffered saline, pH 7.3–7.4

α-Galactosidase from green (unroasted) Santos coffee beans is isolated and purified in its entirety in our laboratory. Purified preparations have a specific activity of 28 units per mg protein (Lowry) at 26°C when assayed at pH 6.5 using 1.25 mM p-nitrophenyl-α-D-galactoside as substrate (Kuo and Goldstein 1983) and a final concentration of 1,500–2,000 units per ml of buffer, pH 5.6. The enzyme is highly stable and can be repeatedly frozen and thawed without loss of activity. α-Galactosidase preparations are devoid of protease and sialidase, and levels of exoglycosidases (other than sialidase), if present, are found only in trace amounts, usually in the range of 0.00006–0.0002% of the activity of α-galactosidase. As such, the purity of our α-galactosidase far exceeds that of currently available commercial preparations. Enzyme-protein homogeneity is also monitored by SDS-acrylamide gel electrophoresis.

4.2.1 Serological Testing of Enzymatically Converted Group O Cells

After enzymatic treatment, cells have completely lost their reactivity with FDA-licensed commercial preparations of human polyclonal anti-B and murine monoclonal anti-B sera (Table 4–1). Other antisera have been used

TABLE 4-1 Reactivity of α-Galactosidase-Treated Human Erythrocytes with Anti-B and Anti-H Reagents[1]

| | Antiserum or Lectin Used | | | |
RBC Treatment	Anti-B Human Polyclonal	Anti-B Murine Monoclonal	Anti-H[2] Murine Monoclonal	Anti-H[2] Lectin
Untreated	12	12	5–8	3–5
Buffer (no enzyme)	12	12	5–8	3–5
α-Galactosidase	0	0	12	12

[1] Scoring system: A numerical system is currently used to score hemagglutination reactions, as it provides greater sensitivity than the old grading system. A comparison of the old and new scoring is as follows:

Old	New	
±	2–4	(microscopic reactions)
1+	5–6	
2+	7–8	
3+	9–10	
4+	11–12	

[2] All group B cells possess some H antigen that has not been converted to type B by galactosyltransferase. The amount of this unconverted H antigen varies somewhat from individual to individual.

from time to time, as will be discussed in Section 4.3.2. In addition, we have also had the opportunity to use exceptionally potent "raw" reagents, courtesy of Hyland Laboratories and Ortho Diagnostic Systems, Inc. These include a high-titer human anti-B serum as well as highly concentrated components of a murine monoclonal antiserum blend. It can also be noted in Table 4-1 that, as reactivity of these cells with anti-B sera decreases, reactivity with anti-H reagents (murine monoclonal anti-H, anti-H lectin) concomitantly increases, as expected, indicating exposure of H antigen or blood group O specificity.

Extensive antigen-profiling of enzyme-treated cells has also been performed as another parameter of membrane change, including tests of antigens A, Rh, MNS, P_1, Lewis, Kell, Duffy, and Kidd. The results showed that those antigens originally present on these cells remained and exhibited no detectable weakening, while those originally absent did not appear after enzyme treatment. Only B antigen and, if present, P_1 antigen, whose activity is also defined by an α-linked terminal galactose residue, are lost after enzyme treatment. (Although the rare antiserum anti-p^k was not available for testing, it is likely that p^k antigenicity is also lost since its specificity, too, is due to galactose in alpha linkage.) While these tests could not encompass all the known human blood group antigens (more than 600 at last count! [Issitt 1985], but many are extremely rare and most are insignificant in transfusion medicine), they did include those antigens considered to be clinically significant from a transfusion standpoint. The results suggest that cryptic antigens were not exposed nor were new antigens being formed from changes in preexisting ones as a result of enzymatic treatment. Enzymatically treated cells also retain their normal sialic acid content as determined by sialic acid analysis. Further, these cells react normally in the polybrene aggregation test, are not agglutinated by reagents which detect sialic acid reduction (*Arachis hypogea* lectin, *Glycine soja* lectin, normal human sera containing anti-T reactivity), and retain their normal complement of sialoglycoprotein antigens (the MNS blood group system).

4.2.2 Other Indices of Red Cell Integrity

Treated cells were then subjected to additional tests which demonstrated both their membrane (structural) and metabolic (functional) integrity, including determination of lysis and maintenance of normal RBC osmotic fragility, adenosine-5'-triphosphate (ATP), 2,3-diphosphoglycerate (2,3-DPG), acetylcholinesterase (AChE), cholesterol, methemoglobin (metHb), and P_{50} levels.

Less than 1% hemolysis is incurred during the enzymatic treatment procedure (Table 4–2). Examination of the cells under the light microscope after treatment with buffer or enzyme reveals them to have normal morphological forms. Furthermore, treated cells do not exhibit any membrane stickiness since they resuspend in the same manner as untreated cells following centrifugation, and they are not agglutinated by their own serum or the sera of A or O individuals. Membrane AChE and cholesterol content are similarly unaffected by either buffer or enzyme treatment, as both are equivalent to levels found in untreated cells.

When membrane osmotic fragility studies are performed, a comparison of the percentage of sodium chloride producing 50% hemolysis of untreated, buffer-treated, or enzyme-treated cells reveals that those treated with enzyme or buffer are comparable to untreated cells and are only slightly more susceptible to osmotic shock (change from 0.42 to 0.47% sodium chloride). Moreover, osmotic fragility is a reversible phenomenon when cells are re-

TABLE 4–2 Membrane Integrity of α-Galactosidase-Treated Human Red Cells

RBC Treatment	Hemolysis (%)	Osmotic Fragility: 50% Hemolysis Point (% NaCl)	Acetylcholinesterase (%)	Cholesterol (%)
Untreated	—	0.42	100	100
Buffer (no enzyme)	0.3	0.47	102	98
α-Galactosidase	0.2	0.47	99	102

placed into an in vivo milieu (plasma or serum). Analysis of the shape of the fragility curve also shows no increased hemolysis occurring at sodium chloride concentrations of 0.6–0.9%, indicating that no part of the cell population has been rendered more fragile by these treatments. From these results it can be concluded that the treatment conditions do not produce any significant membrane alteration, and that these cells have not been prematurely aged.

The functional integrity of α-galactosidase-treated cells has also been established, particularly with respect to the maintenance of normal levels of ATP, 2,3-DPG, metHb, and P_{50} (Table 4–3).

Cellular ATP and 2,3-DPG levels remained greater than 90% of control cells and are clearly uncompromised by the treatment procedure. Similarly, levels of metHb remain low (3%) and P_{50} values are normal. The slight elevation in metHb, while insignificant, is also a reversible phenomenon, like cellular osmotic fragility, because levels rapidly return to normal when cells are replaced into an in vivo environment (plasma or serum). Moreover, the shapes of the oxygen dissociation curves reveal no differences among untreated, buffer-treated, or enzyme-treated cells, further indicating the ability of these cells to transport oxygen efficiently.

We have similarly demonstrated that α-galactosidase-treated cells can be frozen for long-term storage using either a low glycerol-liquid nitrogen ($-196°C$) procedure or a high glycerol-mechanical freeze ($-80°C$) technique with or without metabolic rejuvenation (Lenny et al. 1982a, 1982b). Such frozen-thawed cells retain their metabolic and membrane integrity and are of transfusable quality.

In addition to the tests demonstrating product integrity, we have also evaluated parameters indicative of product purity. α-Galactosidase-treated cells are prepared under aseptic conditions and meet all FDA requirements regarding sterility and apyrogenicity. The efficiency of the wash procedure in removing the enzyme from the final product has also been assessed by testing recipient sera in an ELISA immunoassay procedure for up to five weeks after transfusion for development of antibody to α-galactosidase. No antibody formation to the enzyme has been detected in any recipient and this is discussed further in the section describing in vivo studies.

TABLE 4–3 Metabolic Integrity of α-Galactosidase-Treated Human Red Cells

RBC Treatment	ATP (%)	2,3-DPG (%)	MetHb (%)	P_{50} (mm Hg)
Untreated	100	100	0.6	22.5
Buffer (no enzyme)	92	91	3.2	22.1
α-Galactosidase	95	96	3.4	23.7

4.3 ENZYMATIC CONVERSION OF GROUP B RBC TO GROUP O: IN VIVO STUDIES

Having shown that α-galactosidase-treated RBC maintain their structural and functional integrity in vitro, we proceeded to test their in vivo viability. In these studies we wished to show that 1) the cells would survive normally in the circulation after transfusion to B, A, and O recipients; 2) sufficient B antigen sites had been removed so that the cells would be tolerated in vivo; 3) the cells had not been altered in some manner due to the treatment process in general so as to physically damage them or make them immunogenic; and 4) the enzyme used in treatment had been efficiently washed from the cells before transfusion.

It is commonly known that RBC will not be agglutinated by antisera unless the cells contain a sufficient number and density of antigen sites (Cartron et al. 1974). Generally speaking, RBC containing less than 1,000–2,000 sites/cell are not agglutinable, so the absence of agglutination does not always mean antigen is not present in small quantities. Thus, while α-galactosidase-treated RBC have lost their reactivity with anti-B serum (see Section 4.3.2, for a discussion about other antisera), small amounts of antigen could remain but go undetected. Moreover, it is highly unlikely that enzyme treatment would remove all antigenicity. Whether some sites remain, however, is not of primary concern but rather, whether those remaining are sufficient in number to prevent tolerance in vivo? This question could be answered by monitoring survival time of the treated cells in the circulation, serum anti-B titers of the recipients, and whether treated cells become agglutinable by posttransfusion sera.

Cell survival time in vivo was determined by labeling enzymatically treated cells with chromium-51 (^{51}Cr), administering these cells to study participants, and monitoring the rate of disappearance or radioactivity from the circulation. Such ^{51}Cr-labeling procedures are well established and used routinely in hematology for cell survival and blood volume determinations.

All group A and O individuals possess in their sera a preexisting, or naturally occurring, anti-B antibody, the titer of which varies in different subjects. As such, the anti-B titer of each subject pretransfusion served as the baseline level against which all posttransfusion titers were compared. Titers are determined by testing twofold serial dilutions of recipient sera against 5% suspensions of untreated fresh donor red cells, and results are expressed as the reciprocal of the highest serum dilution that causes macroscopic agglutination. Titration results are often also expressed by assigning a numerical value to each reaction. The total of these values is referred to as the score, which is a semiquantitative but subjective measurement of antibody reactivity. In blood banking protocols, a difference of at least two titer dilutions or 20 or more in score between test samples has been arbitrarily deemed to be significant; under our very stringent detection and scoring conditions, a difference of at least three titer dilutions or of 15–20 in score would be more comparable.

We also investigated the possibility that the treatment process in general might alter the cells in some manner not detectable by the previously described in vitro tests. Such alteration could take two forms: 1) covert membrane damage due to physical manipulation of the cells during the treatment process, or 2) covert membrane perturbation resulting in immunogenicity of these cells. Both types of alteration would be evident in a shortened lifespan of these cells in the circulation, with survival time being reduced in the B recipient as well as in A and O individuals. Cells made immunogenic by the treatment procedure would become agglutinable by the sera of the recipients and could be distinguished from an anti-B reaction, particularly if this also occurred in the B recipient.

Our last consideration focused on the efficiency of enzyme removal from the cells prior to the transfusion; this was measured by screening recipient sera in an ELISA immunoassay procedure for up to five weeks after transfusion to ensure that no antibody to α-galactosidase had developed.

4.3.1 Animal Studies

We began our in vivo viability studies using the gibbon, a nonhuman primate, as a model in which to assess the fate and immune tolerance of small volumes of enzymatically treated cells. Unlike other animal species, including other nonhuman primates (Wiener et al. 1972; Moor-Jankowski and Wiener 1972), gibbon RBC possess a B antigen whose carbohydrate structure is believed to be the same as that of human RBC. [Other animal species, e.g., the rabbit, are known to possess a "B-like" antigen, in that B specificity is indeed due to a terminal α-linked galactose, but the penultimate sugar is not fucose and, thus, the remaining carbohydrate structure is not group O (Eto et al. 1968; Basu and Basu 1973). Moreover, our studies have shown that removal of terminal galactose residues from rabbit RBC by treatment with α-galactosidase exposes a cryptic antigen, resulting in 90–95% destruction of these cells in vivo in less than 24 hours (Lenny and Goldstein 1984).] In our studies with gibbons, human RBC were not used because of interspecies cross-reactivity between other human RBC antigens and gibbon serum, and so, enzymatically treated gibbon RBC were reinfused to gibbons. Additionally in these studies, such treated cells were returned to the donor animal because group O gibbons do not exist and group A animals were not available.

Gibbons were maintained at the Laboratory of Experimental Medicine and Surgery in Primates (LEMSIP) of New York University in Tuxedo, New York. At this early phase of our investigations, enzymatic treatment was performed at either pH 5.7 or 5.8 for 1 to 2 hours. In vitro studies, already described for human cells in the previous section, were also carried out with gibbon erythrocytes and yielded similar results: treatment with enzyme produced normal levels of ATP, 2,3-DPG, metHb, AChE, and cholesterol and normal osmotic fragilities (Lenny 1981).

In vivo survival studies were initiated, using group B and AB gibbons (Lenny and Goldstein 1980). For these studies small volumes of cells (<1 ml) were labeled with ^{51}Cr, treated, and infused to the donor animal. Samples were collected at 30 and 60 min post-infusion, and radioactivity was measured and used as the zero-time 100% survival value. Further samples were taken at appropriate time intervals ranging from 1–21 days. Note that unlike the normal ^{51}Cr half-life (T_{50}) of human red cells, which is about 28 days, gibbon erythrocytes possess a ^{51}Cr T_{50} of approximately 14 days.

As shown in Figure 4–2, greater than 90% of the control or enzyme-treated cells remain in the circulation 24 hours after being returned to the donor animals. The T_{50} of enzyme-treated cells is 17.5 days, which falls within the normal gibbon RBC half-life of 14–18 days. Thus, these results demonstrate that when enzymatically treated cells are returned to the donor, they exhibit normal in vivo survival rates. Figure 4–2 also illustrates that the decay is linear throughout. Further, we were unable to detect the formation of antibody toward enzyme-treated cells, even after returning such cells to the same donor animal three times over 9 months, indicating that under these conditions they were not immunogenic. Additionally, no antibody could be detected to the enzyme, indicating the efficiency of enzyme removal by the wash procedure.

4.3.2 Human Studies

The enzymatic conversion of human group B RBC to group O RBC and the transfusion of these enzymatically converted O cells (ECO RBC) to group B, A, and O normal volunteers have shown much promise. In a series of 11 studies, ECO RBC were administered to normal healthy volunteers in volumes ranging from 1 ml of packed cells to the one-unit level (160–196 ml of packed cells), including second infusions of such cells to two group O participants. A summary of these preclinical studies appears in Table 4–4.

All subjects underwent rigorous testing prior to acceptance into these studies, and both donors and recipients were tested to the same extent. Tests included not only those required by federal, state, and local regulatory agencies regarding donors and donor blood (tests to which recipients were also subjected) but also additional rigorous screening as described here. Each subject was required to provide a full medical history and undergo a physical examination and hematological work-up; all were fully blood-typed, screened for irregular serum antibodies, and tested for hepatitis B surface antigen and antibody, hepatitis B core antibody, alanine aminotransferase (ALT) levels (as an indicator for hepatitis C), human immunodeficiency virus (HIV) antibody, and human T-lymphotropic virus-I (HTLV-I) antibody. Following transfusion, in addition to the monitoring described to

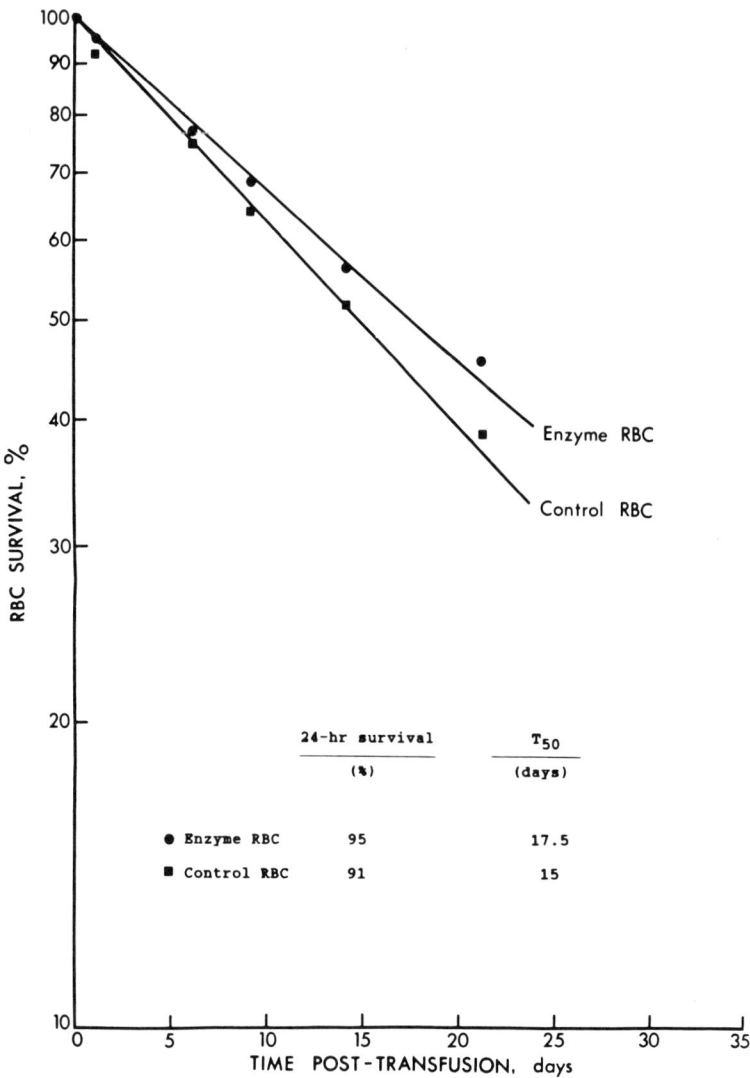

FIGURE 4-2 ^{51}Cr in vivo survival of α-galactosidase-treated gibbon erythrocytes.

determine the fate of the infused cells, hepatitis testing was repeated at one, two, and three months as a precaution against transmission of hepatitis C, and serum antibody screening for the detection of unexpected alloantibodies was repeated through the first five weeks. Subjects receiving volumes of 20 ml or greater remained hospitalized for overnight observation.

The initial preclinical study extended our early work with gibbons by administering small volumes of treated cells to group O and A individuals

TABLE 4–4   [51]Cr In Vivo Survival of α-Galactosidase-Treated Red Cells in Normal Human Volunteers

Study no.	Enzyme Concentration (units/ml RBC)	Recipients' Blood Group	RBCs Infused (ml)	24-Hour Survival (%)[1]	T_{50} (days)[2]
1	630[3]	B, A, O	1 to each	96, 97, 97	33, 30, 31
2	65	B, A, O	2+2+1 to each	88, 88, 86	29, 27, 27
3	75	B, O	20 to each	92, 91	28, 27
4	75	O	160	96	30
5	90	A_1	174	95	36
6	90	A_1	185	99	45
7	90	A_{int}	180	95	36
8	115	O	167	93	35
10[4]	115	O	17	82	27
9	185	O	170	96	40
11[5]	200	O	196	92	36

[1] The FDA standard of acceptability for 24-hour survival is 75%.

[2] 95% of healthy subjects have a T_{50} of 25–37 days (Mollison 1981).

[3] Study no. 1 was done under conditions where the cell hematocrit was low (25%) during the enzyme treatment step. Considerably more enzyme is required to compensate for slow kinetics under these conditions.

[4] The recipient in study no. 8 received a second infusion (17 ml) in study no. 10.

[5] The recipient in study no. 9 received a second full unit in study no. 11.

as well as to the B donor (Goldstein et al. 1982; Goldstein 1983). Prior to transfusion, we performed a full crossmatch between treated cells and participants' sera (including a 10-min, room-temperature, incubation step before the standard 37°C and antiglobulin phases of the test, which we employ as part of our rigorous testing). Then, a suspension of 1 ml of ^{51}Cr-labeled α-galactosidase-treated cells was infused to each of the participants. Samples were withdrawn at 10, 30, and 60 min, and radioactivity was measured, corrected for hemoglobin concentration, and used as the 100% survival value. Subsequent samples were withdrawn at 24 and 48 hours, 6–7 days, and weekly intervals thereafter. As shown in Table 4–4 (study no. 1), both the 24-hour survival and T_{50} results indicated that these cells survived normally in the circulation of all three recipients. No increase in anti-B titer was observed nor did posttransfusion sera agglutinate or lyse treated cells in vitro.

For the second study, again involving A and O recipients as well as the B donor, the experimental protocol called for the transfusion of 2 ml of enzyme-treated cells initially, followed two weeks later with another 2 ml; then a final 1 ml of enzymatically treated cells labeled with ^{51}Cr was given so that a total of 5 ml was transfused to each participant over a four-week time span. Serological testing performed during and for at least five weeks after the three transfusions did not reveal any increase in anti-B titer in the sera of the group O and A participants, nor did the sera of any of the participants agglutinate or lyse freshly prepared enzymatically treated cells. As in the first study, the survival of ^{51}Cr-labeled treated cells in all three participants was normal, further indicating in vivo immune tolerance of enzymatically treated cells (Table 4–4, study no. 2) (Goldstein 1984a, 1984b).

The third and last of the small-volume studies involved the infusion of 20 ml of cells to both the group B donor and a group O recipient (Table 4–4, study no 3). ^{51}Cr RBC survivals were again normal, and no increase in the anti-B titer of the group O recipient was observed.

The ^{51}Cr RBC survival curves constructed from the data obtained in the first three studies are shown in Figure 4–3. The linearity is indicative of normal RBC survival; no downward shift in the slope of the curve could be observed, indicating that enzymatically treated cells were not being selectively destroyed in vivo.

The success of the first three studies prompted us to attempt enzymatic treatment of single whole units of B erythrocytes and to infuse these to either group O or A volunteers (Goldstein et al. 1988). In two of these studies, a small volume (1-2 ml) of α-galactosidase-treated cells was returned to the group B donor, which survived normally (data not shown), since survival of these cells in the original donor had already been established as normal and it was no longer necessary to expose these individuals to radiolabel.

Seven such full-unit studies have been performed, and one group O recipient received a full-unit transfusion twice (see Table 4–4, study nos. 9 and 11) during a period of 4.5 months. Four single units have been infused to group O recipients, and three to group A individuals. RBC survival was determined as before, using 1–2 ml of ^{51}Cr-labeled cells, which were administered immediately after transfusion of the unlabeled unit.

As can be observed in Table 4–4 (studies no. 4–11), RBC survival data were comparable to or better than those obtained in earlier small-volume transfusion studies. Neither immediate nor early destruction of cells was evident, based on blood volume determinations and the use of standards which allow for comparison of expected versus actual counts in vivo. Figure 4–4 illustrates the data obtained from these studies, and, once again, the linearity of these curves is typical of normal RBC survival in the circulation without any selective cell destruction.

As in the previous small-volume studies, serum samples were obtained from the group A and O recipients for at least five weeks after transfusion to monitor anti-B titers and to determine whether posttransfusion sera would agglutinate or lyse ECO RBC in an in vitro crossmatch procedure. The results are summarized in Table 4–5 and 4–6, and described below.

One-unit transfusions of cells treated at an α-galactosidase level of 90 units/ml of packed cells to the three group A recipients produced no sig-

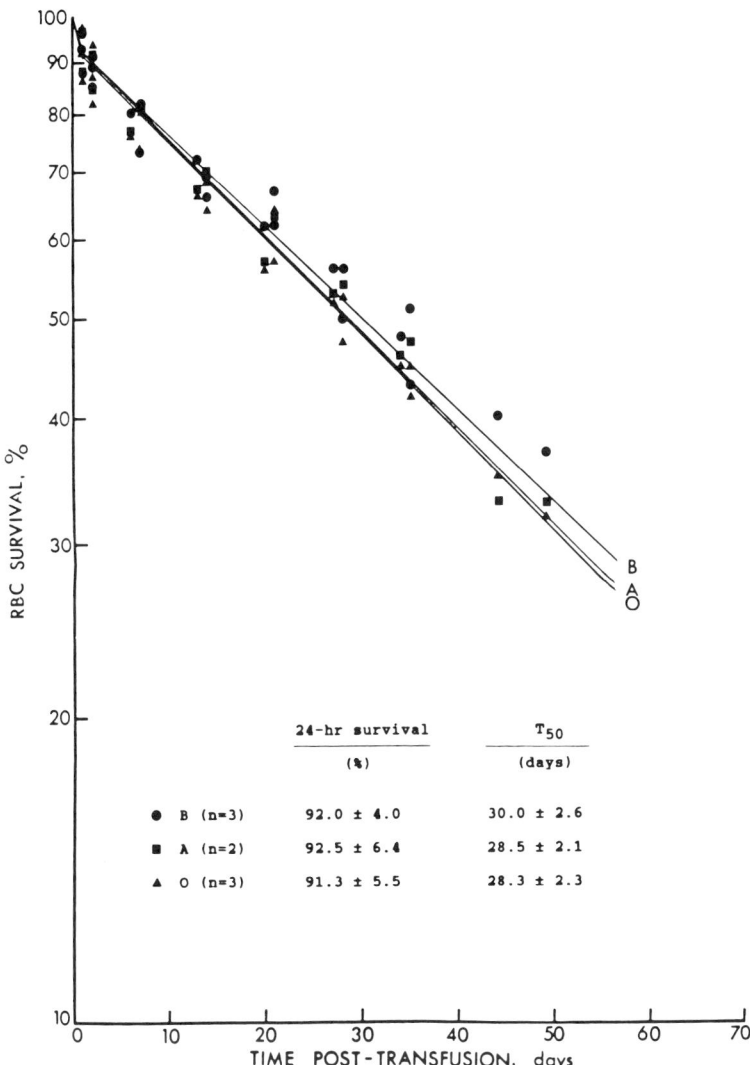

FIGURE 4-3 ^{51}Cr in vivo survival of small volumes of ECO RBC (1 to 20 ml of packed cells) in group B, A, and O normal volunteers.

nificant increase in serum anti-B titer or score. Moreover, posttransfusion sera were unable to agglutinate or lyse enzymatically treated cells when tested in an in vitro crossmatch procedure.

Similarly favorable results were obtained when ECO RBC were transfused to group O recipients, although higher levels of α-galactosidase were needed for cell treatment to maintain the anti-B levels of these recipients

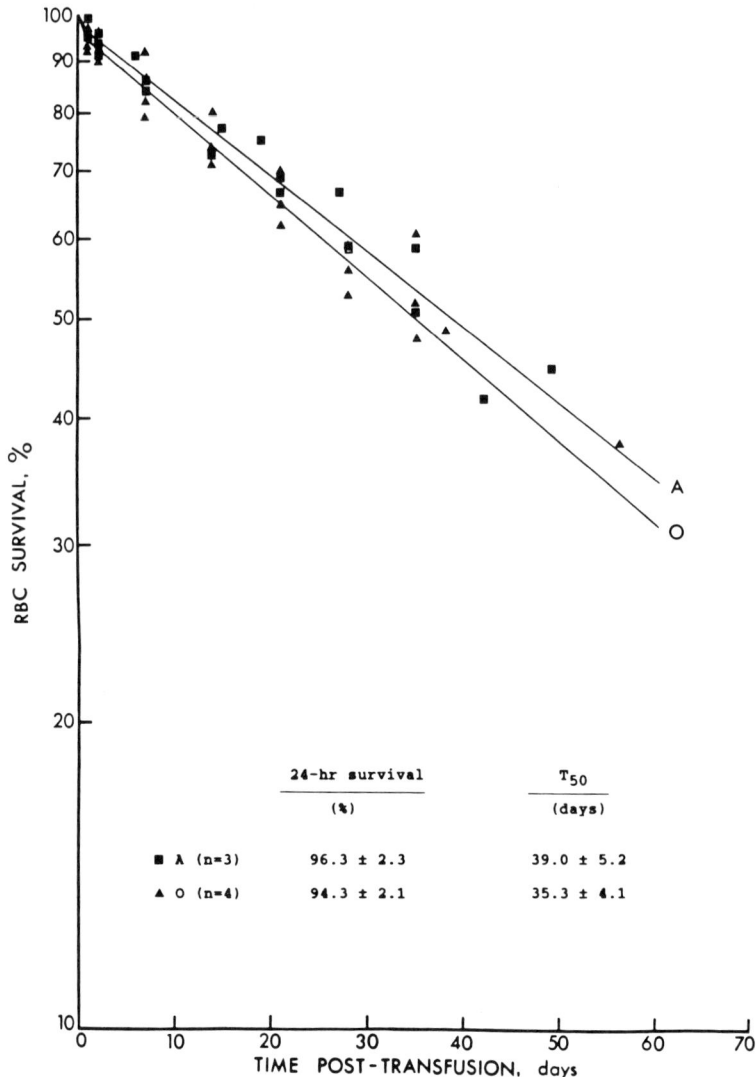

FIGURE 4-4 ^{51}Cr in vivo survival of full-unit volumes of ECO RBC (160–196 ml of packed cells) in group A and O normal volunteers.

at pretransfusion levels. In this series of studies, each of the three group O participants received a full unit of cells that was treated at a different level of α-galactosidase (Table 4–6), and two participants subsequently received second infusions of α-galactosidase-treated cells (see Table 4–7).

The sera of the first group O recipient, who received a full unit of cells treated at an α-galactosidase level of 75 units/ml packed cells (Table 4–6,

TABLE 4-5   Anti-B Antiglobulin Titers and Scores of Three Group A Normal Volunteers Receiving Full Units of ECO RBC[1]

Serum Sample	Group A Recipient No. 1		Group A Recipient No. 2		Group A Recipient No. 3	
	Titer	Score	Titer	Score	Titer	Score
Pretransfusion	128	90	64	76	128	89
Days posttransfusion						
1	64	86	64	70	128	95
2	64	86	64	64	128	93
7	128	89	128	84	256	102
14	64	89	—	—	256	98
21	64	88	64	82	128	91
28	64	88	64	76	128	87
35	—	—	64	80	128	92
42	128	88	—	—	—	—

[1] Group A recipient no. 3 was a weak A intermediate.

TABLE 4-6   Anti-B Antiglobulin Titers and Scores of Three Group O Normal Volunteers Receiving Full Units of ECO RBC[1]

Serum Sample	Group O Recipient No. 1[2]		Group O Recipient No. 2[2]		Group O Recipient No. 3	
	Titer	Score	Titer	Score	Titer	Score
Pretransfusion	64	74	256	95	128	89
Days posttransfusion						
1	64	74	256	95	128	83
2	64	78	256	95	128	84
7	256	106	512	117	256	101
14	512	111	512	117	256	97
21	128	93	512	112	128	90
28	128	88	512	110	—	—
35	128	85	256	98	128	94
79	64	79	—	—	—	—

[1] Recipient no. 1 received cells treated at 75 units/ml of packed RBC, no. 2 at 115 units/ml, and no. 3 at 185 units/ml.
[2] Data from recipients nos. 1 and 2 have been described previously (Goldstein 1989); increases in score are the same as published but titers shown here represent macroscopic determinations.

O recipient no. 1), displayed a three-tube titer increase and a score increase of 37, which peaked at 14 days after transfusion but returned to pretransfusion levels within three weeks. Using thiol reagents (dithiothreitol), the immunoglobulin class of this increased anti-B was identified as IgG. Although anti-B titers were increased in this recipient, such sera were unable to agglutinate or lyse enzymatically treated cells in vitro. These results correlate with the lack of immune destruction in vivo, as can be observed from the normal ^{51}Cr survival results (Table 4–4, study no. 4). No evidence of transfusion reaction was apparent in the recipient, and hemoglobin levels rose after transfusion and remained above pretransfusion levels throughout the 35-day posttransfusion testing period.

The second group O recipient (Table 4–6, O recipient no. 2) received a full unit of cells treated at an α-galactosidase level of 115 units/ml packed cells. Serum anti-B titers were not significantly increased while the score showed an increase of 22 at days 7–14 and returned to pretransfusion levels within three weeks thereafter. In vivo survival of these cells remained normal (Table 4–4, study no. 8). Once again, posttransfusion sera were unable to agglutinate or lyse treated cells in vitro, despite increased anti-B, and these results again correlate with lack of immune destruction in vivo. What was apparent from these two studies, and the first study in particular, is that this modest antibody increase did not adversely affect survival of these cells in vivo.

The third and last group O recipient (Table 4–6, O recipient no. 3) received a full unit of cells treated at an α-galactosidase level of 185 units/ml packed cells. (He subsequently received a second full-unit infusion, described later.) Serum anti-B titers and scores in this recipient showed no significant increase, posttransfusion sera did not agglutinate or lyse treated cells in vitro, and survival of these cells in vivo remained normal (Table 4–4, study no. 9).

The most probable explanation for the modest increase in anti-B in the first two group O recipients is that at lower levels of enzyme (75 and 115 units), sufficient B-antigenic sites remain to induce this slight response, while at higher enzyme levels (185 units), any amounts remaining are below the threshold needed for producing this effect. It is also known that group O individuals tend to exhibit a more potent immune response than do group A. This may be the reason why a similar response was not observed with A recipients, even though the cells they received were treated at the 90-unit level. These results also correlate well with in vitro studies we have more recently done using A-cell-adsorbed human polyclonal anti-A,B serum. Since there is now evidence for the presence of A or A-like antigenic structures on group B cells (Stroup and Treacy 1987; Goldstein et al. 1989), anti-A,B serum was first adsorbed with group A cells to remove anti-A specificity, and it was then reconcentrated to its original potency to eliminate the dilution effects caused by the adsorption procedure. Using this reagent, we are able to detect agglutination with cells treated at lower α-galactosidase levels of 75 and 115 units, whereas cells treated at a level of 185 units are nonreactive.

Our last two studies with normal volunteers (Table 4–4, study nos. 10 and 11) involved second transfusions to two group O recipients who had previously received full-unit volumes. We first selected the individual who had received a one-unit transfusion of cells treated at an enzyme level of 115 units/ml packed cells (Table 4–4, study no. 8), and whose anti-B level had increased by a score of 22 (Table 4–6, O recipient no. 2). Eleven months after the first full-unit transfusion, we administered a small volume of cells (17 ml), also treated at the same enzyme level (115 units/ml packed cells), to determine if this second administration would again evoke an increase in anti-B. As shown in Table 4–7 anti-B titer and score in this individual (O recipient no. 2) did not change, and red cell survival in vivo remained normal (Table 4–4, study no. 10).

Encouraged by these results, we proceeded to transfuse a second full unit to another group O recipient who had received a full-unit transfusion 4.5 months earlier (Table 4–4, first infusion, study no. 9; second infusion, study no. 11). This group O recipient had first received cells treated at an enzyme level of 185 units/ml packed cells and had not produced a significant increase in anti-B (Table 4–6, O recipient no. 3). Following this second transfusion of cells that had been treated at an enzyme level of 200 units/ml packed cells (currently our standard enzyme treatment concentration), anti-B titer and score remained unchanged (Table 4–7, O recipient no. 3) and in vivo red cell survival was normal (Table 4–4, study no. 11). As in past studies, posttransfusion sera were unable to agglutinate or lyse enzymatically treated cells in vitro. Hence, transfusion to a group O recipient of two units of ECO RBC under these conditions, that is, 4.5 months apart and at an enzyme level of 185–200 units, did not produce any increase in anti-B antibody nor any other adverse effects.

We have also monitored the anti-A titers of all group O recipients in these studies since, as mentioned previously, there is evidence for the presence of small amounts of A antigen on group B erythrocytes. We have not

TABLE 4–7 Anti-B Antiglobulin Titers and Scores of Group O Normal Volunteers Receiving Second Transfusions of ECO RBC

	Group O Recipient No. 2[1]		Group O Recipient No. 3[2]	
Serum Sample	Titer	Score	Titer	Score
Pretransfusion	128	87	256	100
Days posttransfusion				
1	128	87	256	99
2	128	87	256	102
7	128	85	256	104
14	128	86	256	99
21	128	88	256	106
28	128	78	256	102
35–38	128	78	256	106

[1] Recipient no. 2 received 17 ml of RBC at 115 units/ml.
[2] Recipient no. 3 received 196 ml of RBC at 200 units/ml.

observed an increase in anti-A titer or score in any of the group O recipients, including the subjects who received two transfusions. In addition, in none of the 11 studies described has antibody to the enzyme been detected in the participants' sera, including the subjects who received two transfusions. This is further confirmation of the efficiency of the washing procedure used to remove enzyme after treatment of the cells.

Finally, to ensure that subjects had not been immunized to other red cell antigens as a result of participation in these studies, all sera were subjected to standard antibody detection tests and were found to be negative. As a further protection for study participants, hepatitis and alanine aminotransferase (ALT) tests were repeated at one, two, and three months posttransfusion, particularly as a precaution against transmission of hepatitis C; all these tests were negative.

Evidence of the in vivo efficacy of ECO RBC has also been documented in those subjects who received full-unit transfusions. Though these individuals were normal healthy volunteers, transfusion of one unit of cells produced a rise in hemoglobin levels ($7.6 \pm 5.5\%$, n = 7) normally observed after a one-unit transfusion. These results suggest that a similar response could be expected in a patient population and that ECO RBC will be efficacious in the treatment of anemia.

In summary, these results demonstrate that enzyme-modified cells survive normally in the circulation of healthy volunteers and, when treated with the appropriate concentration of α-galactosidase (200 units/ml packed RBC), do not effect a clinically significant elevation of naturally occurring, i.e., preexisting, anti-B antibody titers. ECO RBC were well tolerated in all recipients. Vital signs remained stable and evidence of hemolysis (hemoglobinuria) or premature destruction of these cells was absent. Recipients showed no adverse effects to the transfusions in general, nor was there any evidence of refractoriness to or development of any form of new antibody against treated cells. Based upon these promising results in normal healthy volunteers, we will shortly begin Phase 1 clinical trials with ECO RBC.

4.4 ENZYMATIC CONVERSION OF GROUP A RBC TO GROUP O: IN VITRO STUDIES

In contrast to enzymatically converted group B to group O RBC, our studies on the transformation of group A RBC to group O are at the in vitro stage. There are two major reasons for this: 1) alpha-N-acetylgalactosaminidases (A-zymes) are neither as readily available nor abundant in nature as are α-galactosidases (B-zymes); and 2) the structure of the A antigen has been found to be more complex than that of the B antigen or the simplified A structure shown in Figure 4–1.

4.4.1 A-zyme Sources

We have screened a number of A-zymes from various microbial and animal sources and further tested those which appeared promising with synthetic substrates. As is often the case with glycosidases, many will act on synthetic substrates (p-nitrophenyl glycosyl derivatives) or even isolated glycoconjugates but few are capable of cleaving structures on the intact red cell membrane. However, we have isolated and purified an A-zyme from chicken liver which has the ability to cleave terminal α-linked GalNAc residues from the intact red cell membrane (Goldstein et al. 1984).

4.4.2 A Antigen Structure

Contributing to the complexity of the A antigen is the existence of several A subgroups, of which A_1 and A_2 predominate, and the difference in carbohydrate structure among their glycoprotein and glycolipid moieties. (For an excellent review, see Clausen and Hakomori 1989.) It was originally believed that A_1 and A_2 RBC differed quantitatively and possibly qualitatively; it has now been conclusively shown that they also differ qualitatively (Clausen et al. 1985, 1986). Blood group A specificity is carried on both glycoprotein and glycolipid structures by several different types of carbohydrate chains called type 2, type 3, and type 4 (Figure 4–5). More recently, two additional glycolipid A structures have been identified (Clausen et al. 1987) but these exist only in very small amounts on the red cell surface.

The vast majority of A antigens exist as type 2 structures and can be found in both A_1 and A_2 glycoproteins and glycolipids, as well as in A intermediate (A_{int}) cells, whose reactivity with various anti-A reagents classifies these cells somewhere between A_1 and A_2. Of particular relevance is the existence of the type 3 chain structures (Clausen et al. 1985, 1986) referred to as type 3 chain A ("repetitive A") and type 3 chain H ("A-associated H") (see Figure 4–5). Both structures are extensions of type 2 chain A and contain an internal A determinant (α-linked GalNAc). Thus far, these structures have been found only in glycolipid fractions. However, type 3 chain A is believed to be present only in A_1 cells, whereas type 3 chain H, a precursor of repetitive type 3 chain A, is found in A_2 cells. A_{int} cells presumably carry some of both type 3 structures. Type 4 chain A exists in considerably lesser amounts than either type 2 or 3 and has been shown to be present only in A_1 red cells (Clausen et al. 1984).

4.4.3 A-zyme Treatment of A_1, A_2, and A_{int} Red Cells

Treatment of A_1 RBC with our purified chicken liver A-zyme significantly reduces their reactivity with anti-A serum (from an agglutination titer of 512 to one of 8 or 16). Increasing enzyme concentration or treatment time has no further effect on reducing this titer. In contrast, A_2 cells treated under similar conditions lose their reactivity with anti-A serum. Results with A_{int}

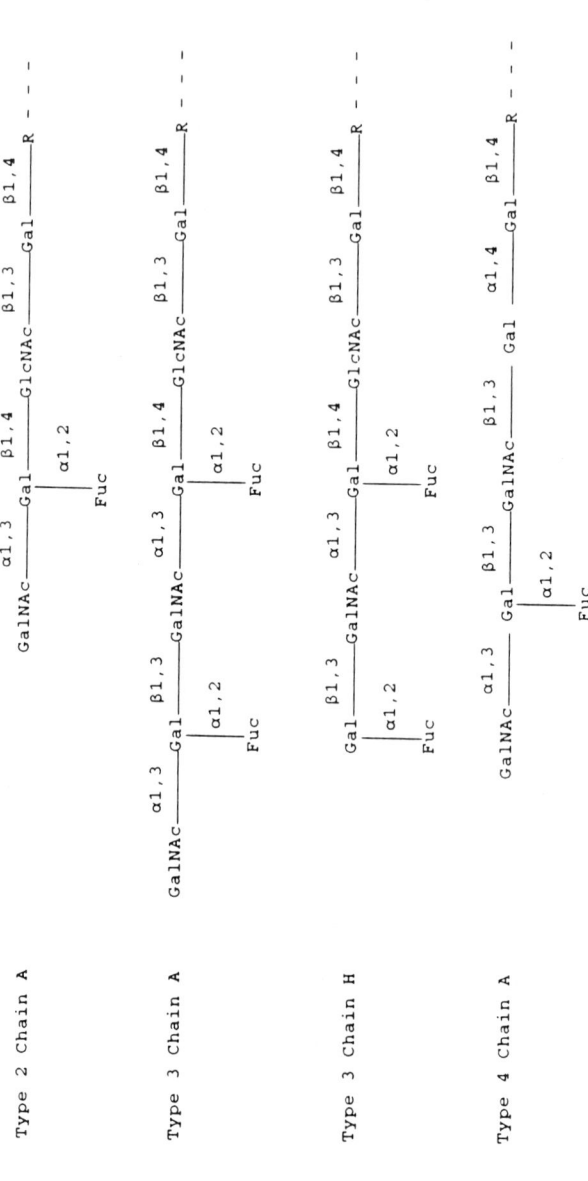

FIGURE 4-5 The four types of carbohydrate structures carrying red cell blood group A and H determinants. R, remainder of the carbohydrate chain, which may be either linear or branched; - - -, either glycoprotein or glycolipid.

cells, again, fall somewhere between those obtained with A_1 and A_2 RBC, and their loss of reactivity with anti-A sera depends on whether they are "weak" or "strong" A intermediates.

We first believed that the inability of the A-zyme to further remove A antigenicity from A_1 and some A_{int} cells was a shortcoming of the enzyme, e.g., steric hindrance was preventing its maximum activity. We now know such is not the case. Using monoclonal antibodies specific for the various group A carbohydrate chains (a generous gift from Dr. Henrik Clausen of the Biomembrane Institute, University of Washington, Seattle, WA) we have been able to demonstrate that our A-zyme efficiently cleaves terminal α-linked GalNAc residues from types 2A, 3A, and 4A chains and that the remaining reactivity of A_1 and some A_{int} RBC with anti-A serum is due to the internal A epitope present in type 3 chain A (see Figure 4–5). Why A_2 cells, which have type 3H structures containing an internal A epitope, become nonreactive with anti-A serum following A-zyme treatment can be explained by the fact that fewer of these structures exist on A_2 cells than do type 3A chains on A_1 cells.

4.4.4   Multiple Enzyme Treatment of Group A RBC

We are taking several approaches to removing the internal A structure from type 3A chains. One approach has been to use an endo-β-galactosidase in combination with our A-zyme. This enzyme, derived from *Bacteroides fragilis*, cleaves Gal to GlcNAc/Glc linkages in linear structures but it is inactive at branch points and at points of attachment to the cell membrane (Scudder et al. 1984). When A_1 RBC are treated with a combination of A-zyme and endo-β-galactosidase, a further reduction in anti-A titer is observed over that seen when A-zyme is used alone, that is, the anti-A titer is reduced from 512 (untreated cells) to 4, but complete loss of reactivity with anti-A serum is not achieved. These results suggest that the larger, type 3A structure is susceptible to endo-β-galactosidase action, as expected, and that the smaller, type 3 chain is responsible for the remaining reactivity of A_1 cells with anti-A serum. Also affected is the larger, type 3H structure (see Figure 4–5) of A_2 cells.

We are continuing to assess the ability of other enzymes, for example, a combination of relevant exoglycosidases or an endoglycosidase, to remove the internal A epitope from blood group A type 3 structures.

4.5   FUTURE PERSPECTIVES: Rh MODIFICATION

Second only to ABO in importance in transfusion medicine is the group of antigens collectively known as Rh. While the Rh system is complex and composed of many antigens, the Rh(D) antigen is considered to be the most significant of the group because it is the most immunogenic. Further, it is the presence of Rh(D) that categorized red cells as Rh positive, and its absence, as Rh negative. Approximately 85% of human red cells are Rh(D) positive.

Major differences exist between the ABO and Rh antigen systems, in that Rh specificities are not carbohydrate-derived but rather are protein in nature and, to date, little is known about their biochemical composition. Further elucidation of Rh structure and orientation within the membrane is ongoing in several laboratories, including our own, and these studies should provide the keys necessary to alter and reduce Rh immunogenicity. To this end, we have recently been able to demonstrate that the extracellular domain of Rh(D) antigen can be attacked by successive treatment of intact red cells with phospholipase A_2 and papain. This can be done under conditions which do not affect the closely related Rh antigens, C/c and E/e (Suyama and Goldstein 1990). These results suggest a unique placement of Rh(D) antigen at the membrane and a possible reason for its superior immunogenicity. We are continuing to pursue the characterization of Rh structure, with our ultimate goal being the production of Rh(D) negative cells from Rh(D) positive ones.

REFERENCES

Basu, M., and Basu, S. (1973) *J. Biol. Chem.* 248, 1700–1706.

Cartron, J., Gerbal, A., Hughes-Jones, N., and Salmon, C. (1974) *Immunol.* 27, 723–727.

Clausen, H., and Hakomori, S.-I. (1989) *Vox Sang.* 56, 1–20.

Clausen, H., Watanabe, K., Kannagi, R., et al. (1984) *Biochem. Biophys. Res. Comm.* 124, 523–529.

Clausen, H., Levery, S., Nudelman, E., Tsuchiya, S., and Hakomori, S.-I. (1985) *Proc. Nat. Acad. Sci.* 82, 1199–1203.

Clausen, H., Levery, S., Kannagi, R., and Hakomori, S.-I. (1986) *J. Biol. Chem.* 261, 1380–1387.

Clausen, H., Levery, S., Nudelman, E., et al. (1987) *J. Biol. Chem.* 262, 14228–14234.

Dejter-Juszynski, M., Harpaz, N., Flowers, H., and Sharon, N. (1978) *Eur. J. Biochem.* 83, 363–373.

Economidou, J., Hughes-Jones, N., and Gardner, B. (1967) *Vox Sang.* 12, 321–328.

Eto, T., Ichikawa, Y., Nishimura, K., Ando, S., and Yamakawa, T. (1968) *J. Biochem.* 64, 205–212.

Finne, J., Krusius, T., Rauvala, H., Hekomaki, R., and Myllyla, G. (1978) *FEBS Lett.* 89, 111–115.

Furukawa, K., and Aminoff, D. (1970) in *Blood and Tissue Antigens* (Aminoff, D., ed.), pp. 415–425, Academic Press, New York.

Gardas, A. (1978) *Eur. J. Biochem.* 89, 471–473.

Ginsburg, V., Hickey, C., Kobata, A., and Sawicka, T. (1971) in *Glycoproteins of Blood Cells and Plasma* (Jamieseon, G.A., and Greenwalt, T.J., eds.), pp. 114–126, Lippincott, Philadelphia.

Goldstein, J. (1983) in *Recent Advances in Hematology* (Hollan, S.R., ed.), pp. 89–97, Hungarian Academic Science, Budapest.

Goldstein, J. (1984a) in *Clinics in Immunology and Allergy*, vol. 4 (de Vries, R.R.P., ed.), pp. 489–502, W.B. Saunders, London.

Goldstein, J. (1984b) in *The Red Cell, Sixth Ann Arbor Conference* (Brewer, G.J., ed.), pp. 139–157, Alan R. Liss, New York.

Goldstein, J. (1989) *Trans. Med. Rev.* 3, 206–212.

Goldstein, J., Siviglia, G., Hurst, R., Lenny, L., and Reich, L. (1982) *Science* 215, 168–170.

Goldstein, J., Hurst, R., Orlando, J., and Grant, A. (1984) *Int. Soc. Blood Transfusion*, p. 86 (Abstracts).

Goldstein, J., Lenny, L., Hurst, R., Benjamin, L., and Jones, R. (1988) *Blood* 72, 277a.

Goldstein, J., Lenny, L., Davies, D., and Voak, D. (1989) *Vox Sang.* 57, 142–146.

Harpaz, N., Flowers, H.M., and Sharon, N. (1975) *Arch. Biochem. Biophys.* 170, 676–683.

Harpaz, N., Flowers, H.M., and Sharon, N. (1977) *Eur. J. Biochem.* 77, 419–426.

Issitt, P. (1985) in *Applied Blood Group Serology*, p. 666, Montgomery Scientific Publications, Miami.

Kabat, E.A. (1970) in *Blood and Tissue Antigens* (Aminoff, D., ed.), pp. 187–198, Academic Press, New York.

Koscielak, J., Miller-Podraza, H., Krauze, R., and Prasek, A. (1976) *Eur. J. Biochem.* 71, 9–18.

Kubo, S.-I. (1989) *J. Forensic Sci.* 34, 96–104.

Kuo, J.-Y., and Goldstein, J. (1983) *Enzyme Microb. Technol.* 5, 285–290.

Lenny, L.L. (1981) Ph.D. thesis, New York University, New York.

Lenny, L.L., and Goldstein, J. (1980) *Transfusion* 20, 618.

Lenny, L.L., and Goldstein, J. (1984) *Transfusion* 24, 87.

Lenny, L.L., Goldstein, J., and Rowe, A. (1982a) *Transfusion* 22, 420.

Lenny, L.L., Goldstein, J., and Rowe, A. (1982b) *Cryobiology* 19, 678–679.

Lenny, L.L., Hurst, R., Goldstein, J., Benjamin, L.J., and Jones, R.L. (1991) *Blood* 77, 1–5.

Levine, P., Robinson, E., Celano, M., Briggs, O., and Falkinburg, L. (1955) *Blood* 10, 1100–1108.

Levy, G.N., and Aminoff, D. (1978) *Fed. Proc.* 37, 1601.

Levy, G.N., and Aminoff, D. (1980) *J. Biol. Chem.* 255, 11737–11742.

Lloyd, K.O., and Kabat, E.A. (1968) *Proc. Nat. Acad. Sci.* 61, 1470–1477.

Mollison, P. (1981) in *A Seminar on Immune-Mediated Cell Destruction*, pp. 45–69, American Association of Blood Banks, Chicago.

Moor-Jankowski, J., and Wiener, A. (1972) in *Pathology of Simian Primates* Part I (Fiennes, R., ed.), pp. 271–317, Karger, Basel.

Morgan, W.T.J. (1947) *Experientia* 3, 257–267.

Oisha, K., and Aida, K. (1976) *Agric. Biol. Chem.* 40, 67–71.

Rovis, L., Anderson, B., Kabat, E.A., Gruezo, F., and Liso, J. (1973) *Biochem.* 12, 1955–1961.

Scudder, P., Hanfland, P., Uemura, K., and Feizi, T. (1984) *J. Biol. Chem.* 259, 6586–6592.

Stroup, M., and Treacy, M. (1987) in *A Scientific Forum on Blood Grouping Serum Anti-A (Murine Monoclonal Blend) Bioclone*, Ortho Diagnostic Systems, Raritan, NJ.

Suyama, K., and Goldstein, J. (1990) *Blood* 75, 255–260.
Watkins, W. (1980) *Adv. Human Genet.* 10, 1–136, 379–385.
Watkins, W., and Morgan, W.T.J. (1954) *Br. J. Exp. Pathol.* 35, 181–190.
Wiener, A., Moor-Jankowski, J., and Socha, W. (1972) *Transplant. Proc.* 4, 101–105.

Chemically Modified and Recombinant Hemoglobin Blood Substitutes

Steven R. Snyder
Joseph A. Walder

The limited stability of whole blood, the need for blood typing, and the risk of blood-borne viral infections have provided the impetus for the development of blood substitutes. In most cases, blood transfusions are given for volume replacement and to enhance oxygen delivery. The current generation of blood substitutes seek only to duplicate these two functions of whole blood. Two oxygen transport systems have been studied most extensively: cell-free hemoglobin and the perfluorocarbons, a class of fully synthetic oxygen carriers (see Chapter 7). In this chapter we review: 1) the basic physiology of oxygen transport by hemoglobin; 2) problems that attend the use of hemoglobin as a blood replacement; 3) adapting hemoglobin for extracellular oxygen delivery by chemical modification of the protein; and 4) use of recombinant DNA techniques to produce hemoglobin and engineer the properties of the protein.

5.1 REGULATION OF THE OXYGEN AFFINITY OF HEMOGLOBIN

The structure and oxygen-binding properties of hemoglobin and of the related oxygen carrier myoglobin have been studied in great detail. Excellent reviews are provided by Dickerson and Geis (1983) and by Bunn and Forget (1986). We shall give only a brief summary here.

Hemoglobin consists of four polypeptide chains, two identical α subunits and two identical β subunits. Each consists of approximately 140 amino acids and has a molecular weight of about 16,000. In each subunit, the porphyrin prosthetic group, heme, is tightly sandwiched between two helices. The iron (Fe) atom at the center of the heme group provides the binding site for oxygen and other ligands such as CO. Only the Fe^{2+} state is capable of binding oxygen. As discussed further below, hemoglobin is susceptible to autooxidation, which results in the formation of methemoglobin (Fe^{3+}). Upon oxygen binding, a distinctive change occurs in the organization of the subunits within the tetramer. This change in quaternary structure is intimately linked to the regulation of the oxygen-binding properties of the protein.

The oxygen-binding curve for hemoglobin in whole blood is shown in Figure 5–1. The curve is sigmoidal, a reflection of the cooperative nature of ligand binding to the protein. As oxygen is bound to hemoglobin, the affinity of the remaining unoccupied sites progressively increases. The com-

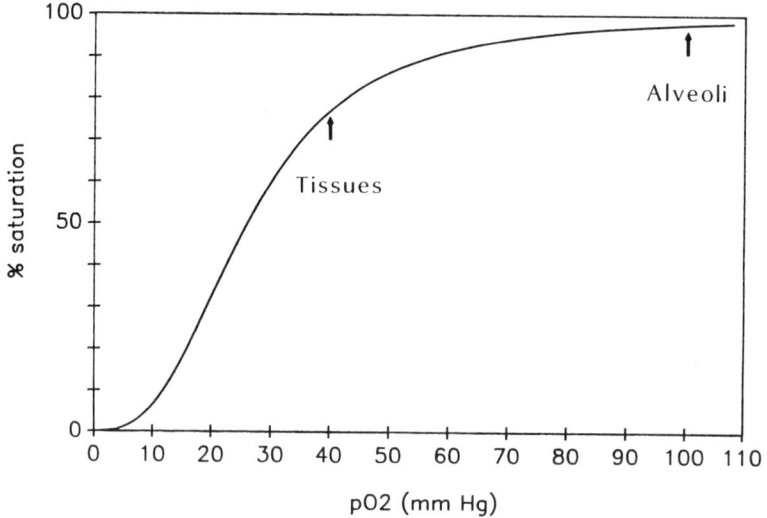

FIGURE 5–1 The oxygen dissociation curve for whole blood. The average pO_2 of tissues (40 mm Hg) and the alveoli (100 mm Hg) are indicated.

munication between the heme groups does not occur by direct site–site interaction. Instead, the effect of ligation is transmitted through the fabric of the protein via the change in quaternary structure alluded to above. To a first approximation, the details of oxygen binding to hemoglobin can be accounted for by the simple stereochemical model first suggested by Monod et al. (1965): Hemoglobin is able to adopt two distinct quaternary structures designated the T and R states. Under deoxygenated conditions, hemoglobin exists almost exclusively in the T conformation. As oxygen is bound, the equilibrium between the two structures progressively shifts to the R form. By the time that three molecules of oxygen are bound, the tetramer is fully in the R state. In this quaternary structure, the affinity for oxygen is increased by about 300-fold compared to the T state. The increase in affinity with occupancy thus arises due to the transition from the low-affinity T structure to the high-affinity R form.

The sigmoidal shape of the oxygen binding curve has important physiological consequences. It allows hemoglobin to become fully saturated with oxygen in the lungs and to offload a substantial fraction of oxygen to the tissues over a narrow range in oxygen tension (see Figure 5–1). The P_{50}, the partial pressure of oxygen required for half-saturation of hemoglobin, is normally about 26 mm Hg for whole blood. At the average partial pressure of oxygen in the tissues, 40 mm Hg, about 25% of the oxygen bound to hemoglobin is liberated. The steep descent of the oxygen-binding curve in this region allows for additional oxygen release without the development of severe tissue hypoxia. This is particularly important for certain organs, such as the heart, in which a much higher fraction of the oxygen bound is normally extracted.

The oxygen-binding properties of hemoglobin are further modulated by three small molecules within the red cell: H^+, CO_2, and 2,3-diphosphoglycerate (2,3-DPG). Each of these so-called allosteric effectors bind preferentially to deoxyhemoglobin (the T state) and have the effect of lowering the oxygen affinity. Under physiological conditions the P_{50} of hemoglobin increases by about 10% for each 0.1-unit drop in pH. This relationship, the Bohr effect, permits greater oxygen release to the tissues in the event that hypoxia, and hence acidosis, develops. CO_2 binds to deoxyhemoglobin by the formation of a carbamate adduct at the N-termini of both the α and β chains:

$$RNH_2 + CO_2 \rightleftharpoons RNHCO_2^- + H^+$$

Approximately 10% of the CO_2 in the blood is transported in this manner. 2,3-DPG binds to deoxyhemoglobin within a cleft between the β subunits. The structure of the complex has been determined by x-ray crystallographic methods (Arnone 1972). 2,3-DPG, a polyanion, is held in place by electrostatic interactions with eight positively charged residues of the β chains. In the transition to oxyhemoglobin, the β chains slide together and largely obliterate the central cavity. As a result the affinity for DPG drops by about 50-fold.

Lack of 2,3-DPG outside of the red cell has a profound effect on the oxygen affinity of hemoglobin and important consequences for its use as a blood substitute. In the absence of 2,3-DPG, the P_{50} of hemoglobin is decreased to about 16 mm Hg as shown in Figure 5–2. The slightly higher pH of plasma (7.4) compared to that within the red cell (7.2) results in a further increase in oxygen affinity. The marked leftward shift in the oxygen-binding curve significantly compromises the offloading of oxygen to the tissues. At the normal tissue partial pressure of oxygen, about 40 mm Hg, the amount of oxygen released is decreased by nearly fivefold. This is the first important problem which must be overcome in order to use extracellular hemoglobin as a blood substitute.

5.2 DISSOCIATION OF THE HEMOGLOBIN TETRAMER

A second, even more serious, problem in the use of hemoglobin as an acellular oxygen carrier arises due to the dissociation of the tetramer to form $\alpha\beta$ dimers:

$$\alpha_2\beta_2 \rightleftharpoons 2\alpha\beta \qquad (1)$$

Under deoxygenated conditions this reaction does not proceed to any sig-

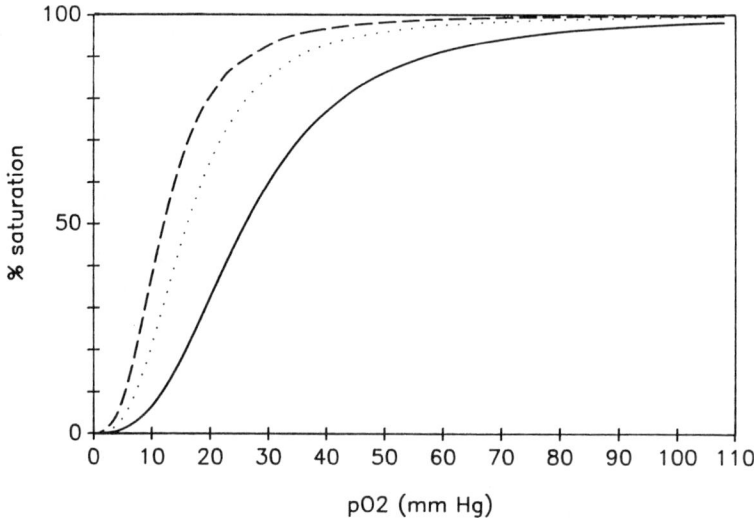

FIGURE 5-2 Oxygen dissociation curves for whole blood (——), hemoglobin in the absence of DPG at pH 7.2 (· · ·), and hemoglobin without DPG at pH 7.4 (– – –).

nificant extent. However, in the oxygenated form, 2–4% of hemoglobin exists as dimers at typical concentrations that would be encountered following transfusion. The molecular weight cutoff for renal filtration of proteins is about 60,000. Hemoglobin dimers having a molecular weight of only 32,000 are very rapidly cleared from the circulation by the kidneys (Bunn et al. 1969; Bunn and Jandl 1969). As this occurs, the equilibrium shown in equation 1 becomes progressively shifted to the right. The intravascular half-life of native hemoglobin is only about 1 hour (Bunn et al. 1969); and the massive hemoglobinuria which occurs poses the risk of renal injury. For hemoglobin to be used as a blood substitute, it is essential that it be modified to maintain the tetramer in an intact form.

5.3 CHEMICAL MODIFICATION OF HEMOGLOBIN

Two classes of cross-linking agents have been described that react with hemoglobin selectively and can be used to block dissociation of the tetramer. Ruth Benesch, Reinhold Benesch, and their colleagues have developed a group of dialdehyde compounds targeted to the 2,3-DPG binding site (Benesch et al. 1975, 1984; Keipert et al. 1989). The parent compound of this series is 2-nor-2-formylpyridoxal phosphate:

It reacts with deoxyhemoglobin to cross-link the lysine 82 residue of one β subunit to the amino-terminal valine residue of the second β chain (Arnone et al. 1977). The reactive aldehyde groups of the molecule are indicated by arrows. The cross-link between the β chains prevents separation of the $\alpha\beta$ dimers. Moreover, the placement of the phosphate group within the 2,3-DPG pocket lowers the oxygen affinity. The P_{50} of the modified hemoglobin (45 mm Hg) is, in fact, substantially greater than that of whole blood (Benesch et al. 1984). Although the level of cooperativity is somewhat reduced, the sigmoidal shape of the oxygen binding curve is retained. Despite these favorable properties, no investigation of this hemoglobin in larger animals has been reported. The scale-up of the preparation of this derivative presents formidable problems. The cross-linking agent itself is difficult to synthesize, and the modified hemoglobin appears to require chromatographic purification.

Our laboratory has worked for some years on a second family of hemoglobin cross-linking agents which also react with amino groups of the protein. The prototype of these reagents is *bis*(3,5-dibromosalicyl) fumarate, and the cross-linking reaction is as follows:

$$R_1\text{---}NH_2 \;+\; H_2N\text{---}R_2 \;\rightarrow\; R_1\text{---}N\overset{H}{\underset{H}{\text{---}}}\overset{O}{\overset{\|}{C}}\text{---}C\overset{H}{=}C\text{---}\overset{O}{\overset{\|}{C}}\text{---}\overset{H}{N}\text{---}R_2 \;+\; 2\;(\text{3,5-dibromosalicyl})$$

This compound was first investigated by Klotz and coworkers as a potential therapeutic agent for sickle cell disease (Walder et al. 1979). We showed that this agent and a series of related analogs react very specifically with oxyhemoglobin to cross-link the β chains between lysine $82\beta_1$ and lysine $82\beta_2$ (Walder et al. 1980; Chatterjee et al. 1982). This has the desired effect of blocking the dissociation of the tetramer, but the oxygen affinity of these derivatives is increased, the opposite effect to the one needed for use of hemoglobin as a blood substitute.

Later we discovered that the reaction of *bis*(3,5-dibromosalicyl) fumarate with deoxyhemoglobin rather than oxyhemoglobin gives rise to a very different product cross-linked between the α subunits (Chatterjee et al. 1986). X-ray crystallographic studies revealed that this cross-link spans the two lysine 99α residues and also blocks dissociation of the tetramer. The hemoglobin remains fully cooperative; despite the presence of the cross-link, it can readily adopt both quaternary structures. In contrast to the β-β cross-linked hemoglobin, this derivative, which we have termed HbXL99α, has a reduced oxygen affinity (Chatterjee et al. 1986). Under physiological conditions, the P_{50} of HbXL99α is 29 mm Hg, resulting in oxygen-transport characteristics that are virtually identical to those of whole blood (Snyder et al. 1987). Although HbXL99α was first isolated as a minor product, we subsequently found that it could be prepared in high yields by carrying out the reaction in the presence of polyanions which bind at the 2,3-DPG site and block modification of the protein within this region (Chatterjee et al. 1986).

HbXL99α has a third very useful property. As a result of the cross-link, the protein is highly heat stable. Under deoxygenated conditions, it can be heated at 70°C for prolonged periods (Estep et al. 1989). This allows for selective denaturation and precipitation of any residual non-cross-linked hemoglobin, a useful purification step, and for complete viral inactivation.

This last feature is extremely important, as it makes possible the use of pooled blood as a source of hemoglobin.

The persistence of HbXL99α and other intramolecularly cross-linked derivatives within the circulation is markedly prolonged compared to unmodified hemoglobin (Snyder et al. 1987; Keipert et al. 1989). The half-life is very dosage and species dependent (Hess et al. 1989). From studies in larger animals, one would expect a plasma half-life for such derivatives in man, at a 50% volume exchange, to be around 20 hours. This is ideal for many applications in which hemoglobin would be used strictly for the purpose of oxygen delivery, such as to perfuse the heart during coronary angioplasty (Rossen et al. 1987) or for organ preservation. It may also be sufficient for large-volume blood replacement for short periods. However, to avoid the necessity for later transfusion with whole blood, or packed red cells, an even longer circulating half-life would be desirable.

Although renal excretion of intramolecularly cross-linked hemoglobins is dramatically reduced, such derivatives remain susceptible to filtration across the somatic capillary beds. The rate of this process can be slowed by further cross-linking the hemoglobin *inter*molecularly to produce higher-molecular-weight species (Sehgal et al. 1983; Keipert and Chang 1985; Snyder et al. 1987). Intramolecular cross-linking remains necessary to block reactions of the following type:

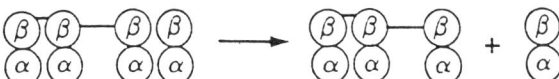

The intravascular half-life of a hemoglobin dimer (128,000 molecular weight) is about twice that of an individual hemoglobin molecule (Snyder et al. 1987). By analogy to normal plasma proteins, an increase in half-life of up to fivefold may be achievable by further polymerization.

A second advantage of intermolecular cross-linking is that a higher hemoglobin concentration can be administered. The normal colloid osmotic pressure (COP) generated by plasma proteins is about 28 mm Hg. This is reached at a hemoglobin concentration of between 8 and 10 gm/100 ml. Although this provides adequate oxygen delivery to the tissues, a greater hemoglobin concentration would be desirable. As the concentration of hemoglobin exceeds 10 gm/100 ml, the COP increases sharply (Moss et al. 1984). Infusion of such a solution would lead to a rapid uptake of water into the vascular space which would tend to cause tissue dehydration, and, of course, dilute the hemoglobin concentration. Intermolecular cross-linking decreases the total number of molecular species present and consequently reduces the COP. With the appropriate level of polymerization, a hemoglobin concentration equal to that of whole blood (14 gm/100 ml) can be used without exceeding the normal COP of plasma (Moss et al. 1984).

Polymerization of hemoglobin has generally been carried out using random cross-linking agents, most commonly, glutaraldehyde. This approach has several disadvantages. Inevitably, a very heterogeneous distribution of products is obtained in which the protein is modified at many different sites on the molecule. There are 48 amino groups in hemoglobin at which glutaraldehyde can react. Such highly modified derivatives may become antigenic, often their oxygen affinity is increased, and usually their susceptibility to autooxidation is enhanced. Moreover the degree of polymerization is difficult to control. Generally, a substantial portion (10 to 20%) of monomeric hemoglobin is produced, as well as a highly polymerized fraction having very aberrant properties. It would be far more desirable to polymerize hemoglobin using specific cross-linking agents that react at a unique site or at a limited number of positions to yield a narrow distribution of molecular weight species.

5.4 PURIFICATION OF HEMOGLOBIN

Early attempts to use hemoglobin as a blood substitute were plagued by nephrotoxicity due to contamination with red cell membrane and cytoskeletal elements. Rabiner and co-workers (1967) were the first to establish the importance of removing red cell stroma. Even low levels of certain membrane phospholipids appear to be toxic in combination with hemoglobin (Feola et al. 1988). Using current ultrafiltration techniques, it is possible to prepare hemoglobin solutions having extremely low phospholipid levels. As in all large-volume parenterals, it is also important to avoid the introduction of bacterial endotoxins.

What level of protein purification is necessary to use hemoglobin as a blood substitute? Hemoglobin represents about 97% of total red cell protein. Removal of red cell membrane and cytoskeletal proteins increases the purity to about 98%. Approximately half of the remaining protein is the enzyme carbonic anhydrase. The heat treatment used in the preparation of Hb-XL99α, alluded to above, not only removes residual non-cross-linked hemoglobin but many other proteins as well. There is no evidence that further purification would be beneficial.

5.5 AUTOOXIDATION OF HEMOGLOBIN

Autooxidation of hemoglobin, the conversion of Fe^{2+} to Fe^{3+} (methemoglobin) in the presence of oxygen, is of concern for two reasons. First, ferric hemes cannot bind oxygen. Moreover, oxygen transport is further compromised by the fact that the remaining ferrous hemes within a tetramer that is partially oxidized have a markedly increased oxygen affinity (Bunn and Forget 1986). Second, reactive oxygen species formed concomitantly with oxidation (superoxide anion, hydrogen peroxide, and the hydroxyl radical) may give rise to toxicity. Autooxidation also limits the storage time for hemoglobin solutions, an important practical consideration in the development of a commercial product.

Hemoglobin autooxidation involves a complex series of reactions. The immediate oxidation product is the superoxide anion, $O_2^-\cdot$, which is formed by a one-electron reduction of molecular oxygen (see eq. 2). The majority of the $O_2^-\cdot$ formed disproportionates to yield H_2O_2 (eq. 3), which can either lead to further hemoglobin oxidation (eq. 4) or react with free ferrous ions (the Fenton reaction, eq. 5) to produce hydroxyl radical, an extremely powerful oxidizing agent.

$$Hbfe(II) + O_2 \rightarrow HbFe(III) + O_2^-\cdot \tag{2}$$

$$H^+ + HO_2\cdot + O_2^-\cdot \rightarrow H_2O_2 + O_2 \tag{3}$$

$$HbFe(II) + \frac{1}{2}H_2O_2 \rightarrow {}^-OH + HbFe(III) \tag{4}$$

$$Fe^{2+} + H_2O_2 \rightarrow Fe^{3+} + {}^-OH + \cdot OH \tag{5}$$

Free metal ions can serve to catalyze these reactions:

$$Fe^{3+} + HbFe(II) \rightarrow Fe^{2+} + HbFe(III) \tag{6}$$

$$Fe^{2+} + O_2 \rightarrow Fe^{3+} + O_2^-\cdot \tag{7}$$

The sum of equations 6 and 7 is equal to equation 2.

Within the red cell, autooxidation is held in check by a NADH-dependent reductase system which continually reduces methemoglobin as it is formed. This enzyme system is not present in plasma, however. At 37°C in physiological salt solutions, hemoglobin oxidizes at a rate of about 4% per hour (see Table 5–1). If this were the case in vivo, it would vitiate the use of hemoglobin as a blood substitute. Fortunately, this does not occur. Following transfusion of native hemoglobin and HbXL99α in the rat, we observed a dramatic decrease in the rate of autooxidation (Snyder et al. 1987). Over a period of 3-5 hours, less than 2% oxidation occurred. Several antioxidant systems present in blood contribute to this effect.

When hemoglobin is incubated in whole blood in vitro, the rate of oxidation is decreased by about 65% compared to simple buffers (Table 5–1). Scavenging of free iron by plasma transferrin can account for about half of this effect. Other metal chelators such as ceruloplasmin may also contribute somewhat. A very important role is played by the erythrocyte. As suggested by the reaction sequence shown above, rapid removal of hydrogen peroxide by sufficiently high concentrations of catalase markedly decreases the rate of hemoglobin autooxidation (Table 5–1). The free concentration of catalase in plasma (about 12 units/ml) is too low to achieve this effect. Erythrocytes, however, provide a very efficient scavenging pathway. The

TABLE 5–1 Hemoglobin Autooxidation Rates In Vitro[1]

Incubation Conditions	Percentage of metHb
Lactated Ringer's[2]	21
Whole blood	9
Transferrin (20 μM)	14
Catalase (4 $\times$ 10^3 U/ml)	11
Catalase (12 U/ml)	21

[1] HbXL99α (human hemoglobin A cross-linked between Lys-99α_1 and Lys-99α_2) was incubated for 5 hours at 37°C, pH 7.4, [Hb] = 0.2 mM tetramer. The initial metHb was 2%. At the end of the incubation period, aliquots were removed from each sample and the % metHb was determined spectrophotometrically. The antioxidants were added to solutions of HbXL99α in lactated Ringer's.

[2] Lactated Ringer's is a physiological salt solution containing 130 mM Na$^+$, 4 mM K$^+$, 1.5 mM Ca^{2+}, 109 mM Cl$^-$, and 28 mM lactate.

rate-limiting step for H_2O_2 removal by red cells is transported across the cell membrane. This occurs by passive diffusion with a rate constant very similar to that for the uptake of H_2O (Snyder, unpublished observations). Once within the red cell, H_2O_2 is degraded rapidly by catalase and glutathione peroxidase. The rate of removal of H_2O_2 by erythrocytes within whole blood corresponds to a catalase activity of 4,000 units/ml. This is only 10-fold lower than the actual enzyme concentration and is sufficient to completely block the component of autooxidation due to H_2O_2 formation (Table 5–1).

Scavenging of H_2O_2 by erythrocytes has another important effect. It enables ascorbate to serve as an effective reducing agent for methemoglobin. To examine the possibility that extracellular methemoglobin might be reduced, we carried out transfusion experiments in the rat with fully oxidized HbXL99α (Snyder et al. 1987). Over a period of 5 hours, the fraction of hemoglobin in the oxidized form decreased to a value of about 20%; 80% of the hemoglobin remaining had been reduced from the Fe^{3+} to the Fe^{2+} state. Reduction of methemoglobin also occurs when it is added to whole blood or plasma in vitro, but to a smaller extent (Table 5–2). Pretreatment with ascorbate oxidase abolishes this effect. The level of reduction that occurs in vitro corresponds to the amount of ascorbate available in plasma, normally between 50 and 100 μM (Table 5–2). To account for the much higher level of reduction of methemoglobin observed in vivo, mechanisms must exist to replenish plasma ascorbate. Regeneration of the reduced form of ascorbate occurs by both dehydroascorbate reductase and semidehydroascorbate reductase within the red cell and other tissues (Halliwell and Gutteridge 1985).

Replenishment of reduced ascorbate alone, however, is not sufficient to account for the extent of reduction of methemoglobin observed in vivo. In

TABLE 5-2   Reduction of methbXL99α In Vitro[1]

Incubation Conditions	Percentage of metHb
Whole blood	89
Plasma	94
Ascorbate oxidase-treated plasma	100
Ascorbate (0.1 mM)[2]	92

[1] Incubations of fully oxidized HbXL99α (100% metHb) were carried out for 2 hours at 37°C, pH 7.4, as described in Table 5-1. Human blood was used for all experiments. Hb concentration = 0.2 mM in each experiment.
[2] Incubation was carried out in lactated Ringer's solution.

TABLE 5-3   Incubation of 50% methbXL99α with 1 mM Ascorbate[1]

Incubation Conditions	Percentage of metHb
Lactated Ringer's	65
Whole blood	32
40% Erythrocyte suspension[2]	20

[1] Incubations of 50% methbXL99α were carried out for 5 hours at 37°C, pH 7.4, Hb concentration = 0.2 mM.
[2] In 50 mM Bis-tris (pH 7.4), 135 mM NaCl.

the presence of excess ascorbate, reduction of methemoglobin in simple buffers under aerobic conditions reaches a limiting value of 50–70% Fe^{3+}. If the initial fraction of methemoglobin is lower than this value, a net oxidation of hemoglobin occurs (Table 5-3). Ascorbate can thus behave both as a reductant and as an oxidant of hemoglobin. The steady-state level of methemoglobin is determined by three reactions: reduction of methemoglobin by ascorbate (eq. 8); the reaction of ascorbate with oxyhemoglobin to yield methemoglobin and H_2O_2 (eq. 9); and further oxidation of hemoglobin by the H_2O_2 so generated (eq. 4).

$$HbFe(III) + AH^- \rightarrow HbFe(II) + AH\cdot \qquad (8)$$

$$2H^+ + HbFe(II)O_2 + AH^- \rightarrow HbFe(III) + H_2O_2 + AH\cdot \qquad (9)$$

Erythrocytes shift the balance between oxidation and reduction of hemoglobin by ascorbate in favor of reduction by rapidly scavenging H_2O_2. Incubation of methemoglobin with 1 mM ascorbate in whole blood or in the presence of washed red cells leads to a level of reduction comparable to that observed in vivo (Table 5-3).

The antioxidant systems present in blood are also likely to be important in preventing cellular injury from the reactive oxygen species formed during

TABLE 5-4   Effect of Physiological Levels of Catalase and Transferrin on Hemoglobin-Induced Arachidonic Acid Peroxidation[1]

Incubation Conditions	Arachidonic Acid Oxidation Products
HbXL99α alone	5.8%
HbXL99α + catalase (4 $\times$ 10^3 U/ml)	2.0%
HbXL99α + transferrin (20 μM)	3.2%

[1] All incubations contained 0.2 mM HbXL99α and ^{3}H-arachidonic acid. Incubations were carried out in lactated Ringer's at 37°C, pH 7.4. After 4 hours, lipids were extracted, and the increase in the percentage of arachidonic acid oxidation products was determined by high-pressure liquid chromatography.

hemoglobin autooxidation. Earlier studies have shown that erythrocytes are able to protect mammalian cells grown in tissue culture from cell death when exposed to high levels of H_2O_2 (Agar et al 1986). Using a very susceptible substrate, arachidonic acid, we have been able to detect low levels of lipid peroxidation initiated by hemoglobin (Table 5-4). Catalase, at a concentration corresponding to the rate of H_2O_2 scavenging of whole blood, decreases this activity by about 65%. Chelation of free Fe by transferrin is also inhibitory.

Although the problems associated with hemoglobin oxidation cannot be discounted, they appear to pose much less of a concern than initially anticipated. Antioxidant systems present in blood markedly retard the rate of hemoglobin autooxidation, provide a pathway for the reduction of methemoglobin, and efficiently quench reactive oxygen species that are formed by hemoglobin oxidation. In clinical applications in which a large volume of blood is exchanged with a hemoglobin solution, supplementation with catalase or other H_2O_2 scavengers, and the use of iron chelators may become necessary.

5.6   RECOMBINANT PRODUCTION OF HEMOGLOBIN

Use of recombinant DNA techniques to produce hemoglobin for use as a blood substitute offers, in principal, several important advantages: potentially an unlimited supply of hemoglobin, a viral-free starting material and the possibility of tailoring the properties of hemoglobin to specific applications. The enormous scale of production needed (50,000 kg, even if only 1 million units were prepared) and the cost efficiency required (about $1/ gm of protein) present formidable challenges, however.

Nagai and co-workers were the first to produce human hemoglobin in *Escherichia coli* (Nagai and Thogersen 1984, 1987). In this system the α and β chains are produced separately as fusion proteins and purified chromatographically under denaturing conditions. The amino-terminal leader sequence is then excised with activated factor Xa, a proteolytic enzyme of the blood-clotting cascade. In the presence of exogenous heme, the subunits can be assembled into a tetramer to yield authentic hemoglobin. Although this provides a useful approach to prepare small quantities of hemoglobin variants for structure-function studies, large-scale production of hemoglobin by this method is not feasible. The cost would be prohibitive, and the process of assembly of the tetramer is too fastidious to scale-up.

At first, it would appear that to overproduce a native heme protein in *E. coli* it would be necessary to engineer the porphyrin biosynthetic pathway as well as to overexpress heme. That this is not the case was first demonstrated by Sligar and colleagues in studies of myoglobin (Springer and Sligar 1987). A synthetic gene for sperm whale myoglobin was constructed with the optimal codons (base triplets) used by *E. coli* for each of the amino acids. Intact myoglobin was produced, comprising up to 10% of total cell protein. Correct folding of the polypeptide chain occurred and the heme group was properly inserted. *E. coli* produced the same heme as found in mammalian myoglobins and hemoglobins. Overproduction of heme presumably occurs in this system due to the relief of negative feedback inhibition of the heme biosynthetic pathway as free heme becomes depleted upon assembly with newly synthesized myoglobin. This accommodating feature of *E. coli* was subsequently exploited to produce hemoglobin by Hoffman and co-workers at Somatogenetics (Hoffman et al. 1989). Coexpression of the α and β subunits within the same cell enables proper assembly of the tetramer to occur and avoids precipitation of the polypeptide chains in inclusion bodies.

As in the case of myoglobin, the α and β chains of hemoglobin produced in *E. coli* retain the methionine residue at the amino terminus. In native hemoglobin, the N-terminal methionine of both subunits is excised. In the mature protein, the α and β chains begin with valine. To create a more authentic amino-terminal sequence, the valine residue can be deleted. This would result in a conservative amino acid substitution of methionine for valine at the N-terminus of both subunits.

Other properties of hemoglobin can also be tailored for its use as a blood substitute by site-directed mutagenesis. More than 400 naturally occurring hemoglobin mutants have been discovered. Of these, 15 have been reported to have a decreased oxygen affinity (Bunn and Forget 1986). It should be possible to construct many additional low-affinity variants by alteration of amino acid residues within the 2,3-DPG binding site, at the interface between the subunits, and within the heme pocket. One synthetic low-affinity mutant has already been described (Nagai et al. 1987). Dissociation of the tetramer could be blocked by the introduction of a disulfide bridge between the $\alpha\beta$ dimers, or by the construction of a single polypeptide chain incorporating both α or both β subunits. It may also be possible to use site-directed mutagenesis to increase the stability of hemoglobin toward autooxidation.

Immunogenicity of mutant hemoglobins is not likely to be a problem. Patients with sickle cell disease repeatedly transfused with normal blood do not develop antibodies against the protein. Even hemoglobins from different mammalian species are not highly immunogenic. The low level of antigenicity of hemoglobin is obviously an important factor for the development of chemically modified derivatives as well.

Single-cell production of hemoglobin presents a paradox when compared to the synthesis of other proteins using recombinant DNA methods. Generally, not only is the yield of the recombinant protein increased when compared to the native source, but the purification is also simplified. This will not be the case for hemoglobin. As noted above, hemoglobin represents about 97% of total red cell protein. This level of production can not be approached in *E. coli*, and certainly not in yeast or a higher eukaryotic cell system. Consequently, a more involved purification scheme will be needed, probably requiring the use of chromatographic methods. It will be difficult for such a process to compete economically with one using donated human blood as the source of hemoglobin.

Because viral contamination can be eliminated by heat treatment, there is no commanding reason to use recombinant methods to produce hemoglobin. It is difficult to estimate the amount of donated blood that will be available for this purpose, but it is probably not less than 3 million units per year. That amount should be adequate through the 1990s. By then, large-scale production of human hemoglobin in transgenic animals should be feasible, if a limitation in the supply of donated blood develops.

Recently, human hemoglobin has been expressed in transgenic mice (Behringer et al. 1989; Ryan et al. 1990). This was accomplished by inclusion of far-upstream regulatory sequences required to turn on the globin genes in erythroid precursors. By the end of the decade it should be possible to achieve this same result in large domestic animals. Methods already exist that would make it possible to ablate the endogenous globin genes to avoid a mixture of different hemoglobins (Capecchi 1989). Of course, it would not be necessary to produce native human hemoglobin. The *E. coli* expression systems described above could be used to design variants having properties tailored for specific applications. These hemoglobins could then be prepared economically and on a large scale using a transgenic host.

5.7 SUMMARY

Several intramolecularly cross-linked hemoglobins having properties useful as blood substitutes have been developed. At least one of these, HbXL99α, is amenable to large-scale production. This hemoglobin, and perhaps other cross-linked derivatives as well, is sufficiently heat stable to achieve complete viral inactivation. This makes it possible to use human blood as a starting material. Preliminary studies on the use of HbXL99α to perfuse the heart during coronary angioplasty appear promising (Rossen et al. 1987). For large-volume blood replacement, a derivative having a longer intravascular retention time would be desirable. The development of more selective cross-linking agents for the polymerization of hemoglobin would be useful for this purpose.

The expression of human hemoglobin in *E. coli* (Nagai and Thogersen 1984, 1987; Hoffman et al. 1989) and in transgenic mice (Behringer et al. 1989; Ryan et al. 1990) has been achieved. The *E. coli* system should prove useful for the design of hemoglobin mutants having specifically tailored properties for use as blood substitutes. Adequate supplies of donated blood will likely be available for at least the next decade for the production of chemically modified hemoglobin derivatives. If the supply of human blood later becomes limiting, large-scale production of human hemoglobin should be feasible in transgenic pigs or cows. The economics of this process could be enhanced by producing other blood proteins of commercial value, e.g., human albumin and factor VIII, in the same animal.

REFERENCES

Agar, A.S., Sadrzadeh, S.M.H., Hallaway, P.E., and Eaton, J.W. (1986) *J. Clin. Invest.* 77, 319–321.

Arnone, A. (1972) *Nature* 237, 146–149.

Arnone, A., Benesch, R.E., and Benesch, R. (1977) *J. Mol. Biol.* 115, 627–642.

Behringer, R.R., Ryan, T.M., Reilly, M.P., et al. (1989) *Science* 245, 971–973.

Benesch, R., Benesch, R.E., Yung, S., and Edalji, R. (1975) *Biochem. Biophys. Res. Commun.* 63, 1123–1129.

Benesch, R., Triner, L., Benesch, R.E., Kwong, S., and Verosky, M. (1984) *Proc. Natl. Acad. Sci USA* 81, 2941–2943.

Bunn, H.F., and Forget, B.G. (1986) *Hemoglobin: Molecular Genetic and Clinical Aspects*, Saunders, Philadelphia.

Bunn, H.F., and Jandl, J.H. (1969) *J. Exp. Med.* 129, 925–934.

Bunn, H.F., Esham, W.T., and Bull, R.W. (1969) *J. Exp. Med.* 129, 909–924.

Capecchi, M.R. (1989) *Science* 244, 1288–1292.

Chatterjee, R., Walder, R.Y., Arnone, A., and Walder, J.A. (1982) *Biochemistry* 21, 5901–5909.

Chatterjee, R., Welty, E.V., Walder, R.Y., et al. (1986) *J. Biol. Chem.* 261, 9929–9937.

Dickerson, R.E., and Geis, I. (1983) *Hemoglobin: Structure, Function, Evolution, and Pathology*, Benjamin/Cummings, Menlo Park, CA.

Estep, T.N., Bobka, E.W., and Ebeling, A.A., et al. (1989) in *Proceedings of the International Symposium on Red Cell Substitutes*, 16–19 May 1989, San Francisco, p. 21.

Feola, M., Simoni, J., Canizaro, P.C., et al. (1988) *Surgery Gynecology and Obstetrics* 166, 211–222.

Halliwell, B., and Gutteridge, J.M.C. (1985) *Free Radicals in Biology and Medicine*, pp. 100–103, Oxford University Press, Oxford.

Hess, J.R., Wade, C.E., and Winslow, R.M. (1989) in *Proceedings of the International Symposium on Red Cell Substitutes*, 16–19 May 1989, San Francisco, p. 16.

Hoffman, S.J., Looker, D.L., Rosendahl, M.S., and Stetler, G.L. (1989) in *Proceedings of the International Symposium on Red Cell Substitutes*, 16–19 May 1989, San Francisco, p. 11.

Keipert, P.E., and Chang, T.M.S. (1985) *Biomater. Med. Dev. Artif. Organs* 13, 1–15.

Keipert, P.E., Adeniran, A.J., Kwong, S., and Benesch, R.E. (1989) *Transfusion* 29, 768–773.

Monod, J., Wyman, J., and Changeux, J.P. (1965) *J. Mol. Biol.* 12, 88–118.

Moss, G.S., Gould, S.A., Sehgal, L.R., Sehgal, H.L., and Rosen, A.L. (1984) *Surgery* 95, 249–255.

Nagai, K., and Thogersen, H.C. (1984) *Nature* 309, 810–812.

Nagai, K., and Thogersen, H.C. (1987) *Methods Enzymol.* 153, 461–481.

Nagai, K., Luisi, B., and Shih, D., et al. (1987) *Nature* 329, 858–860.

Rabiner, S.F., Helbert, J.R., Lopas, H., and Friedman, L.H. (1967) *J. Exp. Med.* 126, 1127–1142.

Rossen, J.D., Snyder, S.R., Marcus, M.L., and Walder, J.A. (1987) *Circulation* 76 (suppl. IV), 27.

Ryan, T.M., Townes, T.M., Reilly, M.P., et al. (1990) *Science* 247, 566–568.

Sehgal, L.R., Rosen, A.L., Gould, S.A., Sehgal, H.L., and Moss, G.S. (1983) *Transfusion* 23, 158–162.

Snyder, S.R., Welty, E.V., Walder, R.Y., Williams, L.A., and Walder, J.A. (1987) *Proc. Natl. Acad. Sci. USA* 84, 7280–7284.

Springer, B.A., and Sligar, S.G. (1987) *Proc. Natl. Acad. Sci. USA* 84, 8961–8965.

Walder, J.A., Zaugg, R.H., Walder, R.Y., Steele, J.M., and Klotz, I.M. (1979) *Biochemistry* 18, 4265–4270.

Walder, J.A., Walder, R.Y., and Arnone, A. (1980) *J. Mol. Biol.* 141, 195–216.

6

Liposome-Encapsulated Hemoglobin: Historical Development of a Blood Substitute

Beth Goins
Alan S. Rudolph
Frances S. Ligler

The use of liposome-encapsulated hemoglobin (LEH) as a blood substitute is an attempt to learn from nature by modeling an oxygen-carrying system after the essential components of a red blood cell (Figure 6–1). Blood transfusions have been used for over a century for resuscitation, and the problems they produce with histocompatibility and infection have become all too obvious. In an attempt to reduce such complications, individual blood components have been transfused, with red blood cells and free hemoglobin administered as oxygen-carrying fluids. Carefully screened red blood cells are effective, either fresh or frozen, but screening is becoming ever more extensive and expensive; and blood supplies free of infective agents are

The authors gratefully acknowledge the financial support of the United States Office of Naval Research, the Naval Medical Research and Development Command, and the National Research Council, of which Dr. Goins is a postdoctoral fellow. The views expressed in this article are those of the authors and not necessarily those of the U.S. Department of Defense.

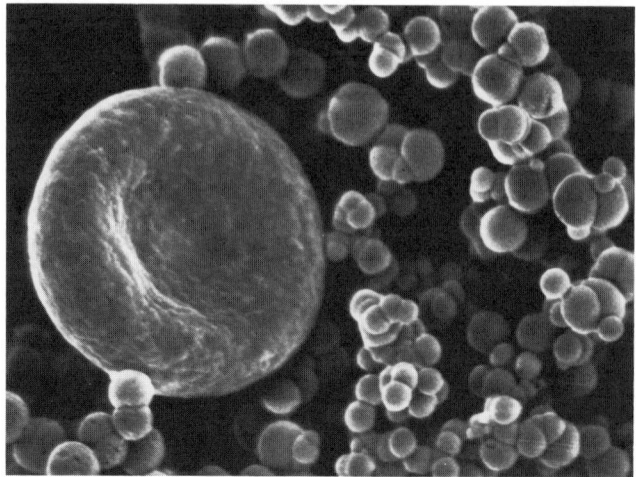

FIGURE 6–1 Scanning electron micrograph of LEH and a normal red blood cell. LEH prepared using a Gaulen homogenizer was mixed with human blood, fixed with glutaraldehyde, and stained using osmium.

limited. Free hemoglobin rapidly dissociates and becomes toxic. Hemoglobin can be cross-linked to stabilize it in vivo, but issues of nephrotoxicity and allergic reactions have not been resolved (for a review, see Odling-Smee and Wilson 1988). Encapsulating the hemoglobin in a liposome protects it from dissociation, maintains its proximity to low-molecular-weight mediators and antioxidants, and extends its circulation time by sequestering it behind a cell-like membrane less quickly recognized as foreign by the immune system.

This chapter discusses the historical development of LEH as an oxygen-carrying resuscitation fluid. Special consideration has been given to the fabrication and biological assessment of LEH, with emphasis on the issues of efficacy and safety. For the curious reader, more detailed reviews describing the in vitro testing (Farmer and Gaber 1987) and biological activity of LEH (Rabinovici et al. 1990c) are available. Finally, some of the future directions for the development of this potential red blood cell substitute are discussed.

6.1 FABRICATION

T.M.S. Chang first proposed the concept of using encapsulated hemoglobin as an "artificial red cell" more than 30 years ago (Chang 1957). Chang originally used synthetic materials (nylons) rather than lipids to encapsulate the hemoglobin. Twelve years later, A.D. Bangham suggested to the British

National Research Development Corporation that hemoglobin could be encapsulated in liposomes to provide a universally compatible blood (Bangham 1983). However, it was not until the late 1970s that laboratory experiments directed toward the encapsulation of hemoglobin in liposomes began (Miller and Djordjevich 1979; Gaber et al. 1983; Hunt and Burnette 1983).

Investigators attempting to bring the concept of an artificial red blood cell into reality faced daunting technological problems which had to be solved prior to testing LEH in significant numbers of animals. First, hemoglobin, which is highly sensitive to oxidation, had to be encapsulated without degradation, and the method of encapsulation had to be amenable to the production of large quantities of LEH. Second, a liposome formulation had to be developed that would be stable during production and storage, and would remain in the circulation for extended periods.

The standard methods of liposome preparation were thoroughly explored for the encapsulation of hemoglobin (Djordjevich and Miller 1980; Gaber et al. 1983; Miller 1981; Olson et al. 1979; Hunt and Burnette 1983; Jopski et al. 1989). For the method to produce a usable product, the encapsulated hemoglobin had to remain in its functional, unoxidized state. Also the final diameter of the liposomes had to be under 1 μm to carry oxygen to the smallest capillaries and to avoid their rapid removal from circulation by the reticuloendothelial system (RES) (Miller 1981). Sonication, detergent dialysis, and reverse-phase methods denatured the hemoglobin. Extrusion through filters or a French press produced small liposomes containing functional hemoglobin, but was only useful for making small quantities. However, extrusion through a Microfluidizer (Beissinger et al. 1986) or a Gaulen homogenizer (Gamble 1988) proved to be successful for the production of multi-liter quantities of LEH, 0.2–0.5 μm in diameter.

The liposome formulations originally used for LEH included natural, unsaturated lipids derived from egg or soybeans, varying amounts of cholesterol, and a negatively charged moiety to prevent aggregation (Olson et al. 1979; Miller 1981; Gaber et al. 1983). Most investigators also included α-tocopherol to reduce lipid peroxidation. The addition of glutathione reduced the oxidation of hemoglobin during manufacture and storage (Stratton et al. 1988). Mediators of hemoglobin function, pyridoxal-5-phosphate or 2,3-diphosphoglycerate, were encapsulated so that oxygen binding and release occurred at pressures equivalent to those in normal red blood cells (Gaber and Farmer 1984).

When LEH made with unsaturated lipids was transfused into rats, it was cleared from the blood in a few hours (Hunt et al. 1985). A.D. Bangham (1983) who originally suggested encapsulating hemoglobin in a liposome, suggested that LEH was unsuitable as an oxygen-carrying fluid because its lifetime in circulation was too short. However, the use of long-chain, saturated phospholipids, such as distearoyl phosphatidylcholine or hydrogenated soy lipid, extended the circulation half-life to 16–20 hours in mice

(Beissinger et al. 1986). Though such long circulation times had been reported for liposomes composed of long-chain, saturated phospholipids, the efficient encapsulation of material into these liposomes required heating the suspension above the phase transition temperature of the lipid (Gregoriadis and Senior 1985). This high temperature of about 60°C quickly denatures hemoglobin. However, inclusion of high concentrations of cholesterol eliminated the phospholipid phase transition and made the fabrication of liposomes possible at room temperature (Mabrey-Gaud 1981).

6.2   IN VIVO STUDIES

The in vivo analysis of LEH has focused on issues important to both LEH efficacy and safety. However, recent studies of the in vivo response to LEH have revealed the need for additional reformulation (Rabinovici et al. 1989; 1990a). Previous modifications to the composition of LEH were primarily in response to problems of production and stability. Thus, much of the recent in vivo work with LEH represents seminal data on the physiological effects of an encapsulated blood substitute. In addition, a large body of data on the effects of large intravenous doses of liposomes will become available through further experimentation, which may be useful for other applications utilizing liposomes as vehicles for drug delivery.

6.2.1   Total Exchange Transfusions

Historically, the first in vivo experiment used to prove efficacy of a blood substitute was to perform a total exchange transfusion of an animal's red blood cells with the artificial oxygen-carrying fluid and document recovery. These are rudimentary experiments which demonstrate that the fluid has sufficient oxygen-carrying capacity to sustain life. They do not, however, address the issue of safety and should not be considered as models for the clinical application of a red blood cell substitute.

Such experiments have been performed with LEH in anesthetized rat models (Djordjevich et al. 1982; Hunt et al. 1985; Ligler et al. 1989). Total isovolemic exchange experiments were performed in which hematocrit levels were brought to 5% or less. Infusion of LEH (in some experiments with 5% serum albumin or 10% Eri-Lyte) during this procedure resulted in the maintenance of baseline blood pressure, heart rate, systemic PO_2, and pH (Djordjevich et al. 1982; Ligler et al. 1989). In other studies, altered hemodynamic parameters were documented, including reduced total peripheral resistance, marginally elevated cardiac output (Djordjevich et al. 1987), and transient increases in blood urea nitrogen (Hunt et al. 1985). Survival of animals infused with LEH in these studies varied from 10–28 hours in one study (Ligler et al. 1989) to 2–5 hours in another (Djordjevich et al. 1982), with gross pathological examination revealing enlargement of the liver and spleen. Further analysis of specific organ function revealed normal creatinine values, normal renal histology, and normal plasma levels of glutamate-pyruvate transaminase (Hunt et al. 1985).

Many of these studies were complicated by the lack of sterility of the preparations, evidenced by signs of septic shock (Djordjevich et al. 1982), so that further analysis of the physiological effects of LEH are difficult. It is clear, however, that these experiments demonstrate the efficacy of LEH. More specifically, when an oxygen-deficient situation presented a lethal challenge for the animal, LEH sustained life by the delivery of oxygen.

6.2.2 Circulation Half-Life

One major determinant of LEH efficacy is its ability to persist in circulation. Not only does the circulation half-life determine the period of efficacy, but it may also define the particular LEH application. For example, it may be that LEH would be most useful in the early phases of trauma, and to minimize side effects, should be cleared from the body quickly. However, longer lifetimes are probably more useful in order to avoid RES blockade and to minimize the need for further transfusions while the body replaces its red blood cells. This latter possibility suggests that issues pertaining to circulation half-life must be considered pertinent to efficacy as well as safety.

Early experiments examining the circulation persistence of LEH revealed that circulation persistence is dose dependent (as is the case with most liposome preparations) as well as formulation dependent (Farmer and Gaber 1987). Injections of moderate amounts of LEH made with distearoyl phosphatidylcholine into the tail vein of mice were cleared with a half-life of 16–20 hours, five times longer than LEH made with dimyristoyl phosphatidylcholine (Farmer and Gaber 1987). Similarly, following exchange transfusion in an anesthetized rat model, a longer circulation half-life (18 hours) was documented for LEH made with hydrogenated soy lipid (Ligler et al. 1989) as compared to 5.8 hours measured for LEH composed of egg phosphatidylcholine (Hunt et al. 1985). Pathological evidence for the removal of LEH from the circulation by the liver and spleen initially came from visible examination of these organs following infusion of LEH. Recent biodistribution data with ^{99m}technicium-labeled LEH infused into the ear vein of anesthetized rabbits support these findings. This study showed that, at 20 hours post-injection, the majority of LEH taken up by the tissues (38% of the total dose) was found in the spleen (15%) and liver (12%) (Phillips et al. 1990). It is interesting to note that, in the larger rabbit, the observed circulation half-life of 36 hours is significantly longer than in the mouse (16–20 hours) using comparable doses equal to 25% of the blood volume (Phillips et al. 1990).

An extensive study to determine the effects of LEH on in vivo RES function has been initiated (S. Weinstock, personal communication). In this study, quantitation and clearance kinetics of RES phagocytic activity in rats treated with three different doses of LEH (4%, 10%, or 25% of blood volume), with liposomes devoid of hemoglobin, and with Krebs-Ringer bicarbonate were measured by colloidal carbon (<0.1 μm) clearance. Preliminary data revealed that initial (10-min) clearance rates in rats given the larger doses (10% and 25%) of LEH or empty liposomes were slower than in the control group at two hours post-treatment. However, by 60 min after LEH or empty liposome administration, the rate of uptake of carbon particles was equivalent to that in control animals (80–95%). In each case, clearance rates were normal at 24 hours and at 2 weeks post-treatment. No significant morphological changes in glutaraldehyde-fixed liver or spleen were detected. These results suggest that the alterations in RES organ function seen following LEH treatment, at the doses studied, are minor and of a transient nature.

6.2.3 Simple Infusions

The monitoring of hemodynamic parameters, following the administration of small doses of LEH in otherwise normal rodents, has been used to document LEH safety. To date, less data have been generated for this purpose than to address LEH efficacy. These studies revealed that small doses of LEH made with hydrogenated soy lipid and administered to a conscious rat produced a number of side effects (moderate increases in blood pressure, heart rate, leukocytosis, and thromboxane and a moderate decrease in platelet count and cardiac output) which diminished over the course of an hour (Rabinovici et al. 1989). Many of these effects were alleviated by a change in the major lipid component of LEH from hydrogenated soy lipid to the synthetic distearoyl phosphatidylcholine (Rabinovici et al. 1990a) or the use of a platelet-activating factor (PAF) antagonist (Rabinovici et al. 1990b). Although the definitive cause of these effects has not been determined, the contaminants present in the hydrogenated soy preparation and the batch-to-batch variability of LEH caused by production methods, which as yet are not completely standardized, were considered to weigh heavily in these results.

6.3 FUTURE DIRECTIONS

Considerable progress has been made toward the development of LEH as a temporary red blood cell substitute. Yet, several key issues remain to be addressed: 1) source of hemoglobin; 2) immunological, pathological, and toxicological characterization; 3) liposomal surface modification to potentially extend the current circulation lifetime; and 4) stability during storage.

Hemoglobin isolated from both human and bovine red blood cells has been used in LEH production and may explain some of the differences in the in vivo data gathered thus far. More recently, human hemoglobin has been cloned and produced by fermentation as a functional protein (Hoffman et al. 1990). Hopefully, this breakthrough will alleviate problems due to shortages of available blood sources, impurities from the isolation procedure, and viral contamination. More importantly, the potential for an immune response to bovine hemoglobin makes the reformulation of LEH with recombinant human hemoglobin a very promising option.

One remaining area which must be explored is the evaluation of the clinical implications following LEH administration. Further characterization of the interaction of LEH with serum components, platelets, and the immune system is crucial. Both short-term and chronic effects of these interactions must be more closely followed at the cellular and biochemical levels as well as correlated with pathological examination. Special attention must be placed on defining the physiological effects associated with any RES impairment following the intravenous injection of large doses of LEH. The possibility of sensitization following multiple doses of LEH must be delineated using LEH containing homologous hemoglobin. In addition, although the individual components of LEH are biodegradable, little clinical information has been gathered to date about the potential toxic effects of the metabolites following LEH removal from circulation. Such a systematic study should now be feasible with the ability to fabricate LEH free of bacterial contamination and in large quantities.

The potential for increasing the in vivo circulation times of LEH by surface modification also merits further study. Although the current formulation meets the original guidelines of a 12-hour half-life set by the National Research Council (1963), reformulation of the product to extend LEH intravascular persistence by 48–72 hours could prevent the administration of a second LEH dose or blood transfusion, thereby limiting the patient's exposure to a blood-borne pathogen. Extended circulation times would also slow the rate of LEH uptake by the RES and reduce the possibility of immunosuppression. For many years, investigators have recognized the difficulty of using liposomes to deliver a drug or oxygen carrier to non-RES cells because of the propensity of the RES to remove circulating liposomes. Since red blood cells are cleared by the RES following enzymatic removal of their sialic acid residues, it was thought that the addition of sialic acid groups to the liposomal surface might extend the circulation time of the liposomes (Geho and Lau 1985). The utilization of gangliosides (sialic acid-containing glycolipids) to enhance the circulation times of liposomes has been demonstrated (Allen 1989; Gabizon and Papahadjopoulos 1988). LEH was modified to include ganglioside G_{M1} as a means to extend the LEH blood retention time (Goins et al. 1990). Ganglioside G_{M1}-containing LEH could be processed with only minor modifications of established procedures, carried oxygen in a similar manner as the original LEH, and showed moderate increases in circulation persistence in mice.

There are several unanswered questions concerning the long-term storage conditions for LEH. As with any pharmaceutical, a sufficient shelf-life must be accomplished to make the product a reality in the marketplace.

Lyophilization is the method of choice for LEH, since it would greatly reduce the mass of material for storage and eliminate the need for refrigeration. The addition of the cryoprotective carbohydrate trehalose during LEH production both preserved liposomal bilayer structure, allowing for a greater retention of hemoglobin following lyophilization, and protected the hemoglobin from oxidative damage (Rudolph 1988). LEH rehydrated following a 3-month shelf-life in the dry state had similar structural and functional properties to the initial LEH preparation (Rudolph and Cliff 1990). Even more encouraging was the fact that there was only a 10% increase over starting methemoglobin levels, indicating that lyophilization is a viable alternative for the long-term preservation of LEH. Further work entails developing lyophilization scale-up procedures and proving lyophilized LEH to be as safe and efficacious upon intravenous administration as LEH stored in a liquid suspension.

In summary, LEH development has come a long way in more than 30 years and, in particular, during the last five years. If the safety of LEH can be demonstrated, LEH will be in use as a human red blood cell surrogate in the near future.

REFERENCES

Allen, T.M. (1989) U.S. patent 4,837,028.

Bangham, A.D. (1983) in *Liposomes* (Ostro, M.J., ed.), p. 15, Marcel Dekker, New York.

Beissinger, R.L., Farmer, M.C., and Gossage, J.L. (1986) *Trans. Am. Soc. Artif. Intern. Organs* 32, 58–63.

Chang, T.M.S. (1957) B.Sc. Honours thesis, McGill University, Montreal, Canada.

Djordjevich, L., and Miller, I.F. (1980) *Exp. Hematol.* 8, 584–592.

Djordjevich, L., Pauli, B., Mayoral, J., and Ivankovich, A.D. (1982) *Anesthesiology* 57, A143.

Djordjevich, L., Kashani. A., Miller, I.F., and Ivankovich, A.D. (1987) *Biorheology* 24, 207–217.

Farmer, M.C., and Gaber, B.P. (1987) *Methods Enzymol.* 149, 184–200.

Gaber, B.P., and Farmer, M.C. (1984) in *The Red Cell: Sixth Annual Ann Arbor Conference* (Brewer, G., ed.), pp. 179–190, Alan R. Liss, New York.

Gaber, B.P., Yager, P., Sheridan, J.P., and Chang, E.L. (1983) *FEBS Letts.* 153, 285–288.

Gabizon, A., and Papahadjopoulos, D. (1988) *Proc. Natl. Acad. Sci. USA* 85, 6949–6953.

Gamble, R.C. (1988) U.S. patent 4,754,788.

Geho, W.B., and Lau, J.R. (1985) U.S. patent 4,501,758.

Goins, B.A., Kestler, K.J., Thourani, V.H., Rudolph, A.S., and Ligler, F.S. (1990) *Biophys. J.* 57, 261a.

Gregoriadis, G., and Senior, J. (1985) in *Targeting of Drugs with Synthetic Systems* (Gregoriadis, G., Senior, J., and Poste, G., eds.), pp. 192–193, Plenum, New York.

Hoffman, S.J., Looker, D.L., Roehrich, J.M., et al. (1990) *Proc. Natl. Acad. Sci. USA* 87, 8521–8525.

Hunt, C.A., and Burnette, R.R. (1983) in *Advances in Blood Substitute Research* (Bolin, R.B., Geyer, R.P., and Nemo, G.T., eds.), pp. 59–69, Alan R. Liss, New York.

Hunt, C.A., Burnette, R.R., MacGregor, R.D., et al. (1985) *Science* 230, 1165–1168.

Jopski, B., Purkl, Y., Jaronis, H.W., Schubert, R., and Schmidt, K.A. (1989) *Biochim. Biophys. Acta* 978, 79–84.

Ligler, F.S., Stratton, L.P., and Rudolph, A.S. (1989) in *The Red Cell: Seventh Ann Arbor Conference* (Brewer, G., ed.), pp. 435–455, Alan R. Liss, New York.

Mabrey-Gaud, S. (1981) in *Liposomes: From Physical Structure to Therapeutic Applications* (Knight, C.G., ed.), pp. 105, Elsevier/North Holland, Amsterdam.

Miller, I.F. (1981) *Chem. Eng. Commun.* 9, 363–370.

Miller, I.F., and Djordjevich, L. (1979) U.S. patent 4,133,874.

National Research Council (1963) *Criteria of Satisfactory Plasma Volume Expanders,* National Academy of Sciences, Washington, DC.

Odling-Smee, G.W., and Wilson, B.G. (1988) in *Blood Substitutes: Preparation, Physiology and Medical Applications* (Lowe, K.C., ed.), pp. 71–86, VCH Publishers, New York.

Olson, F., Hunt, C.A., Szoka, F.C., Vail, W.J., and Papahadjopoulos, D. (1979) *Biochim. Biophys. Acta* 557, 9–23.

Phillips, W.T., Timmons, J.H., Klipper, R., Blumhardt, R., and Rudolph, A.S. (1990) *J. Nucl. Med.* 31, 806.

Rabinovici, R., Rudolph, A.S., and Feuerstein, G. (1989) *Circ. Shock* 29, 115–132.

Rabinovici, R., Rudolph, A.S., and Feuerstein, G. (1990a) *Circ. Shock* 30, 207–219.

Rabinovici, R., Rudolph, A.S., Yue, T.L., and Feuerstein, G. (1990b) *Circ. Shock* 31, 431–445.

Rabinovici, R., Rudolph, A.S., Ligler, F.S., Yue, T.L., and Feuerstein, G. (1990c) *Circ. Shock* 32, 1–17.

Rudolph, A.S. (1988) *Cryobiology* 25, 277–284.

Rudolph, A.S., and Cliff, R.O. (1990) *Cryobiology* 27, 585–590.

Stratton, L.P., Rudolph, A.S., Knoll, W.K., Jr., Bayne, S., and Farmer, M.C. (1988) *Hemoglobin* 12, 353–368.

Medical Oxygen Transport Using Perfluorochemicals

Robert J. Kaufman

Whole blood serves a number of critical functions including delivery of oxygen to tissues, hemostasis, host defense, maintenance of ionic and protein balances, transport of nutrients and hormones, and the removal of metabolic waste products. However, since oxygen deprivation rapidly and irreversibly degrades the function of cells, tissues, and organs, leading to death, the most important function of blood is oxygen delivery to the tissues.

Historically, donated human blood has been the therapeutic of choice for acutely and chronically anemic patients. However, the use of donated blood is not without risks. Blood can be contaminated with an astounding array of infectious agents including HIV, hepatitis virus, cytomegalovirus (CMV), Epstein-Barr virus, and *Brucella abortus*. While blood is currently screened for many pathogens including HIV, it is not possible to identify all pathogens nor to eliminate them with heat sterilization. Recent estimates place the risk of contracting non-A, non-B hepatitis at 7-10% of all patients transfused in the USA with 50% of these patients developing chronic, active hepatitis (Haljamae and Rosenberg 1988).

Immunosuppression after transfusion of homologous blood is also a significant problem and has been linked with increased recurrence of cancer (Blumberg et al. 1985) and with an increased incidence of postoperative infection in certain types of surgical patients (Maetani et al. 1986). In addition to the health-related risks inherent in blood usage, there are other limitations to the use of donated blood. Donor blood must be typed and cross-matched for each patient, resulting in a transfusion delay of a minimum of 20 to 30 min. The use of type O negative blood to circumvent the typing and crossmatching time delay is not without risks.

Blood has a storage lifetime of only 42 days and it must be refrigerated. This makes blood unavailable in many of the situations in which it is most needed, such as in rural trauma incidents, in ambulances and helicopters, on battlefields, and during civilian disasters. Complicating the storage lifetime is the fact that after blood has been stored for a few days, the erythrocyte depletes its 2,3-DPG and oxygen bound to hemoglobin in these cells becomes relatively unavailable. Complete restoration of 2,3-DPG content and oxygen delivery function requires about 12 hours in the circulation. Thus, transfusion of stored red cells does not fully fulfill the oxygen transport function for which it was administered (Huetis et al. 1981).

The search for agents to replace the oxygen transport function of blood has been underway for over 50 years and has centered largely on two approaches: purified hemoglobin derivatives and perfluorochemical (PFC) emulsions. The goals of both PFC and hemoglobin research have been to develop a readily available, disease-free, shelf-stable, safe, cost-effective, universal donor product that would be available when and where it was needed. While neither approach has yet achieved broad clinical success, enormous progress has been made with PFC emulsions over the last five years in understanding and overcoming obstacles to successful product introduction. New medical uses for oxygen transport agents have been identified and explored. Several PFC products are now in various stages of preclinical and clinical evaluation. In late 1989, Fluosol DA-20®, developed and manufactured by the Green Cross Co. (Osaka, Japan), was approved as an oxygen transport agent for distal oxygenation of the myocardium during high-risk coronary balloon angioplasty by the U.S. FDA (Anonymous 1990).

The purpose of this chapter is to review the use of PFCs and their emulsions in medical oxygen transport and to highlight recent developments in the field.

7.1   HISTORY OF PERFLUOROCHEMICALS IN OXYGEN TRANSPORT

PFCs have two important properties that make them attractive for use in biomedical applications. They are extremely inert chemically and biochemically (Simon 1947) and dissolve 15 to 20 times more oxygen than water. PFC oxygen contents can be as high as 35–50 volume % (Gjaldbaek and Hildebrand 1949). Oxygen dissolves in PFCs in accordance with Henry's law and is not chemically bound as it is in hemoglobin. PFC transported oxygen is therefore readily available to the tissues by gradient diffusion.

Sabiston first recognized that PFCs could be exploited for medical oxygen transport. In 1965, he successfully used liquid PFCs as the oxygenator for a cardiopulmonary bypass unit (Howlett et al. 1965). It remained, however, for Leland Clark to galvanize research in this field. Clark, in 1966, demonstrated that mice and young puppies submerged in liquid PFC saturated with oxygen could derive their physiological oxygen requirements via the PFC in their lungs without toxicity (Clark and Gollan 1966). Shortly thereafter, Henry Sloviter discovered that PFCs could be rendered into a plasma-compatible form by emulsification with bovine serum albumin in Kreb's Ringer bicarbonate (Sloviter and Kamimoto 1967). Using this emulsion, he was able to extend the electrical activity in isolated, perfused rat brains far longer compared to rat erythrocytes. Geyer exchanged the blood of rats with an emulsion made from perfluorotributylamine, the surfactant pluronic F-68, and physiological salts. Animals survived on PFC "blood" at high oxygen tensions until sufficient regeneration of RBCs had occurred to support life on room air (Geyer et al. 1968, 1973). The rats developed normally and survived in apparent good health to the end of their normal life expectancy. These total exchange experiments were a graphic demonstration of both the efficacy and the safety of PFCs. The fact that the rats survived with hematocrits as low as 3% demonstrated the physiological gas transport capabilities of PFCs. The safety of emulsion was evident because serum albumin, immunoglobulins, clotting factors, platelets, leucocytes, and erythrocytes, which had all been removed as a consequence of the total exchange, all regenerated, indicating no damage to the liver or the marrow.

Subsequently, clinical research groups initiated animal studies and identified potential clinical applications for PFC oxygen transport agents in fields ranging from shock resuscitation and wound healing to heart attack and cancer therapy. Industrial research resulted in the preparation of the first clinically acceptable emulsion. Fluosol DA-20, by the Green Cross Co.

7.2   DATA ON THE USE OF PFC EMULSIONS

The availability of Fluosol DA-20 as a well-defined, reproducible emulsion led to the generation of consistent toxicity and organ distribution and excretion data on PFCs and their emulsions.

7.2.1   Results in Humans

The most relevant toxicity data on PFCs are those generated on humans in clinical trials involving almost 1,000 patients. Fluosol DA-20 has been given in doses as high as 40 ml emulsion/kg in single doses and as high as 56 ml emulsion/kg in multiple doses. Its side effects are primarily benign and reversible and can be divided into two categories: 1) pulmonary and hemodynamic reactions involving complement activation in response to the pluronic F-68 used as a surfactant in the Fluosol DA-20 and 2) serum enzyme elevations, presumably due to PFC accumulation in Kupffer cells in the liver.

The adverse hematological and pulmonary responses, which effect less than 5% of those treated, include shortness of breath, chest tightness, flushing, hypertension, lower back pain, nausea, and dizziness (Fedor et al. 1988). These reactions have been attributed to pluronic F-68, which causes C3 conversion and generation of C5a-related neutrophil-aggregating activity in a rabbit model (Vercellotti and Hammerschmidt 1982). The adverse reactions are readily controlled by administration of methylprednisolone IV or diphenhydramine IM. Transient neutropenia and thrombocytopenia accompany the reaction to Fluosol DA-20 and are seen in many patients without overt adverse reactions (Tremper et al. 1984).

A summary of adverse reactions to Fluosol DA-20 in the U.S. clinical trials by trial type has been published (Fedor et al. 1988). In 96 anemic patients who received from 2.5 to 55.5 ml emulsion/kg, 13 patients (13.5%) had adverse reactions. In the coronary balloon angioplasty trial only 14 of 297 (4.7%) patients receiving <7 ml emulsion/kg had adverse reactions. In the cancer radiation trials where 97 patients received 463 Fluosol DA-20 exposures with total dosages as high as 56 ml emulsion/kg, 29 patients experienced adverse reactions (29.9%). The higher incidence of events in the cancer patients may be related to the underlying disease condition of the patient. In all cases, the reactions were of short duration (<90 min) and were self-limiting or controlled by diphenhydramine.

Serum enzyme elevations have been observed in some but not all of the anemia clinical trials. Mitsuno (Mitsuno et al. 1982) observed moderate SGOT elevation in eight of 122 patients given 5 to 25 ml/kg of Fluosol DA-20 one week post-infusion. Enzyme elevations persisted in seven of the patients for a month. Waxman reported elevated serum enzymes in one of six patients receiving 20 ml emulsion/kg (Waxman et al. 1984). The patient had elevated alkaline phosphatase, SGOT, SGPT, and lactate dehydrogenase values two days post-infusion but they returned to normal 10 days post-infusion. In the cancer radiation therapy clinical trials, where patients received multiple doses of Fluosol DA-20, 17 of 37 patients had serum enzyme elevations 1.5 to 2.5 times normal including seven with SGOT elevation, eight with SGPT elevation, and 10 with alkaline phosphatase elevation. None of the patients with elevated enzyme levels developed acute symptomatic hepatic problems and all appeared to retain normal liver function (Lustig et al. 1989a). There have been no reports of serum enzyme elevations in the clinical trials of Fluosol DA-20 in coronary angioplasty patients.

RES function in postgastrectomy patients given 1,000 ml of Fluosol DA-20 was assessed by measuring the phagocytic index with iron-chondroitin sulfate (Fujita et al. 1983). Recovery of normal RES function was delayed from one hour after infusion in the Fluosol DA-20 group compared to the control group, although it recovered to the post-operative level on the sixth day after infusion. The results indicate a mild, transient, and reversible depression of RES function in humans. The clinical relevance of the RES depression data is questionable. Fluosol DA-20 has been used in severely hemorrhagic surgical patients, many with GI involvement. These patients would surely be amongst the ones most affected by an increased susceptibility to sepsis. However, clinical experience with Fluosol DA-20 in almost 1,000 patients has not indicated a greater incidence of sepsis or infections in treated patients.

Fluosol DA-20 has produced only slightly toxic symptoms in patients treated with up to 56 ml emulsion/kg. The symptoms have been benign, reversible, and without meaningful functional impairment despite the enormous quantities infused; infusion of 40 ml/kg of Fluosol DA-20 is the equivalent of almost 3 kg of emulsion and over 0.5 kg of PFC. The data strongly suggest that, in general, PFC emulsions may be used safely in patients at therapeutic doses.

7.2.2 Results in Animals

Extensive tissue distribution and toxicity studies have been conducted with Fluosol DA-20 in animals. After intravenous infusion, the PFC concentration of the blood peaks within one hour and then begins a steady, dose-dependent decline. For most PFC emulsions, the blood half-life is about 12-24 hours depending upon dose and species. The PFC particles are cleared from the blood by the cells of the reticuloendothelial system (RES) (Clark et al. 1970; Masuda and Hori 1973; Clark et al. 1975). One week after infusion of a perfluorodecalin emulsion, 30% of the dose of the PFC was retained in the liver, 5% in the spleen, and less than 1% in the bone marrow, lung, adipose tissue, and kidney (Yokoyama et al. 1975a). PFCs leave the body unmetabolized through the lungs (Yokoyama et al. 1975b).

Single-dose toxicity studies with Fluosol DA-20 have been conducted in mice, rats, dogs, and monkeys to evaluate lethality levels and toxicity (Naito and Yokoyama 1981; Lutz and Metzenauer 1980). The intravenous LD_{50} doses for mice and rats are 128 to 144 ml emulsion/kg and 128 to 134 ml emulsion/kg respectively. Fluosol DA-20 increased the liver and spleen weights in rats as much as 150% and 200%, respectively, peaking at four to eight days post-infusion. The effect was found to be dose responsive. The organs returned to normal weights between two and four months post-infusion. The hepatomegaly and splenomegaly began after the accumulation of PFC reached a maximum and continued for two to four days thereafter. Interestingly, functional tests with indocyanine green or the recovery time from pentobarbital-induced sleep reveal a normalized function at the time when liver weight is still rising or at its maximum (Lutz and Wagner 1981). Morphometric analysis of the tissues indicated hypertrophy in the hepatocytes. Autoradiography after eight days of ^{3}H-thymidine infusion revealed a proliferation of nearly all sessile macrophages as well as of particular hepatocytes. These anatomical changes could explain the maintenance of liver function at the time when PFC concentration in the liver is at a maximum (Lutz et al. 1982).

Multiple-dose toxicity studies of Fluosol DA-20 have been conducted with rats and dogs (Naito and Yokoyama 1981) with cumulative doses as high as 150 ml/kg. Dose-responsive depression of weight gain and hematocrit, hepatomegaly, splenomegaly, and elevation of serum enzymes and blood urea nitrogen (BUN) were observed at high doses. All symptoms returned to normal three months after dosing ceased.

There was no mutagenic or teratogenic activity associated with Fluosol DA-20 (Naito and Yokoyama, 1981; Naito et al. 1977).

Lutz evaluated the effect of phagocytosis of PFCs on RES function. He found a reduction in the elimination constant of colloidal carbon occurred in rats treated with Fluosol DA-20 within three hours after injection of 4.4 g Fluosol DA-20/kg. The elimination constant reached a minimum at 6 hours post-infusion and returned to normal at 12 and 24 hours post-infusion. Surprisingly, the elimination constant was depressed again between two and four days post-infusion before returning to normal at eight days post-infusion (Lutz and Metzenauer 1980; Lutz 1983, 1985).

Infusion of Fluosol DA-20 simultaneously with *Escherichia coli* endotoxin decreased the endotoxin LD_{50} dose in mice eightfold. Endotoxin injected several hours post-Fluosol DA-20 infusion was still more toxic than endotoxin infusion alone. Endotoxin injected four days post-Fluosol DA-20 treatment was no more toxic than endotoxin infusion alone. Similar results were obtained if live bacteria were infused into mice simultaneously with or shortly after Fluosol DA-20.

The toxic effects of Fluosol DA-20 observed in animals are reversible hepatomegaly and splenomegaly, moderate and reversible serum transaminase elevation, and moderate and transient depression of the RES. All the effects are benign, particularly considering the enormous amounts of material (>0.5 kg) being introduced intravenously.

7.3 MEDICAL APPLICATIONS AND CLINICAL STUDIES

The list of potential medical applications that has been suggested for PFC-based oxygen transport agents is long and diverse (Table 7–1). Most of these applications have been investigated in animal models; at least four indications have been the subject of clinical studies and one indication has received FDA clearance. This section will focus on the major indications in which sufficient animal and/or clinical data exist to evaluate the therapeutic value of PFCs.

TABLE 7-1   Potential Medical Uses for Perfluorocarbons

Oxygen transport substitute for blood
Tissue preservation in acute myocardial infarct
Oxygenation of ischemic tissue during strokes
Coronary balloon angioplasty
Radiosensitizer for cancer therapy
Cardioplegia for coronary bypass
Priming solution for extracorporeal oxygenation
Liquid ventilation to oxygenate blood during ARDS
Liquid ventilation and lavage for cystic fibrosis
Intraperitoneal perfusion to oxygenate blood during ARDS
Perfusion solution for organ preservation
MRI imaging using ^{19}F
Contrast agent for ultrasound
X-ray contrast imaging
Enhanced oxygenation to accelerate wound healing
Oxygenation during carbon monoxide and cyanide poisoning
Relief of ischemia during sickle cell crises
Dissolution of gas emboli in decompression sickness
Oxygenation of ischemic spinal cord
Ocular applications
Isolated limb perfusion
Drug delivery vehicles

7.3.1 Blood Substitutes

The most common therapeutic application for PFC emulsions has been as
an oxygen transport substitute for red blood cells. Red blood cell substitutes
cover many possible, diverse indications including hemorrhagic shock resuscitation, restoration of blood oxygen content during elective and emergency surgical hemorrhage, priming pumps during coronary artery bypass,
and as an oxygen transport diluent for autologous donation preceding elective surgery.

Geyer's experiments with total exchanged rats provided the first substantial evidence that PFC emulsions could take the place of blood in providing the total oxygen requirements of an animal. Subsequently, exchange
transfusions have been reported in dogs (Suyama et al. 1975) and monkeys
(Ohyanagi et al. 1978; Rosenblum et al. 1985). Monkeys were exchange
transfused with either Fluosol DA-20 or the plasma expander Hespan (HES)
to a hematocrit of <2%. After 6 hours, the survivors were infused with
autologous blood. All of the HES-exchanged monkeys died before the blood
transfusion, while 8 of 10 monkeys given Fluosol DA-20 survived to receive
the blood transfusion, and six of the eight that received blood after 6 hours
survived.

Okada and coworkers studied hemorrhagic shock in mongrel dogs bled
to a mean arterial pressure of 50 mm Hg (Okada et al. 1975). Groups of

dogs were reinfused while breathing 100% oxygen with either Lactated Ringer's (LR) or Fluosol DC. Arterial, venous, and skeletal muscle oxygen tensions were significantly higher in the Fluosol DC group than in the LR group (Table 7-2). Remarkably, the mixed venous oxygen tension of the Fluosol DC group was 78 mm Hg indicating that, in anesthetized, intubated dogs breathing 100% oxygen, the dog's oxygen requirements were largely being met by the PFC with almost no oxygen off-loading by hemoglobin (hemoglobin is greater than 90% saturated at 78 mm Hg).

In studies of hemorrhagic shock in dogs with survival as the endpoint, Ohyanagi and Mitsuno (1975) studied three situations: moderate hemorrhage with PFC or low-molecular-weight dextran (LMD) resuscitation; massive hemorrhage to hematocrit = 10% with PFC or LMD resuscitation; and massive hemorrhage with PFC or LMD resuscitation followed by whole blood infusion. All resuscitations were conducted with the animals breathing 100% O_2 (Ohyanagi and Mitsuno 1975). In all cases, the PFC emulsion produced superior survival to the LMD resuscitation, and when combined with blood given 6 hours later, 100% survival was observed (Table 7-3). The data from these studies offer strong support that interim resuscitation

TABLE 7-2 Hemodynamic and Oxygenation Data in Dogs Resuscitated with Lactated Ringer's and Fluosol DC

Parameter	Lactated Ringer's	Fluosol DC
Cardiac index (I/min/m²)	2.9	2.8
Mixed venous pO_2 (mm Hg)	42	78
Arterial pO_2 (mm Hg)	450	490
Skeletal muscle pO_2 (mm Hg)	25	70
Hb (g/dl)	5.5	6.5
Arterial pCO_2 (mm Hg)	42	47

Data from Okada et al. (1975).

TABLE 7-3 Survival in a Dog Hemorrhagic Shock Model with Various Resuscitation Regimes

Hemorrhage Protocol	Resuscitation Fluid	Hematocrit Preresuscitation	Survival (%)
Moderate	LMD	25 ± 2	40
Moderate	PFC	25 ± 2	75
Massive	LMD	10 ± 2	0
Massive	PFC	10 ± 2	80
Massive	PFC/blood	10 ± 2	100

Data from Ohyanagi and Mitsuno (1975).

by PFC emulsions in environments where blood is not immediately available followed by blood transfusion could make an enormous difference in survival of shock victims.

Makowski compared Fluosol DA-20 and Fluosol DA-35 with 6% HES and whole blood in shock resuscitation (Makowski 1978). Groups of 10 dogs were bled to a pressure of 40 mm Hg and held there for one hour, then reinfused while breathing 100% oxygen with volumes apparently equivalent to the shed blood volume. All four treatments restored mean arterial pressure and cardiac index and increased mean pulmonary artery pressure. Blood and both PFC emulsions restored arterial and venous oxygen contents, but HES did not. All treatments restored oxygen consumption immediately after infusion, however, the PFC emulsion increased oxygen consumption compared to HES or blood almost 50%. Mixed venous oxygen tensions were also higher immediately after perfusion in the PFC-treated animals. In addition, Makowski found that while 80% of the shocked animals reinfused with whole blood or HES died, only 20% of the Fluosol DA-20-treated animals succumbed. The increase in oxygen consumption indicated that PFC emulsions are superior at reducing the oxygen debt that accumulates as a consequence of hemorrhagic shock.

A more recent study by Elliott and coworkers compared Fluosol DA-20 to LR in an animal model of resuscitation. Dogs were bled to and maintained at 60 mm Hg mean arterial pressure (MAP) for 90 min, then 40 mm Hg for 30 min. After the shock period, the animals were given 20 ml/kg of LR, then 42 ml/kg of the experimental fluid, followed by 30 ml/kg of LR. They concluded that Fluosol DA-20 was effective in producing volume expansion, oxygen delivery, and oxygen consumption. They observed an increasing oxygen consumption at 60 min post-infusion similar to previous work and which was statistically significantly greater at 24 hours post-infusion than LR. The authors also found that the PFC contributed as much as 40% to the animals' overall oxygen consumption in the one-hour period post-shock (Elliott et al. 1989).

In summary, the data from the rat, dog, and primate studies on hemodiluted or shocked animals indicate that PFC emulsions can effectively replace volume, deliver adequate oxygen, increase oxygen consumption in animals with an oxygen debt, and markedly increase overall survival rate. The data indicate that PFC emulsions can provide a major portion of an animal's oxygen consumption. The ability of PFCs to increase oxygen consumption in shocked animals may also reflect the superior ability of the small, oxygen-laden particles in the emulsion to traverse and oxygenate previously ischemic microcirculation.

Clinical trials for Fluosol were first reported by Makowski and coworkers in 1978 (Makowski et al. 1978). Terminal and decerebrate patients, 22–33 years of age (victims of severe head injury) were administered between 1,000-1,500 ml of Fluosol DA-20, observed for 24 hours, and then autopsied. Other than transient rises in cardiac output and blood pressure related to the volume loading, the patients exhibited no untoward effects from infusion. Two patients received two 150-ml doses 24 hours apart and did not exhibit sensitization reactions. All patients were autopsied 24 hours post-infusion, and tissue samples were taken and examined histologically. The sections of the liver, spleen, lung, and lymph contained "vacuolated Kupffer cells and macrophages" but were otherwise normal. The other organs were without visible change. These rather unusual clinical trials clearly established the lack of acute toxic responses to Fluosol DA-20 and suggested that it was reasonable to test Fluosol DA-20 in healthy human subjects in a normal Phase I clinical trial.

In Phase I clinical trials of Fluosol DA-20, research employees of the Green Cross Co. received graded doses from 50 to 500 ml of Fluosol DA-20 (Ohyanagi et al. 1979). The subjects receiving 500-ml infusions were phlebotomized 200 ml in advance of infusion. Blood samples were taken at 3, 24, and 48 hours, and 1, 2, 4, 8, and 14 weeks post-infusion and analyzed for complete hematology and clinical chemistry profiles. Throughout the study, all of the subjects' values were within normal range for clinical chemistry, hematology, and hemodynamics. In addition, there were no adverse anaphylactic reactions reported.

Ultimately, Fluosol DA-20 was tested for efficacy in clinical trials in Japan and the USA. In the Japanese trials (Ohyanagi et al. 1984; Mitsuno et al. 1982; Mitsuno and Ohyanagi 1985), which ran from 1979 to 1982, 401 patients received Fluosol DA-20 in dosages of 20 to 30 ml/kg. In 270 patients, Fluosol DA-20 was used instead of blood transfusion for replacement of surgical blood loss or for improvement of acute hemorrhagic anemia. In 131 patients, it was given in an attempt to improve cerebral circulation. These studies do not report on survival nor were the results of Fluosol DA-20-treated patients compared to control groups, leaving the question of efficacy unanswered. The authors determined that Fluosol DA-20 provided about 17% of the tissue oxygen consumption at FiO_2 of 0.5 to 0.6, an amount equal to the plasma contribution. No acute or chronic adverse reactions were observed, except for a transient decline in neutrophils and platelets shortly after infusion. Hemodynamic parameters of patients were either maintained or recovered to normal after treatment with Fluosol DA-20. In patients who died of their diseases, there were no organ abnormalities that could be attributed to Fluosol DA-20. Although PFC could still be detected in some tissues 7 weeks post-infusion, organs analyzed 7 months after treatment were free of PFC.

In the USA, Fluosol DA-20 was tested first on a humanitarian protocol basis where each patient required FDA approval. After six patients received Fluosol DA-20, a medical use protocol was established with strict criteria for patient enrollment. The objectives of this study were to determine clinical safety, hemodynamic, and oxygen transport profiles of Fluosol DA-20. In contrast to earlier reports, fully one-third of the patients experienced an acute reaction to infusion of a 0.5-ml test dose of Fluosol DA-20 which was controlled by treatment with corticosteroids (Tremper et al. 1985). The treated patients had about 3% PFC in their systemic circulation, which contributed about 0.7 volume % O_2 to the total oxygen transported, representing about 25% of the oxygen consumption of the patient. A subsequent report by Gould reported similar results (Gould et al. 1986). The patients in this study were severely anemic with pretreatment hemoglobin levels of about 3 g/dl. They were given up to 40 ml Fluosol DA-20/kg. The Fluosol DA-20 made a significant contribution (Table 7–4) to overall oxygen consumption (28%) and demonstrated the ready availability of the oxygen carried by the PFC phase: 82% of the oxygen dissolved in the PFC was unloaded.

Despite the significant contribution to oxygen consumption noted in the Gould paper, and in contrast to most of the previously published animal studies, the overall level of oxygen delivery and consumption declined in Fluosol DA-20 treated patients (Table 7–5). Gould concluded that Fluosol DA-20 failed to make greater contributions to overall oxygen delivery because of the low volume of PFC in the Fluosol DA-20 formulation and its short intravascular half-life of 24 hours. He speculated that second-generation emulsions with a higher initial concentration of PFC might lead to higher circulating PFC concentrations and better efficacy. He also indicated that a clinical trial where the efficacy of Fluosol DA-20 was evaluated as a short term replacement until red blood cells became available might have a more favorable outcome. Perhaps the most important aspect of this particular clinical trial was the fact that there were no adverse patient reactions to Fluosol DA-20. Gould concluded that Fluosol DA-20 was safe but not effective, largely because of low PFC content.

Despite this outcome, Fluosol DA-20 continues to be utilized on a humanitarian protocol at certain institutions (Spence et al. 1989).

The fact that Fluosol DA-20 failed in clinical trials should not obscure the fact that PFCs did make a significant and reproducibly measurable contribution to oxygen consumption in these severely anemic patients, that side reactions to the product were minimal and controllable and that the overall

TABLE 7–4 Oxygen Dynamics of Three Phases at the Peak Effect of Fluosol DA-20 in Eight Patients[1]

Property	Fluosol DA-20	Plasma	Red Cells
Arterial O_2 content (ml/dl)	0.7 ± 0.1	1.3 ± 0.1	2.8 ± 0.6
Venous O_2 content (ml/dl)	0.2 ± 0.1	0.2 ± 0.1	2.2 ± 0.4
Oxygen unloaded (%)	82 ± 5	82 ± 5	19 ± 5
Contribution to O_2 consumption (%)	28 ± 5	50 ± 5	22 ± 7

[1] Values are given as the mean percent ± SE.
Data from Gould et al. (1986).

TABLE 7–5   Hemodynamics and Oxygen Transport Properties Before and After Fluosol DA-20 Administration in Eight Patients[1]

Property	Before Fluosol DA-20	After Fluosol DA-20[2]
Heart rate (bpm)	117 ± 5	106 ± 4[2]
Mean arterial pressure (torr)	74 ± 6	78 ± 5
Cardiac index (liter/min/m²)	4.5 ± 0.7	4.2 ± 0.7
Hemoglobin (g/dl)	3.0 ± 0.4	2.0 ± 0.4
Total arterial O_2 content (ml/dl)	5.3 ± 0.5	4.8 ± 0.6
O_2 delivery (ml/min/m²)	235 ± 27	197 ± 32
O_2 consumption (ml/min/m²)	109 ± 13	88 ± 11
Partial pressure of arterial O_2 (torr)	356 ± 24	430 ± 19[3]
Partial pressure of mixed venous O_2 (torr)	40.0 ± 3.9	78.2 ± 3
O_2 extraction ratio (%)	46.0 ± 2.5	47.6 ± 3.8

[1] Values are given as the mean percent ± SE.
[2] Data were obtained at peak arterial oxygen content after Fluosol DA-20 administration.
[3] The difference between values before and after Fluosol DA-20 is significant ($p < .05$). Data from Gould et al. (1986).

safety of Fluosol DA-20 was adequate to insure its testing in other clinical indications such as balloon angioplasty and cancer therapy.

7.3.2   Percutaneous Transluminal Coronary Angioplasty (PTCA)

Percutaneous transluminal coronary angioplasty (PTCA) is a procedure that is used to remodel arterial plaque and allow increased coronary artery blood flow. It has gained wide acceptance with over 300,000 angioplasties being performed in the USA annually. In this procedure, a catheter with an inflatable balloon at its tip is threaded into the coronary artery experiencing partial blockage. Once centered in the lesion, the balloon is inflated, compressing the plaque against the vessel wall and increasing the effective diameter of the coronary artery. Current medical practice requires balloon inflation times of 45 sec or longer to insure optimal results, and there is an interest in increasing the inflation time to determine if restenosis (reocclusion of the artery) rates can be reduced. During the period of balloon inflation, the coronary artery is occluded and part of the left ventricle experiences temporary ischemia. Recent studies have shown that balloon inflation times of 20 sec are enough to produce ECG changes, induce ventricular wall motion irregularities, and reduce the left ventricular ejection fraction. Such responses are exacerbated in patients with multivessel disease. In order to increase the time of inflation or to reduce risk in patients with multivessel disease, physicians have attempted to perfuse blood through the central lumen of the angioplasty catheter. Blood perfusion was found to have a number of disadvantages including high viscosity, hemolysis, and the necessity of another arterial access point to provide blood for the perfusion.

In the early 1980s, Spears piloted the use of oxygenated Fluosol DA-20 perfused through the central lumen of the catheter to alleviate symptoms of ischemia during balloon inflation in dogs. The results showed that 8 of 10 dogs perfused with oxygenated Fluosol DA-20 during prolonged balloon inflation times (>5 min) had normal ECGs while controls perfused with oxygenated saline had an ECG injury pattern and increased ventricular ectopy (Spears et al. 1983).

Both Spears (Spears et al. 1984) and Anderson (Anderson et al. 1985) issued preliminary reports on the use of Fluosol DA-20 in humans but the definitive clinical studies were performed by Jaffe and coworkers (Cleman et al. 1986; Jaffe et al. 1988). They used two-dimensional echocardiography to detect the location, extent, and evolution of ischemic contractile dysfunction and quantitatively assessed regional wall motion and ejection fraction by computer analysis of the echocardiograms.

In the clinical trial, 42 symptomatic patients with single lesions of 70%-or-greater stenosis of the coronary artery undergoing PTCA were studied. All patients underwent a preliminary inflation without perfusion. Subsequent inflations were done with either oxygenated LR, oxygenated Fluosol DA-20, or nonoxygenated Fluosol DA-20. The data showed that oxygenated Fluosol DA-20 maintained an ejection fraction identical with the baseline value 45 sec post balloon inflation, while there was 35% reduction in ejection fraction in the controls (Table 7–6). These studies were the basis of the recent FDA approval granted for the use of Fluosol DA-20 in PTCA on high-risk patients (Anonymous 1990).

7.3.3 Myocardial Infarct

Numerous studies have demonstrated that mortality and morbidity in acute myocardial infarct (MI) are directly related to the degree of destruction of the myocardial tissue due to ischemia post infarct. The timeframe of effec-

TABLE 7–6   Ejection Fraction Results with Various Perfusion Treatments

	Number	Baseline Fraction[1]	45-sec Fraction[1]	p Value
Nonperfused	59	57 ± 15	36 ± 14	<.0005
Oxygenated Fluosol DA-20	44	56 ± 14	53 ± 13[2]	Not significant
Lactated Ringer's	10	56 ± 15	38 ± 13	<.0005
Nonoxygenated Fluosol DA-20	7	57 ± 7	37 ± 7	<.0005

[1] Values are given as the mean percent ± SE.
[2] For the 45-seconds nonperfused vs. oxygenated Fluosol, $p > .0001$.
Data from Jaffee et al. (1988).

tive response to coronary artery occlusion has been shown to be short—on the order of 6 hours. After 6 hours, tissue damage is largely irreversible. Therefore, the more rapidly the infarcted myocardium is reoxygenated and reperfused, the more myocardial tissue will be preserved and the final outcome improved for the patient.

PFC emulsions have been proposed as therapeutic agents for preservation of myocardium after an infarct because of their ability to transport oxygen in significant quantities and their small particle size. The small particle size and lowered viscosity of PFC emulsions suggest that they might flow more readily through the long and thin intercapillary connections that make up the collateral circulation in humans, and thus help to oxygenate the myocardium distal to the occlusion (Faithful et al. 1986). These same properties suggest that PFC particles might reperfuse the edematous "no-reflow" vessels resulting from ischemia, bringing in oxygen and removing carbon dioxide in the acidotic tissues functioning under conditions of anaerobic metabolism (Nunn et al. 1983).

Glogar and coworkers assessed the efficacy of PFCs in preservation of ischemic myocardium in dog models (Glogar et al. 1981). Area at risk and collateral flow were measured with technicium (^{99}Tc)-labeled human albumin microspheres. The dogs were hemorrhaged before infusion of PFC emulsion or LR to avoid complications of volume overload. After infusion of the PFC emulsion or a comparable amount of LR, the animals breathed 100% oxygen for 15 minutes prior to permanent occlusion of the left anterior descending (LAD) artery. Dogs were sacrificed 6 hours later. The hearts were sectioned and stained with triphenyltetrazolium chloride (TTC) to determine area of necrosis. The average area at risk (A_R) was about 30% for each group. The area of necrosis (A_N) and the ratio of A_N/A_R were both significantly reduced by about 30% ($p < .01$) in the PFC-treated group compared to both the LR and the control groups.

Nunn and coworkers have examined the effect of Fluosol DA-20 on myocardial salvage in a similar model (Nunn et al. 1983). One hour after ligation of the LAD, Fluosol DA-20 or 0.9% saline was infused with simultaneous withdrawal of blood to a dose of 30 ml emulsion/kg. A group of control dogs that were not exchanged and did not receive 100% oxygen were also included. The animals were sacrificed and the hearts isolated, stained with TCC, and analyzed as above. The ratio of the A_N/A_R is statistically significantly reduced by about 33% in the Fluosol DA-20-treatment group (Table 7–7).

In studies done in pigs, whose collateral circulation is most similar to humans and lacks the variation found in dogs, Faithful et al. (1986) evaluated the ability of Fluosol DA-20 to preserve myocardium during infarct by measuring the change in myocardial oxygen tension in the most hypoxic region of the ischemic myocardium. There were three treatment groups: 1) control, 2) infusion with Fluosol DA-20 after withdrawal of an equal volume of blood 1 hour after LAD ligation, and 3) infusion with the 5% dextran-

TABLE 7-7 Effect of Various Treatments on Myocardial Salvage

	Number	A_N/A_R[1]	A_R/Total Left Ventricle[1]
Group I: control	9	90(56-97) ± 2	23(12-31) ± 2
Group II: saline solution	9	88(80-99) ± 2	18(10-29) ± 2
Group III: Fluosol	9	67(42-80) ± 4[2]	22(14-30) ± 2[3]

[1] Values are given as the mean percent ± SE. The numbers in parentheses represent the range for each group.
[2] For A_N/A_R: III versus I, $p < .01$; III versus II, $p < .001$.
[3] Difference is not significant.
Data from Nunn et al. (1983).

40 in physiological saline after withdrawal of an equal volume of blood 1 hour after LAD ligation. Myocardial oxygen tensions were measured polarographically for each pig through 5 hours post occlusion. Oxygen tension of the ischemic myocardium in the PFC-treated animals increased greatly relative to controls.

Minimization of reperfusion damage by Fluosol DA-20 post MI have been reported by Forman, first using an intracoronary infusion model (Kolodgie et al. 1986), then an intracoronary infusion model with hemodilution (Forman et al. 1987), and finally a systemic infusion model (Bajaj et al. 1989). In the earlier intracoronary model, reperfusion with Fluosol DA-20 after 90 min of ischemia resulted in a 60% reduction in infarct size and improved ventricular function two weeks post infarct. There was no evidence of increased myocardial oxygen tensions in the Fluosol DA-20 group compared to control despite ventilation with 100% oxygen during and for 3 hours after reperfusion. Histologically, there was reduced neutrophil deposition in the microcirculation of the A_R of Fluosol DA-20-treated dogs compared to controls. Electron microscopy revealed capillary obstruction involving endothelial cell protrusions and neutrophil and red cell plugging of the obstructed capillaries in control dogs but not in Fluosol DA-20-treated animals. The authors interpreted this as evidence for improved reflow after the occlusion was eliminated.

In the systemic infusion model (Bajaj et al. 1989), dogs were ventilated with 100% oxygen, and Fluosol DA-20 or LR was infused at 24 ml/kg 1 hour after occlusion of the LAD. Thirty minutes later the occlusion was removed. The dogs were ventilated 3 hours post-infusion with 100% oxygen, then weaned from the ventilator and allowed to recover for 24 hours. The authors measured A_R using monastryl blue and A_N by TCC staining. Regional myocardial blood flow was measured by radiolabeled microspheres. Neutrophil distribution was studied by indium-111 (^{111}In) labeling and reinfusion, and ex vivo neutrophil activation was measured by isolation and zymosan activation. As in previous studies, A_N/A_R was significantly reduced in the Fluosol DA-20 group (24% compared to the LR control) while A_R was similar in both treatment groups at 48–50%. Epicardial blood flow measurements in the central ischemic zone indicated that collateral blood flow variability did not bias the A_N/A_R. Control animals and Fluosol DA-20-treated animals had significantly reduced blood flow in the central ischemic zone during occlusion. One hour post reperfusion, the control animals had significantly reduced myocardial blood flow in the epicardial and endocardial ischemic zone compared to Fluosol DA-20. Based on measurements of ^{111}In counts, there were two- to fourfold increases in neutrophil infiltration into the epicardium, midmyocardium, and endocardium ischemic regions in control dogs compared to Fluosol-DA-20-treated dogs. Non-ischemic regions of the myocardium had similar levels of neutrophils in the LR and Fluosol DA-20-treated groups. Venous neutrophil counts were lowered within 1 hour of Fluosol DA-20 infusion and remained below control counts throughout the reperfusion period. Neutrophils from the Fluosol DA-20 group exhibited lowered ex vivo chemotaxis 1 hour after reperfusion than control groups. Forman (Bajaj et al. 1989) concluded that the primary mechanism by which Fluosol DA-20 preserves myocardium is by reducing neutrophil chemotaxis and adhesion in the ischemic tissue during reperfusion.

In summary, PFC emulsions have been shown to reduce relative infarct size, increase myocardial oxygen tension, and neutralize inflammatory cells during infarct and after reperfusion. Despite the overwhelming evidence of myocardial preservation during MI in animal studies of PFC emulsions from at least four independent labs, there have been no published clinical studies on the use of PFC emulsions to treat acute MI.

7.3.4 Cancer Therapy

Solid tumors possess vascular insufficiencies and blood flow irregularities resulting in significant areas of hypoxic cells, frequently amounting to 10–20% of the total viable tumor cell population (Thomlison and Gray 1955). Hypoxia has long been known to protect cells from the cytotoxic effects of radiation and chemotherapy. The surviving hypoxic cell fraction is generally recognized as capable of reestablishing the tumor and limiting the therapeutic effectiveness of these modalities. There have been numerous efforts to render the hypoxic cell fraction susceptible to radiation and chemotherapy. The development of radiosensitizers such as misonidazole and the use of hyperbaric oxygen (HBO) chambers are just two examples of therapies developed to attack the hypoxic fraction. Neither of these therapies has proven particularly beneficial; misonidazole, which is an effective radiosensitizer in rats, is toxic in humans below the therapeutic dose while HBO has proven cumbersome to test broadly and inconclusive in the limited clinical trials in which it has been tested.

In 1983, Teicher and Rockwell demonstrated that infusion of PFC emulsion in advance of radiation coupled with carbogen (95% oxygen and 5% carbon dioxide) breathing during radiation therapy significantly reduced the surviving fraction of hypoxic cells and the growth rate of implanted rodent tumors (Teicher and Rockwell 1983). Subsequently, Teicher and Rose found that mice bearing tumors from Lewis lung tumor or FSa-II fibrosarcoma treated with Fluosol DA-20 and breathing carbogen had tumor growth delays of up to 30 days compared to rats receiving either radiation alone, radiation and carbogen alone, or radiation and Fluosol DA-20 with air breathing (Teicher and Rose 1984). Others have reported similar results (Song et al. 1986; Rockwell 1985; Lustig and McIntosh 1986).

The mechanism of enhancement of radiation therapy appears to be due to increased oxygenation of the tumor. Song et al. (1987) measured the PO_2 of tumors in mice and verified that tumor oxygenation is markedly increased by the combination of Fluosol DA-20 and carbogen breathing. The results show that tumor oxygenation is increased six-sevenfold in mice breathing carbogen and infused with Fluosol DA-20 over mice breathing room air and twofold over mice breathing carbogen without Fluosol DA-20. Klubes and coworkers (1987) found that Fluosol DA-20 does not increase blood flow through solid tumors in the model they studied. Long has documented by X-ray and histological studies that most implanted and spontaneous animal tumors preferentially accumulated PFC emulsion in macrophages associated with the tumors (Long et al. 1978). What, if any, role this phenomenon plays in radiation therapy enhancement is not clear.

There were two early concerns with the use of Fluosol DA-20 with radiation therapy, one safety and the other technical. The safety issue was the possibility of increased damage to normal tissues as a consequence of using Fluosol DA-20 in combination with radiation therapy, particularly to the bone marrow and stem cells. Mate and Rockwell (1985) found, however, that the viability and radiosensitivity of mouse bone-marrow stem cells was unaltered when Fluosol DA-20 was combined with radiation therapy.

The technical concern centered around the protocol used in the animal experiments and its translation to humans. In most of the animal experiments, dose fractionations were either not used or only three to four fractions were given. This is in contrast to human therapy, where radiation doses are fractionated into 30 to 35 fractions. Fractionation is known to be effective in reducing the hypoxic cell fraction. Thus is was important to determine if the beneficial effect of Fluosol DA-20 was still measurable in a highly fractionated radiation protocol. Moulder and Fish (1988) found that the use of Fluosol DA-20 in mice in combination with a highly fractionated radiation regime of three fractions per week for three to six weeks still significantly decreased the radiation dose necessary to obtain a 50% tumor control rate compared to radiation and carbogen breathing alone.

Clinical trials of Fluosol DA-20 in cancer radiation therapy were first reported in 1986 (Rose et al. 1986). These trials were conducted on 15 patients with Stage III/IV head and neck cancer, who were not candidates for surgical therapy. Radiation was fractionated into about 25 doses of 1.8 Gy over a 5-week period. Fluosol DA-20 was infused at a rate of 8–9 ml/kg on the first day of each week for a total dose of 40–45 ml/kg, and patients breathed 100% oxygen before and during all 25 radiation fractions. Of the 15 patients, 10 had primary and nodal clearance with the longest follow-up post treatment being eight months. The authors noted four cases of acute reactions controllable with diphenhydramine and eight of 15 patients exhibited serum enzyme elevations of two to three times normal, which returned to normal 3 months post therapy. Coagulation times, BUN, creatinine, serum albumin, and bilirubin were unaffected by the treatment. White blood cell counts and hematocrit were slightly depressed but the changes were typical of normal responses to radiation therapy and could not be attributed to Fluosol DA-20. There was some evidence for acceleration of the onset of the mucositis normally caused by the radiation.

Lustig et al. (1989a) have also reported on clinical trials with Fluosol DA-20 in patients with head and neck cancer. In this study, 37 patients were enrolled, and 28 (76%) had complete response (no evidence of primary disease) two months post treatment. The determinant survival, excluding those who died from other causes, was 78% one year post therapy compared to the reported survival rate of 53–62% for the Radiation Therapy Oncology Group. Side effects were reported to be mild and reversible and of about the same frequency, extent and duration as observed in the first clinical trial (Rose et al. 1986). While cancer patients are not considered cured until after five years of complete remission, the initial response to Fluosol DA-20 has been very promising. In addition to the head and neck trials discussed above, clinical trials using Fluosol DA-20 and radiation therapy for treatment of non-small-cell lung cancer (Lustig et al. 1989b) and brain glioma are also underway (Rockwell 1988).

The effects of Fluosol DA-20 and oxygen breathing were also evaluated in conjunction with chemotherapy. Ohyanagi and coworkers found that animals treated with vincristine or spadicomycin and Fluosol DA-20 with oxygen breathing had smaller tumor masses compared to animals treated with the chemotherapeutics alone (Ohyanagi et al. 1983). Subsequently, a number of investigators have reported on the potentiating effect of Fluosol DA-20 on chemotherapy-induced tumor growth delay (TGD) (Teicher and Holden 1987). Fluosol DA-20 with oxygen breathing enhanced the effectiveness of all three major classes of chemotherapeutics: alkylating agents, antibiotics and alkaloids, and antimetabolites (Table 7–8). Despite the large number of papers on the use of Fluosol DA-20 with chemotherapeutics, there have been no reports of clinical studies.

7.3.5 Cardioplegia During Cardiopulmonary Bypass

During cardiopulmonary bypass and cross-clamping, myocardial protection from global ischemia is achieved by the use of chemical cardioplegic arrest and myocardial hypothermia. High potassium in the cardioplegia causes

TABLE 7-8 Effect of the use of Fluosol DA-20 and Oxygen Breathing on the TGD Produced by Various Chemotherapeutic Agents

Drug	Drug Dose (mg/kg)	Treatment Schedule	TGD (days)			
			Drug/Air	Drug/O₂	Drug/Fluosol/Air	Drug/Fluosol/O₂
Alkylating agents						
Melphalan	10	Single dose	2.7 ± 0.3	4.0 ± 0.6	6.5 ± 1.2	9.5 ± 1.3[1]
CDDP	10	Single dose	8.0 ± 1.7	8.2 ± 1.6	5.0 ± 1.6	8.4 ± 2.0
Cytoxan	100	Days 7,9,11,13,15	7.98 ± 0.78	8.96 ± 1.64	6.35 ± 2.14	11.40 ± 3.57
Busulfan	10	Days 7,9,11,13,15	1.10 ± 0.53	2.87 ± 0.98	4.47 ± 0.51	9.26 ± 2.14[1]
Mitomycin	2	Days 7,9,11,13,15	9.03 ± 0.72	3.01 ± 0.98	0.87 ± 0.57	5.33 ± 1.71
BCNU	15	Days 7,9,11,13,15	13.38 ± 1.47	14.13 ± 1.61	18.48 ± 2.78	26.03 ± 2.54[2]
CCNU	100	Days 7,9,11,13,15	3.14 ± 0.42	5.38 ± 1.01	4.88 ± 1.02	12.53 ± 2.25[1]
McCCNU	20	Days 7,9,11,13,15	1.84 ± 0.61	3.25 ± 0.43	5.57 ± 1.78	13.35 ± 2.27[1]
Chlorozotocin	15	Days 7,9,11,13,15	1.56 ± 0.32	1.28 ± 0.42	3.69 ± 0.61	8.34 ± 1.35[1]
Procarbazine	20	Single dose	0.73 ± 0.23	0.21 ± 0.19	2.66 ± 0.87	11.16 ± 2.40[2]
Dacarbazine	400	Single dose	4.14 ± 0.63	6.2 ± 1.5	3.8 ± 0.52	6.8 ± 1.00
Antibiotics and Alkaloids						
Bleomycin	10	Days 6,10,13,16	2.6 ± 0.7	4.69 ± 1.02	5.51 ± 2.26	14.56 ± 2.97[1]
	15	Days 6,10,13,16	3.66 ± 1.0	5.55 ± 1.5	5.86 ± 1.6	16.89 ± 2.94[1]
Vineristine	1.5	Single dose	1.86 ± 0.43	3.14 ± 0.86	2.79 ± 1.01	4.83 ± 1.22
VP-16	10	Days 7–12	1.84 ± 0.33	3.09 ± 0.85	3.07 ± 0.89	4.49 ± 1.22
	15	Days 7–12	4.89 ± 0.66	7.86 ± 0.97	3.11 ± 0.67	11.54 ± 2.14[1]
	20	Days 7–12	6.96 ± 1.0	9.48 ± 1.24	3.95 ± 1.10	21.70 ± 2.68[2]
Antimetabolites						
Methotrexate	0.8	Days 7–12	2.23 ± 0.89	1.43 ± 0.57	0.58 ± 0.53	2.90 ± 0.67
5-Fluorouracil	40	Days 7–10	7.61 ± 0.56	7.72 ± 0.72	9.32 ± 0.88	10.47 ± 1.06

[1] p < .001.
[2] p < .005.
Data from Teicher and Holden (1987).

the heart to cease mechanical and electrical activity. In the cooled and arrested state, maximal protection of the myocardium from ischemia is achieved, and energy requirements are minimized. Other components of the cardioplegia provide for hypertonicity to minimize myocardial edema and buffering to prevent acidosis from anaerobic metabolism. Despite these measures, some metabolic activity remains, with the consequence that energy stores are depleted and conversion to anaerobic metabolism takes place. Energy depletion and anaerobic metabolism result in longer cardiac restart times and slow recovery of ventricular function. The use of blood to provide oxygen to the arrested myocardium has been evaluated and found not to be beneficial at clinically relevant temperatures due to the increased affinity of hemoglobin for oxygen at low temperatures (Magovern et al. 1982).

PFC emulsions have been suggested (Jaffin et al. 1981) as potential additives to cardioplegic solutions. The greater solubility of gases in PFCs at low temperature makes them perfect candidates to maintain oxygen consumption in the arrested, hypothermic heart. Jaffin and coworkers compared a standard cardioplegic solution with oxygenated hyperkalemic PFC emulsion and nitrogenated hyperkalemic PFC emulsion in an isolated rabbit heart model at 23°C for 3 hours of global ischemia and 45 min of normothermic reperfusion. After 3 hours the oxygenated PFC emulsion (FC-43 based)-treated group had 100% of preglobal ischemia myocardial concentration of ATP while the standard cardioplegic and the anoxic PFC emulsion had 62% of control ATP concentration. Forty-five minutes post reperfusion, the oxygenated PFC group had a left ventricular diastolic pressure of 92% of preglobal ischemia myocardial values while the other two solutions had developed pressures of 65% and 68%, respectively. In a related study using the same rabbit heart model and PFC emulsion, the intramyocardial oxygen tension was measured by mass spectrometry. Intramyocardial oxygen tension in globally ischemic, hypothermic hearts were 0.4 mm Hg for the crystalloid group, 1.5 mm Hg for the blood cardioplegia group, and 19.6 mm Hg for the PFC emulsion cardioplegia. Oxygen consumption was 20.4 ml O_2/100 g for the crystalloid, 39.2 ml O_2/100 g for the blood group, and 204 ml O_2/100 g for the PFC emulsion group. Similar studies and results were conducted with Fluosol DA-20. Unfortunately, in these experiments the standard cardioplegia solution was not oxygenated as a control, thereby making it impossible to know if the improvement was due to the PFC or aqueous oxygen content.

Salerno and coworkers evaluated Fluosol DA-20 as a cardioplegic solution in vivo in pigs with hypertrophied myocardium as a model for patients with compromised myocardial function (Novick et al. 1985). The pigs were subjected to 3 hours of global ischemia with cooling to 10–15°C followed by an hour of reperfusion. They compared nonoxygenated crystalloid cardioplegia with oxygenated blood and oxygenated Fluosol DA-20. The results of their study demonstrated that Fluosol DA-20 was superior to nonoxygenated crystalloid cardioplegia but comparable to oxygenated blood

TABLE 7-9 Comparison of the Effects of Standard Cardioplegia, Oxygenated
Blood, and Fluosol DA-20 on Oxygenation and Function

Solution	pO_2 (mm Hg)	ΔO_2 Content	ATP ($\mu mol/g$)	LVDP (% pre)
Nonoxygenated cardioplegia	141	0.6 ± 0.1	16.4 ± 0.7	16
Oxygenated blood	584	5 ± 0.3	18.7 ± 1.3	54
Oxygenated Fluosol DA-20	586	5.8 ± 0.1	19.4 ± 0.6	81

Data from Novik et al. (1985).

for all parameters evaluated including oxygen extraction from the cardioplegia, ATP content, left ventricular diastolic pressure (LVDP), and LVDP dP/dt (Table 7-9). Rousou et al. (1986) conducted a similar study in healthy pigs with similar results.

Menasche, using an isolated rat heart model, compared a PFC emulsion, FC-43, to oxygenated crystalloid. Cardiac output recovery was better 15 min after restart with a PFC cardioplegia compared to oxygenated crystalloid, but the differences were small (100% compared to 90% recovery) and vanished 30 min post reperfusion (Menasche et al. 1983).

Tabayashi and coworkers studied the effect of Fluosol DA-20 compared to oxygenated crystalloid as a cardioplegic solution in dogs subjected to a 120-min global ischemia at an average myocardial temperature of 18.5°C. They compared nonoxygenated crystalloid, oxygenated crystalloid, and oxygenated Fluosol DA-20. They evaluated left ventricular function by sonomicrometry including LVDP, LVDP dP/dt, and stroke volume. Tissue water content was also measured. For all parameters measured, oxygenated cardioplegia were superior to nonoxygenated cardioplegia, and there were no significant differences between oxygenated crystalloid and Fluosol DA-20, in spite of the fact that the Fluosol DA-20 cardioplegia had a greater oxygen content than the oxygenated crystalloid (Tabayashi et al. 1988).

In conclusion, there have been a number of animal studies of the use of PFC emulsions as a cardioplegia for global ischemia during cardiac arrest, but the controls used have often been inadequate to determine the true value of PFC cardioplegia compared to oxygenated crystalloid. To date, PFC emulsions have not been tested clinically as cardioplegia solutions.

7.3.6 Imaging
PFCs have been studied as imaging agents in ultrasound, X-ray, and magnetic resonance imaging (MRI). Long and coworkers investigated brominated PFCs, both neat and emulsified, in the early 1970s, because of the known X-ray opacity of bromides and iodides (Long et al. 1972a, 1972b; Long 1976). They found that brominated PFCs could be instilled intratracheally into an animal's lungs without coughing and inflammatory reactions,

and that it penetrated readily into the alveolar sacs and small respiratory passages and allowed visualization of emphysematous areas of the lung, as well as cavitations and consolidations. Commonly used X-ray contrast agents could not achieve the penetration obtained with PFC and their side reactions were much greater. Long found that neat bromoperfluorocarbons in doses of 1 to 10 ml/kg could be administered orally to image the GI tract. Images were superior to conventional contrast agents in terms of mucosal definition, and typical side effects including diarrhea, caking, and impaction were eliminated. He also found that neat bromoperfluorocarbons could be injected subcutaneously and were sequestered in the lymph nodes, providing an accurate definition of the structure of the node by X-rays (Long et al. 1971).

The fact that PFC emulsions are cleared from the bloodstream by the macrophages of the liver, spleen, and marrow prompted Long et al. (1980) to evaluate bromofluorocarbons as contrast agents for these organs. Hepatic radiopacification was achieved using computed tomography in rats, mice, cats, and dogs by infusing emulsions of perfluorohexyl bromide. The enhancements were visible within 2 hours of infusion of the emulsion and increased over the next 2 to 5 days before dissipating.

Long found that hepatic tumors could be visualized by brominated PFCs due to an apparent migration of PFC-laden macrophages into and around the tumor. A number of tumors have been implanted into mice and tested for concentration of bromoperfluorocarbons including Lewis lung, H-P melanoma, fibrosarcoma, and colon tumors. While tumors varied in their relative uptake of bromoperfluorocarbons, all the tumors tested did take up some bromoperfluorocarbon and became visible with X-rays.

While the cost of the bromoperfluorocarbons compared to barium sulfate and other contrast agents is a significant obstacle to their utilization in X-ray imaging, clinical trials are in progress in France and Ireland using perfluorooctyl bromide emulsions (100% w/v) for lymphatic opacification via subcutaneous injection to detect liver tumors in conjunction with computer-assisted tomography (Anonymous 1988; Mattrey et al. 1988). There have been no published reports on these clinical trials.

The use of PFCs, specifically Fluosol DA-20, to successfully enhance ultrasound images during a myocardial infarct in dogs has been reported by Mattrey using doses of 20 ml emulsion/kg (Mattrey and Andre 1984). A subsequent study by Vessey using the same doses, species, and PFC emulsion revealed no ultrasound enhancement in a blind study (Vessey et al. 1986).

The use of PFCs for use in magnetic resonance imaging has a great deal of potential because the ^{19}F nucleus has a spin of one-half, 100% natural abundance, and a gyromagnetic ratio close to that of protons with a relative sensitivity compared to hydrogen of 0.94. Because fluorine is present in normal tissues at only trace levels, there is no natural background ^{19}F resonance to interfere with the diagnostic signal. However, relatively weak signals are obtained in vivo with ^{19}F imaging due to the flow of blood-borne PFC and the relatively low concentration of ^{19}F atoms achieved in vivo. Sensitivity becomes worse as the field strength increases since most PFCs exhibit multiple peaks at higher fields.

Sloviter (Joseph et al. 1985) partially overcame these problems using a smaller-than-normal surface coil and low field to minimize chemical shift differences. In studies using rats that were exchange transfused 50% with a 40 vol % perfluorodecalin emulsion, a whole body scanner was used to eliminate flow saturation. Despite these modifications, the images were of low resolution and intensity (Joseph et al. 1985). McFarland et al. (1985) used Fluosol DA-20 infused intravenously at doses of 15 to 30 ml/kg in rats with a small-bore magnet and was able to visualize transverse sections of the abdomen after 3 hours of scanning. Filling the stomach and intestine with Fluosol DA-20 also produced an image after one hour of scanning. This image showed essential anatomical detail and was obtained in a reasonable amount of time (McFarland et al. 1985). Geyer obtained images of tumors and RES organs in rats after infusion of two sequential doses of emulsion at 30 ml/kg (Longmaid et al. 1985). Schweighardt cleverly solved the problem of signal intensity reduction due to multiple signals by synthesizing perfluoro-crown ethers, molecules in which all fluorine signals are identical. Signal intensity increased greatly and the required imagable dose was lowered (Schweighardt 1989).

Clark and Ackerman discovered that the relaxation times (T_1) of PFCs change with the oxygen tension of the solution, and they postulated that these changes could be used to form oxygen maps of tissues with different tensions (Clark et al. 1984). They demonstrated the feasibility of oxygen mapping in phantom experiments in vitro. McDonald and coworkers applied this technique in vivo. Using a 30% exchange with an emulsion of perfluorotributylamine (FC-43) in cats, the authors were able to generate an oxygen image of the brain which correlated well with the oxygen maps generated invasively with electrodes. This technique may be useful for localizing regions of relative hypoxia in the brain (Eidelberg et al. 1988).

Clinical trials are underway in the USA using orally administered neat perfluorooctyl bromide in conjunction with MRI to image the GI tract (Anonymous 1988). There are no published reports on these trials.

7.3.7 Liquid Breathing

Liquid breathing using PFCs was the catalyst for an explosion in research on the use of PFCs as oxygen transport agents. While liquid breathing research was rapidly overshadowed by research on emulsions, the field remained active and now promises to make clinical impact. The list of potential applications of "liquid breathing" can be divided into two major groups: liquid ventilation therapies designed to oxygenate a patient with compromised pulmonary function; and lavage therapies which strive to maintain oxygenation while the lung is purged of obstructive material. Liquid ventilation is being investigated as a therapy for premature infant respiratory distress syndrome (RDS), adult respiratory distress syndrome (ARDS), and "respirator lung"—lungs stressed due to prolonged mechanical ventilation. Lavage therapies under investigation include removal of meconium, fibrotic material in cystic fibrosis, and proteinaceous matter in proteinemia diseases.

After Clark and Gollan's famous demonstration of liquid breathing, Modell and coworkers evaluated the feasibility of long-term maintenance of dogs and primates via liquid ventilation (Modell et al. 1970b, 1971). First they determined that, unlike saline lavages, PFC lavages did not extract surfactant from the lungs of dogs nor did they modify their surface tension properties (Modell et al. 1970a). Liquid ventilation of dogs and primates was accomplished with PFC (FX-80, perfluorobutyltetrahydrofuran) for 30 min to 8 hours. Both species became hypercarbic (arterial CO_2 tension of 40 to 80) and acidotic (arterial pH 7.05 to 7.2) as the liquid ventilation progressed but were otherwise well oxygenated. After the PFC had been drained from their lungs, the animals resumed normal air breathing but with significantly depressed arterial oxygen levels (PaO_2) of 45–55 mm Hg. Measurement of the PaO_2 over time indicated a return to normal values about 10 days after liquid ventilation. Gross examination of lungs from dogs serially sacrificed from 1 hour to 10 days after termination of liquid ventilation showed a translucent sheen suggestive of PFC in the dependent alveolar sacs of the 1-hour lungs. Three days after liquid ventilation the translucent areas were less extensive and numerous, and by 10 days only a few lobes had a translucent sheen. Microscopically, the lungs from the 3-hour post liquid ventilation were hyperemic, contained neutrophil exudate in the bronchioles and had some congested alveolar septa filled with numerous intra-alveolar vacuolated macrophages. Ten days after treatment the lungs were virtually normal microscopically. The authors suggested that the hyperemia and inflammatory reactions were due to alveolar distention of liquid ventilation or to an irritant effect of the PFC itself or both. They concluded that the reduced PaO_2, which was readily improved by breathing $FiO_2 = 0.4$, was due to the PFC in the alveolar septa forming a diffusion barrier to oxygen transfer. On balance, the data support the conclusion that liquid ventilation does not cause any adverse morphological, biochemical, or histological effects.

The problems of acidosis and hypercarbia were probably related to inadequate removal of metabolic CO_2. To determine if the hypercarbia and acidosis limited the duration of liquid breathing, four dogs were liquid ventilated for 8 hours with buffer administered to prevent acidosis. All four dogs survived, the acidosis was controlled by the buffer, and the hypercarbia leveled off at a 60–80 mm Hg. While all four dogs survived, two who were given higher-pressure ventilations suffered lung tissue damage, suggesting that pressure control during liquid breathing would be necessary. Moskowitz and Schaffer have subsequently developed ventilators for liquid breathing to solve the problems of CO_2 removal and pressure-induced tissue damage (Schaffer and Moskowitz 1974).

Modell also studied tissue distribution in dogs and primates several years after liquid breathing. The tissues were examined grossly, microscopically, and chromatographically. The results were unremarkable except for the presence of trace amounts (a few milligrams per 100 g of tissue) of PFC in the lungs, liver, and fatty tissue (Modell et al. 1973, 1976).

In summary, liquid ventilation studies with adult dogs and primates resulted in adequate oxygenation during liquid breathing accompanied by acidosis and hypercarbia. After liquid ventilation, PaO_2 levels were depressed for three to 10 days, probably due to PFC in the alveolar sacs. After 10 days and for periods thereafter of up to 3-years post liquid ventilation, lungs were normal grossly, microscopically, and biochemically, with the exception of traces of PFC in various tissues.

In studies with immature and premature animals, the results were markedly different. Rufer and Spitzer studied the use of liquid ventilation with PFC in immature pigs (Rufer and Spitzer 1974). In pigs of 95-days gestation, air ventilation was difficult due to atelectasis, and 75% of the animals died within 15 min. At 100 days gestation, air ventilation was maintained for 90 min. and at 110 days gestation (full term), air ventilation was easily achieved and survival was high. Rufer and Spitzer found that liquid ventilation of the 95-day immature mini-pig could be sustained for over 3 hours and that the animals did not become hypercarbic. A more exciting finding was that compliance in the immature lung after liquid ventilation was improved almost to that of a gestationally mature mini-pig in spite of the fact that lung lavage of the 95-day gestational animals failed to recover any surfactant. It appeared that PFC assumed the role of surfactant in the immature lung, decreased the surface tension of the lung, and restored compliance towards normal during subsequent air breathing. The results suggest that periods of liquid breathing need not be terribly long to have lasting benefit. Schaffer and coworkers found that premature sheep were readily ventilated with PFCs and that peak intratracheal pressures were significantly reduced after liquid ventilation (Schaffer et al. 1976). Schaffer's work also confirmed that hypercarbia and acidosis were not as extreme in the immature lung undergoing liquid ventilation. In a subsequent paper, Schaffer and coworkers found that preterm lambs could be liquid ventilated for 3 hours and that their gas exchange and lung compliance were similar to mature lambs undergoing gaseous ventilation. They believe that this result extends the viability of the preterm lamb to the limit of the pulmonary capillary development rather than that of the pulmonary surfactant system (Schaffer et al. 1983).

Schaffer and coworkers also found that cardiac output is decreased (Lowe et al. 1979) and pulmonary vascular resistance is increased during liquid breathing in preterm lambs (Lowe and Schaffer 1986). Modell and coworkers, on the other hand, found no change in cardiac output in adult dogs (Modell et al. 1970b). The differences in these reports may be due to differential response in premature animals compared to adults, species differences, or methodological differences. If cardiac output decrease and PVR increase are confirmed during liquid breathing, it may have consequences for long-term maintenance via liquid breathing. However, such effects would have no impact on the short-term usage required for the PFC to serve as a temporary surfactant and open the alveoli.

Waldrop (1989) conducted a clinical trial on a 24 to 28-week-old human infant, with 15 min of liquid breathing. Although the baby died, the lungs functioned for 19 hours after the liquid ventilation ceased. Further trials are planned in premature infants.

PFC lavage during liquid ventilation has been studied by several groups. Boren (1970) unsuccessfully attempted to remove carbon particles from the lung by PFC ventilation for 45 min. Calderwood and coworkers compared PFC ventilation for one hour to gas ventilation in dogs instilled intratracheally with 5 ml/kg of corn syrup, a model for pulmonary edema (Calderwood et al. 1973). Immediately after PFC ventilation, PaO_2 improved by 100%, and it remained significantly above the gas-ventilated controls for the 3-hour duration of the experiment. Schaffer compared gas to PFC ventilation on meconium-stained preterm lambs that had clear evidence of meconium in the amniotic fluid, hypoxemia, acidosis, and respiratory distress (Schaffer et al. 1984). Gas ventilation of the lambs on 100% oxygen yielded a PaO_2 of about 80 mm Hg and an alveolar-arterial difference (A-aO_2) of 590 mm Hg. After 90 min of gas ventilation, the animals were switched to PFC ventilation and the PaO_2 improved to 120 mm Hg with an A-aO_2 of only 350 mm Hg. Thirty min after PFC ventilation, the PaO_2 had risen to over 150 mm Hg.

The results indicate that PFC ventilation to achieve lavage and to restore respiratory function seems to be a field with significant therapeutic potential, warranting further research.

7.3.8 Miscellaneous Applications

Numerous other applications for PFCs and their emulsions have been piloted in animal studies. Organ preservation has been demonstrated for the isolated rat brain (Andjus et al. 1967), isolated dog kidney (Berkowitz et al. 1976), isolated rat and guinea pig lung (Sloviter et al. 1970), isolated rat liver (Hall 1975), and isolated rat and dog heart (Toyohira et al. 1978). The preservation times for these organs varied from 2 to 24 hours, histological analysis indicated good retention of anatomical structure, and successful autologous reimplantation was subsequently performed on the dog hearts and kidneys. Intracadaveric preservation of organs was demonstrated in sacrificed dogs perfused with Fluosol DA-20 at 10°C for up to 6 hours with superior maintenance of oxygen consumption compared to LR (Honda

1983). Fluosol DA-20 has been used to perfuse amputated limbs before reattachment, and it proved superior compared to Collin's solution or LR in terms of lower serum enzyme leakage and less scar formation in the skeletal muscle tissue (Takahashi et al. 1987). Despite these successes there have been no clinical trials of PFC emulsions for organ preservation. This may reflect the current status of transplanted organs rather than the efficacy of the treatment. Current practice in all organs but kidney is to harvest, transport, and transplant in 6 hours or less, while keeping the organ in a cold, arrested state with low metabolic demand. Kidneys are handled similarly but can be kept up to 48 hours in the cold. Utilization of PFC emulsions to preserve function goes in the opposite direction from metabolic arrest and would probably require development of a perfusion and gas exchange device to be effective. As organ donation increases in frequency and organ supply exceeds demand, PFCs will play a more prominent role in organ banking and preservation.

PFCs and their emulsions have been successfully employed in treating decompression sickness in rat models at pressures up to 8 atmospheres absolute pressure (ATA). Spiess and coworkers compressed rats to 6.8 ATA, held them there for 30 min, and then rapidly decompressed them over 3–4 min. Post decompression, the rats were randomly infused with either 20 ml/kg of a 10 vol % emulsion of perfluorotributylamine (FC-43) or of 6% HES while breathing 100% oxygen. Only one of 12 animals in the HES group survived for 24 hours, while nine of 12 survived in the FC-43 group. FC-43 maintained mean arterial pressure better after decompression than HES (Spiess et al. 1988). Lutz and Herrmann (1984) demonstrated the same phenomenon with rats pressurized to 8 ATA for 30 min and then decompressed.

Animals were either infused with 50 ml emulsion/kg of Fluosol DA-20 or given no treatment. Seventy-five percent of the Fluosol DA-20 treated animals survived compared to 25% of the controls. These experiments demonstrate that PFCs provide relief from gas emboli produced by decompression at fairly moderate doses of emulsion.

The use of PFC emulsions as therapy for cerebral ischemia has been reported using several animal models with conflicting results. Peerless and coworkers compared the protective effect of Fluosol DA-20 in a feline middle cerebral artery (MCA) occlusion model with mannitol and control groups of animals breathing 100% oxygen. Histologically, the Fluosol DA-20-treated cats had the least tissue damage (Peerless et al. 1981). In a subsequent study using the same animal model, Peerless (1983) compared Fluosol DA-20 (15 ml/kg), 20% mannitol (1.2 g/kg), and saline (15 ml/kg) in permanent and temporary (4-hour) occlusion of the MCA. Infusions were initiated immediately after occlusion of the MCA, and neurological examinations were made at regular intervals. Both groups were sacrificed after 6 hours; the permanent occlusion groups were ischemic for 6 hours while the temporary occlusion groups had 4 hours of ischemia followed by 2 hours of reperfusion.

Neurologically, all of the Fluosol DA-20-treated cats, one-half of the mannitol-treated cats, and <10% of the control animals could stand and walk after 4 hours of ischemia. After 6 hours of ischemia, neurological differences between groups were insignificant. Both Fluosol DA-20- and mannitol-minimized mid-line shift due to cerebral swelling were similar to controls in the permanent occlusion group but there were no differences between the treatments in the temporary occlusion group. Histologically, both Fluosol DA-20- and mannitol-treated permanent occlusion animals exhibited 50% less ischemic neuronal change. There were no histological differences between treatments in the temporary occlusion animals. In a subsequent study with 24-hours occlusion times, Peerless and coworkers found no significant differences between Fluosol DA-20 and control groups (Peerless et al. 1985). Using the same chronic model, Zervas and coworkers found no benefit with Fluosol DA-20 infusion (Kolluri et al. 1986a, 1986b). Perieira and coworkers also found that Fluosol DA-20 was not beneficial in a feline model of temporary occlusion-reperfusion (Perieira et al. 1988). The different results in these studies may be due either to the variability in cat cerebral collateral circulation or to the great difficulty of modeling stroke in animals.

Osterholm and coworkers studied the effects in cats of ventriculosubarachnoid perfusion on the oxygenation of the brain during global ischemia (Osterholm et al. 1983). This model of global ischemia involved bilateral carotid clamping and bleeding to a mean arterial pressure of 30 mm Hg for 15 min. The cats were reperfused intravenously with the withdrawn heparinized blood, and the cerebral ventricles were perfused with either oxygenated PFC nutritional emulsion or a modified Elliott's B solution. In the oxygenated ($PO_2 = 640$ mm Hg) PFC treatment group, 88% of electrocerebral activity was restored compared to complete electrocerebral silence in the group reperfused with the modified Elliott's B solution.

Clinical trials have been conducted in Japan and the USA using Fluosol DA-20 for stroke therapy. In the Japanese study, Oda and coworkers used Fluosol DA-20 in more than 70 patients (Oda et al. 1982). Clinical efficacy was determined by observing changes in neurological function and electroencephalogram activity (EEG) and by measuring regional cerebral blood flow. Patient cerebral ischemia was due to either subarachnoid hemorrhage with subsequent vasospasm or occlusive arterial stenoses. The authors reported clinical improvement in about 65% of the patients from both groups. Nonetheless, it is impossible to assess the value of these data since there were no controls in this study (Oda et al. 1982). The U.S. trial was limited to three patients with vasospasm-induced cerebral ischemia, and the results were inconclusive (Swann et al. 1983). There have been no subsequent clinical reports on the use of PFC emulsions for stroke therapy.

A variety of other applications have been investigated in animals, albeit not extensively. These include the use of PFC emulsions in ARDS therapy (Light et al. 1987), wound healing (Chowdary et al. 1987), carbon monoxide and cyanide poisoning, and sickle cell therapy, and in neat form, as replacements for the vitreous fluid of the eye and as eye-eyelid lubricants (Haidt et al. 1982; Clark et al. 1984).

Investigation of the medical applications of PFC-oxygen-transport agents has been broad both preclinically and clinically. While success has been variable in the past, much of the problem stems from the fact that only a single, low-PFC-content emulsion with surfactant side effects has been widely available. Recently, renewed effort in PFC emulsion research suggests that Fluosol DA-20 will only be the first of many biomedical products based on PFCs that will appear in the near future.

7.4 FUTURE PERFLUOROCHEMICAL PRODUCTS

Fluosol DA-20 was a more than adequate first-generation oxygen transport agent, especially considering the formidable complexities facing the development of a shelf-stable, terminally sterilizable, safe, and effective intravenous emulsion. Thus, the use of Fluosol DA-20 is a major milestone in the history of medical use of PFCs. Nonetheless, Fluosol DA-20 has a number of problems which prevent it from achieving broad medical utility: Fluosol DA-20 has a low PFC content, which is substantially diluted upon infusion, greatly limiting its potential oxygen delivery capability. The major surfactant, pluronic F-68, has been implicated in almost all of the acute side effects of the product, and it produces an emulsion that has to be frozen to remain stable. The anaphylactic reactions and frozen state are major impediments to the use of the product in emergency medicine and pose significant convenience problems in other medical arenas. The intravenous half-life of the PFC particles is short while the tissue half-life of perfluorotripropylamine is long, at 65 days.

7.4.1 Commercial Development

At least four companies are in the development stage with second-generation products: HemaGen/PFC (San Francisco, CA), Green Cross, Alliance Pharmaceutical (San Diego, CA), and Adamantech (Linwood, PA). HemaGen has developed an emulsion based on the use of triglyceride oils to improve surfactant interaction with the PFC, resulting in emulsions with improved stability at PFC contents as high as 70 vol %. Invented by R.F. Shaw, a physician and medical entrepreneur, and L.C. Clark, the original PFC pioneer, these high-PFC emulsions can be sterilized at 121°C for as long as 30 min without degradation or cracking (Shaw and Clark 1987). The mean particle size after sterilization is 0.2 μm by laser light scattering, and the emulsions can be stored for over a year at 4 and 25°C. Particle size does not increase during storage. The viscosity of the emulsion is 7 cP at 37°C, and it does not impair flow through the microcirculation. Animal safety is several times greater than Fluosol DA-20 on a ml PFC/kg basis.

Green Cross Co. recently published information on a second-generation emulsion based on lecithin and perfluoro-N-methyldecahydroisoquinoline (Tsuda and Yokoyama 1989). The emulsion is reported to be only a 10 vol % emulsion and is stable at "cold-room temperature." Perfluoro-N-methyldecahydroisoquinoline reportedly has a short tissue half-life.

Alliance Pharmaceutical has developed an emulsion based on high perfluorooctyl bromide content and using high concentrations of lecithin as the surfactant (Burgan et al. 1987). The formulation was published as a 6% egg yolk lecithin and 100% (wt/vol) perfluorooctyl bromide (about 52 vol %) perhaps including coadditives such as tocopherol. The LD_{50} in mice was reported to be 45 g emulsion/kg or about 32 ml emulsion/kg. Perfluorooctyl bromide was widely investigated as an oxygen transport agent in the 1970s by other investigators including Clark and Yokoyama but was not developed. Long, who had been investigating the utility of bromoperfluorocarbons as radiographic contrast agents for many years, developed the emulsion which is now being evaluated in clinical trials. Total exchanges using the same formulation have been reported. Mean particle size is reported to be 0.2 μm, and the emulsion is reported to be shelf stable at room temperature (Long et al. 1987). Hemodynamic responses in dogs were negligible with this emulsion compared to the well-known hemodynamic collapse in dogs when given Fluosol DA-20 (Mattrey et al. 1989).

Adamantech has developed a lecithin-based emulsion of perfluoromethyladamantane (10 to 20 vol %) which has 0.2-μm particle size, low viscosity, and good shelf stability at 4 or 25°C. Moore has reported that the emulsions are well tolerated by animals (Moore 1987). This emulsion has been tested as a cardioplegia in dogs.

All four of these second-generation products have solved two of the more serious problems with Fluosol DA-20: the acute complement activation caused by pluronic F-68 and the necessity to be stored frozen. All four are in preclinical and clinical studies as of January 1990.

7.4.2 Research on New Materials

The second-generation products now in preclinical and clinical trials are all based on the surfactant lecithin, and as a consequence they have an intravascular half-life comparable to Fluosol DA-20. The major thrust of research in PFC emulsions is toward the development of surfactant/PFC combinations that will remain in the circulation longer.

Schmolka (1975) reviewed the surfactants available through the early years of PFC emulsion research. Since that time, research on surfactants for PFC emulsions has centered mainly in two areas, fluorinated surfactants and analogs of pluronic F-68. Clark and coworkers reported the use of fluorinated surfactants in the early 1970s but the results were not very promising. Schmolka suggested the use of the fluorinated amine oxide XMO-10 as early as 1975 (Schmolka 1975) but it was not until 1982 that Clark reported on the powerful emulsifying properties of this surfactant (Clark et al. 1982). As little as 0.1% of XMO-10 formed stable aqueous emulsions with PFCs. In the case of some structures such as perfluorophenanthrene, the emulsions were transparent and apparently were in the form of microemulsions. Unfortunately, these compounds may be toxic, and they are no longer of much interest.

Delpuech and coworkers have studied perfluoroalkylpolyethylene oxide surfactants and claimed to have made microemulsions of PFCs (Selve et al. 1983). Acyl analogs of these molecules were reported by Fung and coworkers (Fung et al. 1987). Toxicity and stability data for the surfactants and their emulsions have not been reported.

Reiss and coworkers have synthesized pure monoperfluoroacyl esters of 1,4-D-sorbitan but found that they were only moderately surface active (Zarif et al. 1989). Reiss has also prepared perfluoroalkyl glycosides where the perfluoroalkyl chain lengths were six and eight carbons (Greiner et al. 1989).

Reiss reports that perfluorooctylmaltoside forms a stable emulsion with perfluorodecalin and pluronic F-68. The surfactant is nontoxic as a 1% solution in saline at 25 ml/kg. Reiss has recently disclosed two new classes of fluorinated surfactants: C- and N-perfluoroalkyl betaines derived from glycine and perfluoroacyl derivatives of carnitine. Data on their surfactant properties and toxicities have not been published (Bernelin et al. 1989; Nivet et al. 1989).

7.5 CONCLUSION

PFCs have come a long way since Clark's first experiment and the ensuing euphoric period when too much was expected and too little was known. Over the intervening years, the complexities of formulating, manufacturing, testing, and utilizing PFC emulsions have been unraveled, and the first PFC product, Fluosol DA-20, is readily available in the USA. Unfortunately, Fluosol DA-20 proved inadequate for the goals set for it. Nonetheless, it served to define the next level of hurdles that the second-generation of oxygen transport agents will have to clear to achieve widespread use. Fluosol DA-20 was also responsible for the development of a body of knowledge based on the preclinical and clinical efficacy testing of PFC-based products. Medical understanding of PFCs and of their use has been brought into focus by the Fluosol DA-20 experience. The finding that pluronic F-68 was responsible for the undesired acute reactions has led to its removal from second-generation products. The benign and reversible side effects attributable to PFCs are now fairly well understood and are not a threat to organ function at therapeutic doses. Several new formulations have been developed that will expand the uses of PFC emulsions into more applications. Research on third-generation products that will persist significantly longer in the bloodstream is underway, and the first new classes of safe intravenous surfactants have been identified. After a somewhat-long induction period, the use of PFC oxygen transport agents in medicine is poised to enter an exponential growth phase.

REFERENCES

Anderson, H.V., Leimgruber, P.P., Ruobin, G.S., et al. (1985) *Am. Heart J.* 4, 720–726.

Andjus, R.K., Suhara, K., and Sloviter, H.A. (1967) *J. Appl. Physiol.* 22, 1033.

Anonymous (1988) *Otisville BioPharm Annual Report* 4–8.

Anonymous (1990) *FDC Reports* 52, 8.

Bajaj, A.K., Cobb, M.A., Virmani, R., et al. (1989) *Circulation* 79, 645–656.

Berkowitz, H.D., McCombs, P., Sheety, S., et al. (1976) *J Surg. Res.* 20, 595.

Bernelin, R. Le Blanc, M., and Riess, J.G. (1989) in *International Chemical Congress of Pacific Basin,* 17–22 Dec. 1989, abstract no. 662, Honolulu, HA.

Blumberg, N., Agaral, M., and Chuang, C. (1985) *Blood,* 66 (suppl. 1), 274a.

Boren, H.G. (1970) *Fed. Proc.* 29, 1737–1739.

Burgan, A.R., Long, D.M., Mattrey, R.F., et al. (1987) *Biomaterials, Artificial Cells, and Artificial Organs* 15, 367.

Calderwood, H.W., Modell, J.H., Ruiz, B.C., Brogdon, J.E., and Hood, I. (1973) *Anesthesiology* 38, 141–144.

Castro, O., Nesbitt, A.E., and Lyles, D. (1984) *Am. J. Hematology* 16, 15–21.

Chowdary, R.P., Berkower, A.S., Moss, M.L., et al. (1987) *Plastic and Reconstructive Surgery* 79, 98–101.

Clark, L.C., and Gollan, F. (1966) *Science* 152, 1755–1756.

Clark, L.C., Kaplan, S., and Becattini, F. (1970). *J. Thoracic and Cardiovasc. Surg.,* 60, 757–773.

Clark, L.C., Wesseler, E.P., Kaplan, S., et al. (1975) *Fed. Proc.* 34, 1468–1477.

Clark, L.C., Clark, E.W., Moore, R.E., Kinnett, D.G, and Inscho, E.I. (1982) *Prog. Clin. Biol. Res.* 122, 168–180.

Clark, L.C. Ackerman, J.L., Thomas, S.R., et al. (1984) *Adv. Exp. Med. and Biol.* 180, 835–846.

Cleman, M., Jaffe, C.C., and Wohlgelernter, D. (1986) *Circulation* 74, 555–562.

Eidelberg, D., Johnson, G., Barnes, D., et al. (1988) *Magnetic Resonance Med.* 6, 344–352.

Elliott, L.A., Ledgerwood, A.M., Lucas, C.E., et al. (1989) *Critical Care Med.* 17, 166–172.

Faithful, N.S., Erdmann, W., Fennema, M., and Kok, A. (1986) *Brit. J. Anesth.* 58, 1031–1040.

Fedor, E.J., Wisdom, N.L., and Lowe, N.L. (1988) in *Abstract 6th Conference on Chemical Modifiers of Cancer Treatment,* 21–25 March 1988, Paris, pp. 4–14.

Forman, M.B., Puett, D.W., Bingham, S.E., et al. (1987) *Circulation* 76, 469–479.

Fujita, T., Suzucki, T., and Ogawa, R. (1983) *Prog. Clin. Biol. Res.* 122, 265–272.

Fung, B.M., O'Rear, E.A., Afzal, A., et al. (1987) *Biomaterials, Artificial Cells, and Artificial Organs* 15, 411.

Geyer, R.P., Monroe, R.C., and Taylor, K. (1968) in *Organ Perfusion and Preservation* (Norman, J., ed.), pp. 85–97, Appleton, Century and Crofts, New York.

Geyer, R.P., Monroe, R.C., and Taylor, K. (1973) *Fed. Proc.* 32, 927.

Gjaldbaek, J., and Hildebrand, J.H. (1949) *J. Am. Chem. Soc.* 71, 3147.

Glogar, D.H., Kloner, R.A., Muller, J., et al. (1981) *Science* 211, 1439–1441.

Gollan, F., and Clark, L.C. (1966) *The Physiologist* 9, 191.

Gould, S.A., Rosen, A.L., Sehgal, L.R., et al. (1986) *New England J. Med.* 314, 1653–1656.

Greiner, J., Manfredi, A., and Reiss, J.G. (1989) *New J. Chem.* 13, 247–254.

Haidt, S., Clark, L.C., and Ginsberg, R.J. (1982) *Investigative Ophthalmology* 22, 233.

Haljamae, H., and Rosenberg, P.H. (1988) *Acta Anaesthesiol. Scand.* 32 (suppl. 89), 1–3.

Hall, C.A. (1975) *Fed. Proc.* 34, 1513.

Honda, K. (1983) *Prog. Clin. Biol. Res.* 122, 327–330.

Howlett, S., Dundas, D., and Sabiston, D.C. (1965) *Arch. Surg.*, 91, 643–645.

Huetis, D.W., Bove, J.R., and Busch, S. (1981) in *Practical Blood Transfusion,* 3rd ed., pp. 51–61, Little, Brown, and Co., Boston.

Jaffe, C.C., Wohlgelernter, D., and Cabin, H., et al. (1988) *Am. Heart J.* 6, 1156–1164.

Jaffin, J.H., Magovern, G.J., Kanter, R.R., et al. (1981) *Surgical Forum* 13, 290–293.

Joseph, P.M., Yuasa, Y., Kundel, H.S., et al. (1985) *Invest. Radiol.* 10, 504–509.

Klubes, P.S., Hiraga, S., Richard, L.C., Owens, E.S., and Blasberg, R.G. (1987) *Eur. J. Cancer Clin.* 23, 1859–1867.

Kolluri, S., Heros, R.C., Hedley-Whyte, E.T., et al. (1986a) *Surg. Neurol.* 26, 3–8.

Kolluri, S., Heros, R.C., Hedley-Whyte, E.T., et al. (1986b) *Stroke* 17, 976–980.

Kolodgie, F.D., Dawson, A.K., Roden, D.M., et al. (1986) *Am. Heart J.* 112, 1192–1201.

Light, B.M., Perez-Padilla, K., and Kryger, M.H. (1987) *Chest* 91, 444–449.

Long, D.C., Fallano, R., Reiss, J.G., et al. (1987) *Biomaterials, Artificial Cells, and Artificial Organs* 15, 417.

Long, D.M. (1976) U.S. patent no. 3,975,512.

Long, D.M., Nielson, M.D., and Multer, F.K. (1971) *Radiology* 133, 71.

Long, D.M., Liu, M.S., Szanto, P.S., et al. (1972a) *Radiology* 105, 323.

Long, D.M., Liu, M.S., Szanto, P.S., et al. (1972b) *Rev. Surg.* 29, 71.

Long, D.M., Multer, F.K., Greenburg, A.G., et al. (1978) *Surgery* 84, 104–112.

Long, D.M., Lasser, E.C., Sharts, C.M., et al. (1980) *Invest. Radiol.* 15, 242–247.

Longmaid, H.E., Adams, D.F., Nierinckx, R.D., et al. (1985) *Investigative Rad.* 20, 141–145.

Lowe, C., Tuma, R.F., Sevieri, E.M., and Schaffer, T.H. (1979) *J. Appl. Physiol.* 47, 1051–1057.

Lowe, C.A., and Schaffer, T.H. (1986) *J. Appl. Physiol.* 60, 154–159.

Lustig, R.A., and McIntosh, N. (1986) *Prog. Clin. and Biol. Res.* 211, 29–38.

Lustig, R., McIntosh,-Lowe, N., Rose, C., et al. (1989a). *Int. J. Radiation Oncology, Biol. and Phys.* 16, 1587–1593.

Lustig, R.A., Lowe, N., Prosnitz, L., et al. (1989b) *Int. J. Radiation Oncology, Biol. and Phys.* 17, 202.

Lutz, J. (1983) *Prog. Clin. Biol. Res.* 122, 197–208.

Lutz, J. (1985) in *Perfluorochemical Oxygen Transport* (Tremper, K., ed.), pp. 63–93, Little, Brown and Co., Boston.

Lutz, J., and Herrmann, G. (1984) *Pfluger's Archive* 401, 174–177.

Lutz, J., and Metzenauer, P. (1980) *Pfluger's Archive* 387, 175–181.

Lutz, J., and Wagner, M. (1981) *Pfluger's Archive* 394, R14.

Lutz, J., Metzenauer, P., Kunz, E., and Heine, W.D. (1982) in *Oxygen Carrying Colloidal Blood Substitutes* (Frey, R., Beisbarth, H., and Stosseck, K., eds.), pp. 73–81, Zuckschwerdt Verlag, Munich.

Maetani, S., Nishikawa, T., Hirawaka, A., and Tobe, T. (1986) *Ann. Surg.* 203, 275–281.

Magovern, G.J., Flaherty, J.T., Gott, V.L., et al. (1982) *Circulation* 66 (suppl. 1), 60.

Makowski, H. (1978) in *Proc. 4th International Symposium on Perfluorochemical Blood Substitutes,* Kyoto, Japan, 21–22 Oct. 1978, pp. 439–448, Excerptica Medica, Amsterdam.

Makowski, H., Tentshev, P., Frey, R., et al. (1978) in *Proc. 4th International Symposium on Perfluorochemical Blood Substitutes,* Kyoto, Japan, 21–22 Oct. 1978, pp. 47–54. Excerptica Medica, Amsterdam.

Masuda, M., and Hori, M. (1973) in *Proc. 2nd Intercompany Conference on Fluorocarbon,* Osaka, Japan, Igakushobo, pp. 217–228.

Mate, T.P., and Rockwell, S. (1985) in *Abstracts American Society Therapeutic Radiation Oncologists Meeting,* Washington DC. Oct. 1984.

Mattrey, R.F., and Andre, M.P. (1984) *Am. J. Cardiology* 54, 206–210.

Mattrey, R.F., Hilpert, P.L., Long, C.D., et al. (1988) *Critical Care Medicine* 17, 652–656.

Mattrey, R.F., Hilpert, P.L., Long, C.D., et al. (1989) *Critical Care Medicine* 17, 652–656.

McFarland, E.R., Koutcher, J.A., Rosen, B.R., Teicher, B., and Brady, T.J. (1985) *J. Computer Assisted Tomography* 9, 8–15.

Menasche, P., Fauchet, M., Lavergne, A., et al. (1983) *Prog. Clin. Biol. Res.* 122, 363–372.

Menasche, P., Escorsin, M., Birkui, P., et al. (1985) *Am. J. Cadiol.* 55, 830–834.

Mitsuno, T., Ohyanagi, H., and Naito, R. (1982) *Ann. Surg.* 195, 60–69.

Mitsuno, T., and Ohyanagi, H. (1985) in *Perfluorochemical Oxygen Transport* (Tremper, K., ed.), pp. 169–184, Little, Brown and Co., Boston.

Modell, J.H., Gollan, F., Giammona, S.T., and Parker, D. (1970a) *Chest* 57, 263–265.

Modell, J.H., Newby, E.J., and Ruiz, B.C. (1970b) *Fed. Proc.* 29, 1731–1736.

Modell, J.H., Hood, C.I., Kuck, E.J., and Ruiz, B.C. (1971) *Fed. Proc.* 34, 312–320.

Modell, J.H., Calderwood, H.W., Ruiz, B.C., Tham, M.K., and Hood, C.I. (1976) *Chest* 69, 79–81.

Modell, J.G., Tham, M.K., Modell, J.H., Calderwood, H.W., and Ruiz, B.C. (1973) *Toxicology and Applied Pharmacology* 26, 86–92.

Moore, R.E. (1987) *Biomaterials, Artificial Cells, and Artificial Organs* 15, 426.

Moulder, J.E., and Fish, B.L. (1988) *Int. J. Radiation Oncology, Biol. and Phys.* 15, 1193–1196.

Naito, R., and Yokoyama, K. (1981) in *Perfluorochemical Blood Substitutes—Green Cross Technical Information Series No. 7,* pp. 39–50, Green Cross, Osaka, Japan.

Naito, R., Yokoyama, K., and Watanabe, M. (1977) *Artificial Organs* 2, 93.

Nivet, J.B., Le Blanc, M., and Reiss, J.G. (1989) Abstract no. 673, Pacific Basin Chemistry Congress.

Novick, R.J., Stefaniszyn, H.J., Michel, R.P., et al. (1985) *J. Thorac. Cardiovasc. Surg.* 89, 547–566.

Nunn, G.R., Dance, G., Peters, J., and Cohn, L.H. (1983) *Am. J. Cardiol.* 52, 203–205.

Oda, Y., Murata, T., and Uehida, Y. (1982) *Neurol. Surg.* 10, 637–644.

Ohyanagi, H., and Mitsuno, T. (1975) in *Proc. of the Post Congress Symposium, Xth International Congress of Nutrition on Perfluorochemical Blood*, Kyoto, Japan, pp. 21–54.

Ohyanagi, H., Sekita, M., Yokoyama, K., et al. (1978) in *Proc. of the 4th International Symposium on Perfluorochemical Blood Substitutes*, Kyoto, Japan, 21–22 Oct. 1978, pp. 373–389, Excerptica Medica, Amsterdam.

Ohyanagi, H., Toshima, K., Sekita, M., et al. (1979) *Clinical Therapeutics* 2, 306–311.

Ohyanagi, H., Nishijima, M., Usami, M., et al. (1983) *Prog. Clin. Biol. Res.* 211, 315–320.

Ohyanagi, H., Nakaya, S., Okumura, S., and Saitoh, Y. (1984) *Artificial Organs* 8, 110–118.

Okada, K., Kosugi, I., Kawashima, Y., et al. (1975) in *Proc. of the Post Congress Symposium, Xth International Congress of Nutrition on Perfluorochemical Blood*, Kyoto, Japan, pp. 215–224.

Osterholm, J.L., Alderman, J.B., Triolo, A.J., et al. (1983) *Neurosurg.* 13, 381–387.

Peerless, S.J. (1983) *Prog. Clin. Biol. Res.* 122, 353–362.

Peerless, S.J., Ishikawa, R., Hunter, I.G., and Peerless, M.J. (1981) *Stroke* 12, 558.

Peerless, S.J., Nakamura, R., Rodriguez-Salazar, A., and Hunter, I.G. (1985) *Stroke* 16, 38–43.

Perieira, B.M., Weinstein, P.R., and Baena, R.R. (1988) *Neurosurgery* 23, 139–142.

Rockwell, S. (1985) *Int. J. Radiation Oncology, Biol. and Phys.* 11, 97–103.

Rockwell, S. (1988) Personal communication.

Rose, C.M., Lustig, R., McIntosh, N., and Teicher, B. (1986) *Int. J. Radiation Oncology, Biol. and Phys.* 12, 1325–1327.

Rosenblum, W.I., Moncure, C.W., and Behm, F.G. (1985) *Arch. Pathol. Lab Med.* 109, 340–344.

Rousou, J.A., Engelman, R.M., Anisimowicz, L., et al. (1986) *J. Thorac. Cardiovasc. Surg.* 91, 270–276.

Rufer, R., and Spitzer, H.L. (1974) *Chest* 66 (suppl.), 29S–30S.

Schaffer, T.H., and Moskowitz, G.D. (1974) *J. Applied Physiology* 36, 208–213.

Schaffer, T.H., Rubenstein, D., Moskowitz, G.D., and Delivoria-Papadopoulos, M. (1976) *Pediat. Res.* 10, 227–231.

Schaffer, T.H., Tran, N., Bhutani, V.K., and Sivieri, E.M. (1983) *Pediatric Res.* 17, 680–684.

Schaffer, T.H., Lowe, C., Bhutani, V.K., and Douglas, P.R. (1984) *Pediatric Res.* 18, 47–52.

Schmolka, I.R. (1975) *Fed. Proc.* 34, 1449–1453.

Schweighardt, F.K. (1989) U.S. patent no. 4,838,274.

Selve, C., Castro, B., Leempoel, P., et al. (1983) *Tetrahedron* 39, 1313–1316.

Shaw, R.F., and Clark, L.C. (1987) European patent application no. 0 231 091.

Simon, J.H. (1947) *Ind. and Eng. Chem.* 39, 238–341.

Sloviter, H., and Kamimoto, T. (1967) *Nature* 216, 458.

Sloviter, H.A., Yamada, H., and Ogoshi, S. (1970) *Fed. Proc.* 29, 1755.

Song, C.W., Zhang, W.L., Pence, D.M., et al. (1986) *Int. J. Radiation Oncology, Biol. and Phys.* 12, 934–936.

Song, C.W., Lee, I., Hasegawa, T., et al. (1987) *Cancer Res.* 47, 442–446.

Spears, J.R., Serur, J., Baim, D.S., et al. (1983) *Circulation* 68 (suppl. III), 317.

Spears, J.R., Spokojny, A.M., Serur, J.R., et al. (1984) *Am. Heart Assoc. 57th Scientific Sessions,* 12–15 Nov. 1984, Miami Beach, FL, Circulation 69, (suppl. II) p. 245.

Spence, R.K., McCoy, S., Costabile, J., et al. (1989) *Critical Care Med.* 17, S144.

Spiess, B.D., McCarthy, R.J., Tuman, K.J., et al. (1988) *Undersea Biomedical Res.* 15, 31–37.

Suyama, T., Matsumoto, T., Watanabe, M., Hamano, T., and Naito, R. (1975) *Proc. of the Post Congress Symposium, Xth International Congress of Nutrition on Perfluorochemical Blood,* Kyoto, Japan, pp. 225–235.

Swann, K.W., Ropper, A.H., and Zervas, N.T. (1983) *Prog. Clin. Biol. Res.* 122, 399–404.

Tabayashi, K., McKeown, P., Miyamoto, M., et al. (1988) *J. Thorac. Cardiovasc. Surg.* 95, 239–246.

Takahashi, F., Tsai, T., Fleming, P.E., and Ogden, L. (1987) *Plastic and Reconstructive Surgery* 80, 582–590.

Teicher, B.A., and Holden, S.A. (1987) *Cancer Treatment Rep.* 71, 787–789.

Teicher, B.A., and Rockwell, S. (1983) *Am. Assoc. of Cancer Res. Abstr.* 25–28 May 1983, San Diego, CA.

Teicher, B.A., and Rose, C.M. (1984) *Science* 223, 934–936.

Thomlison, R.H., and Gray, L.H. (1955) *Brit. J. Cancer* 9, 539–549.

Toyohira, H., Taira, A., Arikawa, K., et al. (1978) in *Proc. of the 4th International Symposium on Perfluorochemical Blood Substitutes,* Kyoto, Japan, 21–22 Oct. 1978, p. 161, Excerptica Medica, Amsterdam.

Tremper, K.K., Vercellotti, G.M., and Hammerschmidt, D.E. (1984) *Critical Care Medicine* 12, 428–431.

Tremper, K.K., Levine, E.M., and Waxman, K. (1985) in *Perfluorochemical Oxygen Transport* (Tremper, K., ed.), pp. 185–197, Little, Brown and Co., Boston.

Tsuda, Y., and Yokoyama, K. (1989) in *International Chemical Congress of Pacific Basin,* 17–22 Dec. 1989, abstract no. 698, Honolulu, HI.

Vercellotti, G.M., and Hammerschmidt, D.E. (1982) *Blood* 59, 1299–1304.

Vessey, C.G., Armstrong, W.F., West, S.R., et al. (1986) *J. Clin. Ultrasound* 14, 613–618.

Waldrop, M.M. (1989) *Science* 245, 1043–1045.

Waxman, K., Tremper, K.K., Cullen, B.F., and Mason, R. (1984) *Arch. Surg.* 119, 721–724.

Yokoyama, K., Watanabe, M., Yamanouchi, K., et al. (1975a) *Fed. Proc.* 34, 1478–1483.

Yokoyama, K., Yamanouchi, K., Murashima, R., et al. (1975b) *Chem. Pharm. Bull.* 23, 1363–1367.

Zarif, L., Greiner, J., and Reiss, J.G. (1989) *J. Fluorine Chem.* 44, 73–85.

PART II

Plasma Fractions

8

Current Approaches to the Preparation of Plasma Fractions

Johan Vandersande

Plasma protein was first purified in the Harvard Medical School laboratory of E.J. Cohn (Cohn et al. 1946). As a result of the need during World War II for a stable plasma expander to be used under battlefield conditions when large losses of blood were incurred, methods to prepare freeze-dried plasma were developed. However, because of the difficulties encountered during reconstitution of the freeze-dried plasma with sterile water, in the field, Cohn first suggested the use of the plasma protein albumin since it is the protein naturally produced by the body to counteract shock. He suggested that an albumin solution in a final container in a ready-to-use, stable solution could be infused immediately on the battle field to prevent or treat shock.

Requests to commercial pharmaceutical firms by the U.S. government for the production of large quantities of albumin utilizing Cohn's method eventually led to the development of the plasma fractionation industry as we know it today. Later, other scientists, some of whom had been Cohn's coworkers, went on to develop methods to purify other plasma proteins such as immunoglobulins.

Other therapeutically valuable proteins like prothrombin complex are separated in the various plasma fractions produced by the Cohn procedure. The use of cold ethanol and its denaturing effect in the procedure, however, have led to the development of new methods for purifying these proteins, in particular, for labile coagulation factors.

From the many proteins present in plasma, only a few are routinely prepared for therapeutic use today. They are as follows:

Albumin/plasma protein fraction
Antihemophilic factor (factor VIII)
Immunoglobulins
Fibrinogen/fibronectin
Prothrombin complex (factors II, VII, IX, and X)
Activated prothrombin complex concentrates
Antithrombin III

8.1 ALBUMIN

This protein, used in the treatment of shock, is present in a greater concentration than any other protein in plasma. The purification method that Cohn and his coworkers developed is still the basis for most manufacturing methods in use today (Cohn et al. 1946). It is characterized by the following principle: proteins or groups of proteins in plasma are made insoluble by varying five parameters—ethanol concentration, temperature, ionic strength, pH, and protein concentration. Figure 8–1 shows the Cohn method (method 6) as it was described in his publication in 1946.

This procedure is typically carried out in stirred, temperature-controlled tanks equipped with spargers, in order to add buffer and ethanol solutions while avoiding locally high concentrations of such additives and possibly causing denaturation of proteins. Following insolubilization or precipitation, the insoluble proteins are separated from those still in solution by conventional liquid-solid separation techniques such as centrifugation or filtration. The original method required five precipitation and separation steps to reach fraction V, which contains at least 96% albumin. The method also called for the reprecipitation of fraction V in order to reduce the salt concentration of the final albumin fraction. After removal of ethanol and salts from the final fraction by freeze-drying or diafiltration, the albumin is stabilized, sterile filtered, filled, and pasteurized in its final container.

Although many variations of the Cohn procedure have been developed, its principles have remained the same, and it is practiced worldwide in some form or another. In particular, for large-scale fractionation, the advantages of the bacteriostatic effects of ethanol and low temperatures as well as its relatively good yields (around 80%) are unparalleled.

Modifications of the Cohn method have been aimed at reducing volume and combining precipitation steps, such as fractions I and II + III, and IV-1 and IV-4. The goal is usually to reduce capital requirements for processing tanks and pyrogen-free water supply as well as reduce overall operating

LIQUIDS:

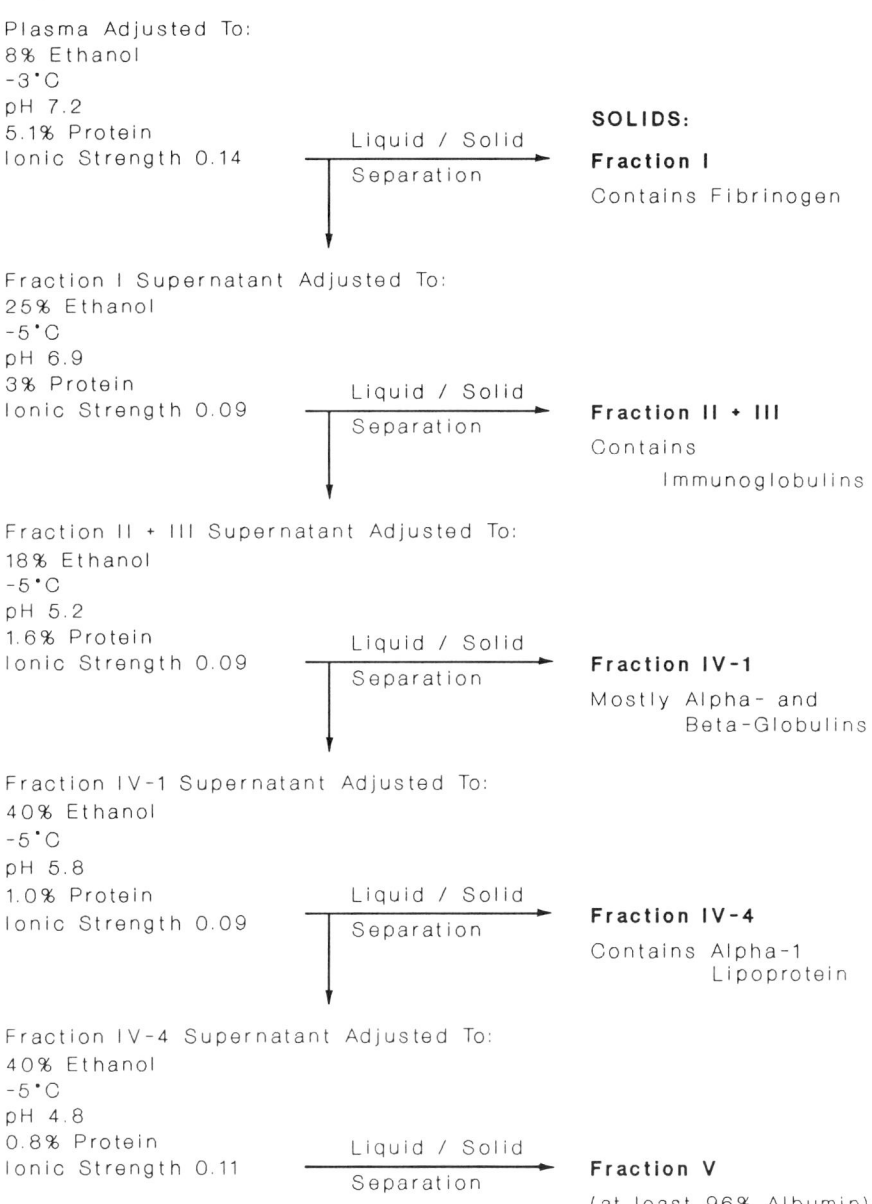

Plasma Adjusted To:
8% Ethanol
-3°C
pH 7.2
5.1% Protein
Ionic Strength 0.14

Liquid / Solid Separation

SOLIDS:

Fraction I

Contains Fibrinogen

Fraction I Supernatant Adjusted To:
25% Ethanol
-5°C
pH 6.9
3% Protein
Ionic Strength 0.09

Liquid / Solid Separation

Fraction II + III

Contains Immunoglobulins

Fraction II + III Supernatant Adjusted To:
18% Ethanol
-5°C
pH 5.2
1.6% Protein
Ionic Strength 0.09

Liquid / Solid Separation

Fraction IV-1

Mostly Alpha- and Beta-Globulins

Fraction IV-1 Supernatant Adjusted To:
40% Ethanol
-5°C
pH 5.8
1.0% Protein
Ionic Strength 0.09

Liquid / Solid Separation

Fraction IV-4

Contains Alpha-1 Lipoprotein

Fraction IV-4 Supernatant Adjusted To:
40% Ethanol
-5°C
pH 4.8
0.8% Protein
Ionic Strength 0.11

Liquid / Solid Separation

Fraction V

(at least 96% Albumin)

FIGURE 8–1 Flow diagram for the purification of plasma fractions using the Cohn process.

expenses. One of the more well-known modifications is the process described by Kistler and Nitschmann (1962).

Chromatographic methods for preparing albumin as reported by Curling (1980), although useful for small-scale purification in the lab, have never developed into widely used industrial-scale processes. Plasma is pretreated in order to remove lipids and adsorbed on a diethylaminoethyl (DEAE) matrix. After desorption and adsorption on a carboxymethyl (CM) matrix, the partially purified albumin is subjected to ultrafiltration and/or gel filtration, formulated, sterile filtered, and filled. Today only a handful of installations, namely small fractionation facilities in Canada, Hungary, and South Africa, are known to be practicing these techniques. The major reasons for this are the proven reliability of the Cohn process, the initial high cost of the chromatographic gel, the large buffer volumes with associated tanks needed, and recoveries that are not better than those obtained by the Cohn process.

8.2   PLASMA PROTEIN FRACTION

This plasma fraction is essentially an albumin product of lower purity. Additional proteins include alpha and beta globulins. The fraction can be obtained by coprecipitating fraction IV-4 with fraction V precipitate. The method, first reported by Hink and coworkers (1957), included a patented solvent-drying technique of the final fraction IV-4 + V that appears to be critical in order to obtain a stable final product, presumably because of lipid removal by the solvent.

8.3   ANTIHEMOPHILIC FACTOR (FACTOR VIII, OR AHF)

This protein, found in minute amounts in plasma, is used in the treatment of hemophilia A. Manufacturers of it have tried to increase the purity of their preparations for several reasons: 1) long-term use of impure factor VIII preparations results in patient exposure to a variety of proteins with undesirable side effects and 2) because factor VIII activity is labile in liquid form, all factor VIII preparations are freeze-dried, requiring good solubility for rapid reconstitution with sterile water immediately prior to use.

The original observation that a small amount (less than 1% by weight) of precipitate forms in frozen plasma when thawed is the basis of all factor VIII preparation methods in use today. Depending on the way plasma is collected, frozen, stored, and thawed, as much as 50% of the theoretical factor VIII activity in plasma (1 activity unit/ml) can be recovered in this crude precipitate.

Purification methods mainly focus on the removal of fibrinogen and immune globulins from this fraction, which is commonly referred to as cryoprecipitate. Purification of cryoprecipitate has been performed using aluminum hydroxide in order to adsorb prothrombin complex. Ethanol and polyethylene glycol fractional precipitation is effective in reducing fibrinogen and fibronectin content. Glycine is used to precipitate factor VIII in order to separate it from the precipitating agents mentioned above. A combination of ethanol and heparin has also been used to obtain a product with a specific activity of 1–2 factor VIII units/mg protein (Bloom 1984). A combination of three precipitating agents (polyethylene glycol, glycine, and sodium chloride) as reported by Ng and coworkers (1986) yields even greater specific activity.

A dramatic breakthrough in factor VIII purification was realized with the application of immunoaffinity chromatography. Factor VIII-specific murine monoclonal antibody, immobilized on a solid matrix packed in a chromatography column, absorbs factor VIII as solutions of cryoprecipitate are pumped through. Subsequent elution yields a factor VIII preparation with a specific activity in excess of 1,500 units/mg. The eluate is subjected to a second chromatography step, an ion exchange step using quaternary aminoethyl (QAE) gel in order to minimize the presence of any monoclonal antibody from the immunoaffinity column (Griffith 1988).

The final bulk solution is sterile filtered, aseptically filled and freeze-dried in final containers.

8.4   IMMUNOGLOBULINS

Immunoglobulins are proteins used to treat immunodeficiencies. Historically these proteins have been purified by cold-ethanol techniques (Oncley et al. 1949) yielding preparations consisting mostly of IgG and traces of IgA and IgM. Figure 8–2 shows the preparation of fraction II (the immunoglobulin fraction) as described by Oncley and coworkers (1949).

Hyperimmune globulin preparations having high titers against specific antigens like tetanus and hepatitis B were used to treat people as well. Stabilized with glycine in a solution containing approximately 16% protein, these products were intended for intramuscular administration. In an effort to increase efficacy by the administration of higher doses, attempts were made to give the product intravenously as early as 1945; however, it turned out that these preparations were rarely tolerated intravenously (Barandun et al. 1962). The reason for this is that most preparations contained aggregates, such as dimers and polymers, that could activate complement without binding to an antigen. Only in the 1980s did preparations of intravenous immunoglobulin first become available and licensed for human use. As a result, the use of the old intramuscular preparations has virtually disappeared.

Although there are several different products for intravenous use on the market today, their purification methods are all aimed at the removal of

Fraction II + III

Dissolved in
20% Ethanol
-5°C
pH 7.2
1% Protein
Ionic Strength 0.005

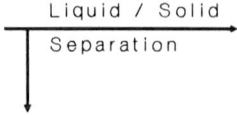

Liquid / Solid
Separation

Solids:

Fraction II + III W

Dissolved in
17% Ethanol
-6°C
pH 5.2
1.2% Protein
Ionic Strenth 0.015

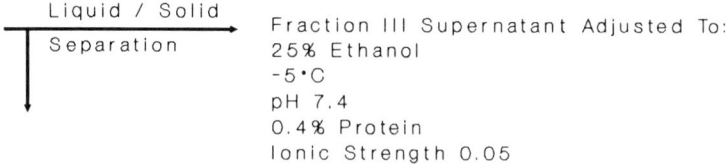

Liquid / Solid
Separation

Fraction III Supernatant Adjusted To:
25% Ethanol
-5°C
pH 7.4
0.4% Protein
Ionic Strength 0.05

Solids:

Fraction III

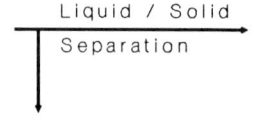

Liquid / Solid
Separation

Solids:

Fraction II
Contains IgG

FIGURE 8–2 Flow diagram for the preparation of fractions II and III using the Oncley method.

aggregates. Likewise virtually all of the methods use fraction II as starting material. Some manufacturers are purifying a solution of fraction II by adding polyethylene glycol (PEG) in increments followed by repeated centrifugation of the solutions in order to remove impurities. Others are subjecting the fraction II solution to ion-exchange chromatography with diethylaminoethyl (DEAE) gels. Another method used is the treatment of fraction II solutions with proteolytic enzymes such as porcine pepsin at low pH, resulting in a product with an anticomplement activity level low enough to make it suitable for intravenous injection. The details of the various methods are generally proprietary but several comparative studies regarding the properties of the various products have been published (Lundblad et al. 1986).

Although the isolation of immunoglobulins directly from plasma by ion-exchange chromatography has never materialized as a large-scale commercial process, it nevertheless does yield a product suitable for intravenous use. After a lipid removal step using dextran sulfate and fumed silica, the plasma is subjected to DEAE sepharose, in order to obtain a separation of crude albumin and IgG. The IgG is further purified by DEAE Cellufine (Bees 1989). The reason for the limited use of these techniques is probably that most manufacturers already have a process that provides them with a crude IgG fraction (fraction I+II+III) as a by-product of their albumin production. Most intravenous immunoglobulin preparations now are freeze-dried in their final container, although at least one manufacturer markets a liquid-formulated product.

8.5 FIBRINOGEN

Historically this protein was recovered from fraction I in the fractionation method as described by Cohn and coworkers (1946). However, because of the high risk of hepatitis associated with the use of this product, manufacturing licenses in the USA were revoked by the FDA in the 1970s. Subsequently, worldwide use of this product has been declining.

Thanks to the advent of viral-inactivation techniques there has recently been a renewed interest in the use and manufacture of fibrinogen, in particular in conjunction with fibronectin and factor XIII, as a sealant for wounds where sutures are impractical. In this particular application fibrinogen is transformed to fibrin by thrombin and $CaCl_2$ (Bagot d'Arc 1989). Lysine-sepharose chromatography has proven particularly useful in the preparation of purified fibrinogen without plasminogen (Matsuda et al. 1972). The final product is sterile-filtered, aseptically filled, and freeze-dried.

8.6 PROTHROMBIN COMPLEX

The group of proteins referred to under this name consists of coagulation factors II, VII, IX, and X and is used in the treatment of hemophilia B (factor IX deficiency).

Cohn fractions III and IV-1 contain these factors, and the complex can be derived from these fractions by adsorption on tricalcium phosphate followed by fractional precipitation with polyethylene glycol (Fekete et al. 1971). Another isolation procedure involves adsorption from plasma by ion-exchange chromatography; DEAE gels are used for this purpose. After adsorption the gel is washed in order to remove unwanted protein and eluted with sodium chloride. This washing procedure results in a lower factor VII content as compared to products derived from Cohn fractions. The eluate can be desalted by gel filtration or diafiltration, although gel filtration is preferred (Brummelhuis 1980). Heparin can be added during the final formulation of prothrombin complex to reduce the risk of activation. The product is sterilized by filtration and freeze-dried in the final container.

8.7 ACTIVATED PROTHROMBIN COMPLEX CONCENTRATES

These concentrates are used to treat patients who have developed inhibitors against coagulation factors. Although they are not often mentioned in the literature, these concentrates represent a significant part of all coagulation products derived from plasma. The factor VIII-inhibitor-bypassing activity in these products has been attributed to a variety of substances, including factor Xa-like activity, but it has not been fully identified yet (Fukui et al. 1985). The concentrates are similar to prothrombin complex concentrates, except some of the clotting factors have been converted to active enzymes.

One method (Eibl et al. 1979) consists of activation of a starting material from plasma or Cohn fraction I supernatant by use of a contact activator such as silica, followed by adsorption onto an ion exchanger.

Another method (Andary et al. 1981) utilizes Cohn fraction IV-1 as starting material. Following adsorption with calcium phosphate, activation with silica, and purification by fractional precipitation with polyethylene glycol, the product is sterile-filtered, filled, freeze-dried, and pasteurized.

8.8 ANTITHROMBIN III

Antithrombin III is a protein in plasma, also called heparin cofactor, or AT III, that is the most recent (1989) of the plasma proteins licensed in the USA for therapeutic use. It is used to treat congenital or acquired deficiencies of antithrombin.

Thanks to the application of affinity chromatography, first reported by Miller-Anderson (Miller-Anderson et al. 1974), large-scale processing with reasonable yield is possible today.

Present in Cohn fraction IV-1, AT III can be purified from this fraction (Hoffman 1989). Fraction IV-1 is suspended and AT III is separated by means of heparin-affinity chromatography; the concentrate is pasteurized for 10 hours at 60°C, and a second affinity-chromatography step is used to separate the AT III from the denatured proteins.

The approach that usually results in highest yields involves adsorption of AT III directly from plasma or fraction I supernatant (Eketorp 1980). Heparin is first immobilized on a solid matrix and then placed in contact with plasma or fraction I supernatant either in batch or continuously by means of a column, thus adsorbing the AT III. The gel is subsequently washed to remove any residual plasma proteins. After elution with sodium chloride, the eluate can be further purified by precipitating the impurities with PEG and by ion-exchange chromatography. Diafiltration is necessary in order to obtain the appropriate ionic strength.

Because AT III is a stable protein, it can be pasteurized in the liquid state, making it a product generally regarded as safe. In spite of its stability, however, the product on the market today is freeze-dried in the final sterile container.

8.9 CURRENT EQUIPMENT AND TECHNOLOGIES USED IN PLASMA FRACTIONATION

Equipment used in plasma fractionation has in many instances been derived from food or dairy processing equipment. Reactor tanks used to precipitate proteins are usually custom built in order to accommodate batch size. They are usually made from a high-grade stainless steel such as 316L and designed to facilitate automated cleaning in place (CIP). The tank interior is typically polished mechanically to a finish of at least 150 grit, although some manufacturers use a higher mechanical finish, followed by electropolishing.

Fractionation facilities built in the 1980s are reminiscent of modern food and dairy plants with high emphasis on environmental controls such as air quality and area segregation by differential air pressure. The fractionation plant of the New York Blood Center is a good example of a facility utilizing Cohn-like fractionation techniques while employing modern process equipment (Figure 8–3).

Most modifications of the original Cohn procedure involve changes in the liquid-solid separation techniques used in order to obtain more complete separation at lower cost. In many facilities the tubular high-speed centrifuges with limited solid-holding capacity of 5–6 kg have been replaced by large sanitary centrifuges holding up to 35 kg, similar to those used in the dairy industry for cream separation. In other cases, centrifuges have been replaced altogether by filtration systems such as pressure-leaf filters or even conventional filter presses. Depending on the fraction being filtered, filter aids like diatomaceous earth and perlites are being used in concentrations up to 2%. One of the critical conditions for a successful liquid-solid separation is the formation of a precipitate consisting of large particles while minimizing the

FIGURE 8-3 The fractionation plant of the New York Blood Center. Courtesy of the New York Blood Center, Inc.

inclusion of mother liquor. Under the right conditions, precipitates with solid contents of up to 38% when separated in centrifuges have been achieved.

High-filtration flow rates and minimal product loss in the filter cakes are a result of optimized precipitation techniques by controlled reagent addition as described by F. Rothstein (in press).

The development of membrane technology for ultrafiltration has dramatically altered many of the conventional process steps in plasma fractionation. Bulk albumin fractions harvested from centrifuges or filters contain up to 60% ethanol and large amounts of salt that were originally removed by a cumbersome reprecipitation followed by freeze-drying or vacuum distillation. Solubilization, followed by diafiltration against pyrogen-free water utilizing an ultrafiltration system with membranes capable of excluding large molecules while passing ethanol and salt molecules, such as described by Vandersande et al. (1982), offers an economically attractive alternative. Likewise there are many applications utilizing ultrafiltration where prior art involved reprecipitation followed by reconstitution. In particular during chromatography, where adjustments to ionic strengths are required in order to achieve adsorption, ultrafiltration processes have become commonplace.

FIGURE 8-4 The immunoaffinity purification plant of Baxter's Hyland division. Courtesy of Baxter Healthcare Corp.

Industrial chromatography columns and associated equipment, having originated in the laboratory, are often mere scale-up versions of lab-size equipment. All too often very large glass columns and bubble traps that are exact replicas of the original lab unit are used. The problem with this is that the original lab equipment was not designed with sanitary considerations in mind, causing problems such as cleanability that could lead to the presence of endotoxin-producing microorganisms, making the products ultimately unsuitable for human use. Today's manufacturer using column chromatography should look for sanitary stainless steel columns and bubble traps, where possible; dedicated buffer tanks that can be cleaned and steamed in place; and a totally enclosed system of fixed sanitary tubing to avoid potential contamination. The immunoaffinity purification plant of Baxter's Hyland division (Glendale, CA), which produces monoclonal-purified factor VIII is an example of a modern manufacturing facility using suitable chromatographic equipment (Figure 8-4).

REFERENCES

Andary, T., Berkebile, L.R., Thomas, W.R., et al. (1981) U.S. patent no. 4,286,056.

Bagot d'Arc, M. (1989) in *Biotechnologie des Proteines du Plasma* (Stoltz, J.F., and Rivat, C., eds.), pp. 101–106, Inserm, Paris.

Barandun, S., Kistler, P., Jeunet, F., et al. (1962) *Vox Sang.* 7, 157–174.

Bees, W. (1989) in *Biotechnologie des Proteines du Plasma* (Stoltz, J.F., and Rivat, C., eds.), pp. 207–215, Inserm, Paris.

Bloom, J.W. (1984) U.S. patent no. 4,478,825.

Brummelhuis, H. (1980) in *Methods of Plasma Protein Fractionation* (Curling, J.M., ed.), pp. 117–128, Academic Press, London.

Cohn, E.J., Strong, L.E., Hughes, W.L., et al. (1946) *J. Am. Chem. Soc.* 68, 459–475.

Curling, J.M. (1980) in *Methods of Plasma Protein Fractionation* (Curling, J.M. ed.), pp. 77–91, Academic Press, London.

Eibl, J., Schwarz, O., and Elsinger, F. (1979) U.S. patent no. 4,160,025.

Eketorp, R. (1980) in *Methods of Plasma Protein Fractionation* (Curling, J.M., ed.), pp. 175–188, Academic Press, London.

Fekete, L.F., and Shanbrom, E. (1971) U.S. patent no. 3,560,475.

Fukui, H., Fujimura, Y., Sugimoto, M., et al. (1985) *Acta Haematologica Japonica* 48, 134–146.

Griffith, M.J. (1988) in *Proceedings of the Symposium on Biotechnology and the Promise of Pure FVIII* (Roberts, H.H., ed.), pp. 69–85, Baxter Healthcare Publications, Brussels, Belgium.

Hink, J.H., Hidalgo, J., Seeberg, V.P., et al. (1957) *Vox. Sang.* 2, 174–186.

Hoffman, D.L. (1989) *Am. J. Medicine* 87 (suppl. 3B), 23–26.

Kistler, P., and Nitschmann, H. (1962) *Vox. Sang.* 7, 414–424.

Lundblad, J.L., Mitra, G., Sternberg, M.M., et al. (1986) *Rev. Infectious Diseases* 8 (suppl. 4), 382–390.

Matsuda, M., Iwanaga, S., and Nakamura, S. (1972) *Thromb. Res.* 1, 619–630.

Miller-Anderson, M., Borg, H., and Anderson, L.-O. (1974) *Thromb. Res.* 5, 439–452.

Ng, P.K., Eguizabal, H.C., and Mitra, G. (1986) *Thromb. Res.* 42, 825–834.

Oncley, J.L., Melin, M., Richert, D.A., et al. (1949) *J. Am. Chem. Soc.* 71, 541–550.

Rothstein, F. (in press) *Protein and Peptide Purification: Process Development and Scale up* (Harrison, R., Jr., ed.), Marcel Dekker, New York.

Vandersande, J., Stryker, M., and Woods, K.R. (1982) *Pharmaceutical Technology* 6, (Nov.) 78–98.

Recombinant Antihemophilic Factors

Kotoku Kurachi

Extensive studies on blood coagulation and its regulatory systems in the past two decades have resulted in the establishment of the basic mechanisms for these systems (Hedner and Davie 1989). The blood coagulation system is composed of two pathways, the intrinsic and extrinsic (Figure 9-1). The complex, multistep, cascade reactions shown are essential for rapid and sufficient fibrin clot formation (by the amplification of reactions), and they are well controlled (by the use of multiple sites for finely tuned regulation). Several protease inhibitors such as antithrombin III are involved in the regulation process (Rosenberg 1989; Bauer and Rosenberg 1989).

Another important regulatory system is the anticoagulant pathway involving protein C, thrombin, and thrombomodulin (Figure 9-2) (Esmon 1987). The blood coagulation cascade and these regulatory systems, therefore, generate just the right amount of fibrin clot in normal conditions. More than 20 factors are known to be involved in these systems. A number of these factors also interact with or are involved in many other physiological

The author thanks Dr. H. Roberts for sharing some of his clinical data. This work is supported in part by research grants from the U.S. National Institute of Health (HL38644) and the Hemophilia Foundation of Michigan.

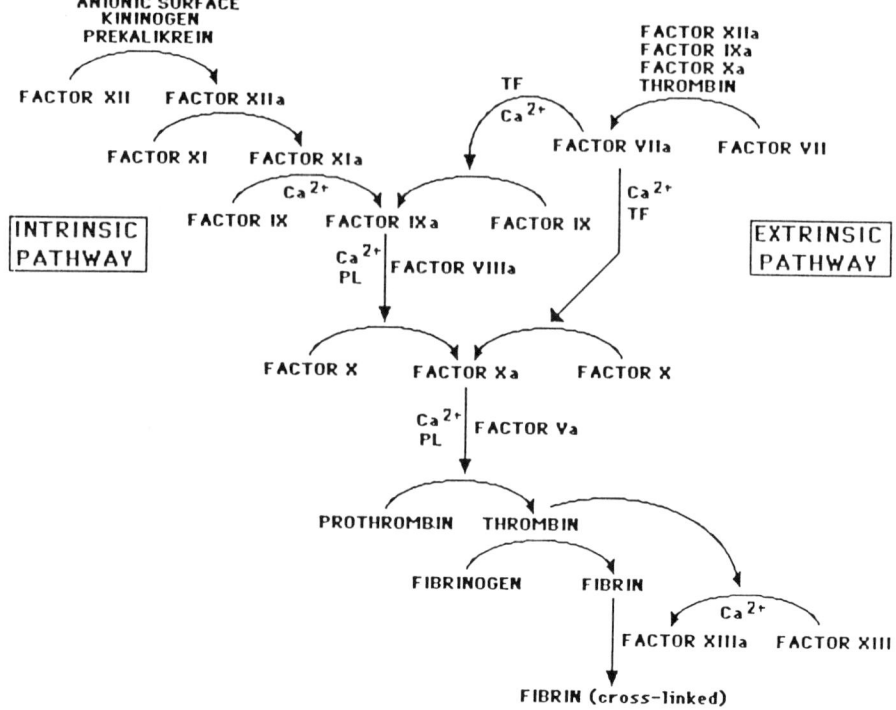

FIGURE 9-1 Diagram of the basic mechanism of blood coagulation. PL, phospholipid; TF, tissue factor.

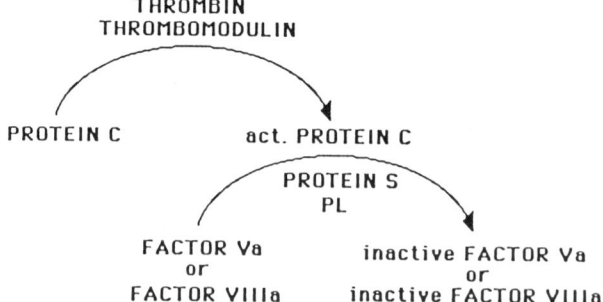

FIGURE 9-2 Diagram of the protein C anticoagulant pathway. PL, phospholipid. Data from Hedner and Davie (1989).

and pathological reactions, such as the complement system and the inflammatory process.

A deficiency in any of these factors results in various blood coagulation disorders, from abnormal bleeding to hypercoagulopathies. The two common abnormal bleeding disorders, hemophilia A (classic hemophilia) and hemophilia B, are due to hereditary disorders of factor VIII or factor IX genes, which are located on the X chromosome. Severe symptoms include frequent joint hemorrhages. A common autosomal bleeding disorder, von Willebrand's disease (vWD), is caused by a deficiency of von Willebrand factor (vWF). Disorders in factors involved in the regulatory systems also result in various coagulopathies.

Bleeding disorders such as hemophilia, as well as coagulopathies due to disorders in regulatory protein factors such as antithrombin III, have been treated primarily by replacement therapies (Luban et al. 1989; Roberts et al. 1989). In these therapies, the blood components, partially purified or as a crude mixture prepared from normal pooled human plasma, are employed. The conventional therapy with crude factor VIII and factor IX concentrates prepared from human plasma has been shown to have several serious side effects and complications. These include possible contamination by hepatitis virus (hepatitis B and non-A, non-B viruses) and HIV-1, alloantibody production, and allergic reactions. Elevated thrombogenic activity was also observed for factor IX concentrates that also contain other vitamin K-dependent proteins. Efforts to employ various processes such as heating and monoclonal antibody-affinity column in the past decade have resulted in highly purified and much safer preparation of these factors free of most of the pathogenic contaminants. These efforts have drastically improved the safety of these preparations. However, possible sporadic contamination by pathogenic viruses still remains a serious risk. Furthermore, difficulties in producing large amounts of these plasma proteins required for replacement therapy are significant problems which need to be solved.

In the past 10 years, we have witnessed the explosive development in molecular genetics and biology. Biotechnology emerging from these disciplines has greatly impacted biology in general as well as medicine. Genetic engineering technology has been intensively applied in the areas of thrombosis and hemostasis (Vehar 1987). To date, all human protein factors (about 20) known to be involved in blood coagulation and its regulation have been cloned for their complementary DNA (cDNA) as well as their genes. The availability of these cDNAs and of genomic DNAs has opened up exciting new avenues for fast and accurate diagnoses as well as replacement therapy for the diseases due to disorders of protein factors involved. The availability of the genetic materials has also opened up the possibility of gene transfer therapies for these congenital disorders.

Recombinant technology promises several major benefits for the treatment of hemophiliac patients, particularly those who require chronic treatment (Luban et al. 1989; Roberts et al. 1989). These benefits include: 1) the possibility of supplying pure and safe proteins in quantity; 2) little or no contamination with pathogenic materials such as hepatitis viruses and human immunodeficiency viruses (HIV-1 and -2); and 3) the possibility of creating modified coagulation factors that may have longer half-lives with improved stability as well as higher biological activity. Cloning of cDNAs and the genes of these protein factors has also facilitated the study of the genetic basis of the diseases at the molecular level. Furthermore, the establishment of an expression system for these proteins has allowed the detailed study of the biochemical properties of these factors, which would otherwise be extremely difficult for some factors (such as factor VIII). The results of these studies will, in turn, allow the development of much better therapies and of more innovative approaches for treating these disorders.

The discussion in this chapter focuses on a few selected factors, especially factors VIII, IX, and VII, the factors which have drawn the most extensive effort so far. Some other factors, including factor XIII, antithrombin III, von Willebrand factor, and protein C, have also been major targets for the application of recombinant technology to be produced in quantity for pharmaceutical purposes. These factors will be briefly discussed at the end of the chapter.

9.1   FACTOR VIII

Factor VIII is a large plasma protein that functions as a cofactor in the activation of factor X by factor IXa in the presence of calcium ions and phospholipids (tenase complex) (Foster and Zimmerman 1989; Vehar et al. 1989) (Figure 9–1). A deficiency of factor VIII results in a common bleeding disorder, hemophilia A. Its prevalence is about one in 10,000, and it is mostly manifested in males due to its gene location on the X chromosome. The isolation of cDNA clones for human factor VIII has allowed the determination of the complete amino acid sequence of this large trace plasma protein (Toole et al. 1986; Gitschier et al. 1984; Vehar et al. 1984). This has opened up an exciting, and otherwise extremely difficult, avenue to study its biochemistry, including its structure–function relationship and precise mechanism of its proteolytic activation (Eaton et al. 1986). Furthermore, the isolation and characterization of its gene (Wood et al. 1984) have facilitated the detailed study of the genetic mechanisms underlying hemophilia A at the molecular level. Factor VIII is synthesized as a single polypeptide chain precursor with a signal peptide of 19 amino acid residues. As shown in Figure 9–3, the mature form of factor VIII (2,332 amino acid residues) is composed of several homologous domain structures (NH_2–A_1 A_2 B A_3 C_1 C_2–COOH) with a heavily glycosylated B domain in the center and two homologous small acidic regions between domains A_1 and A_2, and between B and A_3. Factor VIII is homologous to factor V and to ceruloplasmin (Church et al. 1984) (Figure 9–3). Complementary DNA for factor V, an-

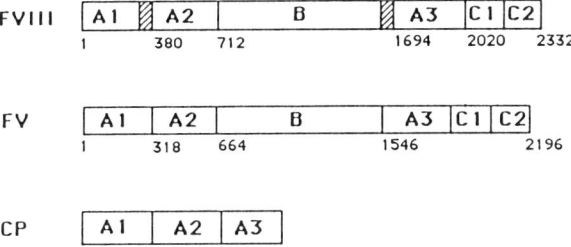

FIGURE 9-3 Schematic drawing of the domain structures of factor VIII (FVIII), factor V (FV), and ceruloplasmin (CP). The shaded regions of FVIII indicate acidic regions; the numbers indicate the amino acid positions where the various domains begin. Data from Vehar et al. (1989).

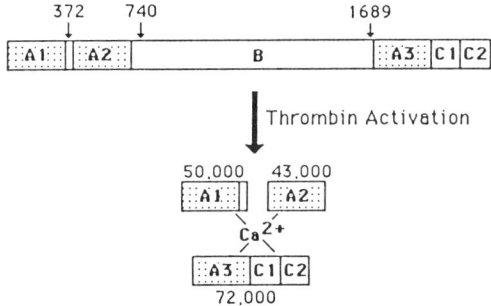

FIGURE 9-4 Schematic drawing of the activation of factor VIII by limited proteolysis with thrombin. (Top) The amino acid positions in factor VIII that are cleaved by thrombin are shown by small arrows. (Bottom) The numbers above and below the three domains in factor VIIIa indicate their sizes in daltons. Data from Vehar et al. (1989).

other large cofactor protein in the coagulation cascade, has been cloned and characterized (Jenny et al. 1987). Upon secretion from the cell, the single polypeptide chain of factor VIII is converted to the two-chain form (carboxy-terminal-derived 80 kDa and amino-terminal-derived 200 kDa) which is the predominant form in the plasma. A further limited proteolysis of the 200-kDa fragment yields activated factor VIII composed of a 50-kDa and 43-kDa amino-terminal-derived fragment, which are associated with the 72-kDa carboxy-terminal-derived fragment in the presence of calcium ions (Eaton et al. 1986; Foster and Zimmerman 1989) (Figure 9-4). During the proteolytic activation process in which thrombin plays a major role, the central, one-third portion (B domain) of the molecule containing most of the carbohydrate chains is freed. Further proteolysis of activated factor VIII by factor Xa, activated protein C, or by thrombin results in the inactivation of factor VIII (Foster and Zimmerman 1989). Kaufman et al. (1988) have shown that the newly synthesized single-chain factor VIII is processed to a two-chain form in the Golgi apparatus. Less than 10% of the total secreted factor VIII is in the intact single-chain form.

The highly purified and biologically active plasma factor VIII needed for clinical use is extremely difficult to obtain in quantity. Cryoprecipitates of plasma used for treatment in the past were heavily contaminated by other proteins, which also cause various side effects. Serious efforts to minimize known pathogenic viruses by means of heating and other methods have apparently succeeded to provide much safer factor VIII concentrates for clinical use (Heldebrant et al. 1985; Piszkiewicz et al. 1985; Levy et al. 1985; Luban et al. 1989; Roberts et al. 1989). Human plasma-derived products, however, are still vulnerable to sporadic contaminations by the known viruses as well as by unexpected viruses, which cannot be inactivated by procedures effective for the known viruses (White et al. 1989; van den Berg et al. 1986).

Recent successful production of recombinant factor VIII has opened up the exciting possibility of providing large quantities of highly pure and safe factor VIII for use in replacement therapy. Production of biologically active recombinant factor VIII has been accomplished by employing expression vectors composed of human factor VIII cDNA (about 9 kb) linked to a strong heterologous promoter, such as adenovirus-2-major late promoter with an SV40 enhancer, in baby hamster kidney (BHK) and chinese hamster ovary (CHO) cells (Wood et al. 1984; Kaufman et al. 1988). The expression level of factor VIII in these cultured cells was low in early attempts. Since then, however, several important factors that significantly effect the level of factor VIII production in cultured mammalian cells (CHO cells) have been identified (Kaufman 1989). These factors include: 1) the important role of vWF in promoting the association of the heavy and light chain of factor VIII upon secretion from the cell and in protecting factor VIII from proteolytic degradation (Kaufman et al. 1988, 1989); 2) poor secretion of newly synthesized factor VIII due to the association with a 78-kDa resident protein (glucose-regulated, immunoglobin-binding protein, GRP78) in endoplasmic reticulum lumen (Dorner et al. 1987, 1988); and 3) a low level of factor VIII mRNA available for translation (Dorner et al. 1989). The coexpression of recombinant factor VIII and vWF in the same CHO cells improved the yield of factor VIII 10- to 20-fold (Kaufman et al. 1989). Dorner et al. (1989) demonstrated that the reduction of GRP78 production by its antisense RNA improved the secretion of proteins which bind to GRP78. The use of inducible promoter to express factor VIII was also shown to be effective in improving factor VIII production level (Dorner et al. 1989). When sodium butyrate was used to enhance the expression, a 19-fold increase of factor VIII mRNA in the cells coexpressing both factor VIII and vWF was observed. These results clearly indicate that recombinant factor VIII production can be significantly improved by manipulating the expression system by means of coexpression, modification of the cells' secretory pathway as well as improved expression vectors.

An exciting aspect of recombinant technology is that it offers a versatile means for modifying the structure of the protein of interest. Various modifications of factor VIII may further improve the production level with an improved half-life as well as the resistance to proteases. For instance, the deletion of the nonessential central portion (B domain) of the molecule has increased the production of biologically active factor VIII up to 10-fold (Toole et al. 1986; Sarver et al. 1987). These modifications coupled with the improvements in expression system described above should lead to the development of an optimized production system for recombinant factor VIII in quantity.

To date, recombinant factor VIII is in advanced clinical trials. The recombinant factor VIII currently used in the clinical trials is of the intact size. It is virtually indistinguishable from the plasma-derived factor VIII in its chemical properties including specific activity (4,000–5,000 U/mg), carbohydrates attached, and $T_{1/2}$ (β-phase of about 13 hours) (White et al. 1989; Macik et al. 1989). No apparent adverse effects were observed in 54 patients treated with the recombinant factor VIII in the advanced clinical trials. Furthermore, no detectable increase in inhibitor activity (alloantibody development) was observed with the recombinant factor VIII preparation employed. These clinical data strongly indicate that all hemophilia A patients will best be treated with recombinant factor VIII preparation in the future, when recombinant technology will be advanced enough so that ample and safe factor VIII can be provided at a reasonable cost.

9.2   FACTOR IX

Factor IX is involved in the middle phase of the blood coagulation cascade (Figure 9–1). A deficiency of factor IX causes an abnormal bleeding disorder, hemophilia B, at a prevalence of about one in 30,000. Factor IX is synthesized as a single, prepro polypeptide chain that undergoes limited proteolysis as well as several post-translational modifications, including γ-carboxylation of the 12 Glu residues in the N-terminal region, β-hydroxylation of Asp 64, carbohydrate chain attachment (both N and O linked), and disulfide bond formation (Hedner and Davie 1989). The mature human factor IX is 415 amino acids in length (Yoshitake et al. 1985). The proleader sequence of 18 amino acid residues in length plays an important role in γ-carboxylation, a post-translational modification as described below. The amino acid sequence of prepro factor IX is shown in Figure 9–5. Upon proteolytic cleavage by factor XIa or factor VIIa, factor IX is converted to the activated form, factor IXa, composed of two chains held together with a disulfide bond. Factor IX is one of the six proteins in the coagulation system which depend on vitamin K for a unique post-translation modification (Hedner and David

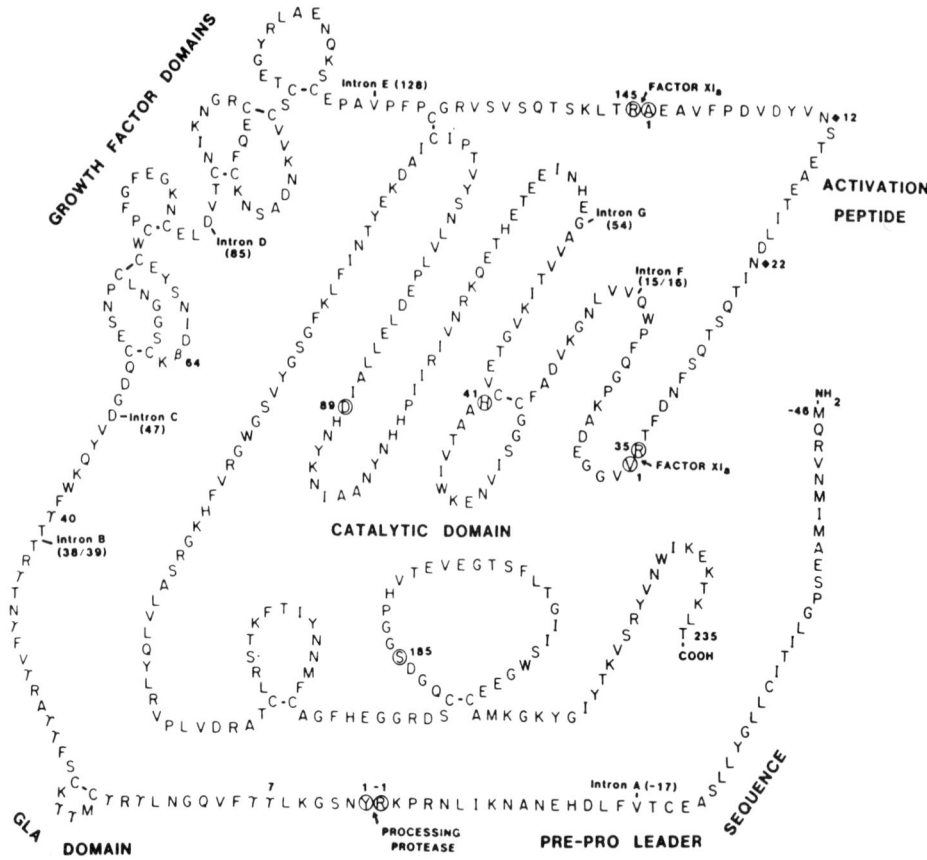

FIGURE 9-5 Amino acid sequence of human factor IX. The intron positions are shown in relation to the amino acid sequence. The sites of proteolytic activation by factor XIa are shown by arrows. β64 is β-hydroxylated aspartic acid. Reproduced with permission from Yoshitake et al. (1985).

1989; Kurachi and Chen 1990). The newly synthesized polypeptide chain of factor IX is γ-carboxylated on the side chain of all 12 glutamic acid residues contained within the amino terminal portion of about 40 amino acid residues (Figure 9-5). Acquiring one more carboxyl group attached to the γ-carbon of the side chain, the glutamic acid residues are changed to γ-carboxylated glutamic acid residues (Gla). This post-translational modification is essential for the biological activities of factor IX. The γ-carboxylated region (Gla region) functions to bind calcium ions. Upon binding calcium ions, the Gla region goes through some significant conformational changes and also serves as the site of binding to phospholipid surfaces provided by activated platelets at vascular lesions. The various post-translational modifications as well as the problems they cause during the production of large quantities of recombinant factor IX are discussed in another chapter in this volume.

Five other vitamin K-dependent proteins (factor VII, factor X, prothrombin, protein C, and protein S) are grossly or partially homologous to factor IX, indicating that they have a common ancestral gene (Kurachi and Chen 1987). Exon-shuffling events apparently played a major role in creating these modern genes from the ancestral genes. Splicing phases of each exon in these genes are conserved, confirming its role in the evolutionary process.

Factor IX gene is located on the X chromosome at q27 distal to hypoxanthine guanine phosphoribosyl transferase and proximal to the long-arm fragile site (Chane et al. 1985; Purrello et al. 1985). Almost all deficiencies in factor IX, therefore, are seen in males; they are rarely seen in females. Human factor IX cDNAs as well as genomic DNA have been isolated and characterized, including the complete nucleotide sequence (about 40 kb in length) (Kurachi and Davie 1982; Yoshitake et al. 1985).

To date, more than 200 abnormal factor IX genes have been studied at the molecular level (Kurachi and Chen 1990). The molecular mechanisms of the factor IX disorders are heterogenous and include single base replacements, deletions, and insertions. A factor IX concentration of less than 1% of the normal level in plasma causes a severe form of abnormal bleeding. This condition can be changed to the mildly affected form if the factor IX concentration level is increased and maintained at 10% or higher.

Factor IX concentrates which have been used for replacement therapy contain various contaminants including other partially activated vitamin K-dependent proteins. These contaminants can cause pathological thrombogenic activity (Luban et al. 1989). Plasma factor IX that has been affinity-purified on monoclonal antibodies is available; however, it is costly to obtain in quantity. Furthermore, these preparations would possibly be contaminated with pathological viruses, despite the various treatments used to eliminate them.

Recombinant factor IX has been produced by mammalian cells such as CHO or BHK cells transfected with expression vectors containing human factor IX cDNA linked to a strong heterologous promoter such as adenovirus-2-major later promoter with an enhancer (SV40). The production of biologically active recombinant factor IX, however, has been hampered by the low efficiency of γ-carboxylation in the cultured cells. The expression levels of factor IX in the early phase of expression experiments were at low μg or sub μg per ml, and only 50–70% of the factor IX polypeptide chains produced was appropriately γ-carboxylated (Busby et al. 1985; Anson et al. 1985; de la Salle et al. 1985). When the expression level was improved to >100 μg/ml, the fraction γ-carboxylated was only 1.5% of the total factor IX molecules produced (Kaufman et al. 1986). Furthermore, the γ-carboxylated fraction of this preparation was found to contain only 6–7 Gla residues per factor IX molecule compared to 12 such residues in the plasma-derived factor IX. This strongly suggests the limited capacity of γ-carboxylation of the expression system employed. Furthermore, γ-carboxylation appears to be a complex reaction which seems to be best optimized when its own propeptide-Gla region sequence was used as a substrate (Berkner et al. 1986). All experimental results reported so far suggest that recombinant factor IX molecules produced by cultured mammalian cells are significantly under-γ-carboxylated. Interestingly, the extent of γ-carboxylation of factor IX seems significantly low while that of factor VII is virtually 100%. One possible reason for this was suggested to be the tyrosine residue at the +1 position in factor IX, where all the other vitamin K-dependent proteins including factor VII have alanine (Berkner et al. 1986). Interestingly, recent mutagenesis experiments indicate that the alanine residue at +1 is apparently favored over the tyrosine residue for more efficient cleavage of the propeptide, and it is indifferent from the tyrosine residue in effecting γ-carboxylation of factor IX (Meulien et al. 1990). Furthermore, replacement of the proline residue at −3 with the valine residue does not effect both propeptide cleavage and γ-carboxylation. Abnormal genes with mutations at −1 and −4, however, are known to effect both propeptide cleavage and γ-carboxylation (Kurachi and Chen 1987, 1990). In this regard, it is interesting to note that the proleader sequence and the Gla region are encoded by one exon. Some fraction of the recombinant factor IX produced by CHO cells was also found as a proform, which indicated that the cleavage between the proleader and mature protein did not take place during secretion (Balland et al. 1988). When the factor IX proleader sequence and Gla region sequence were used to replace the corresponding region of factor VII, only a small fraction (24%) of the factor VII polypeptide produced was found to be active (Berkner et al. 1986). More than 80% of the factor VII polypeptides, however, were precipitated with barium citrate treatment, indicating that these molecules were significantly γ-carboxylated. Some specific Gla residues may be essential for maintaining the biological activity. No information, however, is available as to how many and which specific Glu residues were carboxylated in the factor IX preparation.

These experiments clearly show that the biologically active factor VII can be made with a chimeric vector which contains the factor IX proleader and Gla region sequence. This indicates that the machinery involved in γ-carboxylation is common for various target proteins. These data also suggest that γ-carboxylation requires a set of complex events that may involve the rest of the substrate molecule for maximum efficiency.

Other unique post-translational modifications of factor IX molecule include β-hydroxylation of Asp 64 in the first epidermal growth factor-like domain (Hedner and Davie 1989). The function of this unique amino acid residue is not clear. This residue apparently does not have any significant role in binding to the cell membrane (Derian et al. 1989). Recombinant factor IX produced by various mammalian cells apparently has carbohydrate chains very similar to the plasma factor IX. At the present time, the problems associated with the γ-carboxylation of the amino terminal portion of the mature factor IX appears to be the only major obstacle in producing fully active factor IX in quantity.

No recombinant factor IX sufficient for clinical trials has been produced to date. However, we can expect that the sequence element required for modification will be optimized by using recombinant techniques. Furthermore, the carboxylase and the other components which are involved in γ-carboxylation will be cloned and employed to optimize the modification reaction by coexpression in mammalian cells employed for expressing factor IX.

9.3 FACTOR VII

Factor VII is one of the vitamin K-dependent proteins and has a molecular weight of about 50,000 Da (Hedner and Davie 1989; Kurachi and Chen 1987). Factor VII is involved in the extrinsic pathway of the blood coagulation system (Figure 9–1), and is highly homologous to factor IX (Figures 9–5 and 9–6). Factor VII is activated to factor VIIa by other coagulation factors including factor Xa, XIa, XIIa, and thrombin. Factor VIIa, in turn, proteoloytically activates factor X and factor IX in the presence of tissue factor and calcium ions.

A deficiency of factor VII in human population is very rare. However, its involvement in the extrinsic pathway allows an exciting application of this factor in the treatment of hemophilia by bypassing the steps involving factor VIII and factor IX (Figure 9–1). Namely, factor VIIa may be able to correct abnormal bleeding in hemophilia A or B patients who are deficient in factor VIII or factor IX. In an original clinical trial, the administration of factor VII to patients affected with hemophilia A with antibodies was shown to be effective in stopping the excessive bleeding (Hedner and Kisiel 1983). This application is particularly important for the patients who have developed complications such as antibody against factor VIII or factor IX. The human factor IX cDNA and gene have been isolated and characterized (Hagen et al. 1986; O'Hara et al. 1987). Its gene (12.8 kb in length) is located on chromosome 13 at q34, very close to the factor X locus (de Grouchy et al. 1984; Gilgenkrantz et al. 1986).

Factor VII has been a target of extensive work for production by recombinant technology. Recombinant factor VII was first obtained by employing pDX expression vectors which contain the adenovirus major late promoter linked to the factor VII cDNA and tripartite leader sequence (Berkner et al. 1986). Recombinant factor VII is now produced by Novo Industries (Bagsvaerd, Denmark) and used in clinical trials.

Recombinant factor VII produced by BHK cells is fully γ-carboxylated except for a somewhat lower modification on residue 35, and has essentially the same specific activity as the plasma factor VII preparations (Berkner et

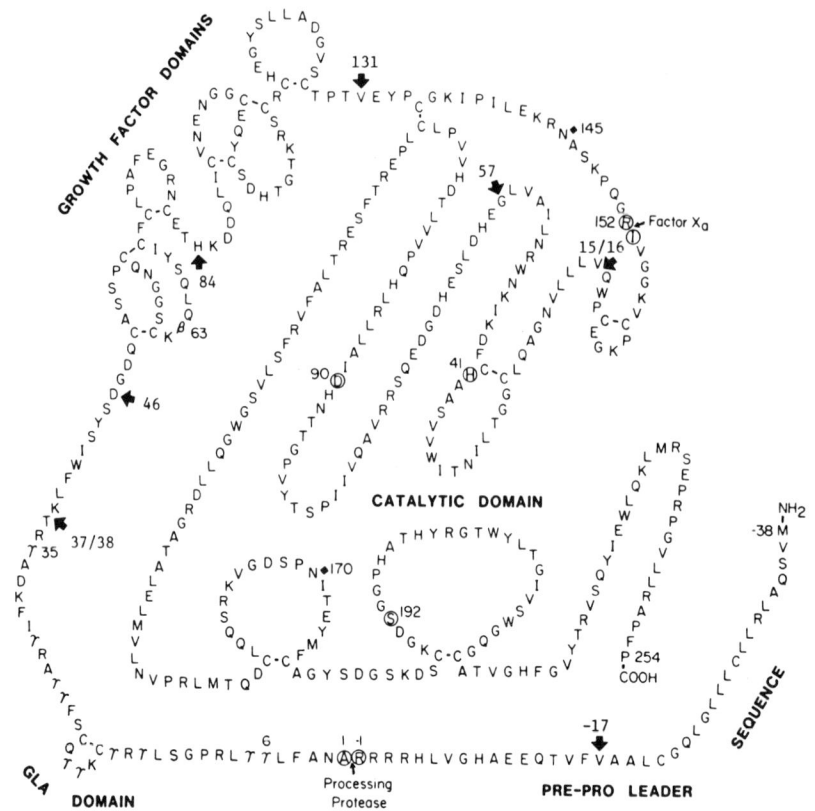

FIGURE 9–6 Amino acid sequence of human factor VII. The intron positions are shown in relation to the amino acid sequence. The sites of proteolytic activation by factor XIa shown by arrows. β64 is β-hydroxylated aspartic acid. Reproduced with permission from Hedner and Davie (1989).

al. 1986; Thim et al. 1988). No other significant differences from the plasma factor VII were found for recombinant factor VII, except for the lower carbohydration at Asn 145 (66% of that found for plasma factor VII). These results strongly indicate that recombinant factor VII produced by heterologous mammalian cells, such as BHK cells, can be used for replacement therapy. Successful uses of recombinant factor VIIa to control bleeding in patients with inhibitor (antibodies) against factor VIII have been reported (Hedner et al. 1988; Macik et al. 1989). For these patients, further supplementation of fresh factor VIII does not improve the condition because of the immediate inhibition of factor VIII by its alloantibodies. In one case (Macik et al. 1989), recombinant human factor VIIa was given to a hemophilia A patient with serious complications including high-titer inhibitor to human factor VIII, allergic reactions to porcine factor VIII aggravated by rapid progression of hematomas and poor venous access (Macik et al. 1989). The administration of recombinant factor VIIa, which consists of nine doses of 60 mg/kg body weight every three to four hours for the first seven doses and every six hours thereafter, followed by regular autoplex therapy, has dramatically improved his condition. The patient remained well for six months after receiving the factor VIIa. In this treatment, no increased turnover of factor VIIa nor thrombi formation were observed. Recombinant factor VIIa was also found to be effective in stopping hemorrhage in both factor VIII- and factor VII-deficient dogs (Brinkhous et al. 1989). In the dog experiment, the drastically increased factor VIIa concentration (7- to 30-fold) also did not result in any increased turnover or thrombosis, indicating that the use of recombinant factor VIIa is an excellent alternative therapy for these conditions. In this experiment, the hemostatic defect in vWD was not corrected, indicating that forced thrombin generation with recombinant factor VIIa does not compensate for the lack of vWF and vWF platelet function. This agrees well with the expected function of vWF in the hemostatic plug formation at the vascular site of injury. Chronic use of recombinant factor VIIa, however, may result in the development of inhibitors (antibodies) in these patients. We need more experience with recombinant factor VIIa in order to test this potential problem.

Use of recombinant factor VIIa also provided valuable insights into the blood coagulation mechanism (Telgt et al. 1989). These data indicate that factor IX activation by factor VIIa is not necessary (or may not play any major role) in order for factor VIIa to exert its hemostatic effect. Factor VIIa requires tissue factors for its activity. This suggests that the administered recombinant factor VIIa may effectively concentrate at the injury site where the tissue factors are exposed. This mechanism of factor VIIa localization works advantageously in replacement therapy. In this regard, however, factor VIIa should not be used in conditions such as massive trauma or sepsis in which a large amount of tissue factors are exposed, until more experience is obtained in its application. The application of factor VIIa for patients with factor VIII or factor IX with inhibitors is one of the clever approaches for solving the complicated problems which would otherwise be very difficult to overcome.

Factor VII has a relatively short half-life (2 to 3 hours). In the future, protein engineering employing recombinant technologies may be able to modify and create a new form of recombinant factor VIIa that has a longer half-life while retaining the same or higher specific activity.

9.4 OTHER RECOMBINANT COAGULATION FACTORS

Besides recombinant factor VIII, factor IX, and factor VII, recombinant proteins of other coagulation and regulatory proteins have been the targets of significant production efforts. These include antithrombin III, protein C, factor XIIIa, as well as von Willebrand factor (vWF).

9.4.1 Antithrombin III

Antithrombin III (ATIII) is a plasma glycoprotein of about 60 kDa (Tollefsen 1989). It functions as a major serine protease inhibitor of the coagulation system by forming a 1:1 complex with most of the serine protease-type coagulation factors including thrombin, factor Xa, factor IXa, factor XIa, and factor XIIa. It, however, does not inhibit factor VIIa and activated protein C. The inhibitory activity of ATIII is augmented several hundredfold by the addition of heparin. ATIII is apparently essential for the control of the activated coagulation factors. Reduced ATIII concentration in the blood circulation causes thrombotic disease at a relatively young age (with a frequency of about 1 in 5,000). It is manifested as an autosomal dominant trait. ATIII concentration levels in patients are no lower than about 50% of the normal control, indicating the importance of this protein in the regulation of blood coagulation. Human cDNAs (about 1.5 kb) and the gene (19 kb) for ATIII have been cloned and characterized (Bock et al. 1982; Chandra et al. 1983; Prochownik et al. 1983, 1985). Its gene is on chromosome 1 at q23–25.

Highly pure, pathogenic contaminant-free preparations of the recombinant ATIII (rATIII) will be particularly useful for those patients with conditions such as DIC (disseminated intravascular coagulation) and deep vein thrombosis. In DIC, for instance, intrinsic ATIII is largely consumed. In this condition the administration of heparin, which normally functions as an anticoagulant by augmenting the ATIII activity, cannot work to control the extensive coagulation in progress unless fresh ATIII is supplemented in replacement therapy.

The expression of rATIII, aiming at its production in quantity, has begun by employing various systems including mammalian cells (COS monkey cells and CHO cells) and yeasts (*Saccharomyces cerevisiae* and *Schizosaccharomyces pombe*) (Stephens et al. 1987; Wasley et al. 1987; Broker et al. 1987). Only a fraction (about 10%) of the rATIII produced, however, was found to be as active as the plasma ATIII. The inactive form of rATIII did not bind to heparin. Slight conformational changes in rATIII are probably responsible for the loss of heparin-binding capacity. More work is required before rATIII can be mass produced.

9.4.2 Protein C

Protein C is a plasma glycoprotein of 56 kDa. It plays a pivotal role in the protein C anticoagulant pathway which involves thrombomodulin (protein C receptor on endothelial cells), protein S (protein C cofactor), and thrombin (Esmon 1987) (see Figure 9-2). The importance of this regulatory pathway of the coagulation cascade has been established by the findings of both protein C and protein S deficiencies in man (Griffin et al. 1981; Bertina et al. 1982; Barbui et al. 1984; Marlar 1985; Brockmans and Conrad 1988). Symptoms of protein C or protein S deficiencies are similar to those of

ATIII deficiency and are inherited as an autosomal dominant trait. Protein C is activated by thrombin bound to its receptor, thrombomodulin, on endothelial cells. Activated protein C than proteolytically inactivates factor VIII and factor V. This reaction is further augmented by protein S, another vitamin K protein. Protein C, therefore, functions antithrombotically. Human cDNAs and the genes for protein C, protein S, and thrombomodulin have been isolated and characterized (Foster and Davie 1984; Beckmann et al. 1985; Foster et al. 1985; Plutkzy et al. 1986). The protein C gene is on chromosome 2. The expression of recombinant activated human protein has recently been reported (Ehrlich et al. 1989).

9.4.3 von Willebrand Factor

von Willebrand factor (vWF) is a large, multimeric glycoprotein (Sadler 1989; Roberts et al. 1989). The human cDNA as well as the genomic DNA for vWF have been cloned and characterized. vWF is synthesized as a single large polypeptide chain composed of a signal peptide (22 amino acid residues), a large proleader sequence (741 amino acid residues), and the mature protein (2050 amino acid residues). The proleader sequence is reported to be involved in the multimerization process of vWF (Verveij et al. 1987). Mature vWF functions as a mediator for platelet adhesion to the subendothelium at the site of vascular lesions, binds factor VIII and functions as a stabilizer of factor VIII in blood circulation. A deficiency of vWF, therefore, shows symptoms similar to both platelet dysfunction as well as the factor VIII: C disorder (hemophilia A). If all degrees of severity are included, vWD appears to affect almost 1% of the population. Clinically significant vWD, however, has a frequency of approximately 1.25 per 10,000, similar to that of hemophilia A. vWD is a very heterogenous disorder that has been classified into three major classes (type I, type II, and type III) with several subtypes. The most common form (type I) is the simple quantitative deficiency of the vWF multimer, likely due to the lowered secretion, and is an autosomal dominant trait. Most type I disorders and a small fraction of type II disorders, can be pharmacologically treated with the vasopressive analog, 1-desamino-8-D-arginine vasopressin (DDAVP). However, some disorders of type I and type II (qualitative disorder with dysfunctional vWF), and all type III (total lack of vWF, recessive trait disorders) are not responsive to treatment with DDAVP, and they require placement therapy with plasma cryoprecipitates. The production of highly purified and safe recombinant vWF, therefore, would be a great benefit to vWD patients. As described above, production of rFVIII may be best achieved when vWF is coproduced from the same mammalian cells employed (Kaufman et al. 1989). Attempts to produce rvWF have begun and we can expect that rvWF will be produced in quantity for the safe treatment of vWD patients in the near future.

9.4.4 Factor XIII

Factor XIII is an enzyme that catalyzes the cross-linking of fibrin monomers to form stable fibrin fibers at the final stage of blood coagulation (Chung and Ichinose 1989). Plasma factor XIII is a heterotetrameric complex of a and b subunits (a_2b_2), while intracellular factor XIII is the homodimer form (a_2). FXIIIa also cross-links fibrin (the α chain) to α_2-plasmin inhibitor, and fibronectin to collagen resulting in the protection of the fibrin clot from premature degradation by plasmin. It also cross-links fibronectin to fibrin (α chain) and to collagen, indicating that factor XIIIa is involved in localizing the fibrin clot to the vascular site of the lesion. Factor XIII is also reported to augment fibroblast growth (Grinnel et al. 1980). These properties strongly indicate that factor XIII is also involved in wound healing (Lorand et al. 1980). The demand for factor XIII to treat factor XIII-deficient patients is low, however, because the factor XIII deficiency is extremely rare (1 in 5 million). The possible use of factor XIII, however, for the purpose of accelerating wound healing is substantial.

Attempts to produce recombinant factor XIII (a subunit) have begun by employing two different yeast (*Saccharomyces cervisiae* and *Schizosaccharomyces pombe*) expression systems (Broker and Bauml 1989; Bishop et al. 1990). Recombinant factor XIIIa produced in these systems was found to be very similar to placental factor XIII in specific activity, activation, and physiochemical properties.

9.5 CONCLUSION

Stable supplies of clean, recombinant, blood coagulation factors and regulatory proteins are essential to keep replacement therapies safe and effective. This is particularly important for hemophilia patients who often require chronic replacement therapies. To date, two recombinant coagulation factors, factor VIII and factor VII, are already being tested in clinical trials. For other factors, significant problems have to be solved before ample recombinant products are available for clinical trials. Progress in recombinant technology, however, is highly promising for soon overcoming these problems. One exciting aspect of the application of recombinant technology is that it also provides us with new approaches for studying the principles of biosynthesis and biochemistry of these proteins. The knowledge obtained in these studies will further augment the optimization of the production of recombinant proteins. The production of recombinant factor VIII is an example of this.

The best therapy for treating hemophilia patients may be the use of the gene transfer approach. Basic studies on somatic cell gene transfer for factor IX as well as factor VIII have already begun. However, it may take many years before this approach is optimized for safety and effectiveness, and ready to be applied to humans.

REFERENCES

Anson, D.S., Austen, D.E., and Brownlee, G.G. (1985) *Nature* 315, 683–685.

Balland, A., Faure, T., Carvallo, D., Cordier, P., and Ulrich, P. (1988) *Eur. J. Biochem.* 172, 565–572.

Barbui, T., Finazzi, G., Mussoni, K., et al. (1984) *Lancet* 2, 819.

Bauer, K.A., and Rosenberg, R.D. (1989) *Am. J. Med.* 87(3B), 39S–43S.

Beckmann, R.J., Schmidt, R.J., Santerre, R.F., et al. (1985) *Nuc. Acids Res.* 13, 5233–5247.

Berkner, K., Busby, S., Davie, E.W., Hart, C., and Insley, M. (1986) in *Cold Spring Harbor Symposium on Quantitative Biology,* vol. LI, pp. 531–541, Cold Spring Harbor Laboratories, New York.

Bertina, R.M., Brackmans, A.W., van der Linden, I.K., and Mertens, K. (1982) *Thromb. Haemost.* 48, 1–5.

Bishop, P.D., Teller, D.C., Smith, R.A., et al. (1990) *Biochem.* 29, 1861–1869.

Bock, S.C., Wion, K.L., Vehar, G.A., and Lawn, R.M. (1982) *Nuc. Acids Res.* 10, 8113–8125.

Brinkhous, K.M., Hedner, U., Garris, J.B., Diness, V., and Read, M.S. (1989) *Proc. Natl. Acad. Sci. USA* 86, 1382–1386.

Brockmans, A.W., and Conrad, J. (1988) in *Protein C and Related Proteins: Biochemical and Clinical Aspects* (Bertina, R.M., ed.), pp. 160–181, Churchill Livingstone, NY.

Broker, M., and Bauml, O. (1989) *FEBS Letter* 248, 105–110.

Broker, M., Ragg, H., and Karges, H.E. (1987) *Biochim. Biophys. Acta* 908, 203–213.

Busby, S., Kumar, A., Joseph, M., et al. (1985) *Nature* 316, 271–273.

Chance, P.F., Dyer, K.A., Kurachi, K., et al. (1985) *Hum. Genet.* 65, 207–208.

Chandra, T., Stackhouse, R., Kidd, V.J., and Woo, S.L.C. (1983) *Proc. Natl. Acad. Sci. USA* 80, 1845–1848.

Chung, P., and Ichinose, A. (1989) in *The Metabolic Basis of Inherited Diseases,* vol. II. (Scriver, C.R., Beaudet, A.L., Sly, W.S., and Valle, D., eds.), pp. 2135–2153, McGraw Hill, New York.

Church, W.R., Jernigan, R.L., Toole, J., Hewick, R.M., and Knopf, J. (1984) *Proc. Natl. Acad. Sci. USA* 81, 6934–6937.

de Grouchy, J., Dautzenberg, M.D., Turleau, C., Beguin, S., and Chavin-Colin, F. (1984) *Hum. Genet.* 66, 230–233.

de la Salle, H., Altenburger, W., Elkain, R., et al. (1985) *Nature* 316, 268–270.

Derian, C.K., VanDusen, W., Przykiecki, C.T., Walsh, P.N., and Berkner, K.L. (1989) *J. Biol. Chem.* 264, 6615–6618.

Dorner, A.J., Bole, D.G., and Kaufman, R.J. (1987) *J. Cell Biol.* 105, 2665–2674.

Dorner, A.J., Krane, M.G., and Kaufman, R.J. (1988) *Mol. Cell. Biol.* 8, 4063–4070.

Dorner, A.J., Wasley, L.C., and Kaufman, R.J. (1989) *J. Biol. Chem.* 264, 20602–20607.

Eaton, D., Rodriguez, H., and Vehar, G.A. (1986) *Biochem.* 25, 505–512.

Ehrlich, H.J., Jaskunas, S.R., Grinnell, B.W., Yan, S.B., and Bang, N.U. (1989) *J. Biol. Chem.* 264, 14298–14304.

Esmon, C.T. (1987) *Science* 235, 1348–1352.

Foster, D., and Davie, E.W. (1984) *Proc. Natl. Acad. Sci. USA* 81, 4766–4770.

Foster, D.C., Yoshitake, S., and Davie, E.W. (1985) *Proc. Natl. Acad. Sci. USA* 82, 4673–4677.

Foster, P.A., and Zimmerman, T.S. (1989) *Blood Rev.* 3, 180–191.

Gilgenkrantz, S., Briquel, M.E., Andre, E., et al. (1986) *Ann. Genet.* 29, 32–35.

Gitschier, J., Wood, W.I., Goralka, T.M., et al. (1984) *Nature* 312, 326–330.

Griffin, J.H., Evatt, B., Zimmerman, T.S., Kleiss, A.J., and Wideman, C. (1981) *J. Clin. Invest.* 68, 1370–1373.

Grinnel, F., Feld, M., and Minter, D. (1980) *Cell* 19, 517.

Hagen, F.S., Gray, C.L., O'Hara, P., et al. (1986) *Proc. Natl. Acad. Sci. USA* 83, 2412–2416.

Hedner, U., and Davie, E.W. (1989) in *The Metabolic Basis of Inherited Diseases*, vol. II, 6th ed. (Scriver, C.R., Beaudet, A.L., Sly, W.S., and Valle, D., eds.), pp. 2107–2134, McGraw Hill, New York.

Hedner, U., and Kisiel, W. (1983) *J. Clin. Invest.* 71, 1836–1841.

Hedner, U., Glazer, S., Pingel, K., et al. (1988) *Lancet* 2 (Nov. 19), 1193.

Heldebrant, C.M., Gomperts, C.K., McDougal, J.S., et al. (1985) *Transfusion* 25, 510–515.

Jenny, R.J., Pittman, D.D., Toole, J.J., et al. (1987) *Proc. Natl. Acad. Sci. USA* 84, 4846–4850.

Kaufman, R.J. (1989) *Nature* 342, 207–208.

Kaufman, R.J., Wasley, L.C., Furie, B.C., Furie, B., and Shoemaker, C.B. (1986) *J. Biol. Chem.* 261, 9622–9628.

Kaufman, R.J., Wasley, L.C., and Dorner, A.J. (1988) *J. Biol. Chem.* 263, 6352–6362.

Kaufman, R.J., Wasley, L.C., Davis, M.V., et al. (1989) *Mol. Cell. Biol.* 9, 1233–1242.

Kurachi, K., and Chen, S.H. (1987) in *The New Dimensions of Warfarin Prophylaxis* (Wessler, S., Becker C.G., and Nemerson, Y., eds.), pp. 67–81, Plenum Publishing Co., New York.

Kurachi, K., and Chen, S.H. (1990) in *Hematologic Disorders in Maternal-Fetal Medicine* (Bern, M.M., and Frigoletto, F., eds.) pp. 449–466, Alan R. Liss, Inc., New York.

Kurachi, K., and Davie, E.W. (1982) *Proc. Natl. Acad. Sci. USA* 79, 6461–6464.

Levy, J.A., Mitra, G.A, Wong, M.F., and Mozen, M.M. (1985) *Lancet* 1, 1456–1457.

Lorand, L., Losowsky, M.S., and Miloszewski, K.J.M. (1980) *Prog. Hemost. Thromb.* 5, 245–290.

Luban, N., Bray, G., Hoyer, L., and Toy, P. (1989) in *Hematology, Educational Program of the American Society of Hematology*, 2–5 Dec. 1989, Atlanta, pp. 68–73, American Society of Hematology, W.B. Saunders Co., Philadelphia.

Macik, B.G., Hohneker, J., Roberts, H.R., and Griffin, A.M. (1989) *Am. J. Hemat.* 32, 232–234.

Marlar, R.A. (1985) *Semin. Thromb. Hemost.* 11, 387–393.

Meulien, P., Balland, A., Lepage, P., et al. (1990) *Prot. Engin.* 3, 629–633.

O'Hara, P.J., Grant, F.J., Haldeman, B.A., et al. (1987) *Proc. Natl. Acad. Sci. USA* 84, 5158–5162.

Piszkiewicz, D., Kingdom, H., Apfelzweig, R., et al. (1985) *Lancet* 2, 1188–1189.

Plutkzy, J., Hoskins, J.A., Long, G.L., and Crabtree, G.R. (1986) *Proc. Natl. Acad. Sci. USA* 83, 546–550.

Prochownik, E.V., Markham, A.F., and Orkin, S.H. (1983) *J. Biol. Chem.* 258, 8389–8394.

Prochownik, E.V., Bock, S.C., and Orkin, S.H. (1985) *J. Biol. Chem.* 260, 9608–9612.

Purrello, M., Alhadeff, B., Esposito, D., et al. (1985) *EMBO J.* 4, 725–729.

Roberts, H., Ginsberg, D., and Gitscher, J. (1989) in *Hematology Educational Program of the American Society of Hematology,* 2–5 Dec. 1989, Atlanta, pp. 55–60, American Society of Hematology, W.B. Saunders Co., Philadelphia.

Romeo, G., Hassan, H.J., Staempfli, S., et al. (1987) *Proc. Natl. Acad. Sci. USA* 84, 2829–2832.

Rosenberg, R.D. (1989) *Am. J. Med.* 87(3B), 2S–9S.

Sadler, J.E. (1989) in *The Metabolic Basis of Inherited Diseases,* vol. II, 6th ed. (Scriver, C.R., Beaudet, A.L., Sly, W.S., and Valle, D., eds.), pp. 2171–2187, McGraw Hill, New York.

Sarver, N., Ricca, G.A., Link, J., et al. (1987) *DNA* 6, 553–564.

Stephens, A.W., Siddiqui, A., and Hirs, C.H.W. (1987) *Proc. Natl. Acad. Sci. USA* 84, 3886–3890.

Telgt, D.S.C., Macik, B.G., McCord, D.M., Monroe, D.M., and Roberts, H.R. (1989) *Thromb. Res.* 56, 603–609.

Thim, L., Bjoern, S., Nicolaisen, E.M., et al. (1988) *Biochem.* 27, 7785–7793.

Tollefsen, D.M. (1989) in *The Metabolic Basis of Inherited Diseases* vol. II, 6th ed. (Schiver, C.R., Beaudet, A.L., Sly, W.S., and Valle, D., eds.) pp. 2207–2218, McGraw Hill, New York.

Toole, J., Knopf, J.L., Wozney, J.M. et al. (1984) *Nature* 312, 342–346.

Toole, J.J., Pittman, D., Murtha, P., et al. (1986) in *Cold Springs Harbor Symposia on Quantitative Biology,* vol. LI, pp. 543–549, Cold Spring Harbor Laboratories, New York.

Vehar, G.A. (1987) *Ann. N.Y. Acad. Sci.* 509, 82–88.

Vehar, G.A., Keyt, B., Eaton, D., et al. (1984) *Nature* 312, 337–342.

Vehar, G.A., Lawn, R.M., Tuddenham, E.G.D., and Wood, W.I. (1989) in *The Metabolic Basis of Inherited Diseases,* vol. II, (Scriver, C.R., Beaudet, A.L., Sly, W.S., and Valle, D., eds.), pp. 2155–2170, McGraw Hill, New York.

van den Berg, W., ten Cate, J.W., Breedervelt, C., et al. (1986) *Lancet* 1, 803–804.

Verveij, C.L., Hart, M., and Pannekoek, H. (1987) *EMBO J.* 6, 2885–2890.

Wasley, L.C., Atha, D.H., Bauer, K.A., et al. (1987) *J. Biol. Chem.* 262, 14766–14772.

White, G.C., McMillan, C.W., Kingdom, H.S., and Shoemaker, C.B. (1989) *New Engl. J. Med.* 320, 166–329.

Wood, W.I., Capon, D.J., Simonsen, C.C., et al. (1984) *Nature* 312, 330–337.

Yoshitake, S., Schack, B.G., Foster, D.C., Davie, E.W., and Kurachi, K. (1985) *Biochem.* 24, 3736–3750.

Recombinant Tissue-Type Plasminogen Activator

Désiré Collen
William F. Bennett

10.1 THE FIBRINOLYTIC SYSTEM

Mammalian blood contains an enzymatic system called the fibrinolytic system that is capable of dissolving blood clots. This system is made up of an inactive proenzyme, plasminogen, which can be converted to the active enzyme, plasmin, by various plasminogen activators. Plasmin is a serine proteinase that digests fibrin to soluble degradation products. Inhibition of the fibrinolytic system occurs at the levels of both the plasminogen activator and the plasmin (Collen and Lijnen 1986; Collen 1987; Bachmann 1987). A diagram of the fibrinolytic system is shown in Figure 10–1.

Plasminogen is a single-chain glycoprotein consisting of 790 amino acids that is converted to plasmin by cleavage of the Arg 560-Val 561 peptide bond (Sottrup-Jensen et al. 1978). Plasminogen and plasmin contain structures called lysine-binding sites that mediate their binding to fibrin and accelerate the interaction between plasmin and its physiological inhibitor α_2-antiplasmin. The lysine-binding sites play a crucial role in the regulation of fibrinolysis (Wiman and Collen 1978a; Collen 1980).

Plasminogen activators are serine proteinases with a high specificity for plasminogen. They hydrolyze the Arg 560-Val 561 peptide bond of plas-

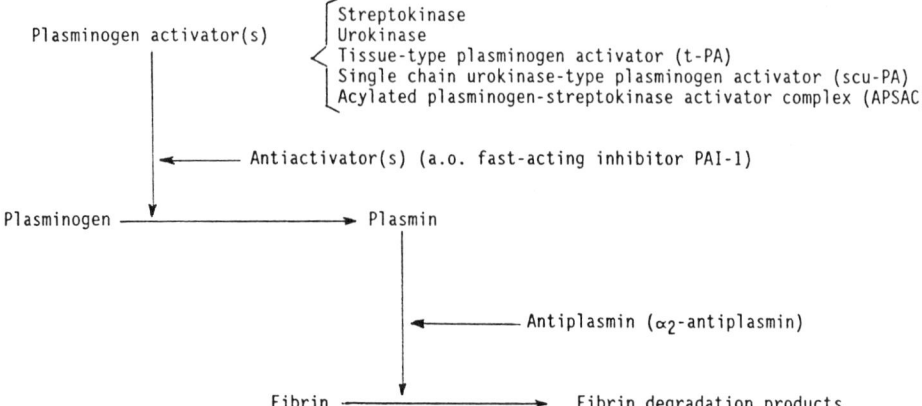

FIGURE 10-1 A diagram of the fibrinolytic system.

minogen yielding the active enzyme, plasmin. The activator streptokinase, however, is a nonenzymatic protein with molecular weight of 47,000 produced by beta-hemolytic streptococci, which activates the fibrinolytic system indirectly. Streptokinase forms a 1:1 stoichiometric complex with plasminogen, which then undergoes a transition exposing an active site in the modified plasminogen moiety, whereby the complex becomes a potent plasminogen activator (Kosow 1975). Acylated plasminogen-streptokinase activator complex (APSAC) is an inactive derivative of the plasminogen-streptokinase activator complex, obtained by acylation of the serine in its active site. APSAC spontaneously reactivates at physiological pH following deacylation (Smith et al. 1981). The plasminogen activator urokinase, originally isolated from urine, is a trypsin-like serine proteinase composed of two polypeptide chains, connected by a disulfide bridge, which activates plasminogen directly to plasmin (White, W.F., et al. 1966).

Two immunologically distinct plasminogen activators have been identified in blood: tissue-type plasminogen activator (t-PA) and single-chain urokinase-type plasminogen activator (scu-PA, pro-urokinase). scu-PA is a single-chain glycoprotein containing 411 amino acids, which is converted to urokinase by hydrolysis of its Lys 158-Ile 159 peptide bond (Holmes et al. 1985). scu-PA has very little activity towards low-molecular-weight substrates but has some intrinsic plasminogen-activating potential; however, it has a catalytic efficiency that is two orders of magnitude lower than that of urokinase (Lijnen et al. 1986). Tissue-type plasminogen activator (t-PA) (see Section 10.2) is a trypsin-like serine proteinase composed of 527 or 530 amino acids (Pennica et al. 1983; Pohl et al. 1984). It occurs either as a single-chain glycoprotein or as a two-chain proteolytic derivative; both forms are enzymatically active (Tate et al. 1987; Petersen et al. 1988). t-PA is a poor plasminogen activator in the absence of fibrin, but it binds specifically to fibrin and activates plasminogen at the fibrin surface several hundredfold more efficiently than in the circulation (Hoylaerts et al. 1982).

The plasminogen activator inhibitor-1 (PAI-1) (Figure 10–1), is a fast-acting inhibitor of t-PA and urokinase (Kruithof et al. 1983, 1984) which normally is at a very low concentration in the blood; however, it may be significantly increased in several disease states, including venous thrombo-embolism and ischemic heart disease. PAI-1 is a serpin and is composed of 379 amino acids with Arg 346-Met 347 at the reactive site (for references see Bachmann 1987).

The plasmin inhibitor α_2-antiplasmin is a glycoprotein of the serine proteinase inhibitor (serpin) superfamily, composed of 452 amino acids with Arg 364-Met 365 as the reactive site (Holmes et al. 1987). α_2-Antiplasmin reacts very rapidly with plasmin, first to form a reversible but inactive complex, which is then slowly converted into an irreversible but active complex (Figure 10–1). The rapidity of the first step of the reaction is dependent on the presence of free lysine-binding sites and a free active center in the plasmin molecule (Wiman and Collen 1978b).

10.2 NONRECOMBINANT TISSUE-TYPE PLASMINOGEN ACTIVATOR (t-PA)

10.2.1 Properties of t-PA

Plasminogen activators have been purified from various sources, including pig heart, hog ovaries, postmortem vascular perfusates, post-exercise blood, and tumor cell lines (for references, see Rijken and Collen 1981). The first satisfactory purification of a human plasminogen activator was obtained from uterine tissue (Rijken et al. 1979). Tissue plasminogen activator, vascular plasminogen activator, and blood plasminogen activator are immunologically identical, but different from urokinase (Rijken et al. 1980). The plasminogen activator found in blood is synthesized and secreted by endothelial cells and is now called "tissue-type plasminogen activator" (t-PA). Human t-PA (melanoma t-PA) has been purified from the culture fluid of a stable human melanoma cell line (Bowes, Roosevelt Park Memorial Institute–7272) in sufficient amounts to permit the study of its biochemical and biological properties (Rijken and Collen 1981; Collen et al. 1982).

Natural t-PA is a serine proteinase with a M_r of about 70,000, composed of a single polypeptide chain of 527 or 530 amino acids with Ser or Gly as the NH$_2$-terminal amino acid (Pennica et al. 1983; Pohl et al. 1984; Vehar et al. 1984). Generally the original numbering system based on 527 amino acids has been used. t-PA is converted by traces of plasmin to a two-chain form by hydrolysis of the Arg 275-Ile 276 peptide bond. Its primary structure is illustrated in Figure 10–2. The NH$_2$-terminal region is composed of several domains with homologies to other proteins: residues 1-43 (F-domain) are homologous to the "finger domains" of fibronectin (Banyai et al. 1983),

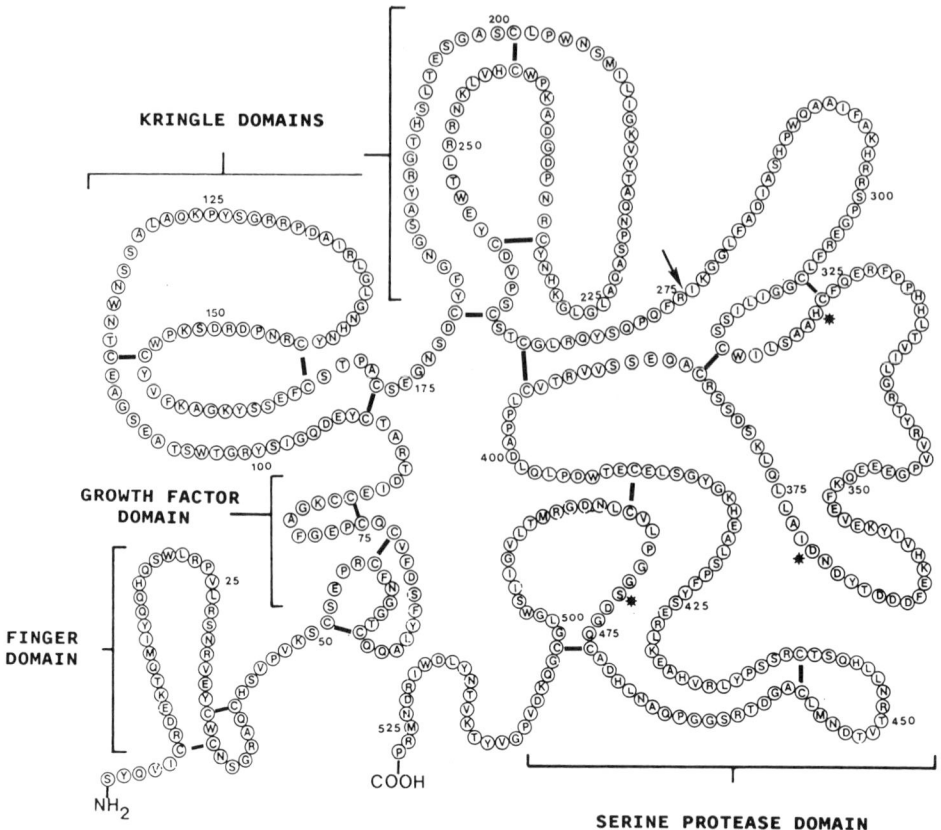

FIGURE 10–2 The primary structure of human tissue-type plasminogen activator (t-PA).

residues 44-91 (E-domain) are homologous to human epidermal growth factor, and residues 92-173 and 180-261 (K_1 and K_2 domains) are both homologous to the "kringle" regions of plasminogen. The region comprising residues 276-527 (P-domain) is homologous to that of other serine proteinases and contains the catalytic site, which is composed of His 322, Asp 371, and Ser 478 (Pennica et al. 1983). t-PA has a specific affinity for fibrin. The dissociation constant of the t-PA-fibrin complex is estimated to be 0.14 μM in the presence of plasminogen (Hoylaerts et al. 1982).

t-PA is relatively inactive in the absence of fibrin, but the presence of fibrin strikingly enhances its activation of plasminogen (Camiolo et al. 1971; Hoylaerts et al. 1982). Kinetic analysis of the activation of native plasminogen (Glu-plasminogen) by t-PA gave the following results: K_m: 9 to 100 μmol/l (Hoylaerts et al. 1982; Rijken et al. 1982; Ranby 1982) and k_{cat}: 0.006 s^{-1} to 0.06 s^{-1} (Hoylaerts et al. 1982; Ranby 1982). In the presence of fibrin,

K_m was reported to be 0.16 to 1.1 μmol/l and k_{cat} was 0.1 to 0.22 s^{-1} (Hoylaerts et al. 1982; Rijken et al. 1982; Ranby 1982). Norrman et al. (1985) showed that there were two phases in the activation of Glu-plasminogen by t-PA in the presence of fibrin; a first phase with K_m 1.05 μmol/l and k_{cat} 0.15 s^{-1} and a second phase with K_m 0.07 μmol/l and k_{cat} 0.14 s^{-1}. These changes and differences in K_m in the presence of fibrin were explained by the exposure of a strong plasminogen-binding site upon initial degradation of fibrin (Suenson et al. 1984). Hoylaerts et al. (1982) found a K_m of 65 μmol/l and a k_{cat} of 0.06 s^{-1} in the absence of fibrin, and a K_m of 0.16 μmol/l with unchanged k_{cat} in the presence of fibrin, corresponding to an increase in catalytic efficiency (k_{cat}/K_m) of t-PA for the activation of plasminogen of about 630-fold.

Miles and Plow (1985) have shown that platelets bind plasminogen and that platelet-bound plasminogen is more sensitive to activation by t-PA. Vaughan et al. (1989) found 120,000 to 290,000 binding sites per platelet with a K_d of 340 to 800 nM. Stricker et al. (1986) reported that incubation with washed human platelets caused a 10- to 50-fold increase in plasminogen activation. These results suggest that platelets may provide a surface for enhanced activation of plasminogen by t-PA. Negatively charged lipids lower the K_m value of plasminogen activation by t-PA 6- to 20-fold, whereas neutral lipids raise the K_m (Soeda et al. 1987) suggesting that plasminogen activation with t-PA may be influenced by the membrane charge. Both plasminogen and t-PA bind specifically to cultured human umbilical vein endothelial cells (Hajjar et al. 1986, 1987) resulting in a three- to fourfold increase of the apparent catalytic efficiency for plasminogen activation. Other authors, however, reported that t-PA binds to endothelial cells via interaction with PAI-1 (Ramakrishnan et al. 1990; Russell et al. 1990).

10.2.2 Gene Structure of t-PA

The gene coding for human t-PA has been localized to chromosome 8 (chromosome 8, bands 8.p.12 → q.11.2) (Rajput et al. 1985; Verheijen et al. 1986a; Yang-Feng et al. 1986). The t-PA gene has been characterized in detail (Ny et al. 1984; Browne et al. 1985; Degen et al. 1986). Thirteen intervening sequences divide the gene into 14 coding regions; the exons range in size from 43 to 914 bp and the introns from 111 to 14,257 bp. Transcription is controlled by a "TATA box" (positions -22 to -29), a "CAAT box" (positions -112 to -116), and a polyadenylation signal.

The complete 2,530-base pair cDNA sequence of mature t-PA (Pennica et al. 1983) contains a single reading frame, beginning with the ATG codon at nucleotides 85 to 87, which is followed, 562 codons later, by a TGA termination triplet at nucleotides 1,771 to 1,773. The aminoterminal serine residue is preceded by 35 amino acids, 20 to 23 of which (residues -35 to -13) probably constitute a hydrophobic signal peptide involved in the secretion of t-PA. The remaining hydrophobic amino acids immediately preceding the start of mature t-PA (residues -14 to -1) may constitute a "pro" sequence similar to that found for serum albumin.

10.2.3 Release and Inhibition of t-PA

Mean t-PA antigen levels in human plasma at rest, measured by immunoassays, range between 3.4 and 6.6 μg/l (for references, see Holvoet et al. 1985). Detectable levels of free, single-chain t-PA were found in plasma of 6 out of 10 healthy subjects (mean value 4.5 μg/l) and increased to 7.6 μg/l after venous occlusion (Holvoet et al. 1987).

The mechanisms regulating the synthesis and secretion of t-PA remain largely unknown. Agents, such as thrombin, that stimulate the release of t-PA from endothelial cells, also stimulate the secretion of PAI-1 (Gelehrter and Sznycer-Laszuk 1986; Hanss and Collen 1987). Dexamethasone increased PAI-1 antigen levels to a greater extent than t-PA antigen levels in rat hepatoma cells and resulted in inhibition of t-PA activity (Gelehrter et al. 1987). Endotoxin stimulates the secretion of PAI-1 activity but not of t-PA antigen (Colucci et al. 1985) whereas histamine causes a secretion of t-PA antigen but not of PAI-1 (Hanss and Collen 1987).

Interleukin-1 stimulates the production of t-PA by human articular chondrocytes (Bunning et al. 1987). Liu et al. (1987) have shown that gonadotropins increase t-PA activity and antigen levels in ovarian cells with a maximum prior to ovulation. This gonadotropin regulation of t-PA activity may play a role in follicle rupture. The increased secretion of t-PA is paralleled by increased steady-state levels of t-PA mRNA (Ny et al. 1987; O'Connell et al. 1987).

The mechanisms involved in the removal of t-PA for the blood are multiple: Clearance by the liver results in a rapid initial half-life of a few minutes both in animals and in humans (see Section 10.2.5). t-PA circulates in plasma in a free active form and as complexes with α_2-antiplasmin, α_1-antitrypsin, C_1-inhibitor (Rijken et al. 1983a; Thorsen and Philips 1984; Lucore and Sobel 1988), and the specific, rapid-reacting inhibitor of t-PA, plasminogen activator inhibitor-1 (PAI-1) (Takada et al. 1989; Alessi et al. 1990).

10.2.4 Mechanism of Action of t-PA

The enhanced activation rate of plasminogen by t-PA in the presence of fibrin has been explained by an increased affinity of fibrin-bound t-PA for plasminogen (Michaelis constant, 65 μmol/l in the absence of fibrin, and 0.16 μmol/l in the presence of fibrin) without significantly influencing the catalytic rate constant of the enzyme (Hoylaerts et al. 1982). However, others have claimed that fibrin influences both the K_m and k_{cat} of the activation of plasminogen by t-PA (Nieuwenhuizen et al. 1988). The kinetic data of Hoylaerts et al. (1982) support a mechanism in which fibrin provides a surface to which t-PA and plasminogen adsorb in a sequential and ordered way yielding a ternary complex. Fibrin essentially increases the local plasminogen concentration by creating an additional interaction between t-PA and its substrate. The high affinity of t-PA for plasminogen in the presence of fibrin thus allows efficient activation on the fibrin clot, while no efficient plasminogen activation by t-PA occurs in plasma.

Free plasmin in the circulation is very rapidly inhibited by α_2-antiplasmin (second-order rate constant, 2 to 4 $\times$ 10^7 $1 \cdot mol^{-1} \cdot s^{-1}$); the rate of this inhibition is strongly dependent on the availability of free lysine-binding sites and a free active site in the plasmin molecule (Wiman and Collen 1978a). Plasmin formed on the fibrin surface has both its lysine-binding sites and its active site occupied and is thus only slowly inactivated by α_2-antiplasmin (half-life of about 10-100 sec, compared to 0.1 sec for free plasmin) (Collen and Lijnen 1986). The fibrinolytic process thus seems to be triggered by and confined to fibrin. On the fibrin surface, single-chain t-PA is quickly converted to a two-chain form during fibrinolysis, but there is little direct evidence that this conversion plays a role in the regulation of fibrinolysis (Rijken et al. 1982).

10.2.5 Pharmacokinetic Properties of t-PA

The turnover of mixtures of ^{125}I-labeled and unlabeled single-chain and two-chain melanoma t-PA (see Section 10.2.1) was studied after intravenous bolus injection into rabbits (Korninger et al. 1981). The fibrinolytic activity of the euglobulin fraction declined rapidly with a half-life of 2 min for the single-chain and 3 min for the two-chain form. The blood radioactivity initially also declined very rapidly. Initial rapid accumulation of radioisotope occurred in the liver, followed by reappearance in the blood. A similar uptake of t-PA by the liver was observed with active site-blocked t-PA. The half-life of t-PA was markedly prolonged after functional hepatectomy. Kuiper et al. (1988) have recently identified two recognition systems for hepatic removal of t-PA via the hepatocytes and via the endothelial cells. Clearance of t-PA from the circulation was also found to be very rapid (initial half-life 1–4 min) in other studies in rabbits and mice (Fuchs et al. 1985; Nilsson et al. 1985; Beebe and Aronson 1986).

Intravenous administration of two-chain melanoma t-PA in seven patients with acute myocardial infarction, at a dose of 6-12 mg over 30 min, resulted in plateau levels of t-PA in plasma of approximately 0.5 μg/ml (Van de Werf et al. 1984b).

10.2.6 Thrombolytic Properties of t-PA

The relative fibrinogenolytic and fibrinolytic properties of human melanoma t-PA and human urokinase have been compared in a system composed of a ^{125}I-labeled-fibrin human blood clot immersed in a plasma milieu (Matsuo et al. 1981a). With t-PA, but not with urokinase, blood clots were dissolved without associated fibrinogen breakdown. In a similar system, non-cross-linked clots were found to lyse more extensively than totally cross-linked clots, and no differences were observed between one-chain and two-chain t-PA (Korninger and Collen 1981).

A summary of the thrombolytic properties of t-PA in animal models of thrombolysis has been reported elsewhere (Collen et al. 1989). In general, the degree of lysis with t-PA was markedly greater than in controls, greater with t-PA than with urokinase or streptokinase, and more fibrin-specific.

In rabbits with experimental pulmonary emboli, t-PA caused thrombolysis at lower doses than urokinase (Matusuo et al. 1981b). Intravenous infusion of melanoma t-PA caused thrombolysis without associated fibrinogen breakdown in dogs with experimental thrombosis of the femoral vein (Korninger et al. 1982). In a preliminary report, Sampol et al. (1983) recanalized thrombosed femoral veins in dogs by infusion of a crude extract of porcine t-PA. In rabbits with jugular vein thrombosis, the extent of thrombolysis by melanoma t-PA was determined mainly by the dose of t-PA and its route of administration, and much less by the age of the thrombus or the molecular form of the activator (Collen et al. 1983). Agnelli et al. (1985) reported that, in rabbits, t-PA provokes much less hemorrhage at equivalent thrombolytic doses than streptokinase. Thrombolysis was obtained within 60 min by intravenous infusion of 5 to 10 μg/kg/min of human melanoma t-PA in dogs with copper-coil-induced coronary artery thrombosis (Bergmann et al. 1983). In addition infusion of t-PA also restored intermediary metabolism and nutritional blood flow without causing systemic fibrinolytic activation.

In 1981, the first patients were treated with t-PA obtained from the Bowes melanoma cell line. Intravenous administration of 7.5 mg t-PA over 24 hours induced complete lysis of a renal and iliofemoral thrombosis in a renal transplant patient without systemic fibrinolytic activation or bleeding (Weimar et al. 1981). However, intravenous infusion of 5-25 mg of t-PA over 24-36 hours did not produce thrombolysis in four patients with deep vein thrombosis (Verstraete and Collen 1985). A pilot study of the use of melanoma cell t-PA was carried out in seven patients with acute myocardial infarction. Coronary thrombolysis in the absence of fibrinogen breakdown was achieved in six of these seven patients (Van de Werf et al. 1984b).

10.3   RECOMBINANT TISSUE-TYPE PLASMINOGEN ACTIVATOR (rt-PA)

10.3.1   Physicochemical Properties of rt-PA

Following the promising initial results obtained with melanoma t-PA, the cDNA encoding human t-PA was cloned and expressed, initially in *Escherichia coli* (Pennica et al. 1983), and subsequently in Chinese hamster ovary

(CHO) cells (Vehar et al. 1986). Initially, the recombinant, tissue-type plasminogen activator (rt-PA) was produced as a two-chain form by a roller bottle process at Genentech Inc. (South San Francisco, CA) and used for early pharmacological and clinical studies (between 1984 and 1986). Subsequently, a single-chain form of rt-PA was manufactured by a large-scale suspension culture production process for commercialization (Activase™, Genentech Inc., South San Francisco, CA; Actilyse™, Boehringer Ingelheim, Ingelheim, Germany). Most clinical experience with rt-PA to date has been obtained with this material. The cDNA of human t-PA has also been cloned and expressed in several other systems (Browne et al. 1985; Kaufman et al. 1985; Harris et al. 1986; Sambrook et al. 1986; Dodd et al. 1986; van Zonneveld et al. 1986a and 1986b; Upshall et al. 1987; Reddy et al. 1987; Markland et al. 1989; Parekh et al. 1989; Sarmientos et al. 1989).

The rt-PA produced by expression of cDNA encoding human t-PA in CHO cells is obtained either as a single-chain molecule consisting of 527 amino acids or as its two-chain derivative generated by cleavage of the Arg 275-Ile 276 peptide bond (Vehar et al. 1986). rt-PA has a highly specific affinity for fibrin. van Zonneveld et al. (1986a and 1986b) have proposed that initial binding of t-PA to fibrin would occur via the finger domain, and in addition via kringle 2 after partial degradation of fibrin and generation of carboxy-terminal lysine. Higgins and Vehar (1987) reported a marked decrease in the dissociation constant for the binding of rt-PA to fibrin upon degradation. These data suggest a mechanism whereby early fibrin digestion by plasmin could accelerate fibrinolysis by increasing the binding of rt-PA and plasminogen. One chain rt-PA binds significantly more to fibrin than two-chain rt-PA (Higgins and Vehar 1987); the stoichiometry of binding is about 1 mol of rt-PA per mol of fibrin monomer.

Spellmann et al. (1989) isolated and characterized 17 carbohydrate structures from rt-PA, which were shown to represent variants and hybrids of either high-mannose or complex carbohydrates, linked to asparagine. No evidence of 0-linked glycosylation was found. rt-PA, like t-PA obtained from melanoma cells, can be separated on lysine-Sepharose in two fractions with different carbohydrate content: type I eluting at lower arginine concentration, contains N-linked carbohydrate on Asn 117, Asn 184 and Asn 448 whereas type II, eluting at higher arginine concentration contains no carbohydrate on Asn 184 (Vehar et al. 1986). Differences in specific activity for these glycoforms from nonrecombinant t-PA have been reported by Wittwer et al. (1989).

The kinetics of the activation of plasminogen by melanoma t-PA and by rt-PA were studied by Zamarron et al. (1984), both in the absence and the presence of CNBr-digested fibrinogen as a soluble cofactor. Michaelis-Menten kinetics applied, with kinetic constants that were very similar to those previously reported for the activation in the presence of solid-phase fibrin (Hoylaerts et al. 1982). The affinity of both enzymes increased markedly in the presence of the soluble cofactor, while the catalytic rate constant did not change significantly.

10.3.2 Pharmacokinetics of rt-PA

The pharmacokinetics of rt-PA were studied following intravenous bolus injection in rabbits (Collen et al. 1984). Fibrinolytic activity in the euglobulin fraction disappeared with a $t_{1/2}$ of 2 min for the single-chain and 3 min for the two-chain form of rt-PA. The initial $t_{1/2}$ of blood radioactivity were also 2 and 3 min respectively, but decline in radioactivity was followed by an apparently slower phase as a result of the reappearance of radiolabeled degradation products in the circulation. In rabbits, the $t_{1/2}$ of rt-PA was inversely correlated to the logarithm of liver blood flow (Bounameaux et al. 1986). Bugelski et al. (1989) recovered radiolabeled rt-PA in the parasinusoidal area of the liver parenchymal cell in rats.

The initial half-life of two-chain rt-PA following infusion in man was 6 min (Verstraete et al. 1985b and 1986; Seifried et al. 1988), similar to that after infusion of melanoma t-PA (Van de Werf et al. 1984b) and to that found in other species: 2 min in rabbits (Collen et al. 1984); 4 min in dogs (Van de Werf et al. 1984a); and 6 min in baboons (Flameng et al. 1985). In patients with acute myocardial infarction, clearance of the single-chain rt-PA produced for commercial use, appears to be 30-40% more rapid than that of the predominantly two-chain roller bottle material initially produced on a pilot scale. However, when the dose of suspension culture rt-PA is increased correspondingly, its thrombus specificity is similar if not better (Garabedian et al. 1987; Mueller et al. 1987). Garabedian et al. (1987) determined that the initial half-life of two-chain rt-PA in blood was 5.3 ± 1.7 min and the terminal half-life was 46 min.

A linear correlation between the infusion rate (4 to 8.2 μg/kg/min) and the mean steady-state plasma concentration of rt-PA (0.52 to 1.4 μg/ml) was observed, although plasma levels of rt-PA varied significantly between patients infused at the same rate (Garabedian et al. 1986). This suggests that no saturation of the clearance mechanism had occurred. These values are in agreement with the estimated plasma concentration of 2.5 μg/ml after rt-PA infusion of 0.5 mg/kg over one hour, assuming a half-life of 6 min. The mean concentration of t-PA found in plasma at rest is 6.6 ± 2 ng/ml (Holvoet et al. 1985); therefore concentrations during treatment with rt-PA are approximately 500 times greater than the physiological concentration.

10.3.3 In Vitro and Animal Studies of rt-PA

The extent of lysis induced by rt-PA of totally cross-linked human or animal plasma clots suspended in autologous plasma varied markedly from one species to the other, the human system being the most susceptible (Lijnen et al. 1984). Systemic activation of the fibrinolytic system in the circulating plasma was minor and dose-dependent in all species.

Using an in vitro clot-lysis system, Fry and Sobel (1988) found that therapeutic concentrations of heparin did not interfere with t-PA binding to fibrin, did not increase activation of free plasminogen, and did not inhibit thrombolysis. These findings suggest that concomitant administration of heparin would not impair t-PA-induced thrombolysis.

Hergrueter et al. (1988) found that infusion of rt-PA prevented thrombosis of an everted femoral arterial segment graft in the rabbit and that rt-PA recanalized thrombosed vessels. The thrombolytic potency of rt-PA was found to be equivalent with that of t-PA purified from conditioned melanoma cell cultures in the rabbit jugular vein thrombosis model (Collen et al. 1984). Gold et al. (1984) found a linear correlation between the rate of infusion of rt-PA and the time to reperfusion in dogs with a coronary artery thrombosis. Flameng et al. (1985) obtained reperfusion by intravenous administration of rt-PA in baboons with coronary artery thrombosis and found a linear correlation between the occlusion time and the infarct size. The thrombolytic effect of rt-PA was found to be sustained beyond its time of clearance from the circulation in the rabbit jugular vein thrombosis model, possibly as a result of binding of rt-PA to fibrin and protection of bound rt-PA from inhibitors (Agnelli et al. 1985).

10.3.4 Clinical Experiences with rt-PA

Thrombotic complications of cardiovascular diseases are a main cause of death and disability, and, consequently, thrombolysis could favorably influence the outcome of such life-threatening diseases as myocardial infarction, cerebrovascular thrombosis and venous thromboembolism.

In the present report, we will briefly review the clinical experience with the use of rt-PA in acute myocardial infarction, pulmonary embolism, deep vein thrombosis, peripheral arterial occlusion, and thromboembolic stroke.

10.3.4.1 Acute Myocardial Infarction.   The role of coronary artery thrombosis in the pathogenesis of myocardial infarction was recognized in the late 1970s (De Wood et al. 1980). It was further recognized that early administration of streptokinase could reperfuse occluded coronary arteries (Rentrop et al. 1984). Reduction in infarct size, preservation of left ventricular function, and reduction in mortality in patients with acute myocardial infarction, have meanwhile been demonstrated for several thrombolytic agents, including rt-PA. The relevant biological properties of thrombolytic agents in terms of their benefit/risk ratio in acute myocardial infarction comprise their efficacy for coronary artery recanalization, their safety, and their effect on left ventricular function and on mortality. At present, insufficient results of comparative clinical trials are available to allow definitive quantitative comparison of the relative efficacy and safety of the various thrombolytic agents. For rt-PA and streptokinase, sufficient information is available from randomized controlled trials or from uncontrolled trials with comparable endpoints, to allow a semiquantitative comparison of their properties. These studies will be briefly summarized below.

Efficacy for Coronary Recanalization.   Two randomized trials have been reported directly comparing the efficacy of streptokinase and rt-PA by angiography 90 min after the start of the infusion (Verstraete et al. 1985a; Chesebro et al. 1987). The frequencies of open coronary arteries at 90 min after the start of the infusion, in the subgroup treated within 3 hours of the onset of symptoms, was 31 of 56 patients (55%) in the streptokinase group and 34 of 42 patients (81%) in the rt-PA group ($p < .01$) (Chesebro et al. 1988). In the subgroup treated at 3-6 hours, the results were 64 of 151 patients (42%) with streptokinase and 109 of 162 patients (67%) with rt-PA ($p < .001$). Overall, the frequency of open arteries at 90 min was 52% higher in patients treated with rt-PA (143 of 204 patients, 70%) as compared to patients treated with streptokinase (95 of 207 patients, 46%). These results indicate that, when measured within a time frame useful for salvaging myocardial tissue, streptokinase is a less efficient agent for coronary thrombolysis, reperfusing at best one out of two occluded arteries, whereas rt-PA is significantly more efficient, reperfusing at least two out of three occluded arteries.

Safety.   Bleeding complications during thrombolytic therapy may be due to the action of the thrombolytic agents on blood coagulation, on the vessel wall, and on the hemostatic plug. In addition, demographic characteristics of the patient and adjunctive therapy with anticoagulant or antiplatelet agents may contribute to bleeding. Quantitative and qualitative evaluation of spontaneous or intervention-related bleeding from results of noncomparable studies is very difficult, especially in association with highly variable frequencies of invasive cardiovascular procedures. Consequently, bleeding complications can only validly be compared in randomized controlled trials. In four small comparative trials with streptokinase and rt-PA, the frequency of bleeding complications was somewhat but not markedly lower with rt-PA than with streptokinase.

Evaluation of the relative risk of cerebral bleeding in association with thrombolytic therapy is most important, but it is unfortunately confounded by several intervening factors. First, acute myocardial infarction in the absence of thrombolytic therapy is associated with an incidence of cerebrovascular accidents, which has been estimated to be approximately 1% in the recent large clinical trials with streptokinase and rt-PA (ISIS-2 1988; Wilcox et al. 1988). Second, the differential diagnosis between hemorrhagic and thrombotic strokes cannot usually be made without the use of CT scans. Third, thrombolytic therapy may convert a thrombotic stroke into a cerebral hemorrhage, without necessarily deteriorating the clinical outcome. Finally, adjunctive therapy may cause an increase of intracranial bleeding above the intrinsic and unavoidable frequency associated with thrombolytic therapy in general and specific thrombolytic agents in particular. Therefore, the definitive answer to the question of the relative frequency of intracerebral bleeding associated with the various thrombolytic agents will require careful comparative studies with very large numbers of patients.

Reduction of Mortality. Mortality reduction in placebo-controlled trials has now been demonstrated for streptokinase, rt-PA, and APSAC (for references, see Collen and Gold 1989). Streptokinase reduces overall mortality at 14-30 days by an estimated 25%. However, in the individual trials, mortality rates in the control groups vary from 6.5-13%, and reductions in early mortality with streptokinase vary from 18-81%. Clearly, the large variability in mortality in the control groups and the impact of streptokinase on mortality are influenced by patient selection, by adjunctive therapy including anticoagulant and antiplatelet drugs, and by mechanical coronary interventions.

Assuming the clinical benefit of thrombolytic therapy in patients with acute myocardial infarction is proportional to the efficacy for coronary thrombolysis, the size of randomized clinical trials required to establish differences between thrombolytic agents can be calculated as follows: On the basis of controlled clinical trials of streptokinase versus placebo, in over 35,000 patients with acute myocardial infarction, the reduction in hospital mortality in the treatment group was found to be approximately 25%. A thrombolytic agent such as rt-PA with a 50% higher efficacy of coronary thrombolysis than streptokinase would thus be anticipated to reduce early mortality by 37.5%. Assuming a control mortality of 9% in the absence of thrombolytic therapy, the mortality with streptokinase treatment would be reduced by 25% to 6.75%, and that with the more potent agent by 37.5% to 5.6%. In order to establish such a difference with a statistical power of 0.8 and a significance level of 0.05, more than 10,000 patients would have to be entered into a randomized trial (D. Finkelstein, personal communication). These numbers illustrate the tremendous task to be undertaken in order to translate efficacy for coronary thrombolysis into reduction of mortality.

Two large, prospective, comparative, clinical trials in patients with acute myocardial infarction—GISSI-2 comparing streptokinase and rt-PA in 20,000 patients; and ISIS-3, comparing streptokinase, APSAC, and rt-PA in 30,000 patients—are presently being carried out. Both trials, in a factorial design, also investigate the effect of heparin. However, heparin is administered by subcutaneous injection and starting only 12 hours and 3 hours after the infusion of thrombolytic agent, respectively. This is most unfortunate, because the recent studies of Bleich et al. (1989) and of Hsia et al. (1990) have reported a significantly lower patency rate, determined by coronary angiography after a mean of 55-59 hours and 7-24 hours respectively, when rt-PA was administered without heparin. Actually the patency rates obtained the rt-PA in the absence of heparin relative to those obtained with rt-PA and heparin were similar to those previously obtained with streptokinase and heparin. On the other hand Sherry (1988) claimed that streptokinase, because of the profound hypocoagulable state which its produces in the blood stream for many hours, does not require adjuvant heparin therapy for coronary recanalization. If the previously established higher reperfusion rates of rt-PA relative to streptokinase require the concomitant use of heparin with rt-PA, but not with streptokinase, the GISSI-2 and ISIS-3 trials, because of their design, can no longer be expected to resolve the crucial question whether efficacy for coronary recanalization translates into clinical benefit. Indeed, with administration schemes that produce comparable rates of recanalization, clinical outcomes are expected to be comparable, unless the reduced viscosity and hypotension induced with streptokinase would confer an additional benefit (Marder and Sherry 1988).

In the meantime, it might well be of interest to review the available data from the small comparative trials with streptokinase and rt-PA (Verstraete et al. 1985a; Chesebro et al. 1987; Magnani 1989; White, H.D., et al. 1989) and with urokinase and rt-PA (Neuhaus et al. 1988; Califf et al. 1989) carried out to date with the simultaneous use of heparin anticoagulation (Table 10–1). Statistical evaluation of the homogeneity of these studies revealed a χ^2 value of 1.34 ($p = 0.85$) allowing metaanalysis of the data. Cumulative in-hospital mortalities were 38/757 (5.0%) in patients random-

TABLE 10–1 In-Hospital Mortality in Randomized Studies with rt-PA Versus Streptokinase (SK) or Urokinase (UK) in Patients with Acute Myocardial Infarction[1]

| | | Mortality Rate | |
| | | rt-PA | SK or UK |
Study	Reference	Treatment	Treatment
ECSG-1	Verstraete et al. (1985a)	3/64	3/65
TIMI-I	Chesebro et al. (1987)	12/157	14/159
New Zealand	White, H.D., et al. (1989)	5/135	10/135
PAIMS	Magnani (1989)	4/86	7/85
TAMI-5	Califf et al. (1989)	8/191	15/190
GAUS	Neuhaus et al. (1988)	6/124	5/121
Total		38/757	54/755
Percent mortality		5.0%	7.2%

The authors are grateful to Alan Hopkins, Genentech Inc., for the metaanalysis of the data shown in this table.
[1] Homogeneity index: χ^2: 1.34, $p = 0.85$. Odds ratio, rt-PA vs. SK/UK: 0.64 (95% CI: 0.4–1.0). The p value of the difference is 0.0523.

ized to rt-PA and 54/755 (7.2%) in patients allocated to the non-fibrin-specific agents streptokinase or urokinase (Table 10–1). The odds-of-death ratio, determined as described by Yusuf et al. (1985) is 0.64, with 95% confidence intervals of 0.4 to 1.0. The difference in mortality between rt-PA and SK/UK has a *p* value of 0.052. These results are derived from small studies, albeit randomized, which were not prospectively designed for a mortality endpoint. However, they agree remarkably well with the values calculated on the basis of the hypothesis that the clinical outcome is primarily determined by the efficacy of the thrombolytic agent, which is estimated to be 50% higher for rt-PA than for streptokinase or urokinase.

10.3.4.2 Deep Vein Thrombosis. Clinical experience with rt-PA in acute deep vein thrombosis is limited; rt-PA infused in a dose of 0.5 mg/kg over 4 hours produced lysis in approximately 60% of patients (Turpie 1987 and 1989). With a rt-PA dose of 100 mg and 50 mg over 8 hours on two subsequent days, comparable results were obtained (Verhaeghe et al. 1989), whereas a continuous infusion of 0.75 to 1.75 mg rt-PA/kg per 24 hours for 2 to 4 days gave better results in a small number of patients (Zimmerman et al. 1988). Two case reports on the use of rt-PA in subclavian vein thrombosis and mesenteric vein thrombosis have recently been reported (Stewart and Mayne 1988; Robin et al. 1988).

10.3.4.3 Major Pulmonary Embolism. In 36 patients with severe pulmonary embolism, 50 mg rt-PA was infused over 2 hours and, if necessary, an additional 40 mg over 4 hours (Goldhaber et al. 1986). Pulmonary angiography revealed a 49% improvement at 6 hours and a decrease in pulmonary artery systolic pressure (Come et al. 1987). In 34 patients with massive recent pulmonary embolism, rt-PA given as a 10 mg bolus followed by 20 mg/hr for 2 hours, caused an improvement of the angiographic score of approximately 15%. A second infusion of 50 mg of rt-PA in 22 patients produced an additional improvement in angiographic severity score of 38% (Verstraete et al. 1988b).

A direct comparison between urokinase given over 24 hours and rt-PA given over 2 hours, was made in 45 patients with pulmonary embolism (Goldhaber et al. 1988). By 2 hours, 82% of rt-PA-treated patients showed clot lysis, compared with 48% of urokinase-treated patients ($p = 0.008$). Improvement in lung scan reperfusion at 24 hours was similar in the two treatment groups. Further studies are necessary to demonstrate whether thrombolytic therapy with rt-PA reduces the morbidity and mortality of pulmonary embolism.

10.3.4.4 Arterial Thromboembolism. Graor et al. (1986) treated 65 patients with peripheral artery or bypass graft thrombosis with intra-arterial administration of 0.05 or 0.1 mg/kg/hr of rt-PA for 2–8 hours. Angiographically proven successful clot lysis was gained in 61 patients, but 76% of these required a secondary procedure. No allergic or adverse reactions occurred, but large hematomas occurred in three patients.

Infusion of rt-PA at a rate of 10 mg/hr into 50 thrombosed femoral and popliteal arteries produced recanalization in 43 of these patients (Verstraete et al. 1988a). Secondary angioplasty led to two reocclusions, and three patients experienced early rethrombosis. A favorable clinical result was thus obtained in 38 patients (76%).

10.3.4.5 Acute Thromboembolic Stroke. Except for a few isolated cases in which rt-PA has been used in patients with recent stroke (Buteux et al. 1988; Henze et al. 1987; Koudstaal et al. 1988), no large-scale trials have been reported. A dose-finding study is underway to determine the optimal dose of rt-PA administered over 60 min (The t-PA Acute Stroke Study Group 1988). In an NIH-sponsored multicenter trial, the incidence of bleeding and early outcome after rt-PA is being investigated (Brott et al. 1988).

Until larger studies to define the efficacy and safety of thrombolytic drugs for acute ischemic stroke are completed, the use of plasminogen activators must remain restricted to experimental protocols.

10.4 MUTANTS AND VARIANTS OF TISSUE-TYPE PLASMINOGEN ACTIVATOR

10.4.1 Deletion Mutants of t-PA
The distinct domains of t-PA appear to be involved in several functions of the enzyme including its binding to fibrin, fast clearance in vivo, plasminogen-activation ability with fibrin specificity (enzymatic properties), and binding to endothelial cell receptors. By means of recombinant DNA technology, the construction, expression, and characterization of deletion mutants of recombinant t-PA lacking one or more of these domains has made possible the investigation of the thrombolytic potential of such molecules and a limited study of their structure-function relationships.

10.4.1.1 Fibrin-Binding Domains. The structures involved in the fibrin binding of t-PA are thought to be mainly on the NH_2-terminal region (A-chain), as evidenced by the intact fibrin affinity of the A-chain isolated after mild reduction of two-chain t-PA (Rijken and Groeneveld 1986; Holvoet et al. 1986). Initially, the finger domain of t-PA (see Figure 10-2), which is homologous with the finger domains of fibronectin, was proposed to be involved in the fibrin-binding of the enzyme (Banyai et al. 1983). This was supported by the demonstration that a deletion mutant of t-PA that encodes only the finger and serine protease domains binds to fibrin (van Zonneveld et al. 1986b). Kagitani et al. (1985), however, have reported that t-PA lacking the finger domain still has significant fibrin affinity. Evidence obtained with additional deletion mutants suggests that binding of t-PA to fibrin is mediated both via the finger domain and via the second kringle region (K_2) (Verheijen et al. 1986b; van Zonneveld et al. 1986a, 1986b). A lysine-binding site (LBS) on kringle 2 (K_2) (amino acids 180–261) is thought to be involved in the interaction of the K_2 domain with fibrin. Competing ligands such as 6-aminohexane (AH) also prevent binding of t-PA to fibrin via K_2 (Verheijen et al. 1987), suggesting the presence of a "AH site" in K_2, similar to that observed in plasminogen (Christensen 1984). This AH site would interact with internal lysine residues in the fibrin matrix, whereas LBS might interact with carboxyl-terminal lysine residues exposed upon initial plasmic digestion of fibrin. However, a t-PA variant lacking only the growth factor domain bound as poorly to fibrin as the finger domain deletion variant (Larsen et al. 1988b; Johannessen et al. 1988). These results suggest either that both amino-terminal domains are required for high affinity to fibrin, or that the polypeptide region connecting the finger and growth factor domains is involved in binding to fibrin, or possibly that the growth factor domain is required for proper folding of t-PA. A t-PA variant lacking both the finger and growth factor domains but containing the kringle regions had intact enzymatic activity but showed moderately (van Zonneveld et al. 1986a, 1986b) or markedly (Larsen et al. 1988b; Johannessen et al. 1988; Kalyan et al. 1988) reduced binding to fibrin. A set of deletion mutants of t-PA, as compared to melanoma t-PA, showed decreased fibrin affinity for all mutants, in the order: t-PA $\gg$ FK$_2$P $>$ FK$_1$K$_2$P $=$ EK$_1$K$_2$P $>$ FP (Burdich et al. 1988). The fibrin affinity of these mutants was found to be directly correlated with the in vitro fibrinolytic activity.

10.4.1.2 Catalytic Domains. The structures required for the enzymatic activity of t-PA are fully comprised within the COOH-terminal region (B-chain). This is evidenced by the intact enzymatic activity of the isolated B-chain, either when separated chemically (Rijken and Groeneveld 1986; Holvoet et al. 1986) or prepared by recombinant DNA technology (Verheijen et al. 1986b; MacDonald ct al. 1986). Thc plasminogen-activating activity of the isolated B-chain, however, is much less stimulated by fibrin (Rijken and Groeneveld 1986; Holvoet et al. 1986; Verheijen et al. 1986b) than is intact t-PA.

Deletion mutants of t-PA have also been used to identify the domains in t-PA that are involved in the stimulation of its plasminogen-activating potential by fibrin. Three groups (Verheijen et al. 1986b; van Zonneveld et al. 1986a, 1986b; Larsen et al. 1988b) have reported that t-PA mutants lacking the finger domain are equally well stimulated by fibrin as is intact t-PA. Mutants lacking the second kringle structure but containing the finger domain show significantly reduced stimulation. These findings suggest that interaction of K_2 with fibrin is primarily responsible for fibrin stimulation.

Johannessen et al. (1988), on the other hand, found that deletion of either the growth factor domain or of both the finger and growth factor domains, also results in significantly reduced stimulation by fibrin. Kalyan et al. (1988) confirmed that a deletion mutant of t-PA lacking both the finger and growth factor domains (K_1K_2P) was less stimulated by fibrin than wild-type t-PA. This contrasts with the similar mutants described by Gething et al. (1988) and by Larsen et al. (1988b), which showed unaltered stimulation by fibrin. This difference may be due to conformational changes in the other domains of the heavy chain caused by the more extensive deletion (residues 2–89) produced by Kalyan et al., as opposed to those of Gething et al. (residues 4–87) or of Larsen et al. (residues 6–86).

Several additional reports on the fibrin affinity of t-PA deletion mutants have added to the existing confusion. Ehrlich et al. (1987) have constructed a t-PA mutant which lacks K_1 and K_2 and in which residue 86 at the carboxy-terminal end of the E domain is linked to residue 262. This mutant had a fibrinolytic activity comparable to that of melanoma t-PA and a similar fibrin affinity. In plasma clot lysis systems in vitro, it appeared to be an efficient thrombolytic agent. These results would suggest that the presence of the finger domain only is sufficient for fibrin stimulation of t-PA. These results thus are in apparent contrast with those of van Zonneveld et al. (1986a, 1986b) who found that the t-PA mutant FEP (deletion of residues 89–254) shows greatly reduced fibrin binding and a reduced stimulation of its plasminogen-activating potential by fibrin.

Gething et al. (1988) reported that t-PA mutants lacking both kringles are not stimulated by fibrin. However, mutants containing only one kringle (either K_1 or K_2) are indistinguishable from one another and from wild-type t-PA, suggesting that K_1 and K_2 would be equivalent in their ability to mediate stimulation of the catalytic activity of t-PA by fibrin. This conclusion is at variance with the results of van Zonneveld et al. (1986a, 1986b) and Verheijen et al. (1986b) which suggest that only K_2 and not K_1 is involved in fibrin binding and stimulation. These discrepancies between the results of different groups indicate that loss of a function by deletion of a domain must be interpreted with care, because observed alterations of function may be due to conformational changes in the remaining part of the molecule rather than to the deletion. It seems appropriate to limit interpretation to the functions which are retained in particular deletion mutants; this provides evidence that the retained function is not associated with the deleted domain.

Finally, a deletion mutant of t-PA has been constructed in which the last three carboxyterminal acids (Met-Arg-Pro) are lacking (Haigwood et al. 1989). The activity of this mutant relative to wild-type t-PA was lower in the absence of fibrin, but displayed about twofold greater stimulation by fibrin than wild-type t-PA. This observation suggests that the carboxy terminus of t-PA might be involved in its interaction with fibrin.

10.4.1.3 Domains Involved in Clearance of t-PA In Vivo.

Animal experiments have indicated that rapid clearance of t-PA (initial $t_{1/2}$, 1-4 min in rabbits and mice) (Korninger et al. 1981; Fuchs ct al. 1985; Krause 1988) occurs almost exclusively via the hepatocytes. Although a receptor for t-PA has not been conclusively identified, the rapid uptake probably involves receptor-mediated endocytosis and lysosomal degradation. Alternatively, it has been suggested that binding of t-PA to endothelial cells may also play a role in its removal from the circulation in vivo (Beebe 1987; Kuiper et al. 1988). By evaluating the clearance rate in rats of isolated A- (aminoterminal) and B- (carboxyterminal) polypeptide chains of human t-PA, evidence was obtained that t-PA is recognized by the liver primarily by the A-chain (Rijken and Emeis 1986).

The determinants involved in the rapid in vivo clearance of t-PA have also been studied with deletion mutants. Larsen et al. (1988a) have studied the pharmacokinetic properties in rats of deletion mutants lacking the finger domain (EK_1K_2P), the growth factor domain (FK_1K_2P), or both (K_1K_2P). The clearance of wild-type t-PA and of FK_1K_2P was biphasic, with an α phase half-life of 0.8 min and 2.1 min, and a β phase half-life of 12 min and 9.2 min respectively. In marked contrast, EK_1K_2P and K_1K_2P each cleared in a single phase with half-lives of 27 and 17 min respectively. These results suggest that clearance of t-PA occurs mainly via the finger domain.

These findings were extended by the construction of glycosylation variants of K_1K_2P, either with the glycosylated Asn 117 residue mutagenized to Gln (K_1K_2P-N117Q), or with the three known glycosylated Asn residues replaced by Gln (K_1K_2P-N117Q, N187Q, N448Q). Following infusion over a 4-hour period in rabbits, t-PA-related antigen disappeared from plasma with an initial $t_{1/2}$ of 25 min for K_1K_2P, 42 min for K_1K_2P-N117Q, and 14 min for the carbohydrate-minus version, as compared to 4 min for natural t-PA (Collen et al. 1988). The specific thrombolytic activity and fibrin specificity of these mutants in rabbits with jugular vein thrombosis were found to be similar to those of natural t-PA (Collen et al. 1988). Upon bolus injection of K_1K_2-N117Q, N187Q, N448Q in dogs with copper-coil-induced coronary artery thrombosis, this mutant was shown to have higher thrombolytic potency than wild-type t-PA (Cambier et al. 1988).

Browne et al. (1988), however, have found that a mutant of t-PA lacking only the growth factor domain (amino acids 51–87) has a greatly prolonged $t_{1/2}$ in a guinea pig model. Johannessen et al. (1988) also found 5- to 10-fold prolonged half-lives in rats and rabbits by deletion of the growth factor domain of t-PA. This suggests that the determinants recognized by hepatic receptors are comprised within the growth factor domain or that deletion mutagenesis has disrupted another structure in t-PA which is involved in its clearance. Kalyan et al. (1988) reported that a deletion mutant lacking both the finger and the growth factor domains (K_1K_2P) is cleared in a monophasic manner in mice with $t_{1/2}$ of 51 min as compared to a biphasic clearance for wild-type t-PA with α-phase half-life of 2 min and a β-phase half-life of 33 min.

In summary, presently available evidence suggests that the structures involved in the rapid hepatic clearance of t-PA are localized in the aminoterminal region, comprising the finger and growth factor domains, but do not allow a more precise identification of the structures within this region.

10.4.1.4 Domains Involved in Binding to Endothelial Cell Receptors. Barnathan et al. (1988a) presented evidence for the existence of two distinct binding sites for rt-PA on cultured human umbilical vein endothelial cells. Binding to the higher-affinity site required an intact catalytic site on the molecule and resulted in inhibition of t-PA by complexation to PAI-1. Binding to the second site occurred with higher capacity but lower affinity, with maintenance of rt-PA activity. K_1K_2P, on the other hand, retains its capacity to bind to cell surface PAI-1, but lacks the determinant required for binding to the high-capacity site on endothelial cells (Barnathan et al. 1988b). Deletion of the finger and growth factor domains of t-PA may thus affect its interaction with cellular receptors and thereby its half-life and distribution in vivo.

10.4.2 Point Mutations of t-PA

Single-chain t-PA is converted to the two-chain form by plasmin at the site of a fibrin clot (Rijken et al. 1982). In addition, some reports (Higgins and Vehar 1987; Tate et al. 1987) claim that one-chain t-PA binds significantly better to fibrin than two-chain t-PA. During fibrinolysis, t-PA would thus be converted to a form with lower fibrin affinity. To clarify the functional consequences of cleavage of t-PA at Arg 275 this residue was converted to glutamic acid or glycine by site-directed mutagenesis (Tate et al. 1987; Petersen et al. 1988). In the absence of fibrin(ogen), these mutants have activity which is 20- to 50-fold lower than that of two-chain t-PA, but in the presence of fibrin, some of the variants were found to have full plasminogen-activating potential without being converted to a two-chain molecule. R275G constructed by Haigwood et al. (1989) and R275C of Higgins et al. (1988a) are exceptions; both had severely attenuated activity in the presence of fibrin. Interestingly R275G did not have low activity when constructed by Higgins et al. (1988b). In vivo, in an arteriovenous shunt model in rabbits and dogs, mutant single-chain and wild-type rt-PA were found to have similar fibrinolytic activity (Eisert and Müller 1988). These experiments thus establish that single-chain t-PA is an active enzyme and not an inactive proenzyme, and suggest that fibrin has a more pronounced stimulatory effect on the activity of single-chain t-PA than on that of two-chain t-PA.

The observation that single-chain t-PA is an active protease opens up a very interesting question as to how this is effected mechanistically. It would also be interesting to find out whether, by mutation, t-PA could be made into a true zymogen, i.e., a protein which is inactive in the single-chain form, and which would become active only after the cleavage of R275.

10.4.3 Resistance to Inhibitors

Only a few site-directed variants of t-PA have been analyzed for resistance to inhibitors. Haigwood et al. (1989) made the interesting observation that variant K277R was resistant to inhibition by PAI-2, while K277I was actually more sensitive than the wild type. Madison et al. (1989) found that a number of mutations in the region from residue 296–304, as well as deletions in this region, conferred resistance to the fast-acting inhibitor, PAI-1. Bennett et al. (1990), have confirmed the latter results, and have suggested that the 296–304 region is the only one in t-PA where PAI-1 resistance can be conferred by point mutations.

10.4.4 Glycosylation Variants of t-PA

If carbohydrates are in any way responsible for the clearance of t-PA, then removal of the glycosylation signal by mutation should produce site-specific carbohydrate-depleted proteins, which would be expected to have prolonged clearance in vivo. In contrast with the domain deletions, however, there is a paucity of data on which to judge the involvement of individual carbohydrate moieties in clearance. Lau et al. (1987) reported having extended the in vivo half-life of t-PA by the site-directed mutation N448Q (our numbering system). However, the authors later reported that the mutant they had studied was not N448Q, but rather N117Q (Lau et al. 1988).

Some of the variants that have been described as lacking glycosylation sites are multiple mutants. Larsen et al. (1989) showed that nonglycosylated t-PA cleared essentially as did wild-type t-PA, and the authors concluded that carbohydrates are not responsible for the rapid clearance of t-PA. This conclusion is supported by the observation of Collen et al. (1984) in which types I and II t-PA, which differ by the presence or absence of a sugar moiety at position 184, are cleared identically in rabbits. Furthermore, when the double mutant N117Q, N184Q was tested in rabbits by Haigwood et al. (1989), the clearance was similar to wild-type. All these results are difficult to reconcile with the extended clearance of N117Q reported by Lau et al. (1987, 1988), and for the moment, the role of carbohydrates in t-PA clearance remains unsettled.

10.5 CONCLUSIONS

Thrombolytic therapy is emerging as the preferred treatment for patients with acute evolving myocardial infarction, resulting in reduced infarct size, preservation of left ventricular function, and reduction in mortality. Recombinant tissue-type plasminogen activator (rt-PA), in the presence of heparin, is a more effective and fibrin-specific coronary arterial thrombolytic agent than streptokinase. Variants of rt-PA have altered enzymatic properties and pharmacological profiles and may constitute improved agents for thrombolytic therapy, with maximized efficacy, minimized side effects, and optimized cost-benefit ratios.

REFERENCES

Agnelli, G., Buchanan, M.R., Fernandez, F., et al. (1985) *Circulation* 72, 178–182.

Alessi, M.C., Juhan-Vague, I., Declerck, P.J., and Collen, D. (1990) *Thromb. Res.* (in press).

Bachmann, F. (1987) in *Thrombosis and Haemostasis 1987* (Verstraete, M., Vermylen, J., Lijnen, R., and Arnout, J., eds.), pp. 227–265, Leuven University Press, Leuven, Belgium.

Banyai, L., Varadi, A., and Patthy, L. (1983) *FEBS Lett.* 163, 37–41.

Barnathan, E.S., Kuo, A., Van der Keyl, H., et al. (1988a) *J. Biol. Chem.* 263, 7792–7799.

Barnathan, E.S., Cines, D., Barone, K., Kuo, A., and Larsen, G.R. (1988b) *Fibrinolysis* 2 (suppl. 1), 28 (abstr. 58).

Beebe, D.P. (1987) *Thromb. Res.* 46, 241–254.

Beebe, D.P., and Aronson, D.L. (1986) *Thromb. Res.* 43, 663–674.

Bennett, W.F., Paoni, N., Botstein, D., et al. (1990) *J. Biol. Chem.* (in press).

Bergmann, S.R., Fox, K.A.A., Ter-Pogassian, M.M., Sobel, B., and Collen, D. (1983) *Science* 220, 1181–1183.

Bleich, S.D., Nichols, T., Schumacher, R., et al. (1989) *Circulation* 80, II–113 (abstr. 455).

Bounameaux, H., Stassen, J.M., Seghers, C., and Collen, D. (1986) *Blood* 67, 1493–1497.

Brott, T., Haley, E.C., Levy, D.E., et al. (1988) *Ann. Emerg. Med.* 17, 1202–1205.

Browne, M.J., Tyrrell, A.W.R., Chapman, C.G., et al. (1985) *Gene* 33, 279–284.

Browne, M.J., Carey, J.E., Chapman, C.G., et al. (1988) *J. Biol. Chem.* 263, 1599–1602.

Bugelski, P.J., Fong, K.L., Klinkner, A., et al. (1989) *Thromb. Res.* 53, 287–303.

Bunning, R.A.D., Crawford, A., Richardson, H.J., et al. (1987) *Biochim. Biophys. Acta* 924, 473–482.

Burdich, M.D., Erickson, L.A., and Schaub, R.G. (1988) *Fibrinolysis* 2 (suppl. 1), 53 (abstr. 120).

Buteux, G., Jubault, V., Suisse, A., and Courtheoux, P. (1988) *Lancet* 2, 1143–1144.

Califf, R.M., Topol, E.J., George, B.S., et al. (1989) *Circulation* 80, II–418 (abstr. 1660).

Cambier, P., Van de Werf, F., Larsen, G.R., and Collen, D. (1988) *J. Cardiovasc. Pharmacol.* 11, 468–472.

Camiolo, S.M., Thorsen, S., and Astrup, T. (1971) *Proc. Soc. Exp. Biol. Med.* 138, 277–280.

Chesebro, J.H., Knatterud, G., Roberts, R., et al. (1987) *Circulation* 76, 142–154.

Chesebro, J.H., Knatterud, G., and Braunwald, E. (1988) *N. Engl. J. Med.* 319, 1544–1545.

Christensen, U. (1984) *Biochem. J.* 223, 413–421.

Collen, D. (1980) *Thromb. Haemost.* 43, 77–89.

Collen, D. (1987) *J. Cell Biochem.* 33, 77–86.

Collen, D., and Gold, H.K. (1989) in *Thrombolysis in Cardiovascular Disease* (Julian, D., Kübler, W., Norris, R.M., et al., eds.), pp. 45–67, Marcel Dekker Inc., New York.

Collen, D., and Lijnen, H.R. (1986) *Crit. Rev. Haemat. Oncol.* 4, 249–301.

Collen, D., Rijken, D.C., Van Damme, J., and Billiau, A. (1982) *Thromb. Haemost.* 48, 294–296.

Collen, D., Stassen, J.M., and Verstraete, M. (1983) *J. Clin. Invest.* 71, 368–376.

Collen, D., Stassen, J.M., Marafino, B.J., et al. (1984) *J. Pharmacol. Exp. Ther.* 231, 146–152.

Collen, D., Stassen, J.M., and Larsen, G. (1988) *Blood* 71, 216–219.

Collen, D., Lijnen, H.R., Todd, P.A., and Goa, K.L. (1989) *Drugs* 38, 346–388.

Colucci, M., Paramo, J.A., and Collen, D. (1985) *J. Clin. Invest.* 75, 818–824.

Come, P.C., Kim, D., Parker, A., et al. (1987) *J. Am. Coll. Cardiol.* 10, 971–978.

Degen, S.J.F., Rajput, B., and Reich, E. (1986) *J. Biol. Chem.* 261, 6972–6985.

De Wood, M.A., Spores, J., Notske, R., et al. (1980) *N. Engl. J. Med.* 303, 897–902.

Dodd, I., Jalalpour, S., Southwick, N., et al. (1986) *FEBS Lett.* 209, 13–17.

Ehrlich, H.J., Bang, N.U., Little, S.P., et al. (1987) *Fibrinolysis* 1, 75–81.

Eisert, W.G., and Müller, T.H. (1988) *Fibrinolysis* 2 (suppl. 1), 52 (abstr. 118).

Flameng, W., Van de Werf, F., Vanhaecke, J., et al. (1985) *J. Clin. Invest.* 75, 84–90.

Fry, E.T.A., and Sobel, B.E. (1988) *Blood* 71, 1347–1352.

Fuchs, H.E., Berger, H., and Pizzo, S.V. (1985) *Blood* 65, 539–544.

Garabedian, H.D., Gold, H.K., Leinbach, R.C., et al. (1986) *Am. J. Cardiol.* 58, 673–679.

Garabedian, H.D., Gold, H.K., Leinbach, R.C., et al. (1987) *Am. Coll. Cardiol.* 9, 599–607.

Gelehrter, T.D., and Sznycer-Laszuk, R. (1986) *J. Clin. Invest.* 77, 165–169.

Gelehrter, T.D., Sznycer-Laszuk, R., Zeheb, R., and Cwikel, B.J. (1987) *Mol. Endocrinol.* 1, 97–101.

Gething, M.-J., Adler, B., Boose, J.-A., et al. (1988) *EMBO J.* 7, 2731–2740.

Gold, H.K., Fallon, J.T., Yasuda, T., et al. (1984) *Circulation* 70, 700–707.

Goldhaber, S., Vaughan, D.E., Markis, J.E., et al. (1986) *Lancet* 2, 886–889.

Goldhaber, S., Kessler, C.M., Heit, J., et al. (1988) *Lancet* 2, 293–298.

Graor, R.A., Risius, B., Young, J.R., et al. (1986) *J. Vasc. Surg.* 3, 115–124.

Haigwood, N., Mullenbach, G.T., Moore, G.K., et al. (1989) *Prot. Engineer.* 2, 611–620.

Hajjar, K.A., Harpel, P.C., Jaffe, E.A., and Nachman, R.L. (1986) *J. Biol. Chem.* 261, 11656–11662.

Hajjar, K.A., Hamel, N.M., Harpel, P.C., and Nachman, R.L. (1987) *J. Clin. Invest.* 80, 1712–1719.

Hanss, M., and Collen, D. (1987) *J. Lab. Clin. Med.* 109, 97–104.

Harris, T.J.R., Patel, T., Marston, F.A.O., et al. (1986) *Mol. Biol. Med.* 3, 279–292.

Henze, T., Boeer, A., Tebbe, U., and Romatowski, J. (1987) *Lancet* 2, 1391.

Hergrueter, C.A., Handren, J., Kersh, R., and May Jr., J.W. (1988) *Plast. Reconstr. Surg.* 81, 418–424.

Higgins, D.L., and Vehar, G.A. (1987) *Biochemistry* 26, 7786–7791.

Higgins, D.L., Lamb, M.C., Young, S.L., et al. (1988a) *Fibrinolysis* 2 (suppl. 1), 123 (abstr. 279).

Higgins, D.L., Lamb, M.C., Holmes, W.E., and Young, S.L. (1988b) *Fibrinolysis* 2 (suppl. 1), 122 (abstr. 278).

Holmes, W.E., Pennica, D., Blaber, M., et al. (1985) *Biotechnology* 3, 923–929.

Holmes, W.E., Nelles, L., Lijnen, H.R., and Collen, D. (1987) *J. Biol. Chem.* 262, 1659–1664.

Holvoet, P., Cleemput, H., and Collen, D. (1985) *Thromb. Haemost.* 54, 684–687.

Holvoet, P., Lijnen, H.R., and Collen, D. (1986) *Eur. J. Biochem.* 158, 173–177.

Holvoet, P., Boes, J., and Collen, D. (1987) *Blood* 69, 284–289.

Hoylaerts, M., Rijken, D.C., Lijnen, H.R., and Collen, D. (1982) *J. Biol. Chem.* 257, 2912–2919.

Hsia, J., Hamilton, W.P., Kleiman, N., et al. (1990) *N. Engl. J. Med.* 323, 1433–1437.

ISIS-2 (Second International Study of Infarct Survival) Collaborative Group (1988) *Lancet* 2, 349–360.

Johannessen, M., Diness, V., Pingel, K., et al. (1988) *Fibrinolysis* 2 (suppl. 1), 30 (abstr. 64).

Kagitani, H., Tagawa, M., Hatanaka, K., et al. (1985) *FEBS Lett.* 189, 145–149.

Kalyan, N.K., Lee, S.G., Wilhelm, J., et al. (1988) *J. Biol. Chem.* 263, 3971–3978.

Kaufman, R., Wasley, L.C., Spiliotes, A.J., et al. (1985) *Mol. Cell Biol.* 5, 1750–1759.

Korninger, C., and Collen, D. (1981) *Thromb. Haemost.* 46, 561–565.

Korninger, C., Stassen, J.M., and Collen, D. (1981) *Thromb. Haemost.* 46, 658–661.

Korninger, C., Matsuo, O., Suy, R., Stassen, J.M., and Collen, D. (1982) *J. Clin. Invest.* 69, 573–580.

Kosow, D.P. (1975) *Biochemistry* 14, 4459–4465.

Koudstaal, P.J., Stibbe, J., and Vermeulen, M. (1988) *Br. Med. J.* 297, 1571–1574.

Krause, J. (1988) *Fibrinolysis* 2, 133–142.

Kruithof, E.K.O., Ransijn, A., and Bachmann, F. (1983) in *Progress in Fibrinolysis*, vol. 6 (Davidson, J.F., Bachmann, F., Bouvier, C.A., and Kruithof, E.K.O., eds.), pp. 365–369, Churchill Livingstone, Edinburgh.

Kruithof, E.K.O., Tran-Thang, C., Ransijn, A., and Bachmann, F. (1984) *Blood* 64, 907–913.

Kuiper, J., Otter, M., Rijken, D.C., and van Berkel, T.J.C. (1988) *J. Biol. Chem.* 263, 18220–18224.

Larsen, G.R., Henson, K., and Blue, Y. (1988b) *J. Biol. Chem.* 263, 1023–1029.

Larsen, G.R., Metzger, M., Henson, K., Blue, Y., and Horgan, P. (1989) *Blood* 73, 1842–1850.

Lau, D., Kuzma, G., Wei, C.M., Livingston, D.J., and Hsiung, N. (1987) *Biotechnology* 5, 953–958.

Lau, D., Kuzma, G., Wei, C.-M., Livingston, D.J., and Hsiung, N. (1988) *Biotechnology* 6, 734.

Lijnen, H.R., Marafino Jr., B.J., and Collen, D. (1984) *Thromb. Haemost.* 52, 308–310.

Lijnen, H.R., Van Hoef, B., and Collen, D. (1986) *Biochim. Biophys. Acta* 884, 402–408.

Liu, Y.-X., Cajander, S.B., Ny, T., et al. (1987) *Mol. Cell. Endocrinol.* 54, 221–229.

Lucore, C.L., and Sobel, B.E. (1988) *Circulation* 77, 660–669.

MacDonald, M.E., van Zonneveld, A.J., and Pannekoek, H. (1986) *Gene* 42, 59–67.

Madison, E.L., Goldsmith, E.J., Gerard, R.D., Gething, M.J.H., and Sambrook, J.F. (1989) *Nature* 339, 721–724.

Magnani, B. (1989) *J. Am. Coll. Cardiol.* 13, 19–26.

Marder, V., and Sherry, S. (1988) *N. Engl. J. Med.* 318, 1585–1595.

Markland, W., Pollock, D., and Livingston, D.J. (1989) *Protein Eng.* 3, 117–125.

Matsuo, O., Rijken, D.C., and Collen, D. (1981a) *Thromb. Haemost.* 45, 225–229.

Matsuo, O., Rijken, D.C., and Collen, D. (1981b) *Nature* 291, 590–591.

Miles, L.A., and Plow, E.F. (1985) *J. Biol. Chem.* 260, 4303–4311.

Mueller, H.S., Rao, A.K., Forman, S.A., et al. (1987) *J. Am. Coll. Cardiol.* 10, 479–490.

Neuhaus, T.L., Tebbe, U., Gottwick, M., et al. (1988) *J. Am. Coll. Cardiol.* 12, 581–587.

Nieuwenhuizen, W., Voskuilen, M., Vermond, A., et al. (1988) *Eur. J. Biochem.* 174, 163–169.

Nilsson, S., Einarsson, M., Ekvärn, S., et al. (1985) *Thromb. Res.* 39, 511–521.

Norrman, B., Wallen, P., and Ranby, M. (1985) *Eur. J. Biochem.* 149, 193–200.

Ny, T., Elgh, F., and Lund, B. (1984) *Proc. Natl. Acad. Sci. USA* 81, 5355–5359.

Ny, T., Liu, Y.-X., Ohlsson, M., et al. (1987) *J. Biol. Chem.* 262, 11790–11793.

O'Connell, M.L., Canipari, R., and Strickland, S. (1987) *J. Biol. Chem.* 262, 2339–2344.

Parekh, R.B., Dwek, R.A., Rudd, P.M., et al. (1989) *Biochemistry* 28, 7670–7679.

Pennica, D., Holmes, W.E., Kohr, W.J., et al. (1983) *Nature* 301, 214–221.

Petersen, L.C., Johannessen, M., Foster, D., Kumar, A., and Mulvihill, E. (1988) *Biochim. Biophys. Acta* 952, 245–254.

Pohl, G., Källström, M., Bergsdorf, N., Wallén, P., and Jörnvall, H. (1984) *Biochemistry* 23, 3701–3707.

Rajput, B., Degen, S.F., Reich, E., et al. (1985) *Science* 230, 672–674.

Ramakrishnan, V., Sinicropi, D.V., Dere, R., et al. (1990) *J. Biol. Chem.* 265, 2755–2762.

Ranby, M. (1982) *Biochim. Biophys. Acta* 704, 461–469.

Reddy, V.B., Garramone, A.J., Sasak, H., et al. (1987) *J. Mol. Biol.* 6, 461–472.

Rentrop, K.P, Feit, F., Blanke, H., et al. (1984) *N. Engl. J. Med.* 311, 1457–1463.

Rijken, D.C., and Collen, D. (1981) *J. Biol. Chem.* 256, 7035–7041.

Rijken, D.C., and Emeis, J.J. (1986) *Biochem. J.* 238, 643–646.

Rijken, D.C., and Groeneveld, E. (1986) *J. Biol. Chem.* 261, 3098–3102.

Rijken, D.C., Wijngaards, G., and Welbergen, J. (1980) *Thromb. Res.* 18, 815–830.

Rijken, D.C., Wijngaards, G., Zaal-De Jong, M., and Welbergen, J. (1979) *Biochim. Biophys. Acta* 580, 140–153

Rijken, D.C., Hoylaerts, M., and Collen, D. (1982) *J. Biol. Chem.* 257, 2920–2925.

Rijken, D.C., Juhan-Vague, I., and Collen, D. (1983a) *J. Lab. Clin. Med.* 101, 285–294.

Robin, P., Gruel, Y., Lang, M., Lagarrigue, F., and Scotto, J.M. (1988) *Lancet* 1, 1391.

Russell, M.E., Quertermous, T., Declerck, P.J., et al. (1990) *J. Biol. Chem.* 265, 2569–2575.

Sambrook, J., Hanahan, D., Rodgers, L., and Gething, M.J. (1986) *Mol. Biol. Med.* 3, 459–481.

Sampol, J., Mercier, C., Houel, F., David, G., and Daver, J. (1983) in *Progress in Fibrinolysis*, vol. 6 (Davidson, J.F., Bachmann, F., Bouvier, C.A., and Kruithof, E.K.O., eds.), pp. 463–466, Churchill Livingstone, Edinburgh.

Sarmientos, P., Duchesne, M., Denefle, P., et al. (1989) *Biotechnology* 7, 495–501.

Seifried, E., Tanswell, P., Rijken, D.C., et al. (1988) *Arzneimittel-Forsch.* 38, 418–422.

Sherry, S. (1988) *J. Am. Coll. Cardiol.* 12, 519–525.

Smith, R.A.G., Dupe, R.J., English, P.D., and Green, J. (1981) *Nature* 290, 132–135.

Soeda, S., Kakiki, M., Shimeno, H., and Nagamatsu, A. (1987) *Biochem. Biophys. Res. Comm.* 146, 94–100.

Sottrup-Jensen, L., Petersen, T.E., and Magnusson, S. (1978) in *Atlas of Protein Sequence and Structure*, vol. 5, suppl. 3 (Dayhoff, M.O., ed.) pp. 91, National Biochemical Research Foundation, Washington, DC.

Spellmann, M.W., Basa, L.J., Leonard, C.K., Chakel, J.A., and O'Connor, J.V. (1989) *J. Biol. Chem.* 264, 14100–14111.

Stewart, A., and Mayne, E.E. (1988) *Lancet* 1, 890.

Stricker, R.B., Wong, D., Shiu, D.T., et al. (1986) *Blood* 68, 275–280.

Suenson, E., Lützen, O., and Thorsen, S. (1984) *Eur. J. Biochem.* 140, 513–522.

Takada, A., Hou, P., Mori, T., and Takada, Y. (1988) *Thromb. Res.* (suppl.) 8, 23–33.

Tate, K.M., Higgins, D.L., Holmes, W.E., et al. (1987) *Biochemistry* 26, 338–343.

The t-PA Acute Stroke Study Group (1988) *Stroke* 19, 134.

Thorsen, S., and Philips, M. (1984) *Biochim. Biophys. Acta* 802, 111–118.

Turpie, A.G.G. (1987) in *Tissue Plasminogen Activator in Thrombolytic Therapy* (Sobel, B.E., Collen, D., and Grassland, E.B., eds.), pp. 131–146, Marcel Dekker Inc., New York.

Turpie, A.G.G. (1989) in *Thrombolysis in Cardiovascular Disease* (Julian, D., Kübler, W., Norris, R.M., et al., eds.), pp. 397–408, Marcel Dekker Inc., New York.

Upshall, A., Kumar, A.A., Bailey, M.C., et al. (1987) *Bio/Technology* 5, 1301–1304.

Van de Werf, F., Bergmann, S.R., Fox, K.A.A., et al. (1984a) *Circulation* 69, 605–610.

Van de Werf, F., Ludbrook, P.A., Bergmann, S.R., et al. (1984b) *N. Engl. J. Med.* 310, 609–613.

van Zonneveld, A.J., Veerman, H., and Pannekoek, H. (1986a) *J. Biol. Chem.* 261, 14214–14218.

van Zonneveld, A.J., Veerman, H., and Pannekoek, H. (1986b) *Proc. Natl. Acad. Sci. USA* 83, 4670–4674.

Vaughan, D.E., Mendelsohn, M.E., Declerck, P.J., et al. (1989) *J. Biol. Chem.* 264, 15869–15874.

Vehar, G.A., Kohr, W.J., Bennett, W.F., et al. (1984) *Biotechnology* 2, 1051–1057.

Vehar, G.A., Spellman, M.W., Keyt, B.A., et al. (1986) in *Cold Spring Harbor Symposia on Quantitative Biology*, vol. LI, pp. 551–562, Cold Spring Harbor Laboratory.

Verhaeghe, R., Besse, P., Bounameaux, H., and Marbet, G.A. (1989) *Thromb. Res.* 55, 5–11.

Verheijen, J.H., Visse, R., Wijnen, J.T., et al. (1986a) *Hum. Genet.* 72, 153–156.

Verheijen, J.H., Caspers, M.P.M., Chang, G.T.G., et al. (1986b) *EMBO J.* 5, 3525–3530.

Verheijen, J.H., Caspers, M.P.M., de Munk, G.A.W., et al. (1987) *Thromb. Haemost.* 58, 491 (abstr. 1814).

Verstraete, M., and Collen, D. (1985) in *Thrombolysis: Biological and Therapeutical Properties of New Thrombolytic Agents* (Collen, D., Lijnen, H.R., and Verstraete, M., eds.), pp. 49–60, Churchill Livingstone, Edinburgh.

Verstraete, M., Bernard, R., Bory, M., et al. (1985a) *Lancet* 1, 842–847.

Verstraete, M., Bounameaux, H., De Cock, F., Van de Werf, F., and Collen, D. (1985b) *J. Pharmacol. Exp. Therap.* 235, 506–512.

Verstraete, M., Su, C.A.P.F., Tanswell, P., Feuerer, W., and Collen, D. (1986) *Thromb. Haemost.* 56, 1–5.

Verstraete, M., Hess, H., Mahler, F., et al. (1988a) *Eur. J. Vasc. Surg.* 2, 155–159.

Verstraete, M., Miller, G.A.H., Bounameaux, H., et al. (1988b) *Circulation* 77, 353–360.

Weimar, W., Stibbe, J., Van Seyen, A.J., et al. (1981) *Lancet* 2, 1018–1020.

White, H.D., Rivers, J.T., Maslowski, A.H., et al. (1989) *N. Engl. J. Med.* 320, 817–821.

White, W.F., Barlow, G.H., and Mozen, M.M. (1966) *Biochemistry* 5, 2160–2169.

Wilcox, R.G., von der Lippe, G., Olsson, C.G., et al. (1988) *Lancet* 2, 525–530.

Wiman, B., and Collen, D. (1978a) *Nature (London)* 272, 549–550.

Wiman, B., and Collen, D. (1978b) *Eur. J. Biochem.* 84, 573–578.

Wittwer, A.J., Howard, S.C., Carr, L.S., et al. (1989) *Biochemistry* 28, 7662–7669.

Yang-Feng, T.L., Opdenakker, G., Volckaert, G., and Francke, U. (1986) *Am. J. Hum. Genet.* 39, 79–87.

Yusuf, S., Collins, R., Peto, R., et al. (1985) *Eur. Heart J.* 6, 556–585.

Zamarron, C., Lijnen, H.R., and Collen, D. (1984) *J. Biol. Chem.* 259, 2080–2083.

Zimmerman, R., Horn, A., Harenberg, J., et al. (1988) *Klin. Wschr.* 66, (suppl. XII), 137–142.

Fibrinogen and Fibrin Formation and Its Role in Fibrinolysis

Birger Blombäck

11.1 INTRODUCTION

Fibrinogen is the soluble protein in blood and tissue extract that, in the presence of thrombin, is transformed into an insoluble network structure called fibrin and is in the form of a gel. Fibrinogen is found in the blood plasma of all vertebrates, and similar gel-forming proteins are also found in several invertebrates, although in many cases there is no chemical relationship between the invertebrate proteins and vertebrate fibrinogen.

More than one hundred years have passed since Denis de Commercy and Olof Hammarsten carried out their pioneering studies of the fibrinogen purified from plasma and of its conversion into fibrin in the presence of thrombin. Since then, we have learned much about the properties of fibrinogen and fibrin, and about fibrin formation in several animal species. Now we are also beginning to understand the processes that govern clot formation and dissolution in blood vessels. Modern molecular biology has

This work was supported by grants from the Swedish Medical Research Council (B91-13X-02475-24B), the Karolinska Institute, and the Tornspiran Foundation, Stockholm, Sweden. The excellent assistance of Sonja Holmqvist, Gerd Sjöberg, and Birgitta Strimme is gratefully acknowledged.

also enabled us to identify the fibrinogen genes and to determine how they are transcribed and translated to functional fibrinogen in the cellular environment. This area of research will aid us to understand normal hemostasis and thrombosis in humans.

This chapter summarizes the most important physicochemical and chemical properties of fibrinogen—the molecular basis of its transformation to fibrin—as well as the processes involved in fibrinolysis. The evolution and biosynthesis of fibrinogen are discussed and finally its importance in health and disease. The reader who requires more information is referred to reviews by Scheraga and Laskowski (1957); Blombäck (1967, 1985); Laki (1968); Doolittle (1973, 1984); Crabtree (1987).

11.2 PHYSICOCHEMICAL PROPERTIES OF FIBRINOGEN

Various physicochemical and chemical properties of fibrinogen are shown in Table 11-1. Fibrinogen is a plasma protein of which more than 90% consists of amino acids. Approximately 3–5% of the mass is made up of covalently linked polysaccharides (Blombäck 1972). The molecule also contains small amounts of ester-linked phosphoric acid and sulfuric acid. The amino acid composition of fibrinogen is similar in several mammalian species (Cartwright and Kekwick 1971).

On the basis of diffusion, sedimentation, and light-scattering measurements, fibrinogen is estimated to have a molecular mass of about 340 kDa (Caspary and Kekwick 1957). Hydrodynamic measurements indicate that the fibrinogen molecule is a rod- or ellipsoid-shaped particle having a length of 40–50 nm and a length-width ratio of 5 to 10 (Scheraga and Laskowski 1957). However, the interpretation of the hydrodynamic data is dependent on the degree of hydration of the molecule. Neutron scattering studies indicate that the molecule may contain as much as 6 g of water per gram of protein (Marguerie and Stuhrmann 1976). This extremely high degree of hydration suggests that the molecule has a swollen-lattice-like structure. Marguerie and Stuhrmann suggested on the basis of their data that the most likely shape of fibrinogen in solution was an oblate disk, although a rodlike structure was also compatible with their data.

As with the hydrodynamic measurements, electron microscopic investigations have failed to yield unambiguous results. Hall and Slayter (1959) found that the molecule was a structure made up of three spherical particles, bound together by a threadlike structure. Köppel (1970) was unable to verify this structure but instead obtained micrographs showing spherical particles which appeared to have the form of a pentagonal dodecahedron. Köppel assumed that the chains of the protein were surface orientated while the inside was filled with water. Using the freeze-etching technique, Bachmann et al. (1975) produced electron micrographs of a structure which may represent the hydrated fibrinogen molecule. The idealized molecule is a cylinder

TABLE 11–1   Physicochemical and Chemical Properties of Fibrinogen

Property	Average Value	Reference
Molecular weight	341,000	Caspary and Kekwick (1957)
Sedimentation coefficient ($S_{20,w}^0$)	7.7–7.9×10^{-13} s	Scheraga and Laskowski (1957); Doolittle (1973)
Diffusion coefficient ($D_{20,w}^0$)	1.8–2.0×10^{-7} cm^2/sec	Scheraga and Laskowski (1957); Doolittle (1973)
Specific viscosity (η)	0.25 dl/g	Scheraga and Laskowski (1957); Doolittle (1973)
Partial specific volume (ν)	0.71–0.72 cm^3/g	Scheraga and Laskowski (1957); Doolittle (1973)
Frictional ratio (f/f^0)	2.34	Marguerie and Stuhrmann (1976)
Molecular volume	3.7×10^3 nm^3	Marguerie and Stuhrmann (1976)
Degree of hydration	6 (g/g protein)	Marguerie and Stuhrmann (1976)
Extinction coefficient ($A_{280}^{1\%}$)	16.25	Blombäck (1958)
Isoelectric point	5.5	Scheraga and Laskowski (1957)
α-helix content	33%	Mihalyi (1965)

with rounded ends (Figure 11-1). It has a length of 45 nm and a width of 9 nm. Irregular forms of the molecule were also observed. The form observed by Bachmann et al. requires that the volume of the hydrated molecule will be seven times greater than the "dry volume" calculated solely from data on the molecular mass and partial specific volume. By applying electron microscopy to a dry preparation of fibrinogen, a different picture of the molecule was produced, which was fairly consistent with the results of Hall and Slayter (1959). Therefore, the model of Hall and Slayter may represent a modified architectural form of fibrinogen produced by dehydration of the native fibrinogen molecule.

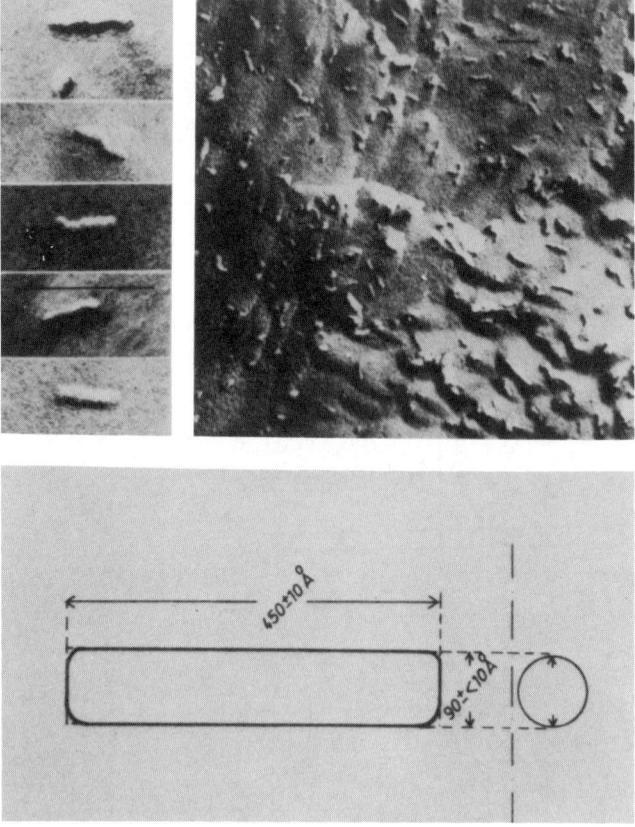

FIGURE 11-1 Top: Electron micrographs of fibrinogen molecules obtained with the freeze-etching technique. The right side shows a typical population of molecules. The left side shows some of the forms present in a population. Bottom: Diagram of the most likely shape and dimensions of the molecule. From Bachmann et al. (1975). Reprinted by permission of John Wiley and Sons, Ltd.

Blombäck and Yamashina (1958) concluded, by quantitative determination of the N-terminal amino acids of fibrinogen, that the molecule is a dimer (Figure 11–2). The two halves each consist of three polypeptide chains: Aα, Bβ, and γ, i.e., the formula for the molecule is (Aα, Bβ, γ)$_2$. Stryer et al. (1963) found in low-angle, X-ray diffraction studies of oriented fibrinogen that the axial repeats in the hydrated molecules were 22.5 nm apart. This number may represent the length of each half-molecule in a dimer of about 45 nm. This indicates that the half-molecules of fibrinogen are joined in the center of the molecule. Tooney and Cohen (1972) crystallized a derivative of fibrinogen lacking a small portion of the carboxy-terminal region of the Aα chain of the molecule. Simulations of electron micrographs of negatively stained crystals suggested that the fibrinogen molecule is 45 nm long and contains several globular domains linked by rodlike regions (Cohen et al. 1983; Weisel et al. 1985).

Fibrinogen in circulating blood appears to consist of a population of several slightly different molecules. By studying the solubility of fibrinogen in ethanol-water mixtures, three types of fibrinogen can be distinguished: one with high, one with low, and one with intermediate solubility (Mosesson et al. 1967). The fibrinogen having low solubility in ethanol-water has been shown to be associated with the plasma fibronectin (Mosesson and Finlayson 1976). Whether the fibrinogen in this complex has properties different from

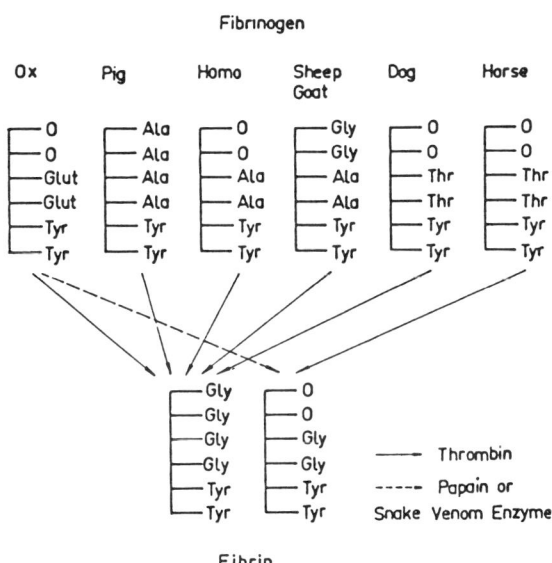

FIGURE 11–2 The six amino-terminal amino acids in fibrinogen and fibrin from various species. The number of chains shown is for a 340-kDa protein molecule. From Blombäck and Yamashina (1958).

those of the remaining plasma fibrinogen has not as yet been fully clarified. The fibrinogen used in most investigations is the form which has intermediate solubility. Fibrinogen having high solubility in ethanol-water has somewhat lower molecular weight. This fibrinogen probably arises as a consequence of enzymatic degradation of fibrinogen as this process has been shown to give rise to products having high solubility (Mosesson et al. 1974; Mosesson 1983). The susceptibility to plasmin, in early stages of fibrinolysis, of terminal segments of the chains of intact fibrinogen has been demonstrated in several studies (Mills and Karpatkin 1970; Gaffney 1971, 1977; Lahiri and Shainoff 1973; Ly et al. 1974a; Nossel 1981). The carboxy-terminal end of the Aα chain of fibrinogen is the first site of attack when the protein is degraded by plasmin. Degradation of the Bβ chain from the amino-terminal end also occurs at a fast rate, while the γ chain is the most resistant to plasmic attack. Amino-terminal analysis of fibrinogen suggests that the heterogeneity may also be explained, in part, by the occurrence of degradation by other endopeptidases than plasmin and by exopeptidases (Blombäck 1967).

Another cause of heterogeneity is the existence of charge heterogeneity in the oligosaccharide and phosphate residues which are bound to the peptide chains (Gaffney 1972). Charge heterogeneity has been demonstrated by chromatography of fibrinogen on DEAE-cellulose matrices (Finlayson and Mosesson 1963; Mosher and Blout 1973). According to Mosher and Blout, fibrinogen is separated into three distinct fractions: the separation is due to the occurrence of two different γ chains in fibrinogen, i.e., γ and γ¹ (Mosesson et al. 1972). The fibrinogen in the first fraction contains γ chains while the fibrinogen in the third fraction contains γ¹ chains. The intermediate fraction contains fibrinogen with both γ and γ¹ chains. Nevertheless, Mosher and Blout (1973) found that even in these chromatographically pure fractions, fibrinogen was heterogeneous. This superimposed charge heterogeneity appeared to be a consequence of differences in numbers of protein-bound phosphate groups. As a result of their experiments, Mosher and Blout suggested that as many as 36 slightly different types of fibrinogen molecules could exist in the blood of any one individual.

11.3 CHAINS AND PROSTHETIC GROUPS OF FIBRINOGEN

Amino-terminal analysis of the fibrinogen molecule from several mammalian species showed that it contained three pairs of polypeptide chains (Blombäck and Yamashina 1958) (Figure 11-2). As mentioned earlier, this suggested that it has a dimeric structure. Following oxidative sulfitolysis of the disulfide bridges, Henschen (1964) was able to isolate the composite polypeptide chains of human and bovine fibrinogen. Similar results have been achieved by other workers following carboxymethylation or oxidation of the disulfide bridges. The molecular mass of the chains denoted Aα, Bβ, and γ was 64, 57, and 47 kDa, respectively for human fibrinogen. Since only three types of chains could be demonstrated in these experiments and since the intact molecule was shown to have six chains, it became clear that the molecule is composed of three polypeptide chains in a paired dimer. The sum of molecular mass for the chains, i.e., 170 kDa, also supported the conclusion that the structure of intact fibrinogen of molecular mass 340 kDa is satisfied by the formula $(A\alpha, B\beta, \gamma)_2$. The half-molecules in this dimer were shown to be covalently linked through symmetrical disulfide bonds at the amino-terminal ends (Blombäck 1969; Blombäck and Blombäck 1972).

As one would expect from the observations carried out on intact fibrinogen, heterogeneity has been demonstrated to exist also in the isolated polypeptide chains of the molecule. At least four variants of the Aα chain differing in chromatographic and/or electrophoretic mobility have been described (Mosesson 1983). The heterogeneity of the Aα chain may be due to differences in charge. The prosthetic groups of ester-bound phosphate are bound partially as serine phosphate in the amino-terminal portion of the Aα chain (Aα Ser 3) (Blombäck et al. 1963). As the Aα Ser 3 residues are only phosphorylated to 30%, one could conclude that the charge heterogeneity is partly due to different extent of phosphorylation of the molecules. In addition, phosphate is also bound to Aα Ser 345 or 346 (Seydewitz et al. 1985). Variation in sulfate content, as was demonstrated for dog fibrinogen, may also contribute to the charge heterogeneity of fibrinogen. Sulfate is bound as tyrosine-O-sulfate in the fibrinopeptide B part of the Bβ chain in several animal species (Krajewski and Blombäck 1968) but this is not the case in human fibrinogen, where a sulfated tyrosine residue is located in some other not-yet-characterized region of the molecule (Jevons 1963).

The carbohydrate portion of human fibrinogen is bound to both the γ and Bβ chains (Gaffney 1972). The carbohydrate chains are of the biantennary type and heterogeneous. They are composed of approximately 11 monosaccharide residues; the chains are branched and built up of mannose, galactose, glucosamine, and sialic acid (Townsend et al. 1982). Sialic acid, or sialic acid and galactose, are situated in terminal positions on the chain branches. Difference in sialylation is the explanation for the charge heterogeneity of the carbohydrate chains. The polysaccharide chains are linked to the polypeptide chain via an asparagine residue. The compound constituting the link between carbohydrate and protein has been isolated in the form of 2-acetamido-1-N-(4-L-aspartyl)-2-deoxy-β-D-glucosylamine (Mester 1969). In the γ chain of human fibrinogen a heterogeneous carbohydrate moiety is bound in this fashion to γ Asn 52 (Iwanaga et al. 1968; Blombäck et al. 1973). The γ chain heterogeneity is partly due to differences in sialylation (Gati and Straub 1978). In the Bβ chain, the carbohydrate chain is linked to Asn 364 (Watt et al. 1979). Here a difference in sialylation also gives rise to two differently charged Bβ chains (Gati and Straub 1978).

The heterogeneity created by difference in sialylation is superimposed on the γ/γ^1 heterogeneity mentioned earlier. The heterogeneity of fibrinogen is thus partly explained by its charge heterogeneity, which exists in its isolated chains. As briefly mentioned earlier, the heterogeneity is also the result of limited proteolytic degradation caused by exo- and endopeptidases. This degradation probably occurs in vivo and preferentially at the carboxy-terminal end of the Aα chain (Mosesson 1983) and the amino-terminal end of the Bβ chain (Nossel 1981).

The heterogeneity may also be traced to the gene level. This possibility is exemplified by the point mutations and deletions which have been described in the various dysfibrinogenemias, such as fibrinogen Detroit (Blombäck et al. 1968b), fibrinogen New York (Liu et al. 1985), and other polymorphisms (Kant et al. 1983).

11.4 PRIMARY STRUCTURE OF FIBRINOGEN

Mainly through the combined efforts of three laboratories, the amino acid sequence of fibrinogen and its disulfide-bonded structure is now known, including the position of most prosthetic groups (Blombäck and Yamashina 1958; Blombäck et al. 1968a, 1969, 1972, 1973, 1974, 1976a; Hessel 1975; Hessel et al. 1979; Gårdlund 1977; Gårdlund et al. 1977; Henschen 1964, 1978, 1983; Henschen and Warbinek 1975; Henschen et al. 1976, 1979, 1983; Henschen and Lottspeich 1977; Lottspeich and Henschen 1977a, 1977b, 1978; Töpfer-Petersen et al. 1979; Doolittle 1981, 1983; Doolittle and Riley 1990; Doolittle et al. 1977a, 1977b, 1979; Bouma et al. 1978; Strong et al. 1985; Wang et al. 1989). With insight into the primary structure of fibrinogen, we can better understand the function of the molecule and of diseases where an abnormal function exists. However, a full understanding of the function of fibrinogen will probably require an elucidation of the tertiary structure by X-ray crystallography. In any event, the unravelling of the primary structure is a prerequisite to solve the puzzle of the tertiary structure in a meaningful way.

Two types of fragmentation have been of fundamental use in solving the primary structure of fibrinogen. The first type of fragments was produced using cyanogen bromide (CNBr), a chemical which reacts with the methionine residues in a protein. In this process methionine is converted into homoserine, and the bond joining the next amino acid residue in a polypeptide is cleaved. In order to place the CNBr-fragments in the intact structure, use is made of fragments produced by fragmentation with plasmin or other proteolytic enzymes. The structures of these will overlap some CNBr fragments and thereby facilitate the deduction of a unique amino acid sequence for the intact chains. Cleavage of human fibrinogen with CNBr results in some 30 fragments with molecular mass ranging from less than 2 to 60 kDa (Blombäck et al. 1968a; Gårdlund et al. 1977). Fibrinogen contains 28-29 disulfide bridges. The largest CNBr fragment, having a molecular mass of 58 kDa, is composed of amino-terminal fragments of the three chains of fibrinogen (Aα, Bβ, and γ). This fragment, which represents approximately 16% of the entire molecule, contains 38% of the disulfide bridges of the molecule and is called the amino-terminal disulfide knot, or N-DSK. N-DSK is a dimeric structure, which was proven by the isolation of symmetrical disulfide-bonded peptides from N-DSK (Blombäck 1969; Blombäck and Blombäck 1972; Blombäck et al. 1976a). The conclusion that N-DSK is a dimer implies also that the fibrinogen molecule is dimeric.

The arrangement of CNBr fragments and of disulfides in the fibrinogen molecule is schematically depicted in Figure 11–3. As can be seen in this diagram, the two halves of the molecule are connected by symmetrical disulfide bridges in the Aα and γ chains. The two halves thus have a twofold symmetry axis. Besides N-DSK, no disulfide fragment from other parts of the molecule have been shown to contain symmetrical disulfides, indicating that the half-molecules in fibrinogen are covalently linked only in the amino-terminal portion of the molecule. From Figure 11–3 it can be seen that the three chains in each half-molecule are also linked by interchain disulfide

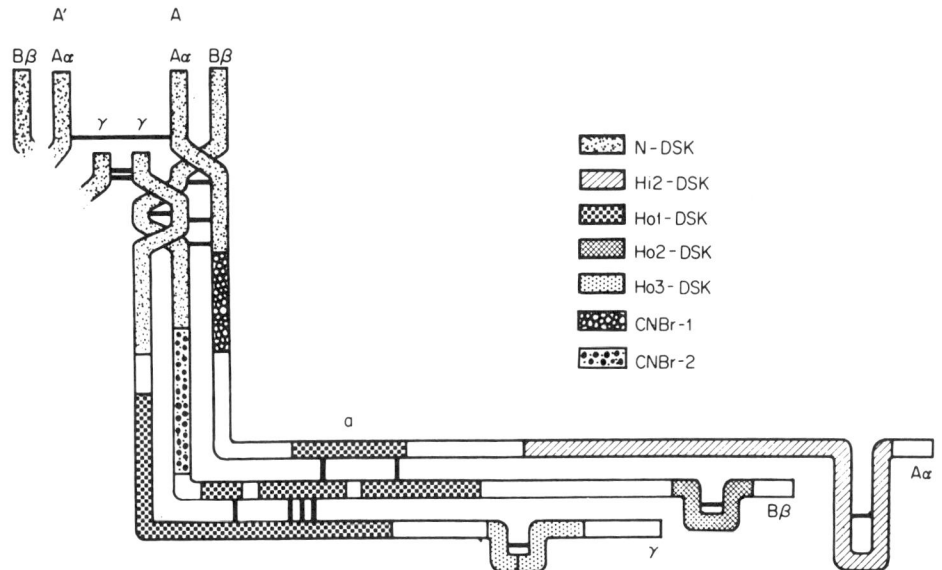

FIGURE 11–3 Schematic representation of a fibrinogen half-molecule with the remaining half-molecule partially indicated. Cyanogen bromide fragments are inserted in the structure. Black bars indicate disulfide bridges, "A" and "a" indicate the location of polymerization domains. Of the corresponding domains ("A′″", "a′″") on the contralateral side, only the location of the A′ domain is indicated. From Blombäck (1979b). Reprinted by permission of John Wiley and Sons, Ltd.

bridges. In this way particularly compact knots are formed both in the amino- and carboxy-terminal portions of molecule.

Fragmentation of fibrinogen by plasmin yields several fragments that have been useful in deduction of the primary structure of fibrinogen. Plasmin has a more narrow specificity than trypsin in the digestion of fibrinogen. Plasmin rapidly cleaves bonds in the carboxy-terminal portion of the Aα chain. Bonds are also cleaved in both the amino- and carboxy-terminal portions of the Bβ chain. The γ chain is also hydrolyzed by plasmin, but at a rate which is considerably slower than that of the other two chains (for review, see Gaffney 1977). The cleavage by plasmin follows a typical pattern, as was shown by Marder et al. (1969). The cleavage first produces fragment X. If digestion is continued a degradation product arises called Y. The final products of the digestion are the so-called "core" fragments, D and E, together with a number of smaller fragments. Fragment Y is of interest since it is produced from fragment X by release of 1 mol of fragment D per mole of fragment X. Hence, fragment Y appears to be composed of 1 mol of fragment E and 1 mol of fragment D. Consequently, it is an intermediate arising from an unsymmetrical cleavage of fragment X. The positions of some plasmin fragments within the fibrinogen molecule are depicted in Figure 11–4. By comparing Figures 11–3 and 11–4, it can be seen that fragment E has a considerable portion of its structure in common with N–DSK (Kowalska-Loth et al. 1973). In contrast to N-DSK, fragment E is devoid of the first 53 amino-terminal amino acid residues in the Bβ-chain and is also missing the carboxy-terminal portion of the γ chain in N-DSK. On the other hand, fragment E has carboxy-terminal extensions of the Aα and Bβ chains which are missing in N-DSK. Based on the structure of these extensions it became possible to isolate CNBr and other fragments providing overlaps between chains in fragment E and the Aα and Bβ chains in fragment D (Gårdlund 1977; Gårdlund et al. 1977; Doolittle et al. 1979).

Most investigators have isolated fragment D as a monomeric structure. Since fibrinogen is a dimer, it follows that there must exist two D fragments per molecule of fibrinogen. Fragment D consists of three chains which have their respective origins in the Aα, Bβ, and γ chains of fibrinogen. Since the γ chain of fragment D has structures in common with the carboxy-terminal portion of the γ chain of N-DSK, alignment between this chain in fragments E and D also became possible (Collen et al. 1975).

In the structure of fibrinogen as finally elucidated the Aα, Bβ, γ, and γ^1 chains were shown to contain 610, 461, 411, and 427 amino acid residues, respectively. Sequence analysis of the Aα chain cDNA revealed a carboxy-terminal extension of 15 amino acid residues. This brings the length of the Aα chain to 625 amino acid residues (Kant et al. 1983). The γ^1 variant is 16 amino acid residues longer than γ and has a different amino acid sequence for 20 carboxy-terminal residues (Wolfenstein-Todel and Mosesson 1981) making it more acidic. It is produced by different splicing of mRNA (Crabtree and Kant 1982a; Fornace et al. 1984). In addition, an intermediate form

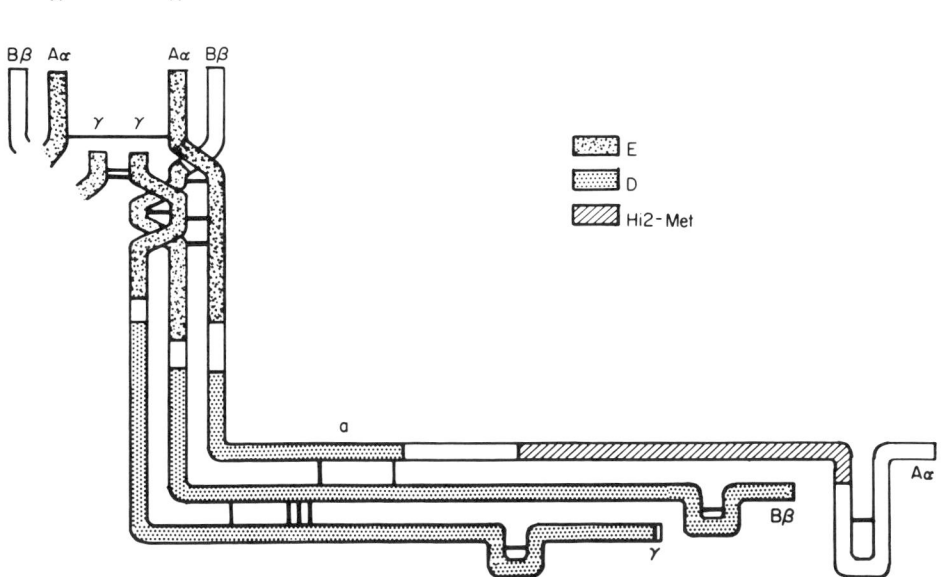

FIGURE 11–4 Schematic representation of a fibrinogen half-molecule with the remaining half-molecule partially indicated. Plasmin fragments are inserted in the structure. E and D indicate classical plasmic fragments. Hi2-Met stands for one of the remaining major fragments isolated from a plasmic digest of fibrinogen. From Blombäck (1979b). Reprinted by permission of John Wiley and Sons, Ltd.

of the γ' chain was reported (Francis et al. 1988). This form is similar to the γ' chain except that the last four residues are missing.

11.4.1 Location of Epitopes in Fibrinogen

Marder et al. (1969) have with the use of antisera raised against fibrinogen demonstrated that the molecule has several antigenic determinants. Some of these determinants have their structural origin in fragment E and fragment D which arise from degradation of fibrinogen with plasmin. Fragments X and Y also have specific determinants and there also exist determinants which are unique for intact fibrinogen.

By using antibodies which have been raised against fragments of fibrinogen, one can elucidate the antigenic structure of the molecule as well as obtain information concerning the location of these determinants (or epitopes) in the three-dimensional structure. Plow and Edgington (1973, 1975) have demonstrated that fragments D and E possess neo-epitopes, which are not expressed in the intact molecule.

Anti-N-DSK cross-reacts to only a small extent with intact fibrinogen and with fragment E, the latter having a considerable portion of its structure in common with N-DSK (Kudryk et al. 1974b). However, after cleavage with CNBr, fragment E reacts to the same extent as N-DSK. This may suggest that the antigenic determinants are not only buried in fibrinogen but also in fragment E. Fragment E in comparison with N-DSK has chain extensions on its Aα and Bβ chains. These extensions are removed on treatment with CNBr. One possible explanation is that the extensions are folded in such a way as to hide the epitopes in fragment E.

In general, antibodies which have been raised against the hydrophobic structures of fibrinogen do not react with the intact fibrinogen molecule. Furthermore, the disulfide bridges in these structures are not available for reduction by thioredoxin in intact fibrinogen (see Section 11.5.4). We suppose, therefore, that the hydrophobic fragments are hidden in the molecule.

11.5 ACTIVATION OF FIBRINOGEN

The fibrinogen-fibrin transformation takes place after removal of two acidic peptides from the amino-terminal portion of the Aα and Bβ chains. The molecular weight of the fibrinopeptides varies between 1.5 and 3 kDa depending on the species from which they have been isolated (for review, see Blombäck 1967, 1979b; Doolittle 1973). Depending on their origin in the chains of fibrinogen the peptides are denoted as either fibrinopeptide A (FPA) or fibrinopeptide B (FPB). The release of these peptides is effected by hydrolytic cleavage, catalyzed by thrombin. Thrombin is a trypsin-like enzyme which has a narrow specificity on protein substrates. Of the few hundred trypsin-sensitive bonds in fibrinogen, only four are preferentially cleaved by thrombin, resulting in release from the fibrinogen molecule of two molecules each of FPA and FPB.

FPA is always released at the faster rate. Polymerization of thrombin-activated fibrinogen molecules usually begins when only small amounts of FPB have been released (Figure 11–5). The release of FPB is greatly accelerated at a later stage, and this acceleration appears to be associated with polymerization of activated fibrinogen (Blombäck et al. 1978, 1979a). FPA from various species shows a marked homology in amino acid sequence (see Blombäck 1979b). This is especially true for structures in close proximity to the thrombin-sensitive arginyl-glycyl bond (Arg 16-Gly 17) in mammalian fibrinogen. The preservation during evolution of structures in FPB is not as evident as is the case with FPA. Thrombin primarily has affinity for structures in the Aα chain, explaining the fast release of FPA. Its specificity with regard to the Aα chain is dependent on a short amino acid sequence around the thrombin-susceptible bond. Thus, it would appear that all of the elements required for recognition by thrombin are present in the first 51 amino acid residues of the Aα chain (Hogg and Blombäck 1974, 1978; Blombäck et al. 1977; Scheraga 1977). These studies of kinetic parameters in thrombin-susceptible substrates have suggested that the se-

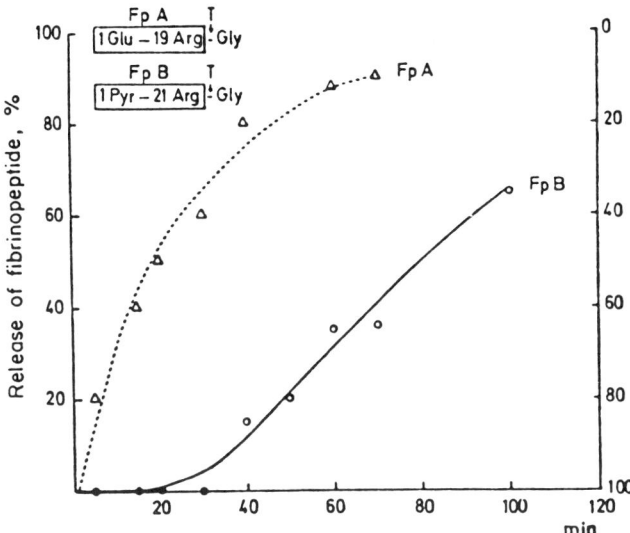

FIGURE 11-5 Release of fibrinopeptides A and B from bovine fibrinogen after addition of thrombin. The susceptible bonds are shown in the upper part of the figure. Note: in human fibrinogen the cleavage of the Aα chain is at Arg 16-Gly 17 and in the Bβ chain at Arg 14-Gly 15. From Blombäck (1979b). Reprinted by permission of John Wiley and Sons, Ltd.

quence Aα 8–23 in human fibrinogen is of prime importance for the fibrinogen-thrombin interaction. The sequence Aα 33–51 is also of some importance in guiding the interaction. The importance of structures residing in FPA (i.e., Phe 9-Arg 16) for thrombin-fibrinogen recognition was already postulated on the basis of comparative structural studies of FPA in different animal species (Blombäck et al. 1966a). It was also suggested that the residue Phe 9 in the natural substrate was in proximity to the Arg 16 residue (Blombäck et al. 1969; Blombäck 1979a). This was based on experiments with synthetic peptides having the Phe residue at different distances from the carboxy-terminal Val-Arg residues. Peptides in which Phe immediately preceded Val-Arg appeared to have the strongest binding to thrombin.

Subsequent elegant kinetic and NMR studies of various synthetic substrates by Scheraga and co-workers have essentially verified this view (Meinwald et al. 1980; Marsh et al. 1982, 1983; Ni et al. 1989a, 1989b). These studies have shown that residues Asp 7, Phe 8, Leu 9, Val 15, and Arg 16 are involved in the interaction of thrombin with peptides derived from the amino-terminal end of the Aα chain of human fibrinogen. The NMR studies suggested binding of thrombin to the Aα 7–16 stretch of peptide. There is a chain reversal within the stretch Asp 7 to Arg 16, such that Phe 8 is brought close to the Arg 16-Gly 17 bond, i.e., the bond split by thrombin in the thrombin-peptide complex (Figure 11–6). Residues Phe 8, Leu 9, and Val 15 of fibrinopeptide A were found to form a hydrophobic cluster which was assumed to fit into the complementary hydrophobic-combining portion of the active site of thrombin. The chain reversal that brings Phe 8 in proximity to Arg 16 is stabilized by hydrogen bonds between Asp 7 and Glu 11 and possibly also Leu 9. In order to form a stable type IIβ turn at Glu 11 and Gly 12, it is necessary that the latter residue be glycine. In an abnormal fibrinogen, fibrinogen Rouen, this position is occupied by valine. NMR-studies of peptides in which only Gly 12 was replaced with valine showed that the conformation of such peptides most likely was changed in such a way that the positioning of Phe 8 in relation to Arg 16 was perturbed (see Figure 11–15 later) and the catalytic efficiency of thrombin was diminished (Ni et al. 1989c).

11.5.1   Polymerization of Activated Fibrinogen

In fibrinogen Detroit (Blombäck et al. 1968b; Mammen et al. 1969) a mutation has taken place at position 19 in the Aα chain, where an arginine in normal fibrinogen has been replaced by serine. This mutation is in an invariable portion of the chain. In fibrinogen Detroit, thrombin releases FPA at almost the same rate as from normal fibrinogen. This suggests that the arginine residue at position 19 in normal fibrinogen does not play a major role in the binding of thrombin to its substrate.

The finding of the Arg 19 → Ser 19 substitution in fibrinogen Detroit first indicated that a polymerization site was located in the amino-terminal part of the Aα chain and that this site in normal fibrinogen was unfolded on release of fibrinopeptide A (Blombäck et al. 1968b; Blombäck 1969, 1970). Earlier studies had shown that polymers formed on release of FPA only by Batroxobin, a snake venom enzyme, had different mass-length ratios than those formed when both FPA and FPB were released (Laurent and Blombäck 1958). The fibrin which is formed on release of FPA has been named fibrin I and that resulting from the release of FPA and FPB has been denoted fibrin II (Blombäck et al. 1978). The light-scattering studies of fibrin polymers by Hantgan and Hermans (1979) also showed a difference between the initial polymer structures of fibrin I and fibrin II. Flow measurements of fibrin gels have also shown that the gel structure of fibrin I is different from that of fibrin II (Blombäck and Okada 1982).

The early findings allowed us to propose that polymerization of activated fibrinogen involved complementary intermolecular interactions of two sets of functional domains named *Aa* and *Bb*, respectively (Blombäck and Blombäck 1972). The *A* domain denotes an amino-terminal domain activated on release of FPA, and *a* denotes a complementary carboxy-terminal domain. The *B* and *b* domains denote similar domains operative on release of FPB.

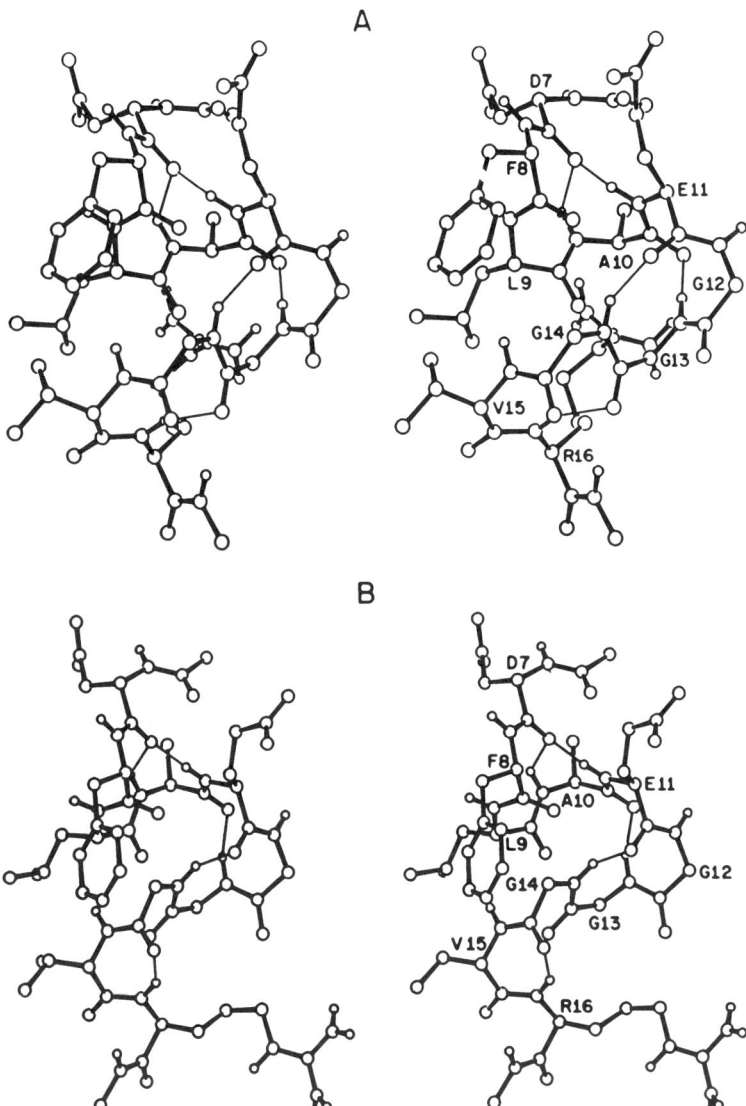

FIGURE 11-6 Stereo drawings of the proposed structure of thrombin-bound FPA fragment (residues 7–16) from normal fibrinogen. (A) Without imposing any hydrogen-bond constraints and (B) with imposition of the hydrogen-bond constraints in the distance geometry calculation. The nonpolar cluster formed by the side chains Phe (8), Leu (9), and Val (15) is clearly seen. The C_a atoms are labeled by the corresponding amino acid residues. Reproduced with permission from Ni et al. (1989b).

Subsequent studies located the *A* domain to N-DSK (Kudryk et al. 1973, 1974a; York and Blombäck 1976). Further evidence for the location of the *A* domain was obtained by Laudano and Doolittle (1978, 1980). Based on the knowledge that the Aα Arg 19 $\rightarrow$ Ser 19 mutation was accompanied by hampered polymerization, they constructed short peptides simulating both the normal and abnormal sequence in this part of the Aα chain. The peptide Gly-Pro-Arg-Pro was shown to inhibit polymerization of activated fibrinogen, but the peptide Gly-Pro-Ser-Pro was not inhibitory. The K_i for the inhibitory peptide was in the millimolar range, i.e., it is a rather weak inhibitor. It is therefore likely that other chain structures in N-DSK participate in creating the *A* polymerization site. Structures in the Bβ and γ chains of N-DSK may also be involved. This suggestion is derived from the fact that thrombin-activated fragment E, which contains all of the structures of N-DSK except Bβ 1–43 and carboxy-terminal residues of the γ chain, binds to a much smaller extent than activated N-DSK to fibrinogen conjugates (York and Blombäck 1976).

The postulated *a* domain was located to plasmin fragment D by Kudryk et al. (1973, 1974a). In fragment D, a polymerization site was further identified in the carboxy-terminal part of the γ chain (Olexa and Budzynski 1981; Varadi and Scheraga 1986; Kuyas et al. 1987a, 1987b). Other structures in fragment D may also be involved in constituting the *a* domain. In agreement with our original postulate it was found that N-DSK from fibrinogen Detroit did not interact with fragment D (Kudryk et al. 1976).

Since fibrinogen is a dimeric molecule, two *A* domains (*A* and *A'*) become activated on release of FPA. Two complementary domains, *a* and *a'*, are also present. These domains are structurally identical but because the half-molecules in fibrinogen have one twofold axis of symmetry, the two halves will not appear identical. On viewing from one direction, we see the front face of the polypeptide chains in one half-molecule, while the corresponding chains in the other half present their back face. This mode of symmetry in the molecule is supported by the chemical investigations which show that the half-molecules of N-DSK (domains *A* and *A'*) are joined by three symmetrical disulfide bridges (Blombäck 1969, Blombäck and Blombäck 1972; Blombäck et al. 1976a, 1976b).

Upon the release of FPA, we postulate that a rearrangement or conformational change takes place in the *A* domains, thereby exposing an amino-terminal structure which is active in polymerization. The activated domains have the ability to interact with the carboxy-terminal domains *a* and *a'*, respectively, in another fibrinogen- or thrombin-activated fibrinogen molecule (i.e., fibrin protomer). The *a* and *a'* domains are functional in fibrinogen as it circulates in blood (Heene and Matthias 1973), but as long as activation of the *A* and *A'* domain has not taken place, polymerization and clot formation do not occur. As each fibrinogen molecule consists of two half-molecules each having two active domains, a polymer can arise via "end-to-end" joining of an activated *A* or *A'* domain in the amino-terminal portion of one molecule to an *a* or *a'* domain in another molecule (Figure 11–7). In this model each molecule in the polymer covers half of the preceding one, as was originally suggested by Ferry (1952).

Based on the postulated location of *A:a* polymerization domains, we would expect that fibrinolytic degradation products inhibit fibrinogen clotting. It is known that fragments X and Y and fragment D inhibit the polymerization of fibrinogen (Budzynski et al. 1967) and that a pronounced bleeding tendency occurs in clinical states where the concentration of these degradation products is high. The presence in blood of fragment D or related fragments is expected to cause the inhibition of the formation of a normal fibrin polymer by interaction with the *AA'* domain of fibrin monomers. Because the fragments are monovalent, no further elongation of the polymers can take place from the N-DSK end of the polymer. In fact, Bang (1964) showed that the fibrin network formed in the presence of fibrinogen degradation products has an appearance which differs from normal.

The release of FPB is slow until, after a lag phase, the release is dramatically accelerated (Figure 11–5). This acceleration appears to be a result of the *BB'* and *bb'* interactions (Blombäck et al. 1978; Blombäck 1979a). The polymerization involving these domains appears to induce an additional conformational strain on the Bβ chain, thereby facilitating the release

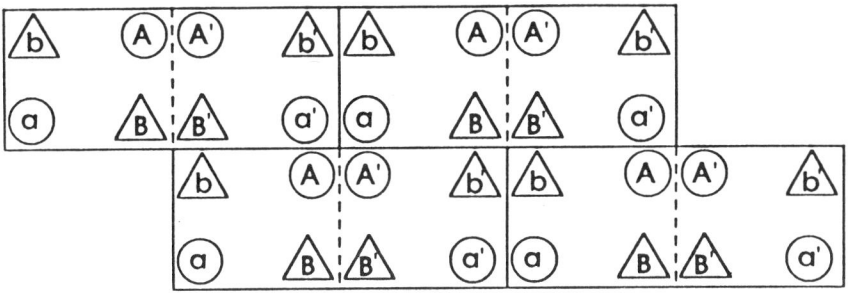

FIBRINOGEN $\xrightarrow{\text{Th}}$ FIBRIN TYPE I $\xrightarrow{\text{Th}}$ FIBRIN TYPE II
(A:a interactions)   (A:a, B:b interactions)
FPA   FPB

FIGURE 11–7 Schematic representation of the sequence of events in protofiber formation. The upper half of the figure shows the possible interactions at sites A:a and B:b. Because of the dimeric nature of fibrinogen, each of these sites is present in duplicate, for example, AA'. The direction of the initial growth of the polymer is different when the A:a or the B:b sites are exclusively activated. The reaction at the bottom summarizes the events which occur during formation of the two types of fibrin. Th, thrombin. Adapted with permission from Blombäck et al. (1978).

of FPB. This is exemplified in experiments with fibrinogen Detroit and Aarhus. In these two fibrinogens, a mutation has occurred at Aα Arg 19, and they both display only slightly slower rates of release of FPA in presence of Batroxobin and thrombin. No polymer formation occurs during this release. The release of FPB occurs at a very slow rate from fibrinogen Detroit and Aarhus (Blombäck 1969; Blombäck and Blombäck 1970; Blombäck et al. 1978, 1988). This slow release was explained as due to the fact that no polymerization took place subsequent to the release of FPA from these abnormal fibrinogens. However, when a critical amount of FPB had been released, polymerization started and the rate of FPB release was accelerated. These findings supported the notion that the release of FPB unfolds a unique set of polymerization sites, i.e., *BB'* and *bb'*. The studies by Olexa and Budzynski (1980) and Shainoff and Dardik (1979) gave additional evidence for the existence of the *B:b* set of polymerization sites in fibrinogen. However, it is also possible that the release of FPB reinforces the polymerization site which is unfolded in normal fibrinogen by the release of FPA (Dyr et al. 1989).

The location of the postulated *B:b* domains have so far not been identified with certainty. The *BB'* sites may, like the *AA'* sites, be located in N-DSK, since when it is activated with Batroxobin it does not bind as efficiently to fibrinogen conjugates as does N-DSK activated with thrombin (York and Blombäck 1976). The *bb'* sites in fibrinogen, and were recently suggested to be located to the carboxy-terminal part of the Aα chain (Hasegawa and Sasaki 1990).

The initial rate of release of FPA is, as mentioned, much faster than that of FPB. The subsequent polymerization event is a fast reaction and stochiometrically related to release of FPA (Blombäck et al. 1984). This means that any set of polymerization sites exposed in normal fibrinogen on release of FPB must be operative at the polymer level. Interaction between such sites may lead to bond formation between the polymers formed subsequent to the release of FPA. The involvement of the *BB'* and *bb'* domains would thus provide a simple means of branching the fibrin fiber to produce an intricate polymer network. Figure 11–7 depicts how these domains may be operative in "end-to-end" and "side-to-side" polymerization and as a consequence, protofibrils are formed. Physical evidence for this type of intermediates in fibrin formation was obtained from hydrodynamic studies by Ferry and his associates (for references, see Ferry 1952) and by electron microscopy (Fowler et al. 1981).

The protofibrils formed on activation of fibrinogen associate to form fibrin fibers, and the latter join into bundles of varying width. The early protofibrils have a width about double that of the fibrinogen molecule. They were shown to be twisted in a helical fashion (Krakow et al. 1972; Williams 1983). This was also postulated by Hermans 1979. Fibrin fibers were also shown to be twisted and were seen to grow to a limiting size of about 100 nm in diameter (Weisel 1986; Weisel and Papsun 1987; Weisel et al. 1987).

The limitation in growth was explained as a consequence of the stretching of the protofibrils near the surface of the fiber. When the amount of energy necessary to stretch a protofibril exceeded the energy available from bonding, the lateral growth would cease.

It would be useful to determine the location of the polymerization domains in the three-dimensional structure of the fibrinogen molecule. The electron micrographs of Bachmann et al. (1975) (Figure 11–1) probably are of the native hydrated molecule. According to Bachmann et al., the average fibrinogen molecule has the shape of a cylinder with rounded ends. The physical domains of the molecule were delineated in some studies (Erickson and Fowler 1983; Weisel et al. 1985), which were partly based on EM investigations of dehydrated fibrinogen molecules, one can distinguish a central globular domain which most likely contains the N-DSK domain. The end regions of the molecule appear to be subdivided into two other globular domains, a proximal-end domain comprising the carboxy-terminal end of the Bβ chain and a distal-end domain comprising the carboxy-terminal part of the γ chain. The carboxy-terminal end of the Aα chains appears to fold back towards the central N-DSK domain, thus creating a separate central domain. The functional domains, *AA'*, are thus presumably located in the central part of the molecule (Figure 11–8). The complementary *aa'* domains are most likely comprised in the two carboxy-terminal globular domains located at opposite ends of the molecule. The latter domains are at least partially present in the plasmin fragment D. The carboxy-terminal parts of the Aα chain, which form a second central domain, may also contain functional structures. These seem not to be involved in protofibril formation, but rather play a role in interactions between fibrin fibers and may thus participate in fiber branching (Weisel and Papsun 1987).

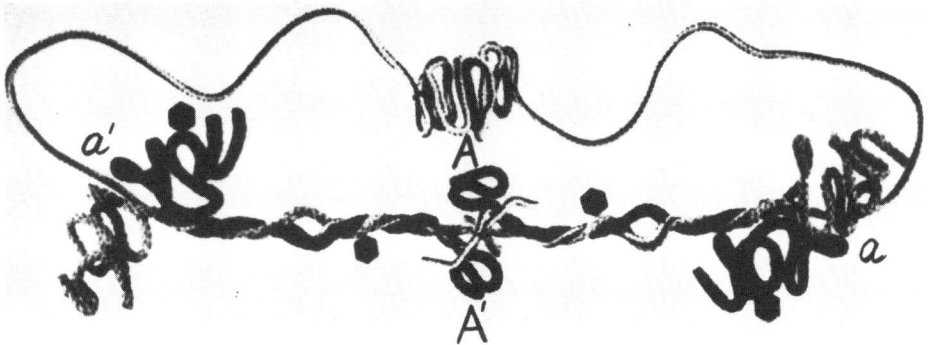

FIGURE 11–8 Schematic three-dimensional model of the fibrinogen molecule. The proposed location of the A,A' and a,a' polymerization domains are indicated. Adapted with permission from Weisel and Papsun (1987).

11.5.2 Fibrin Polymer Stability and Alternative Polymerization Reactions

The fibrin polymer which is formed after activation with thrombin undergoes secondary transformation through factor XIII-catalyzed formation of covalent cross-links between the glutamine and lysine residues in fibrin (Lorand 1972; Doolittle 1973). Factor XIII, a transglutaminase, is present in plasma and platelets, where it exists as a precursor having low or no activity. It is activated by limited proteolysis by thrombin (Buluk et al. 1961). Fibrin polymers stimulate activation of factor XIII by thrombin (Greenberg et al. 1988). The cross-linking involves both γ and α chains in the fibrin protomers. There is one proton-acceptor glutamine residue and one proton-donor lysine residue in each γ chain. In the initial phase of polymerization, these residues in the antiparallel-arranged protomers of the polymer are brought into juxtaposition. In the presence of activated factor XIII and calcium ions, condensation between two γ chains takes place through bond formation between the γ-carbonyl groups of glutamine and the ϵ-amino groups of lysine, with concomitant release of ammonia. The α chain of fibrin undergoes intermolecular cross-linking at several sites which are located in the mid- and carboxy-terminal portion of the chain (Cottrell et al. 1979; Sobel et al. 1988). The resulting stabilized fibrin is insoluble in urea solution and in other solvents which dissolve nonstabilized fibrin or fibrin only cross-linked by γ-γ cross-links.

Thrombin-induced fibrin polymers are the preferred substrates for factor XIII. However, several studies have shown that fibrinogen undergoes a polymerization reaction in the presence of factor XIII (Schwartz et al. 1973; Ly et al. 1974b; Kanaide and Shainoff 1975; Blombäck et al. 1985). This pathway leads to complete incorporation of fibrinogen into polymeric matrices (fibrinogenin) (Blombäck et al. 1985). Factor XIII also catalyzes heteropolymer formation (heteronectin) between fibronectin and fibrinogen in solution (Procyk et al. 1985). The reaction mechanisms in these two polymerization reactions have been elucidated (Procyk and Blombäck 1988). Fibrinogenin formation starts with the introduction of cross-links between the γ chains of fibrinogen units. In this way, linear polymers (dimers, trimers, tetramers, etc.) are formed. In heteronectin formation, the first reaction is between an Aα chain of one fibrinogen molecule and the aminoterminal part of one fibronectin molecule. The heterodimers subsequently combine by introduction of intermolecular cross-links between the γ chains of the heterodimers.

Fibrinogen can also form polymers and gel structures when exposed to shear-stress forces as has been shown by Copley and coworkers (for review see Copley 1988).

11.5.3 Roles of the Disulfide Bridges

Results obtained from investigations using thioredoxin as the reducing agent indicate that certain disulfides in fibrinogen are buried in the intact fibrinogen molecule. It has thus been shown that only 5 out of the 29 disulfide bridges in fibrinogen are rapidly reduced by thioredoxin (Blombäck et al. 1974, 1986). The bonds reduced by thioredoxin include the symmetrical disulfides and the intrachain disulfides in the Aα chains (two per molecule). The symmetrical disulfides which hold together the two half-molecules of fibrinogen seem therefore to be more or less surface exposed and so is the intrachain disulfide of the Aα chain, located in the carboxy-terminal portion of the chain. A fragment containing the latter disulfide is released from the molecule in the early phase of plasmic digestion (Blombäck et al. 1976c). Most of the other disulfide bridges in fibrinogen appear to be buried in core structures in the interior of the molecule. They are certainly of importance for maintaining the conformation of the molecule and as such are kept out of reach of the reducing systems of the organism.

The disulfide bridges in fibrinogen play a fundamental role not only for the fibrinogen's structural integrity but also for the function of its polymerization domains. Thus, it has been shown that thioredoxin-catalyzed reduction of the symmetrical disulfide bridges (Aα 28-Aα 28; γ 8-γ 8, γ 9-γ 9) and of the intrachain disulfide in the Aα chain of fibrinogen caused a total loss of clottability of fibrinogen. On reoxidation, clotting activity returned (Blombäck et al. 1974, 1986). This phenomenon cannot be ascribed to a change in the proteolytic activity of thrombin, because the rate of release of FPA from the reduced fibrinogen is the same as that for the unreduced material. A recent study by Procyk and Blombäck (1990) showed that reduction of the intrachain disulfide bond in the Aα chain, as well as of the symmetrical bond in this chain, did not lead to loss of clotting activity. On the other hand, when, in addition, the intrachain disulfide bond in the carboxy-terminal part of the γ chain was reduced, clotting activity was lost. This study complements the previous study in that it indicates the importance of the symmetrical bonds in the γ chain for the functional integrity of fibrinogen. Furthermore, it suggests that the disulfide-bonded structure in the carboxy-terminal part of the γ chain is important for expression of a polymerization site (presumably *a*) present in this part of the chain (Olexa and Budzynski 1981; Varadi and Scheraga 1986; Kuyas 1987a, 1987b). However, reduction of the intrachain disulfide bond in the γ chain also lead to perturbance of thrombin specificity (Procyk and Blombäck 1990). Instead of preferential cleaving the Arg 16-Gly 17 peptide bond, thrombin also cleaved the Aα Arg 19-Val 20 peptide bond in the reduced fibrinogen. The latter bond is cleaved at a comparatively slow rate and its cleavage does therefore not explain the loss of clotting activity (Procyk et al. 1991).

11.5.4 Fibrin Gel Structure

As mentioned previously, the activated fibrinogen molecules form polymeric structures. The interaction of these structures lead to the formation of a fibrin gel network. The mechanical and spectroscopic properties of fibrin gels have been studied by several investigators. Ferry and Morrison (1947) described properties of gels ranging between the extremes of transparent and opaque. In the transparent gels, the polymer strands are thin and the meshwork fine, whereas in the opaque gels, thick polymer strands are encompassing large liquid spaces (for further references, see Blombäck 1985).

Figure 11–9 shows the typical turbidity pattern after thrombin is added to a fibrinogen solution. After a lag phase, a sudden increase in turbidity occurs. During the lag phase, activation and polymerization takes place but the polymerization is not reflected in any significant increase in turbidity (at 350 nm or above). In the literature, however, this type of turbidity profiles is often referred to as a polymerization curve. This is, however, a misnomer since it is mainly gel formation and lateral growth of strands which are being measured.

One can, by turbidity measurement, obtain a measure of the average fiber mass-length ratio and of the fiber diameter for the hydrated fibers in the fibrin gel (Carr and Hermans 1978; Carr and Gabriel 1980). This is done in plots relating turbidity to the wavelength of light. Extrapolation of turbidity to infinite wavelength, as well as the slope of the plot, gives the fiber mass-length ratio and fiber radius, respectively.

Flow measurements of fibrin gels provide us with a measurement of the porosity, expressed as the Darcy constant (K_s) of the network. The K_s values may be transformed to average fiber mass-length ratio and fiber diameter if the degree of hydration of the fiber is known. In a purified fibrinogen-thrombin system, similar values for fiber dimensions are obtained by both turbidity and permeability measurements (Blombäck et al.

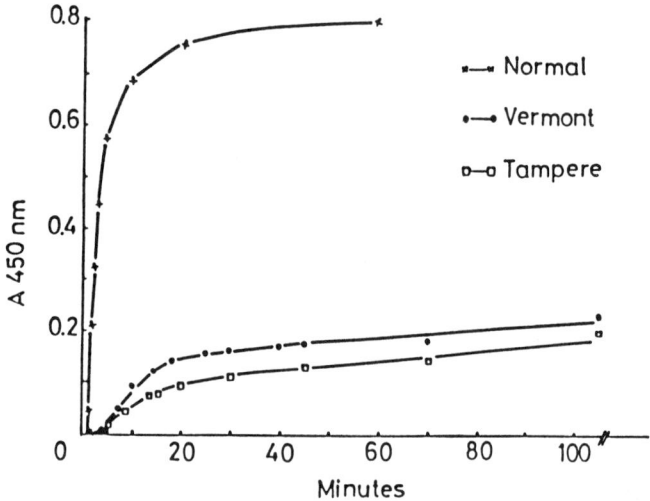

FIGURE 11–9 Turbidity profiles during gel formation for normal fibrinogen, fibrinogen Tampere, and fibrinogen Laconia. Gel formation was with fibrinogen (at 1.5/ml) and thrombin (at 0.35 NIH units/ml) in TNE buffer containing CaCl$_2$ (at 20 mM) and at ionic strength 0.21. Turbidity was measured at 450 nm.

1989). In plasma, however, the two methods give dissimilar results (Blombäck et al. 1990a).

The architectural features of the hydrated gel network can be obtained by scanning confocal three-dimensional microscopy (Blombäck et al. 1989, 1990a). In such studies the matrix of the gel is stained with fluorescein isothiocyanate. This methodology also gives an assessment of porosity and fiber sizes.

Electron microscopic studies of gel matrices may also be performed (Hawn and Porter 1947; Porter and Hawn 1949; Shah et al. 1982; Müller et al. 1984; Mosesson et al. 1987). This methodology requires the removal of water and handling procedures that inevitably distort and compact the gel structure, which is more than 99% water. However, EM offers better optical resolution than light microscopy.

The viscoelastic behavior of fibrin gels can be assessed in studies using rheogoniometers and similar devices to measure deformation (the elastic modulus).

Thrombin-induced gel networks as observed by confocal three-dimensional light microscopy are ordered structures in which circular fiber strands of somewhat varying widths cross each other in different directions in space. The fiber strands often come together at denser nodes, and these nodes may represent nucleation sites. Branching of fibers is infrequent. EM pictures of dehydrated fibrin networks display much less order; and bent, twisted, and branched structures are frequent. These structures are certainly to some extent artifactual as a consequence of dehydration and compaction.

Many factors influence the structure and properties of the gel being formed. We can divide these factors into two groups: kinetic factors and modulating factors. The clotting potential (thrombin concentration) and fibrinogen concentration constitute the kinetic factors and are physiologically the most important in determining the fibrin gel structure (Blombäck and Okada 1982; Blombäck et al. 1989, 1990a).

An increase in thrombin concentration leads to formation of less and less porous gels with thinner fibers and small liquid spaces. Table 11–2 shows permeability and turbidity data for fibrin gels at different thrombin concentrations. Figure 11–10 shows the corresponding structures observed by confocal three-dimensional microscopy.

An increase in fibrinogen concentration (e.g., between 0.5–5 mg/ml) also effects formation of less and less porous gels as in the case of thrombin. However, the fiber diameters do not change much since any decrease is compensated by the increased fibrinogen concentration.

The modulating factors are of two kinds: 1) factors affecting the medium in which clotting occurs, and 2) factors that bind to fibrinogen and/or fibrin polymers and thus influence the formation of the network or of the established network structure after formation.

Among the factors in the first group of modulating factors is the ionic strength. An increase in ionic strength from 0.15 to about 0.22 will cause

TABLE 11-2 Influence of Thrombin Concentration on Clotting Time, K_s, Mass-Length Ratio, and Fiber Diameter[1]

Thrombin (NIH units/ml)	Clotting time (sec)	Permeability Method			Turbidity Method	
		K_s $(cm^2 \times 10^9)$	μ^2 $(\times 10^{-12})$	ϕ^3 (μm)	μ^2 $(\times 10^{-12})$	ϕ^3 (μm)
0.05	531	10.6	36.5	0.198	29.6	0.135
0.21	225	10.4	36.2	0.197	23.0	0.121
0.52	96	5.0	17.4	0.134	17.6	0.104
1.04	60	3.0	10.6	0.107	14.8	0.096
2.02	33	2.7	9.2	0.099	12.5	0.093

[1] Fibrinogen at 1.15 mg/ml was clotted with thrombin at different thrombin concentrations and at ionic strength 0.21.
[2] μ, fiber mass-length ratio in Da/cm.
[3] ϕ, average fiber diameter.

formation of networks of gradually decreasing porosity and fiber thickness. This effect is similar to that seen on increasing the thrombin concentration. However, at ionic strengths above 0.22, swelling of the fiber strands seem to occur and this will lead to a further decrease in porosity. The tightening and certainly the swelling of the fibers may be due to a decrease in protein-protein interaction at increasing high ionic strengths.

The first group of modulating factors also includes plasma proteins like albumin. It is likely that the plasma proteins act by water exclusion. In support of this is the fact that dextran has a similar effect to albumin. Water will thus be withdrawn from the interacting molecules (thrombin and fibrinogen), and less space is available in which polymerization can occur. This will result in coarser and more porous network structures. However, at each protein or dextran concentration, the kinetic factors will to a large extent determine the porosity of the structures formed.

The second group of modulators includes calcium ions and fibronectin. Calcium ions bind to high- and low-affinity sites in fibrinogen (Marguerie et al. 1977). Binding of calcium to the low-affinity binding sites has a dramatic effect on the porosity of the gels being formed (Okada and Blombäck 1983). The porosity and strand width increase with increasing calcium concentration, reaching a saturation level between 10 and 20 mM and a half-maximum value at about the physiological concentration of calcium.

Fibronectin interacts with the fiber strands in the network (Okada et al. 1985). Fibronectin incorporation into fibrin is catalyzed by factor XIII and is a saturable process. As a result the fiber strands will increase in width and the turbidity of the matrix will also increase (Figure 11-11). Since the fractional volume of fibronectin is small, the porosity decreases only slightly.

The parameters of the gel structure as determined by turbidity, permeability, and microscopy may be compared with the viscoelastic behavior of

A

B

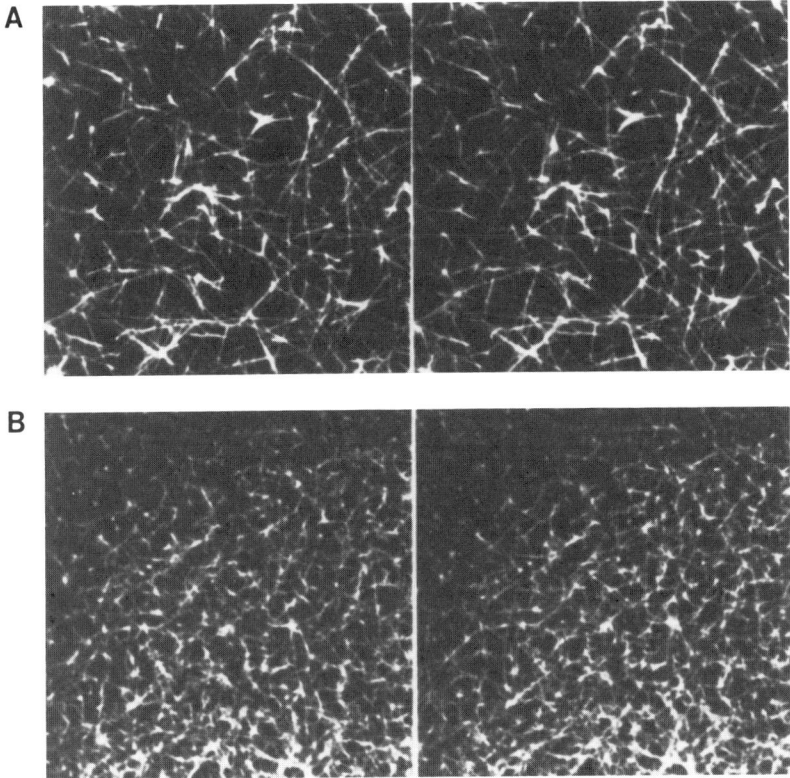

FIGURE 11-10 Three-dimensional micrographs of fibrin gel networks at different thrombin concentrations. The stereo pairs shown are computer reconstructions made from confocal laser microscope recordings. Gel formation was in TNE buffer containing $CaCl_2$ (at 20 mM) of ionic strength 0.21, pH 7.4, and at different thrombin concentrations—conditions: temperature, 24°C; fibrinogen, 1.15 mg/ml. Permeation was with Tris-imidazole-NaCl buffer. (A) Thrombin, 0.05 NIH units/ml. (B) Thrombin, 2.0 NIH units/ml. Field of view: 100 × 100 μm. Reproduced with permission from Blombäck et al. (1990a).

fibrin gels. It is known that the elastic modulus (G') of fibrin gels is dependent on the fibrin concentration of the gel network (Roberts et al. 1974; Carr et al. 1976). Blombäck et al. (1989) showed that clots formed at low fibrin concentrations have a larger porosity and longer fiber strands than gel structures formed at higher fibrin concentrations, and that an increase of these parameters may be related to a decrease in the G' value. Likewise, increasing thrombin concentration has been shown to lead to a slight increase in G' (Roberts et al. 1974), and our work showed that this was accompanied by decreased porosity, strand width, and fiber length. From this comparison,

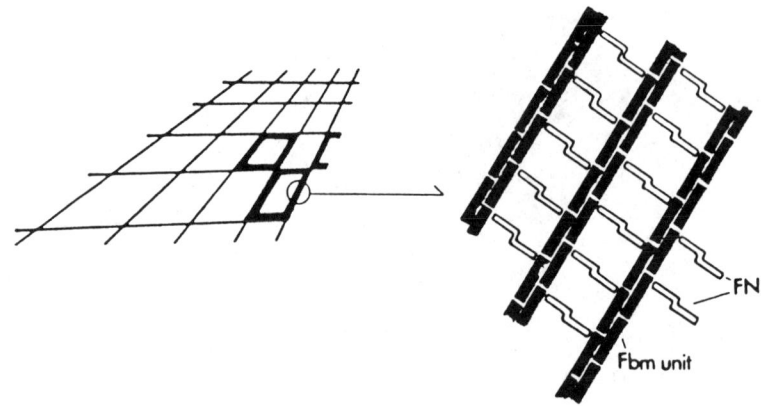

FIGURE 11-11 Schematic representation of the fibronectin arrangement in fibrin strands. Left: a fibrin network that contains some meshes with incorporated fibronectin (heavy lines). Right: a close-up of a portion of the strand structure, showing fibrin polymers in the strand being linked together by fibronectin molecules. FN, fibronectin; Fbm unit, fibrin monomer. Reproduced with permission from Okada et al. (1985).

it appears that there is an inverse relationship between G' and gel porosity, strand width, and fiber length. The elastic modulus drops sharply in clots formed at ionic strengths above 0.15 (Shen et al. 1977a). We showed that the porosity of gels decreases at increasing ionic strengths, but the fiber structure apparently also loosens up through fiber strand swelling. These effects may cause individual fibers of the gel network to extend and have less resistance to stress, or they may allow for slippage between protofibrils of the strands, thus leading to the decrease of the elastic modulus.

Clots made in citrated plasma have been claimed to be more porous and coarser than those formed in a purified fibrinogen thrombin system (Shah et al. 1987). These studies were performed in the absence of calcium, and clotting times of the two systems were not taken into account. We have shown in our previous work that the clotting time of a fibrinogen-thrombin mixture, in the presence of calcium ions, is linearly related to gel porosity (K_s) and thus inversely to the rate of activation (release of fibrinopeptides or thrombin concentration). These relationships also hold at different calcium concentrations (Blombäck and Okada 1982; Okada and Blombäck 1983).

When thrombin is added to recalcified plasma (20 mM calcium) there is, up to certain clotting time, also a linear relationship between the clotting time and the porosity of gels into which all fibrinogen of the plasma is incorporated (Figure 11-12 left side) (Blombäck et al. 1990b). In recalcified plasma, indigenous generation of thrombin from prothrombin occurs after a lag phase even without added thrombin. Addition of thrombin will de-

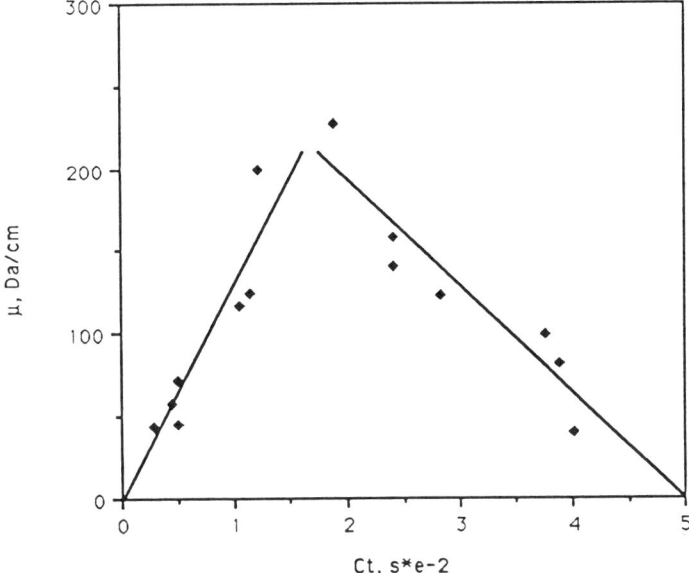

FIGURE 11-12 Fiber mass-length ratio ($\approx$ porosity) of plasma clots versus clotting time (Ct). The plasma (fibrinogen, 2.7 mg/ml) was clotted with different amounts of thrombin in the presence of calcium (at 20 mM). The clots were allowed to mature at room temperature before turbidity or permeation analysis. The data shown are for turbidity measurements. Permeation data gave qualitatively identical results.

crease the length of this lag phase. The generation of thrombin is fast at the end of the lag phase. The right half of Figure 11-12 shows that as the clotting time of recalcified plasma (without thrombin) is approached, porosity decreases. At clotting times approaching the clotting time of recalcified plasma itself, the porosity appears to be determined not only by the added thrombin but also by indigenously generated thrombin or possibly by fibrinogenin formation catalyzed by factor XIII, which may have been activated during the lag phase.

This experiment shows that gel structures in plasma and in a purified fibrinogen-thrombin system can be compared at a given fibrinogen concentration and at a thrombin concentration that gives sufficiently short clotting times (i.e., corresponding to those on the left side of Figure 11-12). Such a comparison shows that the porosity of plasma clots are about 20-fold more porous than those formed in the purified system. This difference is partly due to the presence of albumin in plasma but other components so far not identified may also play a role (Blombäck et al. 1990a).

These data suggest that, within a wide range, the initial rate of activation of fibrinogen also determines the type of structures being formed in plasma. Preliminary analysis of the rate of FPA release at different thrombin concentrations supports this view, at least for gels formed at short clotting times (thrombin concentrations on the left side of Figure 11–12). Fast activation produces a tight network and slow activation produces a porous network structure. What is interesting, however, is that the structure which is formed at the time of clotting is not influenced by the subsequent indigenously generated high level of thrombin, although this is at a time when most of the fibrinogen in the bulk phase is still not activated. We therefore suggest that the first network formed creates a scaffold at the time of clotting and that all subsequently activated fibrinogen molecules in the bulk phase are incorporated into this scaffold. The micrographs of networks formed at different clotting times of plasma are in agreement with this interpretation.

We can only speculate on the nature of the driving force(s) in this intriguing phenomenon. It is possible that upon activation of fibrinogen, the activated species undergo a transient conformational change and that this results in increased rate of diffusion (Larsson et al. 1987). It is also likely that the scaffold has a larger density than the surrounding medium. The combination of increased diffusion of activated fibrinogen and the density heterogeneity created by the scaffold may efficiently drive the deposition of matter to the existing scaffold.

11.6 FIBRIN(OGEN)OLYSIS IN PRESENCE OF PLASMIN

Fibrinogen and fibrin are substrates for a variety of proteolytic enzymes. Extensive proteolysis will make fibrinogen unclottable or will partly or completely lyse any fibrin formed. Plasmin is a proteolytic enzyme that plays a major role in physiological or pathophysiological fibrin(ogen)lysis. This is only partly due to the preferential specificity of plasmin for fibrinogen or fibrin; it is mainly due to the restricted ways in which the enzyme is generated from its precursor plasminogen. This restriction is the result of specific interactions between fibrin(ogen), plasminogen, and various activators. Reviews of this complex issue have been published (see Chapter 10; Wallén 1980, 1987; Collen 1980; Wallén et al. 1990).

Plasminogen, the precursor to plasmin, occurs in two forms in plasma. The main form in plasma is called Glu-plasminogen, since the amino-terminal amino acid is glutamic acid. In the presence of plasmin, this Glu-plasminogen is, by limited proteolysis, transformed to Lys-plasminogen by removal of an amino-terminal peptide segment. As a result lysine appears at the amino-terminal end. The interactions and activation kinetics for these two forms are somewhat different, and this will be addressed later at appropriate places.

Plasminogen binds to fibrinogen but its binding to fibrin appears to be more avid (Thorsen 1975; Cederholm-Williams 1977; Adams Lucas et al. 1983). The binding regions in fibrinogen/fibrin have been shown to be in the fragment D and fragment E (N-DSK) domains (Adams Lucas et al.

1983). This binding enhances the activation of plasminogen and also protects plasmin from binding to the main inhibitor of plasmin in blood, i.e., α_2-antiplasmin and from autolytic degradation. The binding occurs at the lysine-binding sites in plasminogen. These sites have been localized to the five "kringle" structures discovered in the A chain of plasminogen (Sottrup-Jensen et al. 1978). Slight degradation of fibrin by plasmin may increase the affinity even more (Suenson et al. 1984; Tran-Thang et al. 1984).

The activation of plasminogen by the physiologically most important activator, tissue-type plasminogen activator (t-PA), is also restricted by specific interactions. The occurrence of a plasminogen activator in tissues was demonstrated by Astrup and Permin in 1947. Highly purified preparations of t-PA was obtained much later (Cole and Bachman 1977; Rijken et al. 1979; Wallén et al. 1981, 1982). More recently t-PA has been produced by recombinant DNA technology (see Chapter 10; Pennica et al. 1983). t-PA is a single-chain protein (70 kDa) that is synthesized in endothelial cells. It is secreted in response to various stimuli, e.g., thrombin, cytokines. t-PA is transformed into a two-chain (A and B chains) form by plasmin cleavage of a single peptide bond (Wallén et al. 1983). The B chain contains the active center, and the A chain contains sites that facilitate the fibrinolytic process. This chain contains kringle structures and other domains (Jörnvall et al. 1983). t-PA in both the single- and two-chain form binds preferentially to fibrin, and this binding enhances plasminogen activation (Camiolo et al. 1971; Thorsen et al. 1972; Wallén and Wiman 1975). In contrast to plasminogen, the interaction of t-PA with fibrinogen is small. The evidence suggests that t-PA, together with plasminogen, is assembled on fibrin in a ternary complex, i.e., t-PA, plasminogen, and fibrin (Wallén 1977). This leads to stimulation of plasminogen activation by two to three orders of magnitude. In the absence of fibrin, single-chain t-PA has little enzymatic activity in comparison with two-chain t-PA (Norrman et al. 1986). In the presence of fibrin, the enzymatic activity of the two forms is about the same. Slightly degraded fibrin is a better stimulator of t-PA-induced plasminogen activation (Norrman et al. 1985). Degradation mainly occurred at the carboxy-terminal end of the α chain of fibrin by release of a 26-kDa fragment. Preferential binding of both plasminogen and t-PA to "early" degraded fibrin may be of physiological importance, since in relative terms newly formed fibrin will be less rapidly digested than fibrin that has already begun to be degraded.

The t-PA-induced fibrinolysis is not only regulated by fibrin formation but also by the cellular availability and rate of cellular secretion of t-PA. Venous stasis increases secretion and so does stimulation of the endothelium with thrombin at least in in vitro systems.

An inhibitor of t-PA called plasminogen activation inhibitor (PAI) is synthesized in the endothelium and secreted into blood (Chmielewska et al. 1983; Kruithof et al. 1984; Thorsen and Philips 1984; Verheijen et al. 1984). The conditions under which its secretion occurs have not yet been fully clarified. It is found in plasma bound to vitronectin, one of the adhesive proteins in blood (Wiman et al. 1988). PAI reacts avidly with free t-PA and thus will decrease its availability for fibrin. The role of t-PA and its inhibitor is discussed in Section 11.10.

Another activator of plasminogen is urokinase. This activator was first found in urine but has later been identified in several types of epithelial cells of kidney tubulus, urinary bladder, lung, and mammary gland. Urokinase seems to be absent in endothelial cells (Larsson et al. 1984). Urokinase (u-PA) is a two-chain protein (Soberano et al. 1976) but it is synthesized in cells as a single-chain precursor (scu-PA). Scu-PA is present together with u-PA in urine and also in plasma in small amounts (Stump et al. 1986). The activation of plasminogen by u-PA is also stimulated by fibrin, although compared with t-PA, it is far less. The stimulation only concerns one of the two types of plasminogen, i.e., the Glu-plasminogen.

In purified systems, scu-PA has the capacity to activate plasminogen to plasmin (Collen et al. 1984; Gurevich et al. 1984). On the contrary, however, scu-PA does not activate plasminogen in plasma and is not inactivated by inhibitors to u-PA. However, when fibrin is added to the plasma containing scu-PA, a rapid lysis of the fibrin is induced. This interesting phenomenon has been explained as a result of binding of scu-PA to a protein component in plasma that inhibits its activation of plasminogen. Fibrin may counteract this inhibition (Zamarron et al. 1985). Another possibility may be that scu-PA only activates Glu-plasminogen bound to fibrin, since plasminogen in such a complex has a Lys-plasminogen-like conformation (Pannell and Gurevich 1986).

There are other factors in plasma which participate in the activation of plasminogen and thus in the fibrinolytic response. One pathway is elicited by activated factor XII. This pathway is still obscure but it has been suggested that conversion of prekallikrein into kallikrein by factor XII_a is responsible for plasminogen activation. It is not known whether fibrinogen or fibrin stimulates the factor XII-induced activation of plasminogen.

Fibrinogen or rather fragments of fibrinogen play a role in potentiating the action of a nonphysiological activator of fibrinolysis, i.e., streptokinase (SK). Streptokinase binds to human plasminogen and this complex activates plasminogen to plasmin (Robbins and Markus 1978). It has been shown that both fibrinogen and especially an early degradation product of fibrinogen missing the carboxy-terminal part of the $A\alpha$ chain potentiates the action of the SK-plasminogen activator (Takada et al. 1980; Hishikawa-Itoh et al. 1982). The mechanism for this potentiation is not clear although some evidence for formation of a fibrinogen-plasminogen-SK complex has been presented. It appears that the plasminogen in this complex can be activated faster than in the SK-plasminogen complex (Takada et al. 1980).

Fibrin clot lysis is by definition a solubilization of the fibrin network structure. This solubilization does not occur in the earliest phase of digestion. Viscoelastic studies of clot rigidity showed that the cleavage of a bond that connects the carboxy-terminal, two-third segment of the α chain with its amino-terminal part did not change clot rigidity (Shen et al. 1977b). In fact, under certain conditions a slight increase in rigidity was found. The rigidity was rapidly lost when additional bonds in the structure were cleaved. The fact that the initial cleavage of the bond in the α chain did not decrease rigidity suggests that cleavage of this particular bond did not break other important inter- and intramolecular contacts that had been established during fibrin network formation.

The importance of the fibrin network structure for the lysability of a clot has, apart from the above mentioned study, received little attention. As discussed in a previous section, clots may range in structure from extremely tight and rigid ones to coarse and plastic ones. The mechanism of binding of plasminogen and t-PA to fibrin and of the formation of the ternary complex may not be the same on these different structures. The fact that the clot structure has been largely ignored in fibrinolytic studies may explain why the results of different investigators often disagree.

On digestion of cross-linked fibrin with plasmin, fragment D is present in a dimeric form (Gaffney 1977) containing cross-linked γ chains. The detection of this fragment may be useful in the differentiation between the primary fibrinolysis of fibrinogen and the fibrinolysis of fibrin. It appears that in cross-linked fibrin, the α chains are more resistant to plasmin than in non-cross-linked fibrin.

11.7   INTERACTION OF FIBRINOGEN AND FIBRIN WITH IONS, PLASMA PROTEINS, AND CELLS

Calcium ions are indispensable for the functional integrity of fibrinogen (Godal 1960). Conformational or other structural or functional changes of the molecule accompany the removal of calcium (Blombäck et al. 1966c; Ly and Godal 1972/73; Haverkate and Timan 1977). There are three high-affinity binding sites for calcium in fibrinogen (Marguerie et al. 1977). Two of these are located in the carboxy-terminal part of the γ chain (Nieuwenhuizen et al. 1981), and one may be in the amino-terminal disulfide knot (N-DSK) (Nieuwenhuizen et al. 1983). The high-affinity sites are most likely of importance for preserving the integrity of the molecule. This suggestion is based on the finding that EDTA, but not citrate, seriously hampers the polymerization of fibrinogen in presence of thrombin. There are also several low-affinity binding sites for calcium in fibrinogen (Marguerie et al. 1977). Studies of fibrin gel structure at various calcium concentrations suggested that the low-affinity sites are of importance for the alignment of fibrin polymers in the network structure of the gel (Okada and Blombäck 1983). Magnesium ions also seem to interact with fibrinogen and/or fibrin, since this ion counteracts the effect of calcium on the gel structure (Okada and Blombäck 1983).

Fibronectin forms complexes with fibrinogen and activated fibrinogen. These noncovalent interactions are mainly observed at low temperatures (Hörmann 1985) but also at higher temperatures (Okada et al. 1985). A specific interaction between fibronectin and fibrin is catalyzed by factor XIII (Mosher 1975, 1976; Iwanaga et al. 1978; Okada et al. 1985). In this transglutaminase-catalyzed reaction, covalent bonds are introduced between ε-amino groups of the α chains of fibrin (donors) and glutamine residues of fibronectin (acceptors). The acceptor glutamine in fibronectin was identified as residue 3 (from the amino-terminal end) on the basis of its susceptibility to putrescine incorporation in the presence of factor XIII (McDonagh et al. 1981). The donor amine sites in the α chain are probably located in the carboxy-terminal part of this chain (Ehrlich et al. 1983; Sobel et al. 1988). The incorporation is a saturable process in which 1 mol of fibronectin is incorporated per mol of fibrin protomer (Okada et al. 1985). Early fibrin polymers appear not to be the preferred amine donors, but rather the network structure, which is established at the gel point. It is in the strands of this network that fibronectin is incorporated during maturation of the network. It has been suggested (Okada et al. 1985) that in the network, the fibrin polymers of the strands were separated by an array of fibronectin molecules (Figure 11–11). The gel structure thus formed would be expected to become much stronger than a structure merely stabilized by γ- and α-chain cross-links within the polymers of the strands.

Factor XIII also catalyzes the incorporation of α2-antiplasmin into the α chains of a fibrin network (Tamaki and Aoki 1981). It is not known whether the sites for incorporation of fibronectin and α2-antiplasmin, respectively, are the same, and the sites involved in the intermolecular cross-linking of the α chain in fibrin may not be the same as those used for fibronectin and α2-antiplasmin incorporation.

Factor XIII itself binds to fibrin, but this interaction only occurs in the presence of calcium ions (Okada et al. 1985). Factor XIII also binds in a saturable fashion to fibrinogen adsorbed to a surface in the presence of calcium (Silveira et al. 1990). Immunoblotting experiments showed that factor XIII binds to the carboxy-terminal parts of the Aα and Bβ chains of fibrinogen, but surprisingly not to the γ chains (Mary et al. 1987). It is not known whether a specific interaction occurs between factor XIII and fibrinogen in solution. However, the fact that factor XIII copurifies with fibrinogen in different fractionation procedures, including gel chromatography, may indicate noncovalent binding between factor XIII and fibrinogen also in solution.

Fibrinogen is necessary for platelet aggregation to take place in the presence of various agonists, such as thrombin, ADP, epinephrine, and prostaglandins, as was shown by many investigators (for review, see Copley 1979a; Tomikawa et al. 1980a). Binding of fibrinogen to platelets accompanies ADP-induced aggregation in the presence of fibrinogen (Mustard et al. 1978; Marguerie et al. 1979; Tomikawa et al. 1980a). The receptor that is unfolded by the agonists is composed of two glycoproteins and referred to as $GPII_b-III_a$. After induction more than 10,000 receptors are present on the platelet surface (Niewiarowski et al. 1983). The site that binds to the receptors seems, at least partly, to be located in the carboxy-terminal part of the γ chain, since fragments of this part of the fibrinogen molecule inhibit aggregation (Kloczewiak et al. 1983). Other parts of the molecule are also likely to be involved since fragments from the amino-terminal part of fibrinogen (fragment E and N-DSK) also inhibit ADP-induced aggregation in the presence of fibrinogen (Tomikawa et al. 1980a). These fragments both contain a sugar moiety linked to residue γ Asn 52 (Blombäck et al. 1973). In fact, a glycopeptide obtained from fibrinogen was reported to inhibit ADP-induced aggregation in the presence of fibrinogen (Soria et al. 1978).

The divalent nature of fibrinogen (Blombäck and Blombäck 1972) is one possible explanation to the aggregation effect of fibrinogen on exposure of receptor sites. The carboxy-terminal γ chains of the two "arms" of the molecule could bridge gaps between platelets and thus cause aggregation. However, the fact that the dimeric fragments, fragment E and N-DSK, inhibit aggregation in the presence of fibrinogen poses a problem for this explanation (Tomikawa et al. 1980a). However, perhaps there are also structures in the amino-terminal part of fibrinogen bind to platelets.

Platelets adsorb to surface films of fibrinogen (Vroman et al. 1977) and to fibrinogen or fibrin monomer conjugated to sepharose (Tomikawa et al. 1980b). The latter interaction requires calcium ions, but occurs without release of ADP and also in presence of indomethacin. These adsorption processes are most likely due to the exposure of binding sites through conformational changes in fibrinogen as a result of adsorption or conjugation. The binding site on the platelet is not known, but it is probably not identical with the $GPII_b-III_a$ receptor, since agonists like ADP are not involved.

11.8 BIOSYNTHESIS OF FIBRINOGEN

Fibrinogen in mammals is synthesized in the liver and also in megakaryocytes (Leven et al. 1982). The normal fibrinogen concentration in humans is 2–3 g/l plasma. Fibrinogen is also present in platelets (5-10% of the plasma content). The origin of this fibrinogen is disputed. It may be synthesized in megakaryocytes or taken up from plasma by spinocytosis. It has been claimed that a lower-molecular-mass form of fibrinogen is synthesized in megakaryocytes, which is subsequently sequestered into the blood platelets (James et al. 1977). These authors proposed that different structural genes were involved in the production of platelet fibrinogen as compared with plasma fibrinogen. Analysis of platelet fibrinogen by other investigators have shown it to be indistinguishable from plasma fibrinogen (Doolittle et al. 1974). However, other differences than mass between platelet and plasma fibrinogen have been reported (Solum and Lopaciuk 1969; Plow and Edgington 1975). The question of the genetic origin of platelet fibrinogen therefore remains open, especially since some individuals with abnormal fibrinogen in plasma were shown to have apparently normal fibrinogen in their platelets (Soria et al. 1976; Jandrot-Perrus et al. 1979).

The main pool of fibrinogen occurs intravascularly, with a smaller pool extravascularly (Blombäck et al. 1966b). The half-life for human fibrinogen is 3.8–4.9 days. The turnover times show considerable variation in different animal species. The genetic message in somatic cells for fibrinogen synthesis is located on chromosome 4 (Olaisen et al. 1982), or more exactly on its long arm (band 4 q 23-32) (Kant et al. 1985). Three separate, closely linked genes are responsible for transcription and translation of the three fibrinogen chains in humans and rats (Chung et al. 1981; Fornace et al. 1984; Kant et al. 1985; Rixon et al. 1985; Crabtree et al. 1985). The genes are linked in the order $\gamma \rightarrow A\alpha \leftarrow B\beta$. Surprisingly, the direction of transcription of the $B\beta$ gene is reversed as compared to the direction of transcription for the γ and $A\alpha$ genes.

Based on evidence from rats, the fibrinogen genes are coordinately expressed by equal mRNA levels for the three chains under nonstimulated conditions as well as after induction of the acute phase response (Crabtree and Kant 1982b). This is also the case after stimulation with dexamethasone and with hepatocyte-stimulating factor (Otto et al. 1987). However, this may not be true under all conditions, since in chicken hepatocytes under hormone-deprived conditions, unequal synthesis of the three chains at the mRNA level was observed (Plant and Grieninger 1986). Addition of hormones restored balanced expression.

Despite coordinate expression of the genes at the mRNA level, the intracellular levels of the three fibrinogen chains vary in the animal species that have been studied. In human hepatoma cells, the $A\alpha$ and γ chains are in excess over the $B\beta$ chains (Yu et al. 1986). This may also be the case in rabbits (Alving et al. 1982). On the other hand, in rat and chicken hepatocytes, the levels of γ and $B\beta$ chains are in excess over the $A\alpha$ chains (Plant and Grieninger 1986; Hirose et al. 1988). However, in vivo studies in dogs showed no significant difference in de novo synthesis of the three chains of fibrinogen (Kudryk et al. 1982). These conflicting findings may indicate that there is a considerable difference in the cellular assembly mechanisms for fibrinogen in different species.

The intracellular synthesis and assembly of fibrinogen in human hepatoma cells was studied extensively by Redman and co-workers (Yu et al. 1986; Roy et al. 1990). These authors delineated the process from the rough endoplasmic reticulum (ER) to the secreted protein, as shown in Figure 11–13. First, incomplete $B\beta$ chains, together with a pool of free $A\alpha$ and γ chains, accumulate on separate polysomes. In addition to the pools of the $A\alpha$ and γ chains, there are also pools of $A\alpha$-γ complexes. In step 2, the $A\alpha$ and γ chains combine with the $B\beta$ chain, still attached to the polysome. This leads to formations of $B\beta$-$A\alpha$ and $B\beta$-γ complexes (step 3), which are subsequently

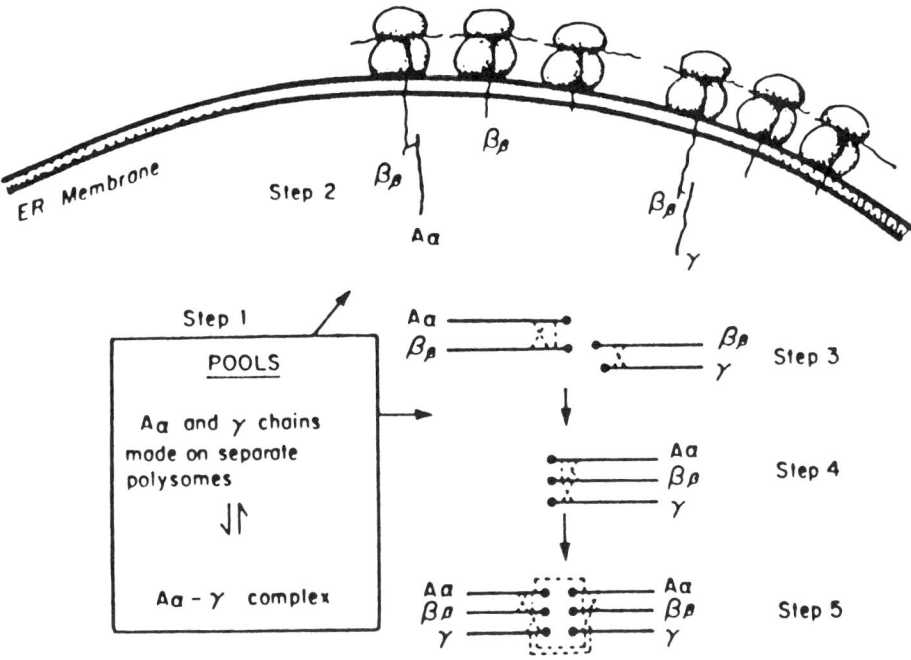

FIGURE 11-13 Intracellular process of fibrinogen synthesis and assembly, starting at step 1 on the left. ER, endoplasmatic reticulum. Adapted with permission from Yu et al. (1986).

released into the lumen of ER. The rest of the assembly process occurs within this organelle. In step 4, a fibrinogen half-molecule is formed by addition of a γ chain to the Bβ-Aα complex and an Aα chain to the Bβ-γ complex.

The final event in the process appears to be the combination of the half-molecules to form dimeric fibrinogen by the formation of symmetrical di-sulfide bonds between the half-molecules. Further processing of the molecule, like glycosylation, is assumed to occur as for other secretory glycoproteins in the Golgi complex. Phosphorylation and sulfylation may also occur in this organelle. For dog fibrinogen, it was shown that these processes did not take place in the ER, although the de novo-secreted fibrinogen was both phosphorylated and sulfylated (Kudryk et al. 1982).

11.9 EVOLUTION OF FIBRINOGEN AND HEMOSTATIC MECHANISMS

An intricate system for hemostasis is present in all vertebrates. In many invertebrates, there are also mechanisms to prevent the loss of body fluids and the entrance of infectious agents, such as tissue contraction, autotomy, cell agglutination, and plasma or lymph coagulation. Even the protozoa have a counterpart to hemostasis called the "surface precipitation reaction" (Needham 1970); however, in the protostomial line of evolution, no gel-forming protein or coagulogen with similarities to vertebrate fibrinogen has so far been identified. In xiposura, e.g., *Limulus* or horseshoe crab species, the coagulogen is present in the hemocytes (Solum 1973). The cells burst and the protein aggregates when in contact with bacterial endotoxins. The coagulogen has a molecular mass of 17 kDa and its amino acid sequence is very different from that of mammalian fibrinogen (Nakamura et al. 1976; Miyata et al. 1983; Iwanaga et al. 1986). In the presence of endotoxin, an enzyme in the hemocytes splits two bonds in the coagulogen. Following this limited proteolysis, aggregation of the remaining protein occurred. Interestingly enough, the enzyme was shown to possess a factor Xa-like specificity for synthetic substrates and vertebrate factor II in the presence of endotoxin (Morita et al. 1985). In another crustacean, the lobster, the gel protein was also shown to be different from vertebrate fibrinogen. In fact, sequence homologies with vitellogenins (precursors of egg-yolk proteins) were shown (Fuller and Doolittle 1971; Doolittle and Fuller 1972; Doolittle and Riley 1990). In the lobster, the coagulogen is not activated by limited proteolysis. Clotting in this species is catalyzed by a transglutaminase (analogous to factor XIII in vertebrate clotting) that makes covalent bonds between the γ-carboxyl groups of glutamine in one coagulogen molecule and the ε-amino groups of lysine in another.

On the deuterostomial line, which encompasses the vertebrates, a putative fibrinogen-like protein has recently been identified in an Echinodermata species, sea cucumber (Xu and Doolittle 1990). This species is closely related to the hemichordata. The evolutionary history of the vertebrate fibrinogen genes appears to have undergone a series of gene duplications, as shown in Figure 11–14. It was suggested that the ancestral gene to the Aα-Bβ-γ chain was duplicated about one billion years ago giving rise to independently evolving Aα chain and Bβ-γ chains. In a more recent duplication the latter gave rise to independently evolving Bβ and γ chains. This reconstruction is based on both phylogenetic and amino acid sequence analyses (Henschen et al. 1983; Doolittle 1983). Thus, the Bβ and γ chains show considerable homology (about 35%) when compared within species and show even more homology when each of them is compared between species. The Aα chain shows the least homology with the other chains and also between species. This evolutionary scenario fits well with the demonstration by recombinant DNA methodology of a putative fibrinogen-like protein in sea cucumber (Xu and Doolittle 1990). This protein shows homologies with the carboxy-terminal parts of the Bβ and γ chains of vertebrate fibrinogen as well as with several other proteins. Sequences corresponding to the amino-terminal parts of these chains could not be identified. No sequence related to the vertebrate Aα chain was reported. These results indicate that the gene duplication leading to independently evolving Bβ and

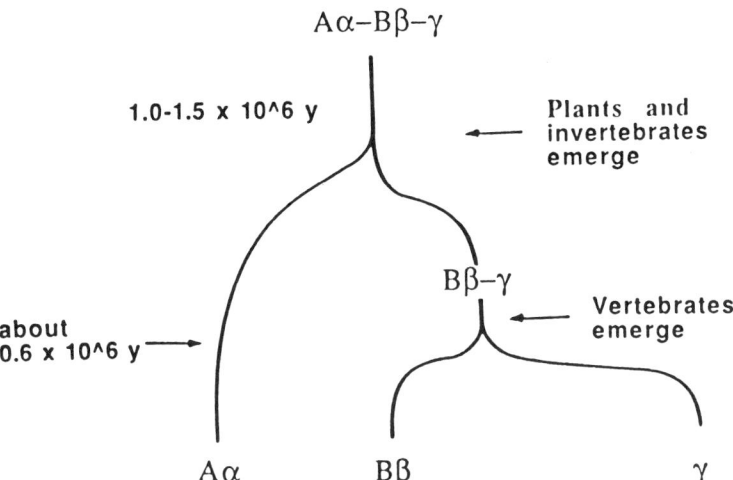

FIGURE 11–14 Proposed evolutionary scheme for the fibrinogen genes. Adapted with permission from Crabtree (1987).

γ chains occurred sometime after separation of this echinodermata species from the rest of the chordate and vertebrate lines, and this may have occurred about 600 million years ago. It is, however, puzzling that one of the important functional domains of vertebrate fibrinogen, i.e., the amino-terminal domain, appeared to be missing in the putative sea cucumber protein. Equally puzzling is the finding "down-stream" from the mRNA of the chicken Aα chain of an "open reading frame" having homologies with vertebrate Bβ and γ chains, as well as with several functionally unrelated proteins (Weissbach and Grieninger 1990). The homologies with other proteins which are functionally not related to fibrinogen should make us cautious about the evolutionary significance of these homologies. We know now that the shuffling of genetic elements between nonrelated genes occurs. Furthermore, the similarities between the sea cucumber and chicken proteins and fibrinogen may only be a reflection of the natural selection of functional domains that may be shared in many evolutionary unrelated proteins. Nevertheless, if these putative proteins are phylogenetically related to vertebrate fibrinogen, this may indicate that vertebrate fibrinogen is unique to all deuterostomia species. Curiously, it may not be present or may have another function in the protostomial line of evolution, since the coagulogens of the crustaceans in the protostomial lineage are functionally and structurally different from vertebrate fibrinogen.

The search for the genetic cradle of the primordial fibrinogen gene will certainly continue. So far, only a few invertebrate species have been investigated. Even in vertebrates we only know the complete sequence of all three chains for a few species. Conserved sequences as well as sections in which a fast flux of amino acid substitutions occurred have been identified (Doolittle 1981). The most extensive phylogenetic studies have been performed on the amino-terminal parts of the Aα and Bβ chains of vertebrate fibrinogens. They are of special interest since they clearly show regions both of rapid flux of amino acid substitutions and of preserved structures (Doolittle and Blombäck 1964; Blombäck et al. 1966a; Blombäck 1970; Söderqvist and Blombäck 1971). The latter are near the thrombin-cleavage site in fibrinogen and suggest the existence of strong selective pressure at this site. The selection pressure on thrombin appears to be more pronounced than on fibrinogen. This is because thrombin from many different vertebrate species has about the same clotting activity regardless of species but fibrinogen's clotting ability differs widely between species (Teger-Nilsson and Blombäck 1974). For instance, there are both slow- and fast-clotting types of fibrinogens. However, thrombin from any species will clot a fast fibrinogen at a fast rate and a slow fibrinogen at a slow rate.

When we compare the hemostatic systems in protostomia and deuterostomia, we find that thrombin-like and factor XIII-like enzymes exist on both lines. The gel-forming proteins are structurally and functionally remarkably different. One may therefore predict that thrombin- and factor XIII-like enzymes were present on the ancestral line leading to protostomia and deuterostomia. The gel-forming proteins on the other hand were possibly selected from different progenitors on the two lines.

11.10   ROLE OF FIBRINOGEN IN HEALTH AND DISEASE

It is generally accepted that the most important function of fibrinogen is that of the structural element in the hemostatic blood clots formed at sites of injured endothelium in the vasculature. However, this does not mean that the importance of fibrinogen is limited entirely to hemostasis. Thus, there is evidence suggesting that fibrinogen/fibrin could be of importance in wound healing and ontogenic processes. Fibrinogen belongs to the acute-phase group of proteins, i.e., the concentration of these proteins increase markedly following infection. The latter situation has given rise to suggestions that fibrinogen may play some role in the organism's defense against infection.

It is conceivable that the major role of fibrinogen in hemostasis is to form fibrin network structures around the platelets which adhere and aggregate around small lesions in the vessel wall. A stable hemostatic plug of platelets would thereby be formed. If fibrin formation fails for some reason, the platelet plug will eventually be swept away and bleeding will restart. Cross-linking of the fibrin formed may also be important for consolidation of the hemostatic plug, since some individuals with a congenital deficiency of factor XIII have severe bleeding disorders.

Increased levels of fibrinogen in the blood are observed in many conditions. In infection, the fibrinogen level in blood is markedly elevated. This is the major cause of the increased red cell sedimentation rate seen in these conditions. In neoplastic diseases, increased fibrinogen levels are often found. In the postoperative state and after trauma, there is also generally an increase in the blood fibrinogen level. The increased fibrinogen level is most likely caused by an increased rate of synthesis. Cytokines, hormones, and other cellular effectors may play a role here.

Increased synthesis of fibrinogen may parallel its increased catabolism. If the rate of catabolism increases more than the synthetic capacity of the organism, the result will be a fall in the level of blood fibrinogen. Such an imbalance may arise in liver diseases. However, this phenomenon is most often seen in states of intravascular coagulation, which must be distinguished from thrombosis, a purely local phenomenon. In intravascular coagulation, fibrinogen catabolism increases because of disseminated deposition of fibrin in the blood vessel system. Uncontrolled activation of fibrinogen may take place following entrance into the blood of substances which promote conversion of prothrombin into thrombin. Intravascular coagulation is a serious condition which, often and paradoxically, manifests itself as a bleeding disorder. It can arise as a complication in several diseases with different pathogenesis. It is found, for example, in *abruptio placentae*, after trauma, in shock, during hemolytic crisis, and in neoplastic disease and the diagnosis is often difficult. The fibrinogen concentration in blood is often quite low but need not necessarily be so. The concentration of other clotting factors is also often reduced. The radioimmunological method for the measurement of FPA (Nossel et al. 1974) is a valuable tool in the diagnosis of intravascular coagulation, since the release of FPA is a measure of the degree of activation of fibrinogen molecules. The level of this peptide in blood is a reflection of fibrin deposition and its intensity. In intravascular coagulation, there is a substantial rise in the level of FPA.

In a rare congenital disease called afibrinogenemia, there appears to be a total lack of fibrinogen in blood. Most of these patients have a bleeding tendency. Thrombin-coagulable fibrinogen cannot be demonstrated in the blood, and no epitopes of fibrinogen are found when examined immunologically. Genetic analysis in a case of afibrinogenemia showed that the genes for fibrinogen (Aα, Bβ, and γ) were apparently present but they were not expressed (Uzan et al. 1984). A point mutation or a small deletion could not be detected in this study. If such changes were present in one gene, processing and expression of the full genetic message may be perturbed.

Congenital hypofibrinogenemia is a more common condition than afibrinogenemia. There is almost no increased bleeding tendency in these patients. It has been suggested that these patients may possibly be heterozygotes with regard to an afibrinogenemia trait.

Dysfibrinogenemia is a condition in which there is a congenital abnormality in fibrinogen structure and function. So far, about one hundred families afflicted with functionally abnormal fibrinogen have been described (for review, see Southan 1988). In regard to the clinical picture, some of these individuals have bleeding symptoms, others have thrombotic tendencies, and the majority have no symptoms at all.

In some of these unusual fibrinogen, mutations or deletions in polypeptide chains have been found. In most cases, the structural defect has not been demonstrated. The abnormal fibrinogens for which a structural change could be shown can be divided into two groups: those having structural changes in the amino-terminal part of the molecule and those with structural changes in the mid-portion or the carboxy-terminal end of the molecule. Most of the point mutations in the amino-terminal part are located within the Aα chain 7–19 sequence. A point mutation, Bβ Arg 14 $\rightarrow$ Cys, has been reported in Fibrinogen Seattle I and Christchurch II (Kaudewitz et al. 1987) and a mutation Bβ Ala 68 $\rightarrow$ Thr in fibrinogen Milan II (Haverkate et al. 1985; Koopman et al. 1990). A deletion (Bβ 9–72) in the amino-terminal part of the Bβ chain was found in fibrinogen New York (Liu et al. 1985). Most of the mutations in the carboxy-terminal part of the molecule are point mutations and are in the γ chain sequence 275–330. In fibrinogen Paris I, abnormal mRNA splicing produces a γ chain with extra carboxy-terminal residues (Mosesson et al. 1976). Point mutations in the carboxy-terminal part of the Bβ chain and the mid-portion of the Aα chain have also been found. Of particular interest is fibrinogen Pontoise (Kaudewitz et al. 1986) in which a point mutation, Bβ 335 Ala $\rightarrow$ Thr, created a new glycosylation site in addition to that at Bβ Asn 364.

The most characteristic abnormality found in coagulation analyses of these mutants is a pronounced lengthening of the "thrombin time" of plasma, i.e., the time needed for the coagulation of the plasma after the addition of thrombin is longer than normal. However, sometimes the clotting time may be only slightly perturbed or it may even be shorter than normal.

The prolongation of clotting time may sometimes be coupled with a slow or with no release of FPA. This is the case for fibrinogen Metz and fibrinogen Petroskey (Henschen et al. 1981; Higgins and Schafer 1981) with Aα Arg 16 $\rightarrow$ Cys and Aα Arg 16 $\rightarrow$ His mutations, respectively. In fibrinogen Milan II (Haverkate et al. 1985; Koopman et al. 1990) with Bβ Ala 68 $\rightarrow$ Thr the release of FPA and FPB by thrombin is also slow. This means that the activation of fibrinogen is hampered, resulting in delayed polymerization and formation.

In other cases, the prolongation of clotting time is not paired with significantly deficient release of FPA and FPB. This is the case for fibrinogen Detroit (Arg 19 $\rightarrow$ Ser; Blombäck et al. 1968b), fibrinogen Aarhus (Aα Arg 19 $\rightarrow$ Gly; Blombäck et al. 1988), fibrinogen Haifa and fibrinogen Bergamo (γ Arg 275 $\rightarrow$ His; Siebenlist et al. 1989; Reber et al. 1986), and fibrinogen Kyoto I (γ Asn 308 $\rightarrow$ Lys; Yoshida et al. 1988). In these fibrinogens, the mutation has caused a perturbation of the polymerization and/or gel formation reactions. In fibrinogen Detroit and Aarhus, the *A* sites, and in Haifa and Kyoto I, the *a* sites, are presumably perturbed.

In some dysfibrinogenemias, the clotting time is not or is only slightly prolonged, and the release of FPA and FPB is normal. In the absence of calcium ions, the clotting time may, however, be drastically prolonged. The gels formed are abnormal in comparison with gels from normal fibrinogen. They display low turbidities and permeability, and confocal three-dimensional microscopy shows abnormal network structures. The gels are rigid and tight. Examples of this type of abnormality are fibrinogen Chapel Hill (Carrell et al. 1983) and fibrinogen Tampere and Lakonia (Blombäck et al. 1990a). Since the clotting times of these fibrinogens are not much prolonged (in presence of calcium ions), it appears that the initial polymerization step proceeds in a normal fashion quantitatively. Abnormal polymers, however, may be formed, and these may interact to form an abnormal network structure. The structural abnormality has not been identified in these fibrinogens. However, in fibrinogen Tampere, there do not appear to be any mutations in the amino-terminal part, and the "hot" spots in the carboxy-terminal part of the γ chain also seem to be normal (Redman 1991).

The question of how the mutations and structural changes in dysfibrinogenemia are related to clinical symptoms is difficult to answer. This is partly due to the fact that in many cases the individuals carrying the abnormal trait are heterozygotes and have a half amount of normal protein (Blombäck and Blombäck 1970). The heterozygotes usually have no or only minor hemostatic defects. The homozygotes, however, often have clinical symptoms. There are exceptions to this rule, i.e., the patient carrying the trait for fibrinogen New York is a heterozygote but showed clinical manifestations. The dysfibrinogenemias with point mutations in the amino-terminal part of the molecule often have bleeding symptoms. Examples of this are found in dysfibrinogenemia Detroit (Aα Arg 19 $\rightarrow$ Ser), Metz (Aα Arg 16 $\rightarrow$ Lys), and Rouen (Aα Gly 12 $\rightarrow$ Val). Bleedings are also associated with dysfibrinogenemia Pontoise, but most of the reported dysfibrinogenemias with carboxy-terminal point mutations are asymptomatic. In dysfibrinogenemia Paris I, having a carboxy-terminal extension of the γ chain (Mosesson et al. 1976), deficient wound healing was reported, possibly related to defective cross-linking by factor XIII.

Several dysfibrinogenemias present a thrombotic disorder. This was the case with fibrinogen New York (deletion Bβ 9–72) and fibrinogen Milan II (Bβ Ala 68 $\rightarrow$ Thr), despite the slow release of fibrinopeptides in these fibrinogens. Deficient fibrinolysis and/or thrombin binding was suggested to be causative (Haverkate et al. 1985; Liu et al. 1986). Dysfibrinogenemia Bergamo (γ Arg 275 $\rightarrow$ His; Reber et al. 1986) was also reported to be associated with a thrombotic tendency. However, other dysfibrinogenemias with the same mutation were reported to be asymptomatic.

In several dysfibrinogenemias with thrombotic diseases, the structural change in the fibrinogen has not been demonstrated. This is the case with fibrinogen Dusard (Soria et al. 1983), where fibrinogen-deficient fibrinolysis was suggested as a cause of the thrombotic conditions (Lijnen et al. 1984). Hampered fibrinolysis may also be part of the pathogenesis in dysfibrinogenemia Chapel Hill (Carrell et al. 1983). In fibrinogen Tampere and Lakonia (Blombäck et al. 1990a) the tendency to form abnormally tight, rigid, fibrin gel structures in vitro may be related to the disease process in vivo.

Fibrinogen Detroit and Aarhus (Blombäck et al. 1978, 1988; Kudryk et al. 1976; Hessel et al. 1986) are examples of the functional consequences which a small change in amino acid sequence may have. The mutations (Aα 19 Arg $\rightarrow$ Ser and Aα 19 Arg $\rightarrow$ Gly, respectively) have occurred in the phylogenetically invariable part of the structure, i.e., the part preserved in the course of mammalian evolution. Evidently, this portion of the Aα chain is critical for polymerization. However, this does not necessarily mean that the affinity of thrombin for this chain is lower than in normal fibrinogen. In fact, it was demonstrated that the rate of FPA release was almost normal. Hence, apparently, the mutation brings about secondary changes in the conformation of fibrinogen in such a way that one of the polymerization domains *(A,A')*, which are normally unfolded after the release of FPA, remains dormant. However, both of these fibrinogens do clot after the release of FPB. As discussed in Section 11.5, this release may activate a second domain *(B,B')*, which, by interaction with a complementary domain, leads to polymerization. This is a slow process and, furthermore, the polymer that eventually forms by this route would be expected to result in a gel structure different from the normal one. In fibrinogen Detroit, Aα Arg 19 is replaced by serine, and, in fibrinogen Aarhus, by glycine. Both fibrinogens show the same functional perturbance when isolated fibrinogen is tested in the presence of thrombin. However, the patients having these two types of abnormal fibrinogen display completely different clinical pictures. In dysfibrinogenemia Detroit, there is a severe bleeding condition, whereas fibrinogen Aarhus is asymptomatic from the hemostatic point of view. The Aα Arg 19 $\rightarrow$ Ser mutation apparently causes abnormal polymerization both in vitro and in vivo in whole blood. The Aα Arg 19 $\rightarrow$ Gly substitution is apparently only expressed in vitro. It is possible that the Aα Arg $\rightarrow$ Ser mutation brings about a more or less irreversible conformational change in fibrinogen. The conformational change resulting from the Aα Arg 19 $\rightarrow$ Gly mutation would be more reversible, especially in vivo. However, factor XIII-induced gel formation of fibrinogen was reported to be normal in dysfibrinogenemia Aarhus, and this may explain the lack of bleeding in the patient (Hessel et al. 1987). Unfortunately, the latter studies were not performed in dysfibrinogenemia Detroit.

Fibrinogen Rouen (Ni et al. 1989c) is an even more powerful example of the dramatic consequences an amino acid substitution can have on the tertiary structure of a protein. In this unusual fibrinogen, with a Gly 12 $\rightarrow$ Val mutation, the normal three-dimensional structure of the thrombin-binding region was disrupted (Figure 11–15) with the result that activation by thrombin was slow.

Thrombosis is one of the major diseases in developed countries, as indicated by the fact that in patients undergoing surgery without prophylactic treatment, 30–40% may develop postoperative thrombosis. Although much progress has been achieved in the fight against thrombosis during the past 20 years, mainly due to the introduction of anticoagulants such as heparin and dicoumarol, thrombosis continues to be a major problem.

In the classical description of venous or arterial thrombi, the role of fibrinogen in formation of the obstructive structure is more or less crucial. In venous thrombi initiated at endothelial lesions, the fibrin network provides the structural framework for the clot in which red cells and other cellular components are trapped. This is called a "red" thrombus. In arterial thrombi, on the other hand, the clot structure is much more compact. Here the clot structure is built up by a mass of platelets in a fibrin network structure. This is the so-called "white" thrombus. The two types of thrombi reflect the physical conditions under which they are formed. On the venous side, where the blood flow is slow, an initial thrombus formed at a site of a lesion in the vessel wall is able to grow in the direction of the bloodstream, because the relatively slow flow in the veins favors a relatively high concentration of procoagulant factors prevailing at some distance from the

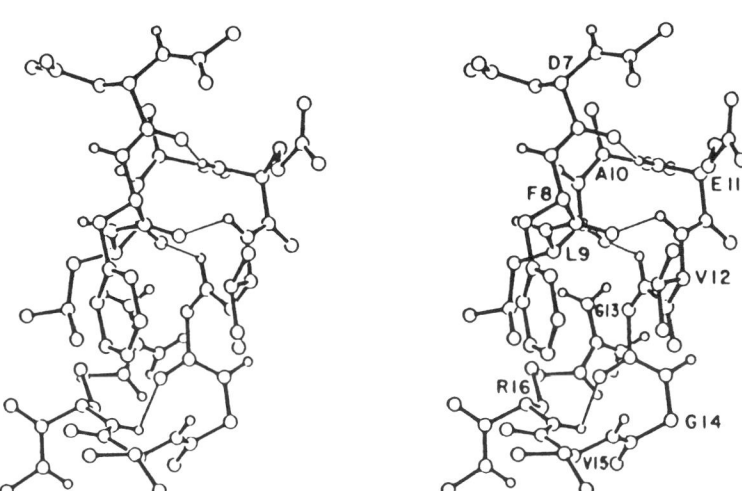

FIGURE 11–15 Stereo drawings of a proposed structure for the thrombin-bound FPA fragment (residues 7–16) from fibrinogen Rouen (Aα 12 Gly $\rightarrow$ Val). The nonpolar cluster formed by the side chains of Phe (8), Leu (9), Val (12), and Val (15) is clearly seen. The C$_a$ atoms are labeled by the corresponding amino acid residues. Compare the structure with proposed structures for the normal peptide given in Figure 11–6. Reproduced with permission from Ni et al. (1989c).

lesion. On the arterial side where the blood flow is rapid, however, the clot will presumably enlarge primarily by aggregation of platelets at the site of lesion, and fibrin formation is restricted to the mass of platelets and its immediate surroundings.

Rudolf Virchow, in the middle of the 19th century, described three factors (the Triad of Virchow) of importance for the development of thrombosis. These were blood flow, blood vessel wall, and the constituents of the blood. A predisposing factor for thrombosis is a slow blood flow, which may be caused by heart disease. Destruction of the endothelial lining of the blood vessels may also be a predisposing factor since it results in extensive platelet deposition which acts as a nucleus for fibrin formation, eventually leading to thrombosis. Finally, constituents of the blood may play a decisive role in determining whether a hemostatic plug remains as such or grows into an obliterating clot structure. Increased levels of coagulation factors may favor thrombus formation merely on an enzyme-kinetic basis. The level of inhibitors of coagulation (e.g., antithrombin III, activated protein C) or factors involved in fibrinolysis may also play a major role in modulating the growth of a thrombus.

We should also consider the consequences that the formation of various types of fibrin may have on the thrombotic process. Recent epidemiological studies have shown that moderately (20–30%) increased levels of fibrinogen and/or factor VII activity carries with it an increased risk for development of ischemic heart disease (IHD) or stroke (Wilhelmsen et al. 1984; Stone and Thorp 1985; Markowe et al. 1985; Kannel et al. 1987; Meade et al. 1986). In several studies, both factors showed an independent association with IHD. In one study (Carlson et al. 1979), the red cell sedimentation rate (ESR) was shown to be a risk factor for IHD. The increased ESR also indirectly links fibrinogen to the process. Our finding that the structure of the fibrin gel is determined by the clotting potential and the fibrinogen concentration acting in concert poses the question whether the common denominator for factor VII activity (clotting potential) and fibrinogen concentration as risk factors in IHD is the architecture of an established clot network rather than the mass of fibrin deposited. Deposited fibrin always occupies a small part (less than 1%) of the clot volume. In a fibrinogen-thrombin system in vitro, the water trapped in the gel network has the volume-filling capacity, and the resiliency of the gel structure determines the occlusive power of such a structure. Tight network structures with small pores and thin fibrin strands are rigid and brittle and break rather than release the water trapped in the clot. An increase in the clotting potential or fibrinogen concentration favors the formation of these types of clots. Elevated levels of factor VII activity and/or fibrinogen in vivo may also favor formation of tight, rigid, space-filling clots at sites of lesions in the vasculature and thereby increase the risk for thromboembolic disease.

Modulators, like proteins and ions, in plasma (see Section 5.4) may play an important role in the organization of fibrin gel structures at lesions in the vasculature. Copley and coworkers (for review see Copley 1979b) have demonstrated that low density lipoproteins (LDL) interact with fibrinogen in the formation of shear-stress induced protein surface layers. It is therefore possible that the formation and quality of the fibrin gel structures are also influenced by certain classes of lipoproteins. LDL is a high risk factor in the development of atherogenesis and thrombosis.

It is also possible that the different architectural forms of fibrin have different affinities for components of the fibrinolytic system, and this may determine how long they prevail in the circulation. The role of the fibrinolytic potential in the thrombotic processes was recently highlighted by Hamsten et al. (1987), who showed that an increased level of the inhibitor of t-PA in blood (i.e., PAI) was a risk factor for IHD.

Finally, much progress has been made in the treatment of thromboembolic diseases. Drugs like heparin and dicoumarol have, during the past few decades, been successfully used in the prophylaxis and treatment of venous thromboembolic diseases. Much less success has been achieved in using these drugs in the treatment and prophylaxis of arterial thrombosis. t-PA was once regarded as the ideal agent for thrombolytic therapy in arterial thrombosis because of its unique property of binding to fibrin but not to fibrinogen. However, the clinical results so far have not matched the early expectations. The dose required is very high due to its extremely short half-life in circulation, which is caused by rapid uptake in the liver (Nilsson et al. 1984; Verstraete et al. 1985). Streptokinase, which seems to lack specific affinity for fibrin, has in fact been shown to be nearly as efficient as t-PA in the lysis of arterial thrombi.

In the future, there may also be a place for other agents in the arsenal of drugs for prophylaxis and treatment of thrombosis; agents which modulate fibrin gel structure are possible candidates for trials; for example dextran, which is a potent modulator of fibrin gel structure, has been shown to be effective in the prevention of thrombosis.

REFERENCES

Adams Lucas, M., Straight, D.L., Fretto, L.J., and McKee, P.A. (1983) *J. Biol. Chem.* 258, 12171–12177.

Alving, B.M., Chung, S.I., Murano, G., Tang, D.B., and Finlayson, J.S. (1982) *Arch. Biochem. Biophys.* 217, 1–9.

Astrup, T., and Permin, P.M. (1947) *Nature* 159, 681–682.

Bachmann, L., Schmitt-Fumian, W.W., Hammel, R., and Lederer, K. (1975) *Macromol. Chemic.* 176, 2603–2618.

Bang, N.U. (1964) *Thromb. Diath. Haemorrh.* 13, 73–80.

Blombäck, B. (1967) in *Blood Clotting Enzymology* (Seegers, W.H., ed.), pp. 143–215, Academic Press, New York.

Blombäck, B. (1969) *Brit. J. Haemat.* 17, 145–157.

Blombäck, B. (1970) in *Symp. Zool. Soc. London no. 27*, pp. 167–187.

Blombäck, B. (1972) in *Glycoproteins, Their Composition, Structure and Function* 5 Part B (Gottschalk, A., ed.), pp. 1069–1081, Elsevier, Amsterdam.

Blombäck, B. (1979a) in *Chromogenic Peptide Substrates* (Scully, M.F., and Kakkar, V.V., eds.), pp. 3–12, Livingstone, Edinburgh.

Blombäck, B. (1979b) in *Plasma Proteins* (Blombäck, B., and Hansson, L.Å., eds.), pp. 223–253, John Wiley & Sons, Chichester, UK.

Blombäck, B. (1985) in *Fibrinogen-Structural Variants and Interactions* (Henschen, A., et al. eds.), pp. 33–42, Walter de Gruyter & Co., New York.

Blombäck, B., and Blombäck, M. (1970) *Nouvelle Revue Française d'Hématol.* 10, 671–678.

Blombäck, B., and Blombäck, M. (1972) *Ann. N.Y. Acad. Sci.* 202, 77–97.

Blombäck, B., and Okada, M. (1982) *Thromb. Res.* 25, 51–70.

Blombäck, B., and Yamashina, I. (1958) *Arkiv. Kemi.* 12, 299–319.

Blombäck, B., Blombäck, M., and Searle, J. (1963) *Biochim. Biophys. Acta* 74, 148–151.

Blombäck, B., Blombäck, M., Gröndahl, N.J., and Holmberg, E. (1966a) *Arkiv. Kemi.* 25, 411–428.

Blombäck, B., Carlsson, L.A., Franzén, S., and Zetterquist, E. (1966b) *Acta Med. Scand.* 179, 557–574.

Blombäck, B., Blombäck, M., Laurent, T.C., and Pertoft, H. (1966c) *Biochim. Biophys. Acta* 127, 560–562.

Blombäck, B., Blombäck, M., Henschen, A. et al. (1968a) *Nature* 218, 130–134.

Blombäck, M., Blombäck, B., Mammen, E.F., and Prasad, A.S. (1968b) *Nature* 218, 134–137.

Blombäck, B., Blombäck, M., Olsson, P., Svendsen, L., and Åberg, G. (1969) *Scand. J. Clin. Lab. Inv.* 24 (suppl. 107), 59–64.

Blombäck, B., Hessel, B., Iwanaga, S., Reuterby, J., and Blombäck, M. (1972) *J. Biol. Chem.* 247, 1496–1512.

Blombäck, B., Gröndahl, N.J., Hessel, B., Iwanaga, S., and Wallén, P. (1973) *J. Biol. Chem. J.* 248, 5806–5820.

Blombäck, B., Blombäck, M., Finkbeiner, W., et al. (1974) *Thromb. Res.* 4, 55–75.

Blombäck, B., Hessel, B., and Hogg, D. (1976a) *Thromb. Res.* 8, 639–658.

Blombäck, B., Hogg, D.H., Gårdlund, B., Hessel, B., and Kudryk, B. (1976c) *Thromb. Res.* (suppl. 8), 329–346.

Blombäck, B., Hessel, B., Hogg, D., and Claesson, G. (1977) in *Chemistry and Biology of Thrombin* (Lundblad, R.L., Fenton, J.W., and Mann, K.G., eds.), pp. 275–290, Ann Arbor Science Publishers, Inc., Ann Arbor, MI.

Blombäck, B., Hessel, B., Hogg, D.H., and Therkildsen, L. (1978) *Nature* 257, 501–505.

Blombäck, B., Okada, M., Forslind, B., and Larsson, U. (1984) *Biorheology* 21, 93–104.

Blombäck, B., Procyk, R., Adamson, L., and Hessel, B. (1985) *Thromb. Res.* 37, 613–628.

Blombäck, B., Adamson, L. Hessel, B., et al. (1986) in *Thioredoxin and Glutaredoxin Systems; Structure and Function* (Holmgren, A., Brändén, C.-I., Jörnvall, H., and Sjöberg, B.-M., eds.), pp. 357–367, Raven Press, New York.

Blombäck, B., Hessel, B., Fields, R., and Procyk, R. (1988) in *Fibrinogen 3. Biochemistry, Biological Functions, Gene Regulation and Expression* (Mosesson, M.W., et al., eds.), pp. 263–266, Elsevier Science Publishers, B.V. (Biomedical Division), Amsterdam.

Blombäck, B., Carlsson, K., Hessel, B., et al. (1989) *Biochim. Biophys. Acta* 997, 96–110.

Blombäck, B., Banerjee, D., Carlsson, K., et al. (1990a) in *Fibrinogen, Thrombosis, Coagulation, Fibrinolysis* (Liu, C.Y., and Chien, S., eds.), Plenum Publishing Corp., New York (in press).

Blombäck, B., Fatah, K., and Hessel, B. (1990b). Unpublished observations.

Blombäck, M., Blombäck, B., and Holmqvist, H. (1976b) *Thromb. Res.* 8, 567–577.

Bohonus, V.L., Doolittle, R.F., Pontes, M., and Strong, D.D. (1986) *Biochemistry* 25, 6512–6516.

Bouma, H., III, Takagi, T., and Doolittle, R.F. (1978) *Thromb. Res.* 13, 557–562.

Budzynski, A.Z., Stahl, M., Kopeć, M., et al. (1967) *Biochim. Biophys. Acta* 147, 313–323.

Buluk, K., Januszko, T., and Olbromski, J. (1961) *Nature* 191, 1093–1094.

Camiolo, S.M., Thorsen, S., and Astrup, T. (1971) *Proc. Soc. Exp. Biol. Med.* 138, 277–280.

Carlson, L.A., Böttiger, L.E., and Åhfeldt, P.E. (1979) *Acta Med. Scand.* 206, 351–360.

Carr, M.E., and Gabriel, D.A. (1980) *Macromolecules* 13, 1473–1477.

Carr, M.E., and Hermans, J. (1978) *Macromolecules* 11, 46–50.

Carr, M.E., Shen, L.L., and Hermans, J. (1976) *Anal. Biochem.* 72, 202–211.

Carrell, N., Gabriel, D.A., Blatt, P.M., Carr, M.E., and McDonagh, J. (1983) *Blood* 62, 439–447.

Cartwright, T., and Kekwick, R.G.O. (1971) *Biochim. Biophys. Acta* 236, 550–562.

Caspary, E.A., and Kekwick, R.A. (1957) *Biochem. J.* 67, 41–48.

Cederholm-Williams, S.A. (1977) *Thromb. Res.* 11, 421–423.

Chmielewska, J., Rånby, M., and Wiman, B. (1983) *Thromb. Res.* 31, 427–436.

Chung, D.W., Rixon, M.W., MacGillivray, R.T., and Davie, E.W. (1981) *Proc. Natl. Acad. Sci. USA* 78, 1466–1470.

Cohen, C., Weisel, J.W., Phillips Jr., G.N., et al. (1983) *Ann. N.Y. Acad. Sci.* 408, 194–213.

Cole, E.R., and Bachmann, F.W. (1977) *J. Biol. Chem.* 252, 3729–3737.

Collen, D. (1980) *Thromb. Haemost.* 43, 77–89.

Collen, D., Kudryk, B., Hessel, B., and Blombäck, B. (1975) *J. Biol. Chem.* 250, 5808–5817.

Collen, D., Stassen, J.M., Blaber, M., Winkler, M., and Verstraete, M. (1984) *Thromb. Haemost.* 52, 27–30.

Copley, A.L. (1979a) *Folia Haematol. Leipzig* 106, 732–764.

Copley, A.L. (1979b) *Thromb. Res.* 14, 249–263.

Copley, A.L. (1988) *Biorheology* 25, 377–399.

Cottrell, B.A., Strong, D.D., Watt, K.V., and Doolittle, R.F. (1979) *Biochemistry* 18, 5405–5410.

Crabtree, G.R. (1987) in *The Molecular Basis of Blood Diseases* (Stamatoyanno-poulus, G., Nienhuis, A.W., Leder, P., and Majerus, P.W., eds.), pp. 631–661, W.B. Saunders Co., Philadelphia.

Crabtree, G.R., and Kant, J.A. (1982a) *Cell* 31, 159–166.

Crabtree, G.R., and Kant, J.A. (1982b) *J. Biol. Chem.* 257, 7277–7279.

Crabtree, G.R., Comeau, C.M., Fowlkes, D.M., Malley, J., and Kant, J.A. (1985) *J. Mol. Biol.* 185, 1–19.

Doolittle, R.F. (1973) *Adv. Protein Chem.* 27, 1–109.

Doolittle, R.F. (1981) *Scientific American* 245, 126–135.
Doolittle, R.F. (1983) *Ann. N.Y. Acad. Sci.* 408, 13–27.
Doolittle, R.F. (1984) *Ann. Rev. Biochem.* 53, 195–229.
Doolittle, R.F., and Blombäck, B. (1964) *Nature* 202, 147–152.
Doolittle, R.F., and Fuller, G.M. (1972) *Biochim. Biophys. Acta* 805–809.
Doolittle, R.F., and Riley, M. (1990) *Biochem. Biophys. Res. Comm.* 167, 16–19.
Doolittle, R.F., Takagi, T., and Cottrell, B.A. (1974) *Science* 185, 368–370.
Doolittle, R.F., Cassman, K.G., Cottrell, B.A., Friezner, S.J., and Takagi, T. (1977a) *Biochemistry* 16, 1710–1715.
Doolittle, R.F., Cassman, K.G., Cottrell, B.A., et al. (1977b) *Biochemistry* 16, 1703–1709.
Doolittle, R.F., Watt, K.W.K., Cottrell, B.A., Strong, D.D., and Riley, M. (1979) *Nature* 280, 464–468.
Dyr, J.E., Blombäck, B., Hessel, B., and Kornalik, F. (1989) *Biochim. Biophys. Acta* 990, 18–25.
Ehrlich, P.H., Sobel, J.H., Moustafa, Z.A., and Canfield, R.E. (1983) *Biochemistry* 22, 4184–4192.
Erickson, H.P., and Fowler, W.F. (1983) *Ann. N.Y. Acad. Sci.* 408, 146–163.
Ferry, J.P. (1952) *Proc. Natl. Acad. Sci.* 38, 566–569.
Ferry, J.D., and Morrison, P.R. (1947) *J. Am. Chem. Soc.* 69, 388–400.
Finlayson, J.S., and Mosesson, M.W. (1963) *Biochemistry* 2, 42–46.
Fornace, A.J., Cummings, D.E., Comeau, C.M., Kant, J.A., and Crabtree, G.R. (1984) *J. Biol. Chem.* 259, 12826–12830.
Fowler, W.E., Hantgan, R.R., Hermans, J., and Erickson, H.P. (1981) *Proc. Natl. Acad. Sci. USA* 78, 4872–4876.
Francis, C.W., Müller, E., Henschen, A., Simpson, P.J., and Marder, V.J. (1988) *Proc. Natl. Acad. Sci.* 85, 3358–3362.
Fuller, G.M., and Doolittle, R.F. (1971) *Biochemistry* 10, 1311–1315.
Gaffney, P.J. (1971) *Nature New Biol.* 230, 54–56.
Gaffney, P.J. (1972) *Biochim. Biophys. Acta* 263, 453–458.
Gaffney, P.J. (1977) *Br. Med. Bull.* 33, 245–252.
Gati, W.P., and Straub, P.W. (1978) *J. Biol. Chem.* 253, 1315–1321.
Godal, H.C. (1960) *Scand. J. Clin. Lab. Invest.* 12, (suppl. 53), 1–20.
Greenberg, C.S., Achyuthan, K.E., Rajagopalan, S., and Pizzo, S.V. (1988) *Arch. Biochem. Biophys.* 262, 142–148.
Gurevich, V., Pannell, R., Louie, S., et al. (1984) *J. Clin. Invest.* 73, 1731–1739.
Gårdlund, B. (1977) *Thromb. Res.* 10, 689–702.
Gårdlund, B., Hessel, B., Marguerie, G., Murano, G., and Blombäck, B. (1977) *Eur. J. Biochem.* 77, 595–610.
Hall, C.E., and Slayter, H.S. (1959) *J. Biophys. Biochem. Cytol.* 5, 11–17.
Hamsten, A., de Faire, U., Walldius G., et al. (1987) *Lancet* 2, 3–9.
Hantgan, R.R., and Hermans, J. (1979) *J. Biol. Chem.* 254, 11272–11281.
Hasegawa, N., and Sasaki, S. (1990) *Thromb. Res.* 57, 183–195.
Haverkate, F., and Timan, G. (1977) *Thromb. Res.* 10, 803–812.
Haverkate, F., Koopman, J., Kluft, C., et al. (1985) in *Fibrinogen. Structural Variants and Interactions* vol. 3, (Henschen, A., Hessel, B., McDonagh, J., and Saldeen, T., eds.), pp. 223–236, Walter de Gruyter and Co., Berlin.
Hawn, C.V.Z., and Porter, K.R. (1947) *J. Exp. Med.* 86, 285–292.
Heene, D.L., and Matthias, F.R. (1973) *Thromb. Res.* 2, 137–154.

Henschen, A. (1964) *Ark. Kemi.* 22, 375–396.

Henschen, A. (1978) *Hoppe-Seyler's Z. Physiol. Chem.* 359, 1757–1770.

Henschen, A. (1983) *Thromb. Res.* (suppl. 5), 27–39.

Henschen, A., and Lottspeich, F. (1977) *Hoppe-Seyler's Z. Physiol. Chem.* 358, 1643–1646.

Henschen, A., and Warbinek, R. (1975) *Hoppe-Seyler's Z. Physiol. Chem.* 356, 1981–1984.

Henschen, A., Lottspeich, F., Sekita, T., and Warbinek, R. (1976) *Hoppe-Seyler's Z. Physiol. Chem.* 357, 605–608.

Henschen, A., Lottspeich, F., Töpfer-Petersen, E., and Warbinek, R. (1979) *Thromb. Haemost.* 41, 662–670.

Henschen, A., Southan, C., Soria, J., Soria, C., and Samama, M. (1981) *Thromb. Haemost.* 46, 103 (abst. 314).

Henschen, A., Lottspeich, F., Kehl, M., and Southan, C. (1983) *Ann. N.Y. Acad. Sci.* 408, 28–43.

Hermans, J. (1979) *Proc. Natl. Acad. Sci. USA* 76, 1189–1193.

Hessel, B. (1975) *Thromb. Res.* 7, 75–87.

Hessel, B., Makura, M., Iwanaga, S., and Blombäck, B. (1979) *J. Eur. Biochem.* 98, 521–534.

Hessel, B., Stenbjerg, S., Dyr, J., et al. (1986) *Thromb. Res.* 42, 21–37.

Hessel, B., Adamson, L. Procyk, R., et al. (1987) *Brit. J. Haemat.* 66, 355–361.

Higgins, D.L., and Schafer, J.A. (1981) *J. Biol. Chem.* 256, 12013–12017.

Hirose, S., Oda, K., and Ikehara, Y. (1988) *Biochem. J.* 251, 373–377.

Hishikawa-Itoh, Y., Sugie, I., Kato, H., and Iwanaga, S. (1982) *J. Biol. Chem.* 92, 1129–1140.

Hogg, D.H., and Blombäck, B. (1974) *Thromb. Res.* 5, 685–693.

Hogg, D.H., and Blombäck, B. (1978) *Thromb. Res.* 12, 953–964.

Hörmann, H. (1985) in *Hematology, Plasma Fibronectin, Structure and Function*, vol. 5, (McDonagh, J., ed.) pp. 99–120, Marcel Dekker Inc., New York.

Iwanaga, S., Blombäck, B., Gröndahl, N.J., Hessel, B., and Wallén, P. (1968) *Biochim. Biophys. Acta* 160, 280–283.

Iwanaga, S., Suzuki, K., and Hashimoto, S. (1978) *Ann. N.Y. Acad. Sci.* 312, 56–73.

Iwanaga, S., Morita, T., Miyata, T., Nakamura, T., and Aketagawa, J. (1986) *J. Protein Chemistry* 5, 255–268.

James, H.L., Ganguly, P., and Jackson, C.W. (1977) *Thromb. Haemost.* 38, 939–954.

Jandrot-Perrus, M., Mosesson, M.W., Denninger, M.H., and Ménaché, D. (1979) *Blood* 54, 1109–1116.

Jevons, F.R. (1963) *Biochem. J.* 89, 621–624.

Jörnvall, H., Pohl, G., Bergsdorf, N., and Wallén, P. (1983) *FEBS Letter* 156, 47–50.

Kanaide, H., and Shainoff, J.R. (1975) *J. Lab. Clin. Med.* 85, 574–597.

Kannel, W.B., Wolf, P.A., Castelli, W.P., and D'Agostino, R.B. (1987) *J. Am. Med. Assoc.* 258, 1183–1186.

Kant, J.A., Lord, S.T., and Crabtree, G.R. (1983) *Proc. Natl. Acad. Sci. USA* 80, 3953–3957.

Kant, J.A., Fornace, Jr., A.J., Saxe, D., et al. (1985) *Proc. Natl. Acad. Sci. USA* 82, 2344–2348.

Kaudewitz, H., Henschen, A., Soria, J., and Soria, C. (1986) in *Fibrinogen—Fibrin Formation and Fibrinolysis*, vol. 4 (Lane, D.A., Henschen, A., and Jasani, M.K., eds.), pp. 91–96, Walter de Gruyter, Berlin.

Kaudewitz, H., Henschen, A., Pirkle, H., et al. (1987) *Thromb. Haemost.* 58, 515 (abstr. 1901).

Kloczewiak, M., Timmons, S., and Hawiger, J. (1983) *Thromb. Res.* 29, 249–255.

Koopman, J., Grimbergen, J., Lord, S.T., Haverkate, F., and Mannucci, P.M. (1990) *Xth Fibrinogen Workshop*, 28–30 June 1990, Rouen, France (abstr.).

Köppel, G. (1970) *Thromb. Diath. Haemorrh.* (suppl. 39), 71–73.

Kowalska-Loth, B., Gårdlund, B., Egberg, N., and Blombäck, B. (1973) *Thromb. Res.* 2, 3–8.

Krajewski, T., and Blombäck, B. (1968) *Acta Chem. Scand.* 22, 1339–1346.

Krakow, W., Endres, G.F., Siegel, B.M., and Scheraga, H.A. (1972) *J. Mol. Biol.* 71, 95–103.

Kruithof, E.K.O., Tran-Thang, C., Ransijn, A., and Bachmann, F. (1984) *Blood* 64, 907–913.

Kudryk, B., Reuterby, J., and Blombäck, B. (1973) *Thromb. Res.* 2, 423–450.

Kudryk, B.J., Collen, D., Woods, K.R., and Blombäck, B. (1974a) *J. Biol. Chem.* 249, 3322–3325.

Kudryk, B., Reuterby, J., and Blombäck, B. (1974b) *Eur. J. Biochem.* 46, 141–147.

Kudryk, B., Blombäck, B., and Blombäck, M. (1976) *Thromb. Res.* 9, 25–36.

Kudryk, B., Okada, M., Redman, C.M., and Blombäck, B. (1982) *Eur. J. Biochem.* 125, 673–682.

Kuyas, C., Haeberli, A., and Straub, P.W. (1987a) *Thromb. Haemost.* 58, 22 (abstr. 86).

Kuyas, C., Sigrist, H., and Straub, P.W. (1987b) *Thromb. Haemost.* 58, 267 (abstr. 977).

Lahiri, B., and Shainoff, J.R. (1973) *Biochim. Biophys. Acta* 303, 161–170.

Laki, K. (1968) in *Fibrinogen* (Laki, K., ed.) pp. 1–398, Marcel Dekker Inc., New York.

Larsson, L.-I., Skriver, L., Nielsen, L.S., et al. (1984) *J. Cell. Biol.* 98, 894–903.

Larsson, U., Blombäck, B., and Rigler, R. (1987) *Biochim. Biophys. Acta* 915, 172–179.

Laudano, A.P., and Doolittle, R.F. (1978) *Proc. Natl. Acad. Sci. USA* 75, 3085–3089.

Laudano, A.P., and Doolittle, R.F. (1980) *Biochemistry* 19, 1013–1019.

Laurent, T.C., and Blombäck, B. (1958) *Acta Chem. Scand.* 12, 1875–1877.

Leven, R., Schick, P.K., and Budzynski, A. (1982) *Blood* 65, 501.

Lijnen, H.R., Soria, J., Soria, C., Collen, D., and Caen, J.P. (1984) *Thromb. Haemost.* 51, 108–109.

Liu, C.Y., Koehn, J.A., and Morgan, F.J. (1985) *J. Biol. Chem.* 260, 4390–4396.

Liu, C.Y., Wallén, P., and Handley, D.A. (1986) in *Fibrinogen–Fibrin Formation and Fibrionolysis* vol. 4 (Lane, D.A., Henschen, A., and Jasani, M.K., eds.), pp. 79–89, Walter de Gruyter, Berlin.

Lorand, L. (1972) *Ann. N.Y. Acad. Sci.* 202, 6–30.

Lottspeich, F., and Henschen, A. (1977a) *Hoppe-Seyler's Z. Physiol. Chem.* 358, 1521–1524.

Lottspeich, F., and Henschen, A. (1977b) *Hoppe-Seyler's Z. Physiol. Chem.* 358, 935–938.

Lottspeich, F., and Henschen, A. (1978) *Hoppe-Seyler's Z. Physiol. Chem.* 359, 1451–1455.

Ly, B., and Godal, H.C. (1972/73) *Haemostasis* 1, 204–209.

Ly, B., Kierulf, P., and Arnesen, H. (1974a) *Thromb. Res.* 5, 301–314.

Ly, B., Kierulf, P., and Jakobsen, E. (1974b) *Thromb. Res.* 4, 509–522.

Mammen, E.F., Prasad, A.S., Barnhart, M.I., Au, C.C., and Schwandt, V. (1969) *J. Clin. Invest.* 48, 235–249.

McDonagh, R.I.P., McDonagh, J., Petersen, T.E., et al. (1981) *FEBS Lett.* 127, 174–178.

Marder, V.J., Shulman, N.R., and Carroll, W.R. (1969) *J. Biol. Chem.* 244, 2111–2119.

Marguerie, G., and Stuhrmann, H.B. (1976) *J. Mol. Biol.* 102, 143–156.

Marguerie, G., Chagniel, G., and Suscillon, M. (1977) *Biochim. Biophys. Acta* 490, 94–103.

Marguerie, G.A., Plow, E.F., and Edgington, T.S. (1979) *J. Biol. Chem.* 254, 5357–5363.

Markowe, H.L., Marmot, M.G., Shipley, M.J., et al. (1985) *Brit. Med. J.* 291, 1312–1314.

Marsh, H.C., Jr., Meinwald, Y.C., Lee, S., and Scheraga, H.A. (1982) *Biochemistry* 21, 6167–6171.

Marsh, H.C., Jr., Meinwald, Y.C., Thannhauser, T.W., and Scheraga, H.A. (1983) *Biochemistry* 22, 4170–4174.

Mary, A., Achyuthan, K.E., and Greenberg, C.S. (1987) *Biochem. Biophys. Res. Comm.* 147, 608–614.

Meade, T.W., Mellows, S., Brozovic, M., et al. (1986) *Lancet* 2, 533–537.

Meinwald, Y.C., Martinelli, R.A., van Nispen, J.W., and Scheraga, H.A. (1980) *Biochemistry* 19, 3820–3825.

Mester, L. (1969) *Bull. Soc. Chim. Biol.* 51, 635–648.

Mihalyi, E. (1965) *Biochim. Biophys. Acta* 102, 487–499.

Mills, D., and Karpatkin, S. (1970) *Biochem. Biophys. Res. Commun.* 40, 206–211.

Miyata, T., Hiranaga, M., Umezu, M., and Iwanaga, S. (1983) *Ann. N.Y. Acad. Sci.* 408, 651–654.

Morita, T., Nakamura, T., Miyata, T., and Iwanaga, S. (1985) in *Bacterial Endotoxins Structure. Biomedical Significance and Detection with the Limulus Amebocyte Lysate Test*, pp. 53–64, Alan R. Liss Inc., New York.

Mosesson, M.W. (1983) *Ann. N.Y. Acad. Sci.* 408, 97–113.

Mosesson, M.W., and Finlayson, J.S. (1976) *Prog. in Hemostasis and Thrombosis* 3, 61–107.

Mosesson, M.W., Alkjaersig, N., Sweet, B., and Sherry, S. (1967) *Biochemistry* 6, 3279–3287.

Mosesson, M.W., Finlayson, J.S., and Umfleet, R.A. (1972) *J. Biol. Chem.* 247, 5223–5227.

Mosesson, M.W., Galanakis, D.K., and Finlayson, J.S. (1974) *J. Biol. Chem.* 249, 656–664.

Mosesson, M.W., Amrani, D.L., and Ménaché, D. (1976) *J. Clin. Invest.* 57, 782–790.

Mosesson, M.W., DiOrio, J.P., Müller, M.F., et al. (1987) *Blood* 69, 1073–1081.

Mosher, D.F. (1975) *J. Biol. Chem.* 250, 6614–6621.

Mosher, D.F. (1976) *J. Biol. Chem.* 251, 1639–1645.

Mosher, D.F., and Blout, E.R. (1973) *J. Biol. Chem.* 248, 6896–6903.

Mustard, J.F., Packham, M.A., Kinlough-Rathbone, R.L., Perry, D.W., and Regoeczi, E. (1978) *Blood* 52, 453–466.

Müller, M.F., Ris, H., and Ferry, J.D. (1984) *J. Mol. Biol.* 174, 369–384.

Nakamura, S., Takagi, T., Iwanaga, S., Niwa, M., and Takahashi, K. (1976) *Biochem. Biophys. Res. Commun.* 72, 902–908.

Needham, A.E. (1970) *Symp. Zool. London no. 27*, pp. 19–44.

Ni, F., Konishi, Y., Frazier, R.B., Scheraga, H.A., and Lord, S.T. (1989a) *Biochemistry* 28, 3082–3094.

Ni, F., Meinwald, Y.C., Vasquez, M., and Scheraga, H.A. (1989b) *Biochemistry* 28, 3094–3105.

Ni, F., Konishi, Y., Bullock, L.D., Rivetna, M.N., and Scheraga, H.A. (1989c) *Biochemistry* 28, 3106–3119.

Nieuwenhuizen, W., van Ruijven-Vermeer, I.A.M., Nooijen, W.J., et al. (1981) *Thromb. Res.* 22, 653–657.

Nieuwenhuizen, W., Vermond, A., and Hermans, J. (1983) *Thromb. Res.* 31, 81–86.

Niewiarowski, S., Kornecki, E., Budzynski, A.N., Morinelli, T.A., and Tuszynski, G.P. (1983) *Ann. N.Y. Acad. Sci.* 408, 536–555.

Nilsson, T., Wallén, P., and Mellbring, G. (1984) *Scand. J. Haematol.* 33, 49–53.

Norrman, B., Wallén, P., and Rånby, M. (1985) *Eur. J. Biochem.* 149, 193–200.

Norrman, B., Pohl, G., Jörnvall, H., and Wallén, P. (1986) *Eur. J. Biochem.* 59, 7–13.

Nossel, H. (1981) *Nature* 291, 165–167.

Nossel, H.L., Yudelman, L., Canfield, R.E., et al. (1974) *J. Clin. Invest.* 54, 43–53.

Okada, M., and Blombäck, B. (1983) *Thromb. Res.* 29, 269–280.

Okada, M., Blombäck, B., Chang, M.D., and Horowitz, B. (1985) *J. Biol. Chem.* 260, 1811–1820.

Olaisen, B., Teisberg, P., and Gedde-Dahl, Jr., T. (1982) *Human Genet.* 61, 24–26.

Olexa, S.A., and Budzynski, A.Z. (1980) *Proc. Natl. Acad. Sci.* 77, 1374–1378.

Olexa, S.A., and Budzynski, A.Z. (1981) *J. Biol. Chem.* 256, 3544–3549.

Otto, J.M., Grenett, H.E., and Fuller, G.M. (1987) *J. Cell. Biol.* 105, 1067–1072.

Pannell, R., and Gurevich, V. (1986) *Blood* 67, 1215–1223.

Pennica, D., Holmes, W.E., Kohr, W.J., et al. (1983) *Nature* 301, 214–221.

Plant, P.W., and Grieninger, G. (1986) *J. Biol. Chem.* 261, 2331–2336.

Plow, E., and Edgington, T.S. (1973) *J. Clin. Invest.* 52, 273–282.

Plow, E., and Edgington, T.S. (1975) *J. Biol. Chem.* 250, 3386–3392.

Porter, K.R., and Hawn, C.V.Z. (1949) *J. Exp. Med.* 90, 225–232.

Procyk, R., and Blombäck, B. (1988) *Biochim. Biophys. Acta* 967, 304–313.

Procyk, R., and Blombäck, B. (1990) *Biochemistry* 29, 1501–1507.

Procyk, R., Adamson, L., Block, M., and Blombäck, B. (1985) *Thromb. Res.* 40, 833–852.

Procyk, R., Kudryk, B., Callender, S., and Blombäck, B. (1990) *Xth Fibrinogen Workshop,* 28–30 June 1990, Rouen, France (abstr.).

Procyk, R., Medved, L., Kudryk, B., and Blombäck, B. (1991) *Biochemistry* (submitted).

Reber, P., Furlan, M., Henschen, A., et al. (1986) *Thromb. Haemost.* 56, 401–406.

Redman, C. (1991) Unpublished observations.

Rijken, D.C., Wijngaards, G., Zaal-de Jong, M., and Welbergen, J. (1979) *Biochim. Biophys. Acta* 580, 140–153.

Rixon, M.W., Chung, D.W., and Davie, E.W. (1985) *Biochemistry* 24, 2077–2086.

Robbins, K.C., and Markus, G. (1978) in *Fibrinolysis* (Gaffney, P.J., and Balkuv-Ulutin, S., eds.), pp. 61–75, Academic Press, New York.

Roberts, W.W., Kramer, O., Rosser, R.W., Nestler, F.H.M., and Ferry, J.D. (1974) *Biophys. Chem.* 1, 152–160.

Roy, S.N., Mukhopadhyay, G., and Redman, C.M. (1990) *J. Biol. Chem.* 265, 6389–6393.

Scheraga, H.A. (1977) in *Chemistry and Biology of Thrombin* (Lundblad, R.L., Fenton, II, J.W., and Mann, K.G., eds.), pp. 145, Ann Arbor Science Publishers, Ann Arbor, MI.

Scheraga, H.A., and Laskowski, Jr., M. (1957) *Adv. Protein Chem.* 12, 1–131.

Schwartz, M.L., Pizzo, S.V., Hill, R.L., and McKee, P.A. (1973) *J. Biol. Chem.* 248, 1395–1407.

Seydewitz, H.H., Kaiser, G., and Witt, I. (1985) in *Fibrinogen. Structural Variants and Interactions,* vol. 3 (Henschen, A., Hessel, B., McDonagh, J., and Saldeen, T., eds.), pp. 121–132, Walter de Gruyter Co., Berlin.

Shah, G.A., Ferguson, I.A., Dhall, T.Z., and Dhall, D.P. (1982) *Biopolymers* 21, 1037–1047.

Shah, G.A., Nair, C.H., and Dhall, D.P. (1987) *Thromb. Res.* 45, 257–264.

Shainoff, J.R., and Dardik, B.N. (1979) *Science* 204, 200–202.

Shen, L.L., Hermans, J., McDonagh, J., and McDonagh, R.P. (1977a) *Am. J. Physiol.* 232, H629–H633.

Shen, L.L., McDonagh, R.P., McDonagh, J., and Hermans, J. (1977b) *J. Biol. Chem.* 252, 6184–6189.

Siebenlist, K.R., Mosesson, M.W., DiOrio, J.P., et al. (1989) *Thromb. Haemost.* 62, 875–879.

Silveira, A., Hessel, B., Blombäck, B., et al. (1990) Unpublished observation.

Sobel, J.H., Thibodeau, C.A., and Canfield, R.E. (1988) *Thromb. Haemost.* 60, 153–159.

Soberano, M.E., Ong, E.B., Johnsson, A.J., Levy, M., and Schoellman, G. (1976) *Biochim. Biophys. Acta* 445, 763–773.

Solum, N.O. (1973) *Thromb. Res.* 2, 55–70.

Solum, N.O., and Lopaciuk, S. (1969) *Thromb. Diath. Haemorrh.* 21, 428–440.

Soria, J., Soria, C., Samama, M., Poirot, E., and Kling, C. (1976) *Pathol. Biol.* 24, (suppl.), 15–17.

Soria, J., Soria, C., Bertrand, O., and Samama, M. (1978) *Biochem. Biophys. Res. Comm.* 82, 442–450.

Soria, J., Soria, C., and Caen, J.P. (1983) *Brit. J. Haem.* 53, 575–586.

Sottrup-Jensen, L., Claeys, H., Zajdel, M., Petersen, T.E., and Magnusson, S. (1978) in *Progress in Chemical Fibrinlysis* vol. 3 (Davidson, J.F., Rowan, R.M., Samama, M.M., and Desnayers, P.C., eds.), pp. 191–209, Raven Press, New York.

Söderqvist, T., and Blombäck, B. (1971) *Naturwissenschaften* 58, 16–23.

Southan, C. (1988) in *Fibrinogen, Fibrin Stabilisation and Fibrinolysis* (Francis, J.L., ed.), pp. 65–99, E. Horwood Ltd, Chichester, UK.

Stone, M.C., and Thorp, J.M. (1985) *J. Royal College of Gen. Practitioners* 35, 565–569.

Strong, D.D., Moore, M., Cottrell, B.A., et al. (1985) *Biochemistry* 24, 92–101.

Stryer, L., Cohen, C., and Langridge, R. (1963) *Nature* 197, 793–794.

Stump, D.C., Thienpont, M., and Collen, D. (1986) *J. Biol. Chem.* 261, 1274–1278.

Suenson, E., Lützen, O., and Thorsen, S. (1984) *Eur. J. Biochem.* 140, 513–522.

Takada, A., Mochizuki, K., and Takada, Y. (1980) *Thromb. Res.* 19, 485–492.

Tamaki, T., and Aoki, N. (1981) *Biochim. Biophys. Acta* 661, 280–286.

Teger-Nilsson, A.-C., and Blombäck, B. (1974) *Thromb. Res.* 5, 223–233.

Thorsen, S. (1975) *Biochim. Biophys. Acta* 393, 55–65.

Thorsen, S., and Philips, M. (1984) *Biochim. Biophys. Acta* 802, 111–118.

Thorsen, S., Glas-Greenwalt, P., and Astrup, T. (1972) *Thromb. Diath. Haemorrh.* 28, 65–74.

Tomikawa, M., Iwamoto, M., Söderman, S., and Blombäck, B. (1980a) *Thromb. Res.* 19, 841–855.

Tomikawa, M., Iwamoto, M., Olsson, P., Söderman, S., and Blombäck, B. (1980b) *Thromb. Res.* 19, 869–876.

Tooney, N.M., and Cohen, C. (1972) *Nature* 237, 23–25.

Töpfer-Petersen, E., Lottspeich, F., and Henschen, A. (1979) *Thromb. Haemost.* 41, 671–676.

Townsend, R.R., Hilliker, E., Li, Y.T., Laine, R.A., Bell, W.R., and Lee, Y.C. (1982) *J. Biol. Chem.* 257, 9704–9710.

Tran-Thang, C., Kruithof, E.K., and Bachmann, F. (1984) *J. Clin. Invest.* 74, 2009–2016.

Uzan, G., Courtois, G., Besmond, C. et al. (1984) *Biochem. Biophys. Res. Comm.* 120, 376–383.

Varadi, A., and Scheraga, H.A. (1986) *Biochemistry* 25, 519–528.

Verheijen, J.H., Chang, G.T., and Kluft, C. (1984) *Thromb. Haemost.* 51, 392–395.

Verstraete, M., Bounameaux, H., de Cock, F., Van de Werf, F., and Collen, D. (1985) *J. Pharmacol. Exp. Ther.* 235, 506–512.

Vroman, L., Adams, A.L., Klings, M., et al. (1977) *Ann. N.Y. Acad. Sci.* 283, 65–76.

Wallén, P. (1977) in *Thrombosis and Urokinase* (Sherry, S., and Paoletti, R., eds.), pp. 91–102, Academic Press, London.

Wallén, P. (1980) in *Fibrinolysis* (Kline, D., and Reddy, K.N.N., eds.), pp. 1–24, CRC Press, Boca Raton, FL.

Wallén, P. (1987) in *Fundamental and Clinical Fibrinolysis* (Castellino, F.J., et al., eds.), pp. 1–18. Elsevier Science Publishers B.V. (Biomedical Division), Amsterdam.

Wallén, P., and Wiman, B. (1975) in *5th Congress on Thrombosis and Haemostasis* 21–26 July 1975, Paris. pp. 348 (abstr. 331).

Wallén, P., Bergsdorf, N., and Rånby, M. (1982) *Biochim. Biophys. Acta* 719, 318–328.

Wallén, P., Rånby, M., Bergsdorf, N., and Kok, P. (1981) in *Progress in Fibrinolysis,* vol. 5 (Davidson, J.F., Nilsson, I.M., and Åstedt, B., eds.), pp. 16–23, Churchill, Livingstone, Edinburgh.

Wallén, P., Pohl, G., Bergsdorf, N., et al. (1983) *Eur. J. Biochem.* 132, 681–686.

Wallén, P., Cheng, X., and Ohlsson, P.I. (1990) in *Fibrinogen, Coagulation, Thrombosis, Fibrinolysis* (Liu, C.Y., and Chien, S., eds.), Plenum Publishing Corp, New York (in press).

Wang, Y.Z., Patterson, J., Gray, J.A., et al. (1989) *Biochemistry* 28, 9801–9806.

Watt, K.W.K., Takagi, T., and Doolittle, R.F. (1979) *Biochemistry* 18, 68–76.

Weisel, J.W. (1986) *Biophys. J.* 50, 1079–1093.

Weisel, J.W., and Papsun, D.M. (1987) *Thromb. Res.* 47, 155–163.

Weisel, J.W., Nagaswami, C., and Makowski, L. (1987) *Proc. Natl. Acad. Sci. USA* 84, 8991–8995.

Weisel, J.W., Stauffacher, C.V., Bullit, E., and Cohen, C. (1985) *Science* 230, 1388–1391.

Weissbach, L., and Grieninger, G. (1990) *Proc. Natl. Acad. Sci.* 87, 5198–5202.

Wilhelmsen, L., Svärdsudd, K., Korsan-Bengtsen, K., et al. (1984) *New Engl. J. Med.* 311, 501–505.

Williams, R.C. (1983) *Ann. N.Y. Acad. Sci.* 408, 180–193.

Wiman, B., Almquist, Å., Sigurdardottir, O., and Lindahl, T. (1988) *FEBS Letter* 242, 125–128.

Wolfenstein-Todel, C., and Mosesson, M.W. (1981) *Biochemistry* 20, 6146–6149.

Xu, X., and Doolittle, R.F. (1990) *Proc. Natl. Acad. Sci. USA* 87, 2097–2101.

York, L., and Blombäck, B. (1976) *Thromb. Res.* 8, 607–618.

Yoshida, N., Terukina, S., and Okuma, M. (1988) *J. Biol. Chem.* 263, 13848–13856.

Yu, S., Kudryk, B., and Redman, C. (1986) in *Fibrinogen—Fibrin Formation and Fibrinolysis*, vol. 4 (Lane, D.A., Henschen, A., and Jasani, M.K., eds.), pp. 3–13, Walter de Gruyter Co., New York.

Zamarron, C., Lijnen, H.R., and Collen, D. (1985) *Thromb. Haemost.* 54, 102 (abstr. 604).

12

Fibrinogen-Fibrin: Preparation and Use of Monoclonal Antibodies as Diagnostics

B.J. Kudryk
A. Bini
S.F. Rosebrough
T.F. Schaible

Monoclonal antibodies (MoAbs) specific for human fibrin(ogen) or for some of its many degradation products have been reported in the literature since the early 1980s. One of the major aims of these studies has been the identification of MoAbs* that could not only serve as useful clinical tools but also as probes for identifying conformational changes that may accompany the fibrinogen-to-fibrin transition or those that may result from the interaction of fibrin(ogen) with other proteins or certain cells. Noninvasive methods to determine the existence and precise anatomical location of established and/or propagating thrombi would be of obvious clinical importance. In

Some of the work reported here was supported by grant number HL24230 from the U.S. National Heart, Lung and Blood Institute, the National Institutes of Health; NATO Collaborative Research Grant (CRG-890569); and Centocor, Inc., Malvern, Pennsylvania. The authors wish to thank Raffaella Bertazzi, Tellervo Huima, Laura Pampallona, Anna Rohoza, and Daniela Spadano for their excellent help in the preparation of this work.
* See the list of abbreviations and terms at the end of this chapter.

this regard, at least six different laboratories have identified fibrin- or fi-
brin(ogen)-split product-specific MoAbs, which are now being tested—some
are already in late-phase clinical trials—as potential clot-imaging agents.
Assays developed with a number of different fibrin(ogen)-specific MoAbs
have been shown useful in monitoring both the in vivo balance between
thrombin and plasmin activity and the efficacy of thrombolytic therapy. The
availability of a good number of neoepitope-specific antibodies has made
possible the development of some screening procedures for the detection of
hemostatic disturbances that are both extremely sensitive and quite rapid
to perform. In addition, since such neoepitope-specific antibodies fail to
react with the parent fibrinogen molecule, assays with these MoAbs can now
be performed using nonprocessed human plasma. Fibrin-specific and other
MoAbs have been used in the ABC immunoperoxidase technique for the
localization and molecular characterization of fibrin(ogen) and its proteolytic cleavage products in tissues obtained from patients with atherosclerosis, renal disease, and malignancy. In this chapter, we describe the preparation, specificity, and use of a number of different fibrin(ogen)-specific
MoAbs identified in our own laboratory and that of other investigators.

12.1 FIBRINOGEN-TO-FIBRIN TRANSITION AND FIBRIN(OGEN)OLYSIS

Fibrinogen is a large (M_r 340,000), symmetrical blood glycoprotein composed of three pairs of nonidentical polypeptide chains (Aα chain, M_r 66,000,
610 residues; Bβ chain, M_r 54,000, 461 residues; γ chain, M_r 48,000, 411
residues). The six chains are held together by 29 disulfide bonds, and the
protein contains no free sulfhydryl groups. Three of the 29 disulfide bridges,
the so-called symmetrical disulfides, unite the two halves of fibrinogen and
thereby form the dimeric protein. Cleavage of fibrinogen with CNBr produces about 30 fragments, the largest of which is called the amino (NH$_2$)-
terminal disulfide knot (N-DSK). A slightly smaller fragment called (T)N-
DSK can be derived from fibrin II or cross-linked fibrin II by CNBr cleavage.
Cleavage of fibrinogen or non-cross-linked fibrin II with plasmin results in
a number of small fragments and also two large core products called fragment D and fragment E. The CNBr-generated fragments N-DSK and (T)N-
DSK are structurally related to the plasmin-derived fragment E. When fully
cross-linked fibrin II is cleaved with plasmin, fragment E and fragment D-
dimer remain as core degradation products. In fragment D-dimer, each core
unit is derived from a different fibrin II molecule, which had been cross-
linked by factor XIIIa during fibrin stabilization. By electron microscopy,
fibrinogen appears as a rodlike, trinodular (or three-globule) structure about
450 Å long. The central and the outer nodules have been designated as the
E and D domains, respectively (for review, see Crabtree 1987).

The thrombin-catalyzed release of fibrinopeptides (two of FPA and two of FPB) from the NH_2-terminal portions of both the $A\alpha$ and $B\beta$ chains of the fibrinogen E domain results in the formation of fibrin monomer. Identification of blood levels of FPA and FPB may serve as an index for in vivo thrombin activity. Once formed, fibrin monomers spontaneously begin to polymerize and, when the process is complete, a firm gel results. As already mentioned, fibrin gels can be further stabilized by factor XIIIa-mediated cross-linking of adjacent monomers. The fibrinolytic system, specifically the active enzyme plasmin, is capable of dissolving blood clots (fibrinolysis) and thereby restoring vascular integrity. Since plasmin can also degrade most blood proteins, thrombolytic therapy with any of the available plasminogen activators often leads to a very significant decrease in blood fibrinogen levels (fibrinogenolysis). A number of intermediate split products are generated when both fibrin II and fibrinogen are degraded by plasmin. Due to cross-linking, the intermediate soluble fragments (called X-oligomers by some) derived from fibrin II are much more complex than those which originate from fibrinogen. The terminal core fragments of fibrin II (fragment E and fragment D-dimer) and fibrinogen (fragment D and fragment E) can be distinguished by the presence or absence of FPA (in the E domain) and monomeric or dimeric form of the γ chain (in the D domain).

12.2 MONITORING BLOOD LEVELS OF FIBRIN(OGEN) DEGRADATION PRODUCTS

It has been almost 20 years since Nossel and his collaborators described a polyclonal antibody-based assay for FPA (Nossel et al. 1971). Since that time, many investigators have tried to devise specific, sensitive and rapid analytical methods for measuring in vivo concentrations of fibrin(ogen) degradation products. The rationale for development of such assays, principally immunoassays, is that they would be very useful not only in the diagnosis but also in the treatment of impaired hemostasis. Due to varying degrees of cross-reactivity with intact fibrinogen, only a limited number of assays using polyclonal antisera have given meaningful data. With the advent of hybridoma technology, a number of laboratories have attempted to prepare MoAbs for clinical use. Below we will identify some of the already available antibodies and also discuss analytical methods developed with each. At the outset, however, it should be pointed out that, to date, only limited clinical experience has been obtained with all but a few of these MoAb-based assays.

12.2.1 Immunoassays Using MoAbs Specific for Epitopes Associated with or Released from the E Domain

Table 12–1 lists 14 monoclonal antibodies specific for epitopes associated with the E domain. In terms of clinical utility, a significant number (10/14) of these MoAbs have the very favorable property of not cross-reacting with

TABLE 12-1 Properties of MoAbs Specific for Epitopes in the E Domain

Antibody	Isotype	Epitope Structure	Reactive with Fibrinogen	Reference
ESF9	?	"Free" FPA	No	Dawes et al. 1985
ESF1	?	"Free" FPA	No	Dawes et al. 1985
Anti-FPA	?	FPA-containing fragments	Yes	Shainoff et al. 1986
8C2-5	IgG1,κ	"Free" FPA	No	Kudryk et al. 1989b
D21-7	IgG1,κ	"Free" FPA	No	Kudryk et al. (in preparation)
Y18	IgM,κ	FPA-containing fragments	Yes	Koppert et al. 1985
fbn-17	IgM	Aα 17-22	No	Scheefers-Borchel et al. 1985
1-8C6	IgG2a,κ	Bβ 1-42	Yes	Kudryk et al. 1983
64C5	IgG1,κ	Bβ 15-21	No	Hui et al. 1983
59D8	IgG1,κ	Bβ 15-21	No	Hui et al. 1983
55D10	IgG1,κ	Bβ 15-21	No	Hui et al. 1983
LIAB-6	IgM	Fragment E	Yes	McCabe et al. 1984
T2G1	IgG1,κ	Bβ 15-21	No	Kudryk et al. 1984
FDP-14	IgG1,κ	Bβ 54-118	No	Koppert et al. 1986

the parent (i.e., intact fibrinogen) molecule. The availability of these so-called neoepitope-specific MoAbs makes possible the development of assays in which intact plasmas can serve as the test sample. Clearly, intact plasma cannot be used as the test sample in any assay where the primary antibody is one of the four listed in Table 12–1 that can react with both intact fibrinogen and its corresponding degradation product. However, as has already been demonstrated with antibody Y18 (Hoegee-de Nobel et al. 1988), some of these so-called panspecific MoAbs, when used as tagging reagents in "capture/tag" assays (see below) may also prove to be very valuable clinical tools.

As shown in Table 12–1, six MoAbs have been identified that can react with "free" FPA and/or FPA-containing fragments. Antibodies ESF9 and ESF1 were the first FPA-specific MoAbs to be identified (Dawes et al. 1985). Antibody ESF9 has been used in a RIA with labeled FPA and results have shown that an excess of fibrinogen could not significantly inhibit binding of the ligand to the antibody. In a RIA using antibody ESF1, only slight cross-reactivity was obtained with FPA derived from Rhesus monkey fibrinogen. This result is very interesting in that Rhesus monkey FPA contains 16 amino acid residues like the human peptide, but it differs in structure by one residue (Aα Ser 3 in human, Aα Thr 3 in monkey). Our own group has isolated the two remaining neoepitope-specific MoAbs to FPA which are identified in Table 12–1. An ovalbumin conjugate of the human FPA homologue Aα 7-16 served as the immunogen for the preparation of antibody 8C2-5 (Kudryk et al. 1989b). As clearly demonstrated in Figure 12–1A, the latter probe is so specific for "free" FPA that it even fails to react with the smallest known FPA-containing peptide (Aα 1-21) present in blood (Weitz et al. 1986). This antibody also reacts with dog FPA but not with des Arg (Aα 1-15) or shorter (e.g., Aα 1-14, Aα 1-13, Aα 1-12, etc.) peptide homologues of human FPA. The latter result as well as the data depicted in Figure 12–1 strongly suggest that Aα Arg 16, with a free carboxyl group, is an essential component of the epitope that interacts with antibody 8C2-5.

In an effort to prepare other FPA-specific antibodies, we have also utilized both FPA-ovalbumin and FPA-mouse albumin conjugates as well as various FPA polymers—prepared without a carrier protein—as immunogen. Following sensitization with FPA-ovalbumin (actually the N-tyrosyl derivative of human FPA was used for conjugation to the carrier protein), mouse spleen cells were fused with an immunoglobulin nonsecreter myeloma (P3X63Ag8.653). One hybridoma cell line has been isolated which secretes antibody D21-7. As with antibody 8C2-5, D21-7 binds ELISA plates coated with ovalbumin conjugates of FPA or of N-tyrosyl FPA, but it fails to react with either human fibrinogen, the Aα chain, or other FPA-containing fragments of fibrinogen.

Using antibodies 8C2-5 and D21-7, we have developed several immunoassays for measuring human FPA levels directly in plasma. Figure 12–2 shows the standard dose-response curves for FPA using HRPO-con-

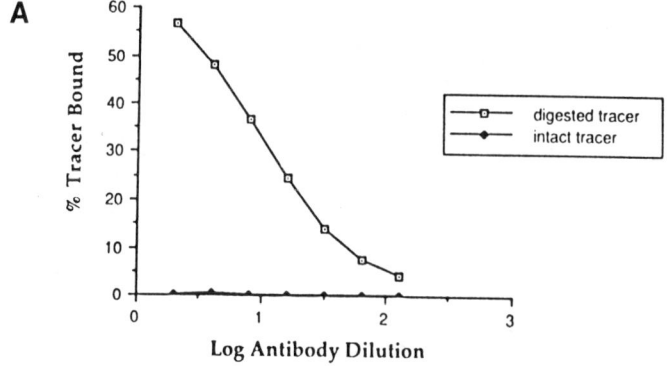

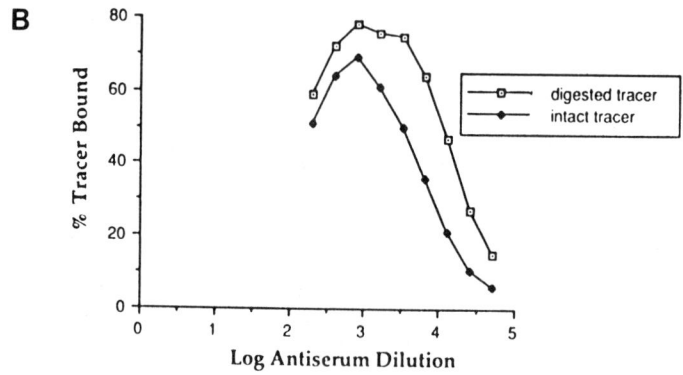

FIGURE 12–1 (A) Binding RIA using various dilutions of MoAb/8C2-5 mixed with a standard amount of ^{125}I-Tyr-Aα 1-21 ligand before and after digestion with thrombin. (B) For comparison, binding RIA results obtained with a polyclonal rabbit anti-FPA sera (R-33) and ^{125}I-Tyr-Aα 1-21 before and after thrombin digestion are also shown. Antiserum R-33 was prepared by Nossel and his colleagues (Nossel et al. 1976) and has been shown to cross-react extensively with intact fibrinogen and FPA-containing fibrinogen fragments. Reproduced with permission of Grune & Stratton, Inc., from Kudryk et al. (1989b).

jugates of each antibody. In these experiments, dilutions of FPA were made in an ELISA-compatible buffer. It should be pointed out that the ELISA inhibition curves shown in Figure 12–2 are identical when normal human plasma or buffer containing 5 mg/ml fibrinogen is used as diluent (data not shown). By comparison with antibody 8C2-5, the assay using D21-7 is significantly less sensitive in its ability to measure "free" FPA. Since only sparse data dealing with quantitative immunoassays using antibodies ESF9 and ESF1 have appeared (Dawes et al. 1985; Prowse 1986), it is presently impossible to compare the sensitivity obtained with our MoAbs with the

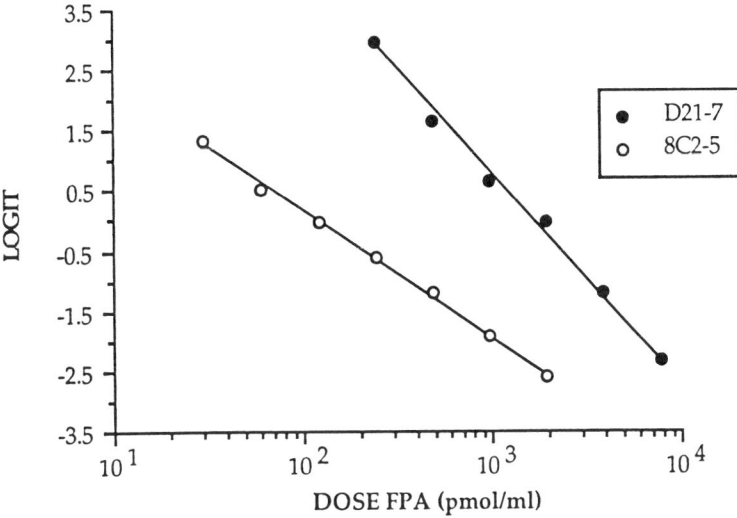

FIGURE 12–2 ELISA-determined standard dose-response curves of reactivity between human FPA and enzyme (HRPO) conjugates of antibodies 8C2-5 and D21-7. In these experiments, plates were coated with ovalbumin conjugates of the N-tyrosyl derivative of human FPA. Inhibition was determined as follows: $(A/A_0) \times 100$ where A is absorbance in the presence of FPA, and A_0 is absorbance in control wells (buffer with no FPA added). Response data were linearized using logit transforms.

two other neoepitope-specific FPA probes. Although the FPA EIAs using both of our antibodies are quite rapid, easy to perform, and use only very small amounts of easily prepared and stable reagents, the assay sensitivity is much lower than that observed with polyclonal FPA antisera such as R-33 (Nossel et al. 1976). However, the unique specificity of both MoAbs makes them potentially useful. For example, assays with the latter antibody and incompletely ethanol-extracted clinical plasma samples will never result in an overestimate of FPA levels as has frequently been observed with methods using antibodies which cross-react even slightly with intact fibrinogen. The question of sensitivity particularly of the 8C2-5-based assay, can be addressed by simply using larger sample volumes and/or more concentrated ethanol-extracted clinical material. The availability of both D21-7 and 8C2-5 also offers the opportunity for development of a "capture/tag" type EIA with a less specific probe. In such an assay, antibody D21-7 or 8C2-5 could serve as the "capture" probe, and an HRPO-labeled second antibody, directed to a different epitope on FPA, could be the "tag" reagent. The latter antibody could be panspecific, that is, one that would react with fibrinogen as well as with FPA-containing fragments. A necessary requirement for the tag antibody is that it be directed to an epitope other than the one with which either D21-7 or 8C2-5 react. Binding of two different antibodies to a peptide the size of FPA would not be unique. A recent report has described two antibodies that can bind simultaneously to a 24-residue peptide. It appears that the outer boundaries of the two epitopes are separated by only three amino acids (Jackson et al. 1988).

Two panspecific antibodies (anti-FPA and Y18) reactive with either FPA or FPA-containing fragments are also listed in Table 12–1. Antibody anti-FPA has been useful for probing the molecular nature of fibrin(ogen) oligomers in extracts derived from atherosclerotic aortas (Valenzuela et al. 1988). (More details on the use of this probe are given in Section 12.3.) It would be interesting to see if antibody anti-FPA and D21-7 (or 8C2-5) can simultaneously bind FPA. Antibody Y18 was prepared from a fusion experiment using spleen cells from an animal previously immunized with fragment Y derived from human fibrinogen. This antibody reacts similarly with fibrinogen, fragments X and Y, N-DSK, Aα chain, as well as Aα 1-51. Since all immunoreactivity with Y18 is abolished when any of the antigens just identified are digested with thrombin, the epitope to which this probe is directed must reside in or around the enzyme-sensitive Aα Arg 16-Gly 17 bond. Other data has led to the speculation that the NH$_2$-terminal portion of FPA as well as Aα Arg 16 are essential components of the epitope to which antibody Y18 is directed (Koppert et al. 1985). Antibody Y18 has proven to be a useful tag probe in two different EIAs. In one of these, an assay for primary fibrinogenolysis products, the capture antibody is FDP-14 (see Table 12–1). The latter is specific to a neoepitope in the fragment E moiety and will bind all fragments E, irrespective of whether they are derived from fibrinogen or fibrin. In the assay, antibody FDP-14 will capture all fragments E from plasma; however, HRPO-labeled Y18 will tag only those which are derived from fibrinogen but not fibrin. The assay is rapid and sensitive (lower detection limit <0.025 μg/ml) with a coefficient of variation of 3–8% (Koppert et al. 1987). In the other assay, the capture antibody (antibody G8) is directed to the COOH-terminal 150-amino acid segment of the Aα chain. In this assay, antibody G8 will bind only fibrin(ogen) but not derivatives (e.g., fragments X and Y) that are missing the COOH-terminal region of the Aα chain. Since HRPO-labeled Y18 will tag fibrinogen and products that are derived from fibrinogen, this assay can be used for estimating plasma fibrinogen concentration. It has been shown that, unlike the functional fibrinogen assay (clotting rate), the G8 capture/Y18 tag EIA is not disturbed by the presence in plasma of fibrin(ogen) degradation products, heparin, EDTA, or oxalate (Hoegee-de Nobel et al. 1988).

Albumin conjugates of a synthetic hexapeptide whose structure was identical to that of Aα 17-22 were used as the immunogen for preparation of antibody fbn-17 (Müller-Berghaus et al. 1985; Scheefers-Borchel et al. 1985). This peptide was selected because it includes the new NH$_2$-terminus of the α chain of fibrin I or II, and it was assumed that it would encompass a cleavage-site-specific epitope of fibrin I and II. Antibody fbn-17 reacts completely with the hexapeptide before and after conjugation to albumin as well as with fibrin I and II. Very importantly, this same antibody also reacts with plasma samples containing soluble fibrin I or II. Antibody fbn-17 has been used for development of an EIA which can detect plasma fibrin concentrations down to about 14 ng/ml. Four MoAbs directed to the same cleavage-site-specific epitope of fibrin II are also identified in Table 12–1. Three of these (64C5, 59D8, and 55D10) were prepared using hemocyanin conjugates of a synthetic heptapeptide whose structure was identical to that found at the new NH_2-terminus (Bβ 15-21) of the β chain of fibrin II (Hui et al. 1983). Antibody T2G1 was prepared by our group using the fibrin-derived fragment (T)N-DSK as immunogen (Kudryk et al. 1984). Recent studies have shown that antibodies 59D8 and T2G1 are virtually indistinguishable (Kudryk et al. 1989a). As will be discussed in Section 12.4, these same two MoAbs are currently being used in clinical studies as clot imaging agents. An EIA has been developed with antibody T2G1 that measures plasma levels of the fibrin-derived peptide Bβ 15-42. Ethanol-extracted plasma samples serve as test material for this high sensitivity assay which has a lower detection limit of 5 pmol/ml. The panspecific antibody 1-8C6 was also isolated by our group (Kudryk et al. 1983). The immunogen used for preparation of this antibody was the fibrinogen-derived fragment N-DSK. Studies conducted with many different fragments derived from fibrin(ogen) suggest that the latter antibody is directed to an epitope in or around the thrombin-susceptible Bβ Arg 14-Gly 15 bond. T2G1- and 1-8C6-based EIAs using ethanol extracts of patient samples are useful in clinical investigations dealing with the very early events in fibrino(geno)lysis. For example, a study using both assays and samples from myocardial infarction patients undergoing thrombolytic therapy demonstrated a marked increase in blood levels of both peptides, particularly Bβ 1-42, following administration of streptokinase. Such results confirm that, in addition to fibrinolysis, there is significant degradation of fibrinogen following infusion of the plasminogen activator (Eisenberg et al. 1986). In a more recent study, the 1-8C6-based EIA as well as a fragment D-dimer assay were used to differentiate the fibrin specificity of two doses of t-PA (100 vs 150 mg administered over 6 hours) in patients with acute myocardial infarction (Eisenberg et al. 1988). A conclusion drawn from this work was that use of such EIAs should facilitate development and implementation of optimal dose regimens of t-PA or other activators of the fibrinolytic system.

Table 12–1 also lists two fragment E-specific antibodies. As already discussed, the neoepitope-specific antibody FDP-14 (Koppert et al. 1986) is used in an assay that measures primary fibrinogenolysis products. This same antibody also serves as a capture probe in several other assays. In one, which measures the sum of products of fibrinogenolysis and fibrinolysis present in plasma, the tag probe is a polyclonal rabbit anti-FDP (Koopman et al. 1987). The other assay measures but does not discriminate between cross-linked and non-cross-linked fibrin degradation products present in plasma. The tag antibody (designated DD-13) in the latter assay reacts very strongly with the D-dimer but much less so with various fragments D derived from fibrinogen (Koppert et al. 1988). Antibody LIAB-6 is a product of a fusion experiment using spleen cells of an animal immunized with urine collected from a patient with bladder cancer. The antibody reacts with native fibrinogen and most but not all fragment E species derived from a late-stage digest of fibrinogen. Using this antibody in combination with an enzyme-linked polyclonal goat anti-human fibrinogen probe, a sensitive capture/tag EIA has been developed which may be useful in monitoring bladder cancer patients for tumor recurrence after surgery (McCabe et al. 1984).

12.2.2 Diagnostic Value of MoAbs Reactive with D Domain or Other Regions of Fibrin(ogen)

A number of groups have prepared MoAbs specific for neoepitopes associated with fragment D of fibrin(ogen) or fragment D-dimer of cross-linked fibrin. Antibody 231-12B is a low-affinity probe that has a very restricted, if not absolute, specificity for fragment D. The antibody was prepared by fusing myeloma cells with spleen cells from a rat immunized with fibrinogen fragment D. In a RIA with labeled fragment D, 50% displacement was obtained with about 22 μg/ml "cold" fragment D. A 10-fold more sensitive EIA has been described using 231-12B as capture and the panspecific probe 280-4A as tag (Kennel and Lankford 1983). Several neoepitope-specific MoAbs have been prepared using purified D-dimer as immunogen (Matsumoto et al. 1985). Antibody 03204 can discriminate quite well between the immunogen and fragment D. A sensitive, simple, and rapid assay called DALIA has recently been developed with $F(ab')_2$ isolated from a different antibody (03202), which also resulted from the latter study. In this method, which most probably measures X-oligomers rather than the smaller degradation products like D-dimer, both agglutinated and nonagglutinated particles are counted by a Coulter Counter, and the level of agglutination is expressed as the ratio of the total volume of antigen-bound and free latex. Intact plasma can be used as test material, and the results have shown that the mean concentration of FDP in the plasma of normals (n = 98) was 0.69 μg/ml and in DIC patients (n = 142) it was 13.24 μg/ml (Sakai et al. 1988). A mixture of early FbDP was used in the preparation of antibody F2C5 which reacts with fragment D, and reacts much more with D-dimer but not with fibrinogen, fragments X and Y, or early FDP. Such results suggest a preferential specificity of the antibody for fibrin-degradation products. An agglutination immunoassay using F2C5-coated latex beads has been developed, and, since this probe does not react with intact fibrinogen, the assay offers the possibility of direct, rapid, and accurate measurement of total fragment D and D-dimer in unprocessed plasma (Mirshahi et al. 1986).

In terms of sensitivity, the latter is similar to the capture/tag assay using antibody 231-12B (Kennel and Lankford 1983).

Antibodies recognizing epitopes on D-dimer but not on fibrinogen or the monomeric fragment D have been prepared by several groups. Rylatt and collaborators were first to identify two such probes, designated DD-1C3 and DD-3B6 (Rylatt et al. 1983). As immunogen, these investigators used lysates of cross-linked fibrin but, just prior to fusion, the sensitized mice were boosted with purified D-dimer. Following fusion, a number of hybridomas secreting antibodies reactive with D-dimer were identified. As expected, most reacted with fibrin(ogen) and all fragments containing the D domain. One of these panspecific MoAbs and DD-3B6 have been used in several highly sensitive capture/tag assays developed by Rylatt's group. The DD-4D2 (panspecific) capture/DD-1C3 tag EIA has a lower limit of sensitivity for D-dimer of about 10 ng/ml. Using this assay, significant levels of "D-dimer equivalent"—this term is used since, in addition to D-dimer, the neoepitope-specific probe recognizes a number of related molecular species present in a digest of cross-linked fibrin—were found in serum of patients (Elms et al. 1983). However, since it is well known that substantial amounts of fibrin(ogen) degradation products are removed when blood is allowed to clot (Elms et al. 1986; Gaffney and Perry 1985; Nieuwenhuizen 1988), the latter assay using serum as test material will almost always result in an underestimation of degradation products. The DD-3B6 capture/DD-4D2 tag EIA allows for plasma to be used as test material. As expected, the D-dimer equivalent in plasma of most patients assayed by the latter EIA greatly exceeded the serum values (Whitaker et al. 1984).

The DD-3B6-based capture/tag EIA has been used in many laboratories to screen the plasma of patients predisposed to thrombosis or those undergoing thrombolytic therapy. Elevated levels (mean = 593 ng/ml) of D-dimer equivalent were observed in plasma of women (n = 12) with DVT showing no signs of PE (Hafter et al. 1985). Much higher values (mean = 3,337 ng/ml) were obtained in pregnant women (n = 6) who developed DVT with pulmonary embolism. Four obstetrical patients with severe DIC had D-dimer values of 50,500–390,000 ng/ml. The D-dimer value for healthy young females (n = 23) was about 50 ng/ml. In this same study, patients with ovarian cancer (n = 22) were screened for both D-dimer and the tumor marker CA 125. A distinct correlation among levels of the D-dimer and the tumor marker and tumor progression was reported. This same assay was used to screen plasma of men who had been discharged from the hospital following a myocardial infarct. Antigen levels in the patient population were significantly higher when compared to that found in age-matched controls. It was suggested that the results may reflect a global increase in thrombotic tendency in this patient population (Rogers et al. 1986). A capture/tag RIA with these same antibodies has been employed in studies whose aim was to clarify the role of coagulation and fibrinolysis in renal disease (Kamitsuji et al. 1986). Urinary levels of D-dimer equivalent were increased in patients with active glomerular lesions such as crescentic glomerulonephritis. In patients with membranous nephropathy and focal and segmental sclerosis/ hyalinosis, there was a correlation between proteinuria and D-dimer equivalent. It was concluded that the ratio of these two parameters may be of value in assessing active glomerular disease. The DD-3B6-based capture/ tag EIA and a RIA for the platelet α-granule protein, platelet factor 4, were used to screen plasma of patients with various degrees of uremia (Gordge et al. 1989). Previous studies had shown that the kidney does not participate in the clearance of either marker. Raised levels of both antigens were found in patients with CRF. Since intravascular coagulation may be a universal feature of ARF, very high levels of D-dimer equivalent (mean = 2451 ng/ ml in ARF vs mean = 244 ng/ml in CRF) were observed in this patient group. As mentioned above, EIAs for both Bβ 1-42 and D-dimer equivalent have been used to differentiate the fibrin specificity of two different doses of t-PA (Eisenberg et al. 1988). While neither regimen of t-PA showed significant difference in plasma values of D-dimer equivalent, much higher values of Bβ 1-42, indicative of more intense fibrinogenolysis, were observed with the higher dose of t-PA.

In addition to the capture/tag EIA, Rylatt's group has also developed a more rapid but less sensitive agglutination assay (Elms et al. 1986; Hillyard et al. 1987). In this procedure, dilutions of plasma are mixed with polystyrene beads previously coated with the neoepitope-specific antibody DD-3B6. Results have shown that plasmas containing about 250 ng/ml D-dimer equivalent fail to agglutinate the latex particles. The assay can detect 2 μg pure D-dimer or D-dimer/fragment E complex per ml but not 500 μg of either fibrinogen fragments X, D, or E per ml (Greenberg et al. 1987). Data obtained to date with the DD-3B6-based immunoassays (both the EIA and latex agglutination) confirm that they indeed are very useful clinical tools for monitoring impaired hemostasis. Another capture/tag EIA for D-dimer equivalent using two different MoAbs (MA-15C5 and MA-8D3) has also been described (Declerck et al. 1987); however, clinical experience with this assay is still limited. The same can be said for an assay using two antibodies (IgM isotype) specific for different epitopes on the complex X-oligomer fraction derived from cross-linked fibrin (Gaffney et al. 1988a and 1988b).

12.3 CHARACTERIZATION OF FIBRIN(OGEN) ANTIGENS IN TISSUES

The presence of fibrin(ogen)-related antigens in vascular and extravascular space has been described in a large number of diseases such as atherosclerosis, renal disease, malignancy, and inflammation (Dvorak 1986; Dvorak et al. 1983; Hancock and Atkins 1985; Haust et al. 1965; Kao and Wissler 1965; Rickles and Edwards 1983; Vassalli and McCluskey 1964b; Woolf and Carstairs 1967). To date, histological detection of fibrin has relied mainly on histochemical methods or on immunohistochemistry using poly-clonal antisera. Routine stains, such as Lendrum and PTAH, are based on nonspecific chemical reactivity with the fibrin(ogen) molecule (Lendrum et al. 1961; Luna 1968), and the reaction does not occur in clots prepared from purified protein (Gitlin and Craig 1957). In previous studies, the use of immunofluorescence with polyclonal antisera to fibrinogen led to difficulties in interpretation because of cross-reactivity with both fibrinogen and fibrin (Dvorak et al. 1983; Hancock and Atkins 1985; Paronetto and Koffler 1965; Zacharski et al. 1983). Hybridoma technology has allowed the development of antibodies that react specifically with epitopes of fibrinogen, fibrin, or with their various degradation products of possible pathophysiological im-portance. The identification and distribution of the different molecular forms in which fibrin(ogen) and its degradation products are present in diseased tissues may provide important information on the pathogenesis of the disease process.

Monoclonal antibodies 1-8C6 (Kudryk et al. 1983), T2G1 (Kudryk et al. 1984), and GC4 (Kudryk et al. 1985, 1988, 1989a) have been utilized as probes for the molecular characterization and distribution of fibrinogen/fibrin I, fibrin II, and fibrin(ogen) degradation products in human tissues (Bini 1987; Bini et al. 1988, 1989a, 1989b). These same antibodies have also been used to study tissues from patients with malignant pleural mesothe-lioma (Wojtukiewicz et al. 1989), SCCL (Wojtukiewicz et al. 1990a), and colon cancer (Wojtukiewicz et al. 1990b). Antibody 1-8C6, along with mon-oclonal murine anti-FPA as well as polyclonal antifibrinogen-chain anti-bodies have also been used to probe the molecular weight distribution and fibrinopeptide content of fibrinogen-related protein in atherosclerotic aortas (Valenzuela et al. 1988). Some parallel studies have used other monoclonal antibodies to distinguish fibrinogen from D-dimer and cross-linked fibrin in renal disease (Deguchi et al. 1989; Takemura et al. 1987) and in ovarian tumors (Wilhelm et al. 1988).

12.3.1 Atherosclerotic Lesions

Our investigation stemmed from results of previous studies on the molecular characterization of fibrin(ogen)-related material in human thrombi and ath-erosclerotic plaques (Bini 1987; Bini et al. 1986, 1987). The latter showed that three different molecular forms of fibrinogen (i.e., the intact molecule, fibrin I, and fibrin II) were present in atherosclerotic vessels, and that in-creased amounts of fibrin II were associated with progression and severity of the lesion. Moreover, the major component of fibrin(ogen)-related ma-terial present in human thrombi was identified as fibrin II, as previously postulated (Nossel 1981). Those data, obtained by RIA, suggested that the presence of fibrin(ogen)-derived material in atherosclerotic vessels was not just related to incorporation of mural thrombi in which fibrinogen and fibrin II are not present in comparable amounts (Bini ct al. 1986). Therefore, we wanted to identify the distribution of the different molecular forms of fibrin(ogen) in the vessel wall. Monoclonal antibodies 1-8C6, T2G1, and GC4 were used with the ABC technique to study the distribution of fibrinogen/fibrin I (reactivity with 1-8C6), fibrin II (reactivity with T2G1), and fibrin(ogen) degradation products (reactivity with GC4) in a large number of normal, atherosclerotic, and other diseased tissues (Bini et al. 1989a, 1989b).

The thrombi and atherosclerotic plaques examined were all specimens obtained from vascular and cardiothoracic surgery at the Columbia Presbyterian Hospital (New York, NY) and were histologically classified. Our observations on these samples could be summarized as follows:

1) The presence of fibrinogen/fibrin I and fibrin II in each type of lesion was as previously detected by RIA. Increased fibrin II was associated with the more severe atherosclerotic plaques. In addition, the use of antibody GC4 allowed the detection of degraded fibrin(ogen), i.e., fragment D/D-dimer-related antigen, directly in the vessel wall. Antibody reactivity varied for each type of lesion, as shown in Table 12-2.

2) The distribution varied for each type of lesion and was different for fibrinogen/fibrin I, fibrin II, and fragment D/D-dimer. Essentially, four patterns of distribution were observed: In early plaques, fibrinogen/fibrin I was detected in the intima and on the luminal surface (Figure 12-3A) and around foam cells and macrophages (Figure 12-4A). Fibrin II was present on the endothelium in one coronary artery (Figure 12-3B), and in small flecks around smooth muscle cells and macrophages. Fragment D/D-dimer-related antigen was rarely observed in these lesions. In fibrous plaques, fibrinogen/fibrin I was distributed in short threads or flecks around smooth muscle cells and macrophages (Figure 12-5) and on the luminal surface in areas of thrombus. Fibrin II was distributed in long threads in the intima and media, around foam cells, and in areas of thrombus. Fragment D/D-dimer-related antigen was seen in areas of large fibrin deposits (Figure 12-5C). In more advanced lesions, fibrinogen/fibrin I and fibrin II were largely distributed in loose connective tissue and around foam cells, macrophages, and cholesterol crystals, and calcium deposits. Long threads and bundles of fibrin

TABLE 12-2 Summary of Antibody Reactivity of Human Atherosclerotic Lesions

Tissue Sample[1]	MoAb Reactivity Type		
	1-8C6	*T2G1*	*GC4*
Normal wall	2/12	1/12	—
Early plaques	6/7	3/7	1/7
Fibrous plaques	5/6	4/6	4/6
Advanced plaques	7/8	7/8	6/8

[1] Tissues were classified by histological criteria. Adapted from Bini et al. (1989a) by permission of the American Heart Association, Inc.

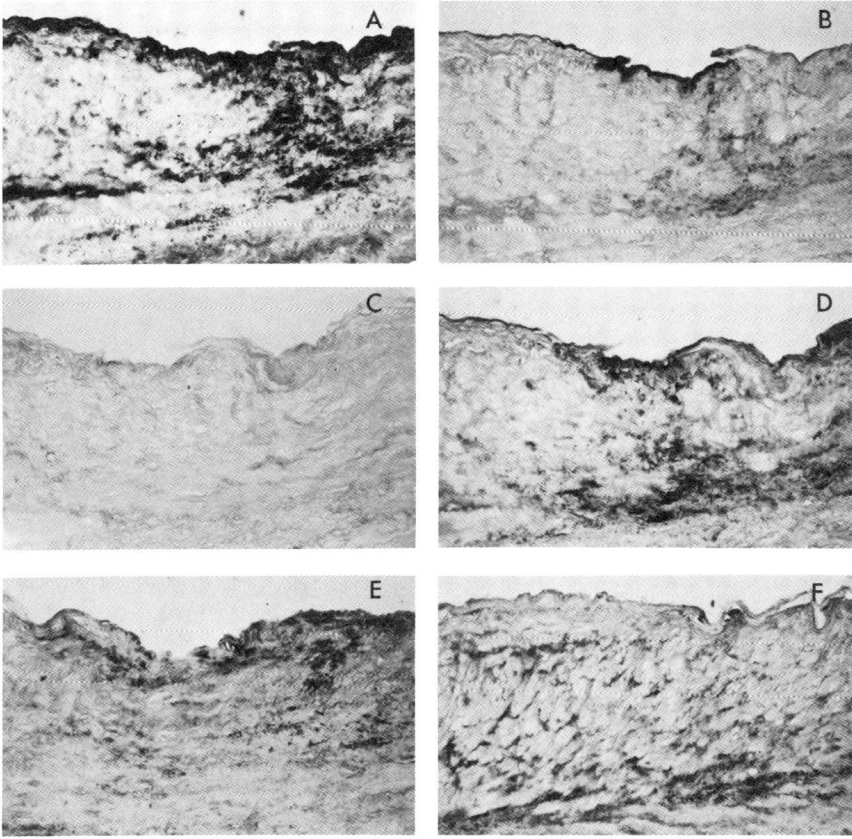

FIGURE 12-3 Micrographs of an early lesion of the right coronary artery. (A) Fibrinogen/fibrin I is present both on and beneath the luminal surface. (B) Some fibrin II is present on the lumen. (C) Fragment D/D-dimer is not detected in this lesion. (D) Polyclonal antiserum to fibrinogen shows distribution in the same area stained by the monoclonals. (E) Albumin follows the distribution of fibrinogen. (F) Staining for IgGs shows a different distribution from albumin. Reproduced by permission of the American Heart Association, Inc.; from Bini et al. (1989a).

II and fragment D/D-dimer-related antigen were observed parallel to the lumen, suggesting incorporation of mural thrombi (Bini 1987; Bini et al. 1987, 1989a). Moreover, cross-sections of coronary arteries showed that successive stages of lesion are present along the lumen of the same vessel (Bini et al. 1989a). These observations suggested further that the presence of fibrin-derived material in atherosclerosis is not only related to thrombus formation and to altered permeability of the endothelium (Smith 1977; Smith et al. 1976), but might also be due to an active interaction of fibrinogen with monocyte-derived cells or vascular wall cells.

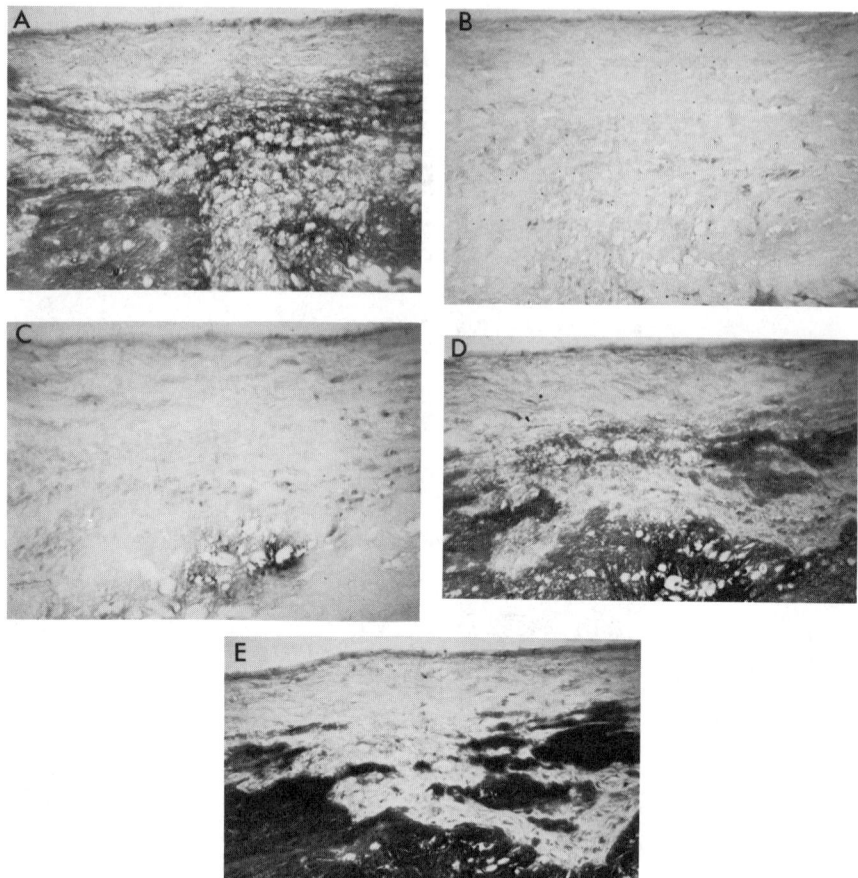

FIGURE 12-4 Micrographs of an early lesion in the left coronary, close to its branching from the ascending aorta, from a heart transplant recipient. (A) A cluster of foam cells is observed under an apparently normal intimal surface surrounded by fibrinogen/fibrin I. (B) Only traces of fibrin II are present in the same lesion. (C) A focus of lysis with degradation products is clearly visible. (D) The areas detected by the polyclonal antiserum to fibrinogen includes the areas detected by the monoclonals. (E) Staining for albumin is very intense in the deeper intima layer, but not on the lumen. Reproduced by permission of the American Heart Association, Inc.; from Bini et al. (1989a).

In another study using both polyclonal and monoclonal antibodies, extracts from atherosclerotic intimal tissue have been prepared for immunoelectrophoretic assessment of molecular-weight distributions and fibrinopeptide content of fibrinogen deposits. Both saline and SDS extracts contained monomeric and cross-linked oligomeric fibrinogen-related anti-

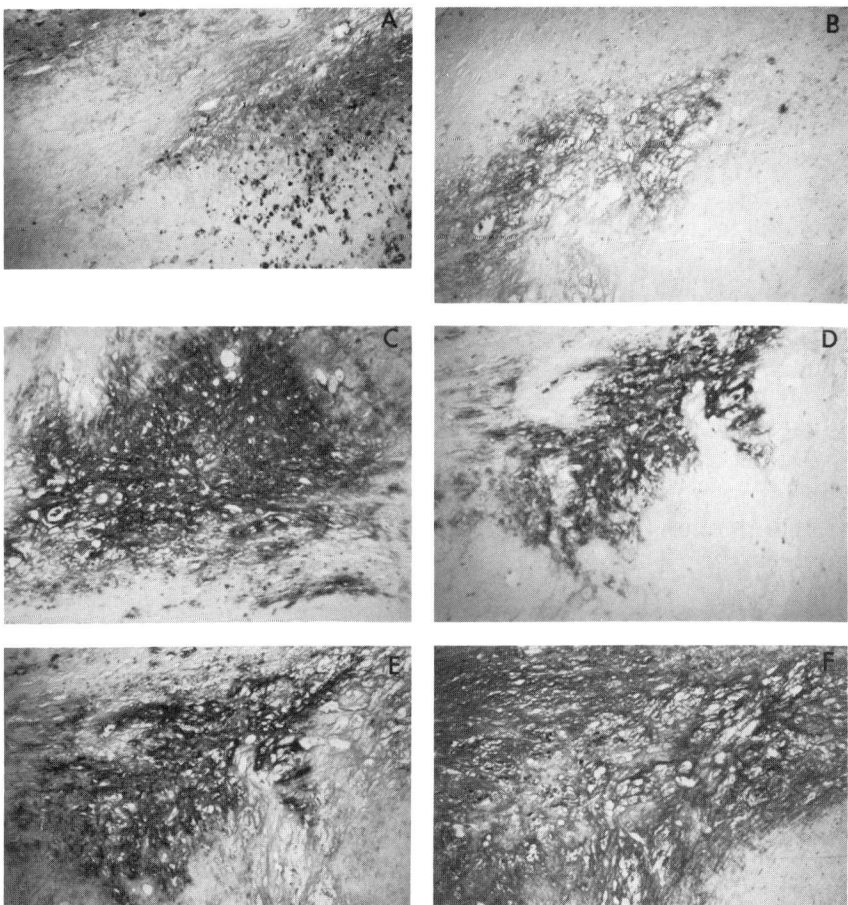

FIGURE 12-5 Micrographs of a fibrous plaque from a popliteal artery. (A) In this area of the lesion, fibrinogen/fibrin I are distributed in small flecks suggesting an association with smooth muscle cells and macrophages. (B) In the loose connective tissue of the same lesion, fibrinogen/fibrin I are distributed in longer threads and more diffuse staining is observed. (C) A large area of fibrin II is present in the same site in the tissue and around foam cells. (D) A large focus of degradation products is visible. (F) Albumin is widely distributed in the same area. Reproduced by permission of the American Heart Association, Inc.; from Bini et al. (1989a).

gen with FPA and FPB substantially intact (i.e., cross-linking of fibrinogen without prior release of fibrinopeptides and formation of fibrin II). The cross-links involved the Aα chain much more than the γ chain. Taken together, these results suggest that an alternate pathway to polymerization, one not depending on the coagulation mechanism and, therefore, not involving factor XIIIa, may contribute to the deposition process in tissues. For example, it is possible that tissue transglutaminase, released following tissue damage, is responsible for oligomer formation. The presence of dimeric forms of fibrinogen in many human plasma samples may be due to leaching from the intimal tissue (Valenzuela et al. 1988).

12.3.2 Normal and Complicated Pregnancy—Renal Disease

The development and possible application of this approach to the study of fibrin deposits in other tissues was tested on a large number of samples obtained by surgical or needle biopsy for diagnostic purposes. Normal human placenta at term was studied as a positive control (Bini et al. 1989b). Perivillous and intravillous deposits of fibrinogen/fibrin I, fibrin II, and fragment D/D-dimer-related antigen was observed in all placentas studied (Figure 12–6). It is known that fibrin deposits in the placenta increase progressively during pregnancy although the mechanism of its formation as well as its role is unknown. We found that placental areas containing fibrinogen/fibrin I and fibrin II were much more extensive than those con-

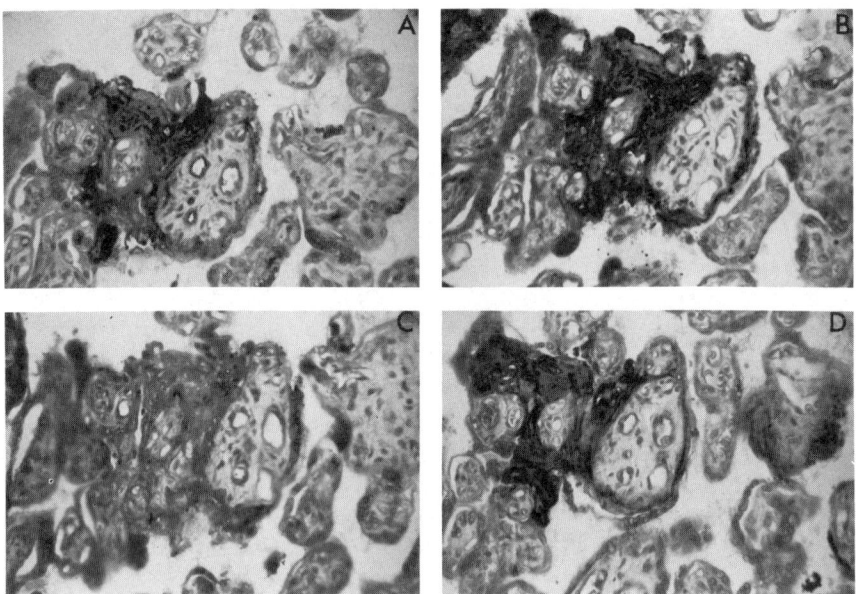

FIGURE 12–6 Micrographs of fibrin deposits in human placenta. Intense perivillous and intervillous distribution of fibrinogen/fibrin I (A) and fibrin II (B) is shown. No villi capillari stain is observed in (B). (C) Fragment D/D-dimer is visible in smaller areas. (D) Staining with the polyclonal antiserum to fibrinogen does not distinguish the areas detected by the monoclonal antibodies. Adapted with permission from Bini et al. (1989b).

taining the fragment D/D-dimer-related antigen. These results suggest that the rate of fibrin formation is increased in this tissue. Fibrin II deposits and fragment D/D-dimer-related antigen were absent from villi stem vessels, although they were present in the mesenchymal stroma and along the throphoblast basement membrane (Figure 12–6). This seems to indicate that maternal rather than fetal fibrinogen is involved in the formation of fibrin in the placenta. Further studies on the identification of fibrin(ogen)-related antigen in the human placenta may provide useful information on the progression of both normal and complicated pregnancy.

Systemic disorders of intravascular coagulation frequently involve damage to the kidney, and a large number of "primary" renal diseases may be complicated by coagulation. The susceptibility of the renal vascular bed to coagulation process can be attributed to a variety of factors, and the kidney is usually the primary target of vascular damage in eclampsia/pre-eclampsia, postpartum hemolytic uremic syndrome, thrombotic thrombocytopenic purpura, and the Shwartzman reaction. Moreover, the presence of immunological reactions in the glomerular capillary wall in a variety of glomerulonephritis (e.g., lupus nephritis or anti-GBM disease) may cause local coagulation in the glomerular capillary lumen. Accumulation of fibrin seems to play a major role in the pathogenesis of experimental and human glomerular disease itself, whatever the initial triggering mechanisms may be (Hancock and Atkins 1985; Vassalli and McCluskey 1964a, 1964b), and there is a broad correlation between the amount of fibrin deposited and the histologic damage. We applied the ABC technique using antibodies 1-8C6, T2G1, and GC4 to study the molecular composition of fibrin(ogen)-derived deposits in renal diseases of vascular and nonvascular origin (Bini et al. 1988). The patients were divided into three groups: one with microangiopathies, one with primary glomerulonephritis, and one with lupus nephritis of various types. This study showed that fibrin formation and lysis occur at different levels of the renal vasculature in both systemic disorders of coagulation and in many primary glomerulopathies of diverse etiology (Figure 12–7). Fibrinolysis was greater in glomerular capillaries than in larger vessels, and in microangiopathies more than in other renal diseases. These observations suggest that damage to the renal endothelium in kidney diseases of different origin might all involve the formation and lysis of fibrin deposits (Bini et al. 1988; Bini, A., D'Agati, V., Pirani, C., et al. in preparation). Recent studies have shown that low-molecular-weight fibrin(ogen)-degradation products enhance the release of growth factors from cultured endothelial cells (Lorenzet et al. 1988). It is possible that such a mechanism can contribute to mesangial proliferation. Moreover, following severe damage to the glomerular capillary wall, passage of fibrin(ogen)-related antigen in Bowman's space could stimulate the parietal epithelial cells to proliferate and form cellular crescents. More detailed studies on lupus nephritis are now in progress, and these may help us understand the coagulative and fibrinolytic processes which contribute to the development and progression

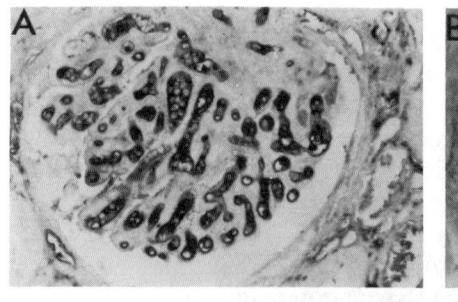
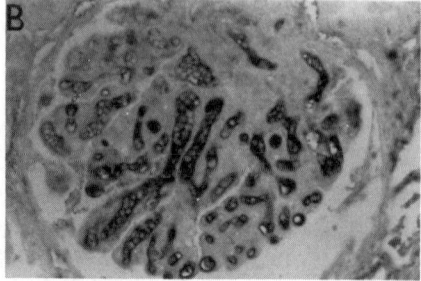
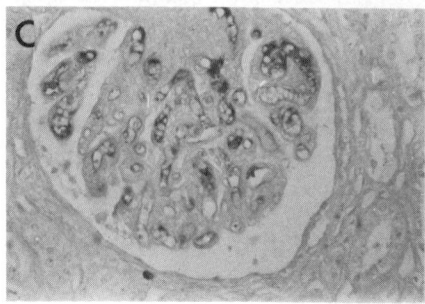

FIGURE 12-7 Micrographs of fibrin(ogen) deposits in a case of hemolytic uremic syndrome using MoAbs 1-8C6, T2G1, and GC4. (A) Fibrinogen/fibrin I is extensively present along the endothelial aspect and in the lumens of most glomerular capillaries. (B) Fibrin II has a similar distribution and intensity of staining. (C) Fragment D/D-dimer is visible in the glomerular capillary lumens of the same glomerulus. However, the staining is much weaker and more focal. Data from Bini et al. (1989b).

of kidney disease. The use of the described technique may provide important information on the development of new therapeutic strategies in the treatment of human renal diseases and could be applied routinely to selected renal biopsies to monitor the efficacy of a given therapeutic regimen.

12.3.3 Malignancies

A large number of clinical and experimental studies have shown an association between blood coagulation and malignant disease (Donati et al. 1981; Rickles and Edwards 1983). Recent immunohistochemical studies have shown the presence of coagulation factors and macrophage antigens in tumor stroma (Adany et al. 1988, 1989). In solid tumors, the malignant cells are dispersed in tumor stroma which primarily consists of connective tissue, new blood vessels, inflammatory cells (mainly macrophages and lymphocytes), and a fibrin matrix (Dvorak 1986; Dvorak et al. 1983; Folkman 1985; Form et al. 1984; Liotta et al. 1983). The formation of fibrin in tumors can be the result of activation of coagulation by procoagulants of the tumor cells themselves (Donati et al. 1984), and/or by activation of infiltrating macrophages (Colvin and Dvorak 1975; Edwards and Rickles 1984; Lorenzet et al. 1983). Previous studies have shown that although there is no correlation between the amount of fibrin deposited in tumors and their degree of malignancy, the pattern and distribution of fibrin in tumor stroma is characteristic and constant for a given type of tumor (Dvorak 1986). The differences in the local balance between fibrin deposition and degradation could reflect the predominance of procoagulant or fibrinolytic activity in various tumors (Dano et al. 1985).

Studies on the identification of fibrinogen/fibrin I, fibrin II, and fibrin(ogen) degradation products in solid tumors with differing stroma content have been reported. Monoclonal antibodies (1-8C6, T2G1, GC4) as well as rabbit antibodies to a host of coagulation factors—including but not limited to tissue factor, factors VII, X, V, "a" subunit of factor XIII, and proteins S and C—have been used to study the mechanism of coagulation activation in mesothelioma (Wojtukiewicz et al. 1989a), colon cancer (Wojtukiewicz et al. 1989b), and SCCL (Wojtukiewicz et al. 1990). Whereas no consistent pattern of reactivity was observed using any of these probes with tissue specimens of patients with malignant pleural mesothelioma, samples from patients with SCCL and colon cancer, processed simultaneously, reacted significantly with some of these antibodies. SCCL tumor cells reacted with antibodies to tissue factor, factor V, and proteins C and S. Stroma from the same patient group stained for factor XIII, fibrinogen, and fibrin (Wojtukiewicz et al. 1990). In contrast, specimens from patients with colon cancer showed no evidence of factors VII and X, good reactivity with antibodies to factor XIII and fibrinogen, but virtually no fibrin II deposits in tumor stroma (Wojtukiewicz et al. 1989b). The conclusion drawn from these studies is that the absence of components of coagulation/fibrinolysis may account for the low incidence of metastatic dissemination in mesothelioma. These results and data for other tumor types should provide more detailed information on the presence of fibrin in tumors, on its role on tumor growth and metastasis, and on the use of anticoagulant or fibrinolytic agents in cancer therapy.

12.4 MONOCLONAL ANTIBODIES AS THROMBUS IMAGING AGENTS

Thrombus detection, via gamma-camera imaging of localized radiolabeled, fibrin-specific, monoclonal antibodies, has proven useful for both experimental and clinical applications. Table 12–3 lists all the antibodies which have been used in experimental and clinical studies. The use of certain components of the coagulation/fibrinolysis system (e.g., radiolabeled fibrinogen, plasmin, plasminogen, urokinase, streptokinase, fibrin fragment E1, platelets, thrombospondin, and t-PA) as thrombus imaging agents has not been as successful owing to their low uptake rate in the thrombus or their

TABLE 12–3   Antibodies Used in Experimental and Clinical Thrombus Imaging Studies

Antibody	Isotype	Epitope Associated with	Epitope Structure	Reference[1]
64C5	IgG1,κ	E domain	Bβ 15-21	Hui et al. 1983
59D8	IgG1,κ	E domain	Bβ 15-21	Hui et al. 1983
DD-3B6	IgG1	D domain	?	Rylatt et al. 1983
T2G1	IgG1,κ	E domain	Bβ 15-21	Kudryk et al. 1984
NIBn123	IgM	X-oligomers from cross-linked fibrin	?	Gaffney et al. 1988a
15C5	IgG1,κ	D domain	?	Declerck et al. 1987
8D3	IgG1,κ	D domain	?	Declerck et al. 1987
Y22	IgG1,κ	D domain	?	Wasser et al. 1989
GC4	IgG1,κ	D domain	?	Kudryk et al. 1988

The references listed are only those that describe, for the first time and in some detail, the preparation and characterization of each antibody.

ability to accumulate only in fresh (<24 hours) or growing thrombi (Koblik et al. 1989; Oster and Som 1989). Antibodies directed against some of these components have had similar limitations and results, i.e., antibodies specific for fibrinogen (Spar et al. 1965), platelets (Oster and Som 1989; Oster et al. 1985; Peters et al. 1986; Som et al. 1986), activated platelets (Palabrica et al. 1989), and thrombospondin (Koblik et al. 1989).

12.4.1   Animal Model Studies

Antibody T2G1 is directed to an epitope on the new NH_2-terminal end (Bβ 15-21, GHRPLDK) of the human fibrin β chain (Kudryk et al. 1984). Since this identical sequence is also present on the dog fibrin β chain, antibody T2G1 cross-reacts completely with dog fibrin and therefore this animal has served as a useful in vivo thrombus model. Imaging with radiolabeled antibody T2G1 has proven successful for both fresh and aged experimentally-induced thrombi. We initially investigated the use of ^{131}I-labeled T2G1 for imaging fresh canine thrombi (about 3 hours old at the time of antibody injection) and compared it to ^{125}I-labeled nonspecific polyclonal murine IgG concurrently injected (Rosebrough et al. 1985). Radiobioassay results showed preferential thrombus uptake of T2G1 (about 15 times more than control), suitable thrombus concentrations (0.25% dose/gram), and target-to-background ratios of 9 (thrombus/blood) and 240 (thrombus/muscle) for positive images in all dogs. Although positive results were obtained at 24 and 48 hours after injection, earlier images were not definitive due to the large blood pool of the antibody (Figure 12–8) and, consequently, the low thrombus-to-background ratios. The antigenic site for T2G1 has a potential imaging disadvantage because it may be split from the clot during the early

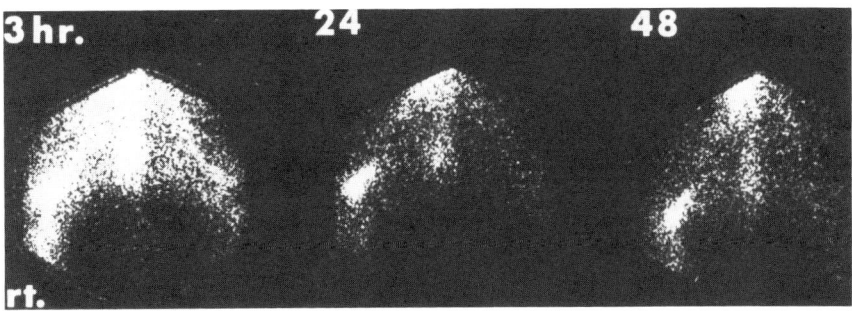

FIGURE 12-8 Gamma camera images of an experimentally induced thrombus in the femoral vein of a dog using ^{131}I-labeled T2G1. The camera was positioned over the lower trunk and upper legs. At three hours after antibody injection, images are negative due to large blood pool of antibody. After 24 and 48 hours, images were positive (right leg, left side of each image) due to higher thrombus-to-background ratios.

stages of plasmin-induced fibrinolysis. To ascertain the effect of thrombus age on T2G1 localization, thrombi were aged from 1–5 days before antibody injection and imaging performed 48 hours thereafter (Rosebrough et al. 1987). As was found for fresh thrombi, these images were positive, indicating adequate antigen concentrations for successful radioimmunoimaging even in 7-day-old thrombi (formed 5 days prior to antibody injection and images obtained 2 days after antibody injection).

In order to improve image quality and provide positive images in shorter time, the use of antibody fragments was investigated. There are many advantages with using the F(ab′)₂ and Fab fragments; these include: 1) their small size speed their diffusion into the extravascular fluid, and, for the Fab and Fab′ fragments, excretion via the glomerulus increases circulatory clearance and may provide higher thrombus-to-background ratios; 2) the elimination of nonspecific binding of antibody by Fc receptors in the liver, spleen, and on leukocytes; 3) they may be less immunogenic due to their smaller size and the removal of the Fc fragment; 4) less potential Fc-mediated reactions such as complement activation and immune complex disease. We have investigated the use of [131]I- and ^{111}In-labeled F(ab′)₂ and Fab fragments of T2G1 for thrombus imaging in the dog (Grossman et al. 1987; Rosebrough et al. 1988) and concluded that ^{111}In-labeled F(ab′)₂ was the best thrombus imaging agent.

Another neoepitope-specific monoclonal antibody, designated GC4, has also been developed by our group. This antibody does not bind ELISA plates coated with fibrinogen, fibrin monomer, or any of the three component chains of fibrinogen, fibrin, or cross-linked fibrin. In competition experiments, antibody GC4 fails to react with native fibrinogen in solution and reacts quite poorly with intact digests derived from either fibrinogen, fibrin, or cross-linked fibrin. However, this antibody does react with purified high-molecular-weight-plasmin-digest products of either human or dog fibrinogen, fibrin, or cross-linked fibrin, as well as with some but not all human or dog D and D-dimer fragments (Kudryk et al. 1985, 1988, 1989a).

Since antibody GC4 binds plasmin-digested dog fibrin, its utility as a thrombus imaging agent was directly compared to T2G1 in the canine thrombosis model. For successful imaging, an antibody must retain its immunoreactivity after radiolabeling. Many monoclonal antibodies lose immunoreactivity after radiolabeling or during coupling reactions. Prior to their use as imaging agents in the canine model, the immunoreactivity of T2G1 and GC4 was determined by affinity chromatography on fibrin-coupled Sepharose and plasmin-digested fibrin-coupled Sepharose, respectively. T2G1 was insensitive to direct labeling with [131]I- and ^{111}In-labeling after coupling with DTPA cyclic dianhydride. Immunoreactivity (>75%) was determined by extrapolation to the ordinate from a double reciprocal plot of total/bound radiolabeled antibody versus amount of antigen-gel added per tube (Lindmo et al. 1984; Rosebrough et al. 1989b). Immunoreactivity of GC4 radiolabeled with ^{131}I and ^{111}In was similarly measured. GC4 was insensitive to iodination; however, unlike T2G1, it lost immunoreactivity (Rosebrough and Maley 1990a) when high concentrations of DTPA were present (Figure 12–9). With the use of low concentrations of DTPA (DTPA/GC4 molar ratio of ~1.0), GC4 immunoreactivity was retained, and the

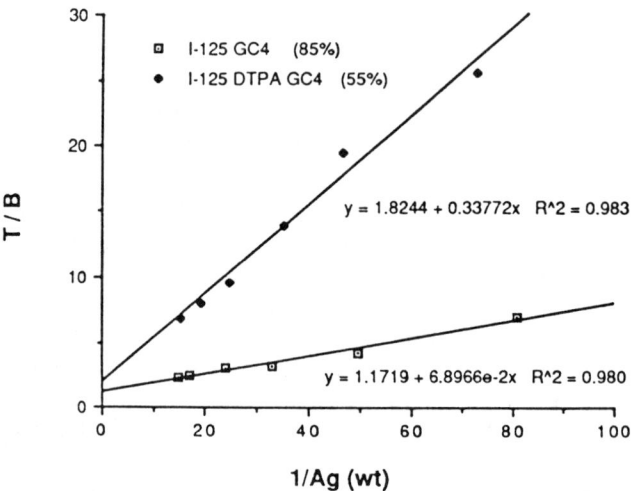

FIGURE 12–9 Double reciprocal plots of ^{125}I-labeled/GC4 versus amount of antigen-Sepharose gel added per tube. Immunoreactivity was determined by extrapolation to the ordinate. Antibody/GC4 exhibited high immunoreactivity (85%) when labeled with ^{125}I, but lost immunoreactivity (55%) when the same preparation was labeled with ^{111}In using high concentrations of DTPA.

radiolabeling yield and specific activity were sufficient for imaging studies. In dogs with 3-hour-old thrombi, GC4 localization was about 50% less than that obtained with T2G1. Conversely, with 3-day-old thrombi, GC4 localization was about two times more than that obtained with the other probe. Nevertheless, radiolabeled GC4 yielded positive 24- and 48-hour images in all dogs regardless of thrombus age. To more closely mimic the clinical setting, dogs containing 1- and 3-day-old thrombi were treated with therapeutic doses of heparin initiated at least 3 hours prior to antibody injection. In these dogs, the thrombus localization ratio of GC4/T2G1 approached 3, thrombus-to-blood ratios were above 20, thrombus concentrations (% dose/gram) were above 0.5%, and images were positive 4 hours after antibody injection. Thus, antibody GC4 may be better suited than T2G1 for thrombus imaging and may prove ideal for clinical use. Theoretically, GC4 may be more advantageous than T2G1 since clinical thrombi are often 24 hours old (i.e., may be "aged" or partly degraded thrombi) and prospective patients to be evaluated by immunoscintigraphy are often undergoing heparin therapy (Rosebrough et al. 1989a, 1990b).

In other canine model studies, the imaging properties of ^{99m}Tc-labeled T2G1 Fab' were compared with those of a ^{111}In-labeled Fab fragment of a similar fibrin-specific monoclonal antibody, designated 59D8 (Hui et al. 1983). Although thrombus uptake and the time required to visualize the "hot spot" were not significantly different for the two probes, the ^{99m}Tc-labeled T2G1 fragment gave slightly higher thrombus/blood and thrombus/muscle ratios and also showed a faster blood disappearance rate (Knight et al. 1989). In EIAs, antibody Y22 shows strong reactivity with human, rabbit, rat, sheep, and dog fibrin and only weak binding to fibrinogen. This antibody seems to be directed to a conformation-dependent epitope in the D domain of fibrin and, after labeling, binds to forming as well as preformed plasma clots in a plasma environment. Imaging studies using ^{99m}Tc-labeled Y22 (intact antibody) showed hot spots in both the abdomen and jugular vein (sites where thrombi were experimentally induced or deposited) of rabbits 18 hours after injection of antibody (1.0 mg, 37 MBq) and about 21 hours following thrombus induction/deposition. In the clot-containing jugular vein, clot/blood ratios at 45 min and 4 and 18 hours were 1.2, 4.9, and 7.7, respectively. Induced clots in the inferior vena cava of Wistar rats were visualized 4 hours after injection of ^{99m}Tc-labeled Y22 (Wasser et al. 1989). Papain-generated and ^{123}I-labeled Fab fragments prepared from two D-dimer specific antibodies (15C5 and 8D3, antibodies are against nonoverlapping epitopes on D-dimer) have also been used for imaging studies in the rabbit experimental thrombosis model. Since neither antibody cross-reacts with rabbit fibrin or its degradation products, nonocclusive human plasma clots were formed in the external jugular vein. Following thrombus induction and "aging" (for either 1, 24, or 72 hours), equimolar mixtures of the two labeled Fab fragments were injected. The vein segment/blood ratios at 24 hours was 6.6 for animals with nonocclusive clots and 1.5 in animals that underwent the same surgical procedure without clot induction (Holvoet et al. 1989). Imaging human fibrin clots formed in rabbit external jugular vein using both the D-dimer-specific probe DD-3B6 (Rylatt et al. 1983) and an IgM antibody (NIBn 123) specific for X-oligomers (Gaffney et al. 1988a) has been reported recently. In these studies, both antibodies (intact) were ^{131}I-labeled. Despite some differences in reactivity of the two probes as observed by in vitro methods, both gave comparable jugular thrombus/heart ratios and discernable images 18 hours post injection (Tymkewycz et al. 1989).

12.4.2 Thrombus Imaging in Patients

Several reports have presented preliminary data on the use of antifibrin antibodies as clot imaging agents in patients with DVT. Two of the nine antibodies (59D8 and T2G1) listed in Table 12–3 have been used in clinical trials. Both ^{111}In-labeled F(ab')₂ T2G1 and ^{99m}Tc-labeled Fab' T2G1 are currently in clinical trials with encouraging preliminary results. However, published reports have appeared only for studies employing the 59D8 antibody.

Immunoscintigraphic studies with the 59D8 antifibrin antibody have utilized the Fab fragment labeled with ^{111}In via DTPA chelate. This method provides over 90% incorporation of the radionuclide onto the antibody fragment. Generally, studies using ^{111}In-labeled 59D8 antifibrin (Jung et al. 1989; Lusiani et al. 1989; Alavi et al. 1990) have enrolled patients with signs or symptoms of deep-vein thrombosis (i.e., swelling, pain, sensation of heat in the lower extremities), and have had deep-vein thrombosis confirmed by contrast venography performed either before or after the antifibrin study. All patients have been injected via a peripheral vein with 0.5 mg of the Fab fragment labeled with 2 mCi of [111]In. Antifibrin images of the lower extremities have been acquired at several time intervals post-injection, out to 24 hours post-injection.

Two studies (Jung et al. 1989; Lusiani et al. 1989) enrolled patients both with and without documented deep-vein thrombosis, allowing the determination of diagnostic sensitivity and specificity. In their respective overall patient population, Jung et al. (1989) reported a sensitivity of 84% and a specificity of 81% in a total of 52 patients (31 patients with DVT, 21 without), while Lusiani et al. (1989) reported sensitivity of 78% and a specificity of 92% in a total of 30 patients (18 patients with DVT, 12 patients without). In the study by Alavi et al. (1990), only a few of the patients had negative venograms, which did not permit a determination of diagnostic specificity. These investigators also compared specific anatomic sites for localization of deep-venous thrombi by contrast venography or by antifibrin imaging in 28 patients with positive contrast venograms. There were 53 positive sites on the venograms, and 56 on the antifibrin images, 41 of which matched. All of the results reported above were obtained in images acquired 2–4 hours after antifibrin injection; images acquired beyond this time, up to 24 hours, did not further improve overall diagnostic accuracy.

These preliminary studies also showed that factors related to age and location of the deep-venous thrombus and heparinization of the patient appear to affect the diagnostic accuracy of antifibrin scintigraphy. When Jung et al. (1989) analyzed a subgroup of patients with symptoms for less than 10 days, sensitivity improved to 92%, compared to 84% when all patients were considered. Similarly, Lusiani et al. (1989), in analyzing a subgroup with symptoms less then 30 days, reported a sensitivity of 100%. Therefore, to the extent that duration of clinical symptoms represents the age of deep-venous thrombi, sensitivity of antifibrin scintigraphy is better for acute thrombi.

Both Jung et al. (1989) and Alavi et al. (1990) have also examined the ability of ^{111}In-labeled antifibrin to detect deep-vein thrombi in different regions of the leg. In the study by Jung et al. (1989), sensitivity and specificity were 92% and 96% in the calf, 82% and 83% in the popliteal (knee) region, 63% and 96% in the thigh, and 18% and 100% in the pelvis. Alavi et al. (1990) also observed a decline in the capability of antifibrin to detect deep-venous thrombi in the thigh, but only in patients receiving heparin. These results suggest a diminished sensitivity of ^{111}In-labeled antifibrin in detecting proximal deep-venous thrombosis. Because of the larger size of thrombi in the thigh, these results could represent a slower penetration of the antibody into the central portion of the thrombus. Whether extending the time of performance of scintigraphy could improve the sensitivity for proximal deep-venous thrombi needs to be investigated (Jung et al. 1989).

Preliminary results from Alavi et al. (1990) suggest that heparinization may impair the detection of proximal deep-venous thrombi by ^{111}In-labeled antifibrin. In 19 nonheparinized patients, antifibrin successfully identified all five thrombi in the thigh that were confirmed by contrast venography in this group of patients. However, in 14 heparinized patients, only 7 of 13 thrombi in the thigh were detected by antifibrin. In the study by Jung et al. (1989), 79% of the patients received heparin, yet a high overall sensitivity was observed. A heparin effect on the uptake of antifibrin by the thrombus could be due to the prevention of new fibrin-formation, while ongoing endogenous fibrinolysis would remove some of the epitope from the clot, resulting in a net loss of sites for binding. Furthermore, endogenous fibrinolysis may result in the release of the antifibrin-labeled fibrin from the surface of the intraluminal thrombus (Alavi et al. 1990). However, the preliminary results suggest that, although heparin anticoagulation may reduce sensitivity for detecting individual sites of thrombus, the overall disease detection rate may be unaffected by heparin. Certainly, further studies are required.

Recent studies have also been undertaken with ^{99m}Tc-labeled T2G1 antifibrin in patients with clinical signs and symptoms of deep-venous thrombosis. This technique employs the Fab' fragment of the antibody to which

[99mTc] is bound at an incorporation of greater than 90%. The labeling procedure can be accomplished in less than 15-30 minutes. Figure 12–10 shows a contrast venogram and an antifibrin scan in a 43-year-old woman with symptoms of deep-venous thrombosis of 3-days duration in the right calf, knee, and thigh. The venogram shows a lateral view of the knee and distal femur and demonstrates definite intraluminal filling defects corresponding to areas of acute deep-venous thrombosis in the popliteal and superficial femoral vein. The antifibrin scan is a posterior view of the knee region and distal thigh obtained 4 hours post-injection and demonstrates intense focal uptake of the antibody in areas corresponding to thrombus in the deep veins.

In summary, preliminary studies of antifibrin scintigraphy in the diagnosis of DVT have shown promising results. Overall sensitivity and specificity are high, but appear to diminish for aged clots and in more proximal locations of the lower extremity. Diagnostic images can almost always be obtained within 2-4 hours after injection of the antibody. Further studies are required in a broadly defined patient population to fully evaluate di-

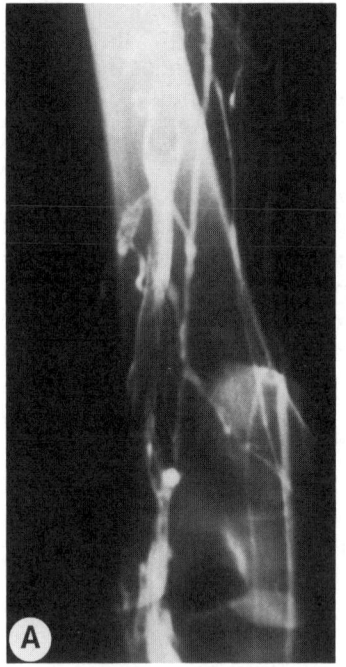 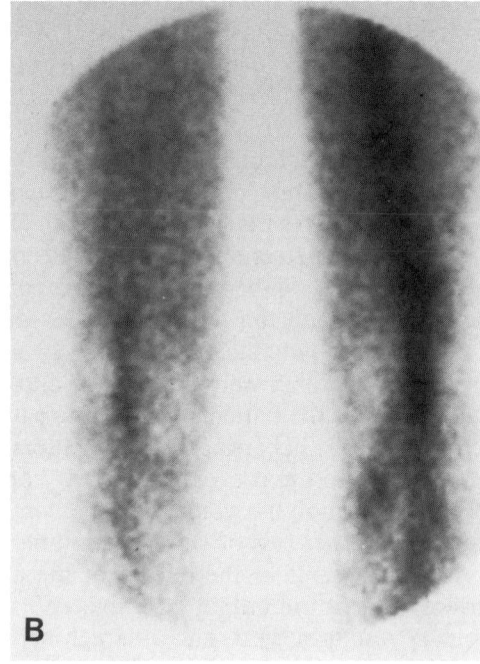

FIGURE 12–10 (A) A contrast venogram and (B) a [99mTc]-labeled T2G1 antifibrin scan in a 43-year-old woman symptomatic for DVT in the right leg. The venogram is a lateral view of the knee and distal femur, and the antifibrin scan is a posterior view of the knee and distal thigh. The areas of intense focal antibody uptake correspond to the locations of acute venous thrombosis in the venogram.

agnostic performance and to ascertain the ultimate clinical utility of this technique.

ABBREVIATIONS AND TERMS

ABC	Avidin-biotin complex immunoperoxidase technique
ARF	Acute renal failure
CNBr	Cyanogen bromide
CRF	Chronic renal failure
DALIA	Distribution analyzing latex immunoassay
DIC	Disseminated intravascular coagulation
DTPA	Diethylenetriaminepentaacetic acid
DVT	Deep-vein thrombosis
ELISA	Enzyme-linked immunoabsorbant assay
EIA	Enzyme immunoassay
FbDP	Fibrin degradation products
FbgDP	Fibrinogen degradation products
FDP	Fibrin(ogen) degradation products
Fibrin I	Fibrin (des FPA) prepared by clotting fibrinogen with batroxobin or similar enzymes
Fibrin II	Fibrin (des FPA/des FPB) prepared by clotting fibrinogen with thrombin or similar enzymes
FPA	Fibrinopeptide A ($A\alpha$ 1-16, M_r 1536)
FPB	Fibrinopeptide B ($B\beta$ 1-14, M_r 1578)
Fragment D	COOH-terminal fragment (M_r 93,000) of fibrinogen or non-cross-linked fibrin obtained by digestion with plasmin in buffers containing $CaCl_2$
Fragment D-dimer	COOH-terminal fragment (M_r 186,000) of cross-linked fibrin obtained by digestion with plasmin in buffers containing $CaCl_2$
Fragment E	The NH_2-terminal fragments (M_r 50,000) obtained by plasmin digestion; the formula for the predominant species is given by ($A\alpha$ 20–78, $B\beta$ 54–122, γ 1–53)$_2$
Fragment X	Early plasmin degradation product (M_r 225,000 to 333,000) of fibrinogen
Fragment Y	Early plasmin degradation product (M_r 150,000 to 170,000) of fibrinogen
GBM	Glomerular basement membrane
HRPO	Horseradish peroxidase
MoAb	Monoclonal antibody
N-DSK	Amino-terminal disulfide knot ([$A\alpha$ 1-51, $B\beta$ 1-118, γ 1-78]$_2$, M_r 58,000) of fibrinogen obtained by cleavage with CNBr

PE	Pulmonary embolism
PTAH	Phosphotungstic acid hematoxylin
RIA	Radioimmunoassay
SCCL	Small cell carcinoma of the lung
SDS	Sodium dodecyl sulfate
(T)N-DSK	Amino-terminal disulfide knot ([Aα 17-51, Bβ 15-118, γ 1-78]$_2$, M$_r$ 52,000) of fibrin II or cross-linked fibrin II obtained by cleavage with CNBr
t-PA	Tissue-type plasminogen activator
X-oligomers	High-molecular-weight cross-linked fibrin degradation products

REFERENCES

Adany, R., Szegedi, A., Ablin, R.J., et al. (1988) *Thromb. Haemostas.* 60, 293–297.

Adany, R., Kappelmayer, J., Berenyi, E., et al. (1989) *Thromb. Haemostas.* 62, 850–855.

Alavi, A., Palevsky, H., Gupta, N., et al. (1990) *Radiology* 175, 79–85.

Bini, A. (1987) Ph.D. thesis, Columbia University College of Physicians and Surgeons, New York.

Bini, A., Sobel, J., and Kaplan, K.L. (1986) in *Fibrinogen and Its Derivatives* (Müller-Berghaus, G., Scheefers-Borchel, U., Selmayr, E., and Henschen, A., ed.), pp. 173–177, Elsevier Science Publishers, Amsterdam.

Bini, A., Fenoglio, J., Jr., Sobel, J., et al. (1987) *Blood* 69, 1038–1045.

Bini, A., D'Agati, V., Pirani, C., et al. (1988) *J. Histochem. Cytochem.* 36(7a), 938.

Bini, A, Fenoglio, J.J., Jr., Mesa-Tejada, R., et al. (1989a) *Arteriosclerosis* 9, 109–121.

Bini, A., Mesa-Tejada, R., Fenoglio, J.J., Jr., et al. (1989b) *Lab. Invest.* 60, 814–821.

Colvin, R.B., and Dvorak, H.J. (1975) *J. Exp. Med.* 142, 1377–1390.

Crabtree, G.R. (1987) in *The Molecular Basis of Blood Diseases* (Stamatoyannopoulos, G., Nienhuis, A.W., Leder, P., and Majerus, P.W., ed.), pp. 631–661, W.A. Saunders Company, Philadelphia.

Dano, K., Andreasen, P.A., Grondahl-Hansen, J., et al. (1985) *Adv. Cancer Res.* 44, 139–266.

Dawes, J., Drummond, O., Micklem, L.R., et al. (1985) *Thromb. Haemostas.* 54, 41.

Declerck, P.J., Mombaerts, P., Holvoet, P., et al. (1987) *Thromb. Haemostas.* 58, 1024–1029.

Deguchi, F., Tomura, S., Yoshiyama, N., et al. (1989) *Nephron* 51, 377–383.

Donati, M.B., Poggi, A., and Semeraro, N. (1981) in *Coagulation and Malignancy. Recent Advances in Blood Coagulation* (Poller, L., ed.), pp. 227–259, Churchill Livingstone, Edinburg.

Donati, M.B., Semeraro, N., and Gordon, S.G. (1984) in *Hemostatic Mechanism and Metastasis* (Honn, K.V., and Sloane, B.F., eds.), pp. 62–71, Martinus Nijhoff, Boston.

Dvorak, H.F. (1986) *N. Engl. J. Med.* 315, 1650–1659.

Dvorak, H.F., Senger, D.R., and Dvorak, A.M. (1983) *Cancer Metastasis Rev.* 2, 41–73.

Edwards, R.L., and Rickles, F.R. (1984) in *Hemostatic Mechanism and Metastasis* (Honn, K.V., and Sloane, B.F., eds.), pp. 342–354, Martinus Nijhoff, Boston.

Eisenberg, P.R., Sherman, L.A., Jaffe, A.S., et al. (1986) *Circulation* 74 (part 2), II-245.

Eisenberg, P.R., Sobel, B.E., and Jaffe, A.S. (1988) *Circulation* 78, 592–597.

Elms, M.J., Bunce, I.H., Bundesen, P.G., et al. (1983) *Thromb. Haemostas.* 50, 591–594.

Elms, M.J., Bunce, I.H., Bundesen, P.G., et al. (1986) *Am. J. Clin. Pathol.* 85, 360–364.

Folkman, J. (1985) *Adv. Cancer Res.* 43, 175–203.

Form, D.M., VanDeWater, L., Dvorak, H.F., et al. (1984) *JNCI* 73, 1207–1214.

Gaffney, P.J., and Perry, M.J. (1985) *Thromb. Haemostas.* 53, 301–302.

Gaffney, P.J., Creighton, L.J., Perry, M.J., et al. (1988a) *Br. J. Haematol.* 68, 83–90.

Gaffney, P.J., Creighton, L.J., Callus, M., et al. (1988b) *Br. J. Haematol.* 68, 91–96.

Gitlin, D., and Craig, J.M. (1957) *Am. J. Pathol.* 33, 267–283.

Gordge, M.P., Faint, R.W., Rylance, P.B., et al. (1989) *Thromb. Haemostas.* 61, 522–525.

Greenberg, C.S., Devine, D.V., and McCrae, K.M. (1987) *Am. J. Clin. Pathol.* 87, 94–100.

Grossman, Z.D., Rosebrough, S.F., McAfee, J.G., et al. (1987) *Radiographics* 7, 913–921.

Hafter, R., Schröck, R., von Hugo, R., et al. (1985) *Scand. J. Clin. Lab. Invest.* 45, 137–144.

Hancock, W., and Atkins, R. (1985) *Semin. Nephrol.* 5, 69–77.

Haust, M.D., Wyllie, J.C., and Moke, R.H. (1965) *Exp. Mol. Pathol.* 4, 205–216.

Hillyard, C.J., Blake, A.S., Wilson, K., et al. (1987) *Clin. Chem.* 33, 1837–1840.

Hoegee-de Nobel, E., Voskuilen, M., Briët, E., et al. (1988) *Thromb. Haemostas.* 60, 415–418.

Holvoet, P., Stassen, J.M., Hashimoto, Y., et al. (1989) *Thromb. Haemostas.* 61, 307–313.

Hui, K.Y., Haber, E., and Matsueda, G.R. (1983) *Science* 222, 1129–1132.

Jackson, D.C., Poumbourios, P., White, D.O., et al. (1988) *Mol. Immunol.* 25, 465–471.

Jung, M., Kletter, K., Dudczak, R., et al. (1989) *Radiology* 173, 469–475.

Kamitsuji, H., Whitworth, J.A., Dowling, J.P., et al. (1986) *Am. J. Kidney Dis.* 7, 452–455.

Kao, V.C.Y., and Wissler, R.W. (1965) *Exp. Mol. Pathol.* 4, 465–479.

Kennel, S.J., and Lankford, P.K. (1983) *Clin. Chem.* 29, 778–781.

Knight, L.C., Maurer, A.H., Ammar, I.A., et al. (1989) *Radiology* 173, 163–169.

Koblik, P.D., DeNardo, G.L., and Berger, H.J. (1989) *Sems. Nucl. Med.* 19, 221–237.

Koopman, J., Haverkate, F., Koppert, P., et al. (1987) *J. Lab. Clin. Med.* 109, 75–84.

Koppert, P.W., Huijsmans, C.M.G., and Nieuwenhuizen, W. (1985) *Blood* 66, 503–507.

Koppert, P.W., Koopman, J., Haverkate, F., et al. (1986) *Blood* 68, 437–441.

Koppert, P.W., Kuipers, W., Hoegee-de Nobel, E., et al. (1987) *Thromb. Haemostas.* 57, 25–28.

Koppert, P.W., Hoegee-de Nobel, E., and Nieuwenhuizen, W. (1988) *Thromb. Haemostas.* 59, 310–315.

Kudryk, B., Rohoza, A., Ahadi, M.,et al. (1983) *Mol. Immunol.* 20, 1191–1200.

Kudryk, B., Rohoza, A., Ahadi, M., et al. (1984) *Mol. Immunol.* 21, 89–94.

Kudryk, B., Rohoza, A., Ahadi, M., et al. (1985) *Thromb. Haemostas.* 54, 275.

Kudryk, B., Rohoza, A., Ahadi, M., et al. (1988) in *Fibrinogen 3—Biochemistry, Biological Functions, Gene Regulation and Expression* (Mosesson, M.W., Amrani, D.L., Siebenlist, K.R., and DiOrio, J.P., ed.), pp. 129–132, Elsevier Science Publishers B.V., Amsterdam.

Kudryk, B.J., Grossman, Z.D., McAfee, J.G., et al. (1989a) in *Monoclonal Antibodies in Immunoscintigraphy* (Chatal, J.-F., ed.), pp. 365–398, CRC Press, Boca Raton, FL.

Kudryk, B., Gidlund, M., Rohoza, A., et al. (1989b) *Blood* 74, 1036–1044.

Lendrum, A.C., Fraser, D.S., Slidder, S.W., et al. (1962) *J. Clin. Pathol.* 15, 401–413.

Lindmo, T., Boven, E., Cuttitta, F., et al. (1984) *J. Immunol. Methods* 72, 77–89.

Liotta, L.A., Rao, C.N., and Barsky, S.H. (1983) *Lab. Invest.* 49, 636–649.

Lorenzet, R., Peri, G., Locati, D., et al. (1983) *Blood* 62, 271–273.

Lorenzet, R., Sobel, J.H., Bini, A., et al. (1988) *Circulation* 78, 394.

Luna, L.G. (1968) *Manual of Histologic Staining Methods of the Armed Forces Institute of Pathology*, pp. 81–86, McGraw-Hill Book Co., New York.

Lusiani, L., Zanco, P., Visona, A., et al. (1989) *Angiology* 40, 671–677.

Matsumoto, T., Nishijima, Y., Teramura, Y., et al. (1985) *Thromb. Res.* 38, 297–302.

McCabe, R.P., Lamm, D.L., Haspel, M.V., et al. (1984) *Cancer Res.* 44, 5886–5893.

Mirshahi, M., Soria, J., Soria, C., et al. (1986) *Thromb. Res.* 44, 715–728.

Müller-Berghaus, G., Scheefers-Borchel, U., Fuhge, P., et al. (1985) *Scand. J. Clin. Lab. Invest.* 45, 145–151.

Nieuwenhuizen, W. (1988) *Blut* 57, 285–291.

Nossel, H.L. (1981) *Nature* 291, 165–166.

Nossel, H.L. Butler, V.P., Jr., Wilner, G.D., et al. (1976) *Thromb. Haemostas.* 35, 101–109.

Nossel, H.L., Younger, L.R., Wilner, G.D., et al. (1971) *Proc. Natl. Acad. Sci. USA* 68, 2350–2353.

Oster, Z.H., and Som, P. (1989) *AJR* 152, 253–260.

Oster, Z.H., Srivastava, S.C., Som, P., et al. (1985) *Proc. Natl. Acad. Sci. USA* 82, 3465–3468.

Palabrica, T.M., Furie, B.C., Konstam, M.A., et al. (1989) *Proc. Natl. Acad. Sci. USA* 86, 1036–1040.

Paronetto, F., and Koffler, D. (1965) *J. Clin. Invest.* 44, 1657–1664.

Peters, A.M., Lavender, J.P., Needham, S.G., et al. (1986) *Br. Med. J.* 293, 1525–1527.

Prowse, C.V. (1986) *Vox Sang.* 50, 65–70.

Rickles, F.R., and Edwards, R.L. (1983) *Blood* 62, 14–31.

Rogers, S., Sweetnam, P.M., Perry, M.J., et al. (1986) *Thromb. Res.* 43, 389–393.

Rosebrough, S.F., Kudryk, B., Grossman, Z.D., et al. (1985) *Radiology* 156, 515–517.

Rosebrough, S.F., Grossman, Z.D., McAfee, J.G., et al. (1987) *Radiology* 162, 575–577.

Rosebrough, S.F., Grossman, Z.D., McAfee, J.G., et al. (1988) *J. Nucl. Med* 29, 1212–1222.

Rosebrough, S.F., McAfee, J.G., Grossman, Z.D., et al. (1989a) *J. Nucl. Med.* 29, 747.

Rosebrough, S.F., McAfee, J.G., Grossman, Z.D., et al. (1989b) *J. Immunol. Methods* 116, 123–129.

Rosebrough, S.F., and Maley, B.L. (1990a) *Antibody Immunoconjugates Radiopharmaceuticals* 3(2), 137–149.

Rosebrough, S.F., McAfee, J.G., Grossman, Z.D., et al. (1990b) *J. Nucl. Med.* 31, 1048–1054.

Rylatt, D.B., Blake, A.S., Cottis, L.E., et al. (1983) *Thromb. Res.* 31, 767–778.

Sakai, Y., Maeda, M., Takei, F., et al. (1988) *Thromb. Res.* 50, 469–479.

Scheefers-Borchel, U., Müller-Berghaus, G., Fuhge, P., et al. (1985) *Proc. Natl. Acad. Sci. USA* 82, 7091–7095.

Shainoff, J.R., Valenzuela, R., Gonda, S., et al. (1986) *BioTechniques* 4, 120–128.

Smith, E.B. (1977) *Am. J. Pathol.* 86, 665–674.

Smith, E.B., Alexander, K.M., and Massie, I.B. (1976) *Atherosclerosis* 23, 19–39.

Som, P., Oster, Z.H., Zamora, P.O., et al. (1986) *J. Nucl. Med.* 27, 1315–1320.

Spar, I.L., Goodland, R.L., and Schwartz, S.I. (1965) *Circulation Res.* 17, 322–329.

Takemura, T., Yoshioka, K., Akano, N., et al. (1987) *Kidney International* 32, 102–111.

Tymkewycz, P.M., Creighton, L.J., Gascoine, P.S., et al. (1989) *Thromb. Res.* 54, 411–421.

Valenzuela, R., Shainoff, J.R., Lucas, F.V., et al. (1988) in *Fibrinogen 3—Biochemistry, Biological Functions, Gene Regulation and Expression* (Mosesson, M.W, Amrani, D.L., Siebenlist, K.R., and DiOrio, J.P., ed.), pp. 313–316, Elsevier Science Publishers B.V., Amsterdam.

Vassalli, P., and McCluskey, R.T. (1964a) *Am. J. Pathol.* 45, 653–666.

Vassalli, P., and McCluskey, R.T. (1964b) *Ann. NY Acad. Sci.* 2, 1052–1062.

Wasser, M.N.J.M., Koppert, P.W., Arndt, J.W., et al. (1989) *Blood* 74, 708–714.

Weitz, J.I., Landman, S.L., Crowley, K.A., et al. (1986) *J. Clin. Invest.* 78, 155–162.

Whitaker, A.N., Elms, M.J., Masci, P.P., et al. (1984) *J. Clin. Pathol.* 37, 882–887.

Wilhelm, O., Hafter, R., Coppenrath, E., et al. (1988) *Cancer Res.* 48, 3507–3514.

Wojtukiewicz, M.Z., Zacharski, L.R., Memoli, V.A., et al. (1989a) *Thromb. Res.* 55, 279–284.

Wojtukiewicz, M.Z., Zacharski, L.R., Memoli, V.A., et al. (1989b) *Thromb. Haemostas.* 62, 1062–1066.

Wojtukiewicz, M.Z., Zacharski, L.R., Memoli, V.A., et al. (1990) *Cancer* 65, 481–485.

Woolf, N., and Carstairs, K.C. (1967) *Atherogenesis* 5, 373–386.

Zacharski, L.R., Schned, A.R., and Sorenson, G.D. (1983) *Cancer Res.* 43, 3963–3968.

In Vivo and In Vitro Regulation
of Blood Cell Production

Hematopoietic Stem Cell Processing and Storage

David Ciavarella

The most important clinical application of hematopoietic stem cell (HSC) biology is bone marrow transplantation (BMT), perhaps more precisely referred to as HSC transplantation, since bone marrow is no longer the sole source of transplantable HSC. Despite enormous gains in the science of transplantation, BMT very much retains its empirical side. Many key scientific issues are unresolved, especially the nature and regulation of the totipotential HSC, which still eludes positive identification, the role of non-hematopoietic elements (both donor and recipient) in engraftment, and the immunobiology of the graft versus host reaction, which remains as a critical problem in allogeneic BMT. Important allied issues include the relative scarcity of human leucocyte antigen (HLA)–compatible donors for patients in need of allogeneic transplants, the problem of tumor contamination of autologous grafts, and the need for more active, less toxic chemotherapeutic agents. The indications for BMT, especially autologous grafting, are also constantly being expanded and refined.

Thus, the processing and storage of HSC, which are the topics of this chapter, do not represent limiting factors in the growth of HSC transplantation. Nevertheless, improvements in this area have and will continue to contribute to advances in the clinical usefulness of this burgeoning therapy. In this chapter, I will discuss the collection, processing, and storage of HSC, but not the underlying concepts of hematopoiesis. Similarly, the use of colony-forming unit assays (CFUs) and hematopoietic growth factors will be discussed in the context of this chapter, but full details of the methods and applications of these assays and proteins are beyond the scope of this review and author. The reader is referred to other discussions of these topics (Fauser and Messner 1979; Groopman et al. 1989; Gabrilove 1989).

13.1 HEMATOPOIETIC STEM CELL COLLECTION AND PROCESSING

13.1.1 Dosage Considerations

The ability to restore hematopoiesis in a patient treated with myeloablative chemotherapy or radiotherapy remains the best and only proof that the infused bone marrow (BM) or peripheral-blood (PB) nucleated cell preparation contains early HSC. The question of the dose of nucleated cells needed to ensure engraftment is still unsettled, although much empirical evidence in animals and humans has been acquired. There are three widely accepted albeit not completely interchangeable ways of expressing cell dose. These are number of nucleated cells (NC), number of mononuclear cells (MNC), and number of CFUs, usually CFU–granulocyte/macrophage (GM) or CFU–granulocyte/erythrocyte/macrophage/megakaryocyte (GEMM) (also referred to as CFU-MIX), or CFU-culture (C) in older studies. Most often, the numbers are expressed per kilogram of patient body weight, although CFUs are often reported as number per 10^5 or 10^6 MNC or per milliliter of blood.

Nucleated cell counts encompass all nucleated marrow or peripheral blood cells, including those mature or near-mature cells such as bands, granulocytes, and orthochromic erythroblasts whose regulation is clearly very different than HSC. The minimally accepted dose for transplantation is 2×10^8 NC per kg. Others (Storb et al. 1977; Lasky et al. 1982) use 3×10^8 NC per kg as their standard dose. Some investigators distinguish between autologous and allogeneic grafts. For example, Gorin (1986) states that 0.5 $\times$ 10^8 NC per kg is the lower threshold for autologous BMT.

Mononuclear cell counts are perhaps a more precise measure, since it is widely held that HSC are mononuclear cells. The number of MNC in the marrow is approximately 20–30% of the number of NC (Williams 1983), depending upon the differential cell counts accepted a standard and how certain cells, e.g., promyelocytes, are classified. If one assumes a MNC/NC ratio of 0.20 and a minimum transplantable dose of 2×10^8 NC per kg, then a minimum MNC dose is 0.4×10^8 MNC per kg. This is about 2.5–4 $\times$ 10^9 MNC for the average adult patient. A safer dose, assuming 3×10^8 NC per kg and a MNC/NC ratio of 0.30, would be 0.9×10^8 MNC per kg, or about 5–8 $\times$ 10^9 MNC for the average patient.

A final measure of cell dose is that of CFUs infused. Data using burst-forming unit–erythroid (BFU-E), CFU-C, CFU-GM, and CFU-GEMM (Spitzer et al. 1980; Atkinson et al. 1985; Hartmann et al. 1985; Torres et al. 1985; Stiff et al. 1987a; Rowley et al. 1987; Visani et al. 1988) have been published, with most data available for CFU-GM. Theoretically, one could argue that CFU-GEMM, being a more primitive cell, would be a better choice.

Empirical data on the number of CFUs needed for engraftment have been generated, but, in order to provide comparisons to NC and MNC data, we can first consider the concentration of these culturable cells in the blood and marrow. The variability in the assay systems, e.g., the use of agar versus methylcellulose, and the variability among laboratories (To et al. 1986) have resulted in a wide range of published concentrations. McCarthy and Goldman (1984) summarized published results on the numbers of HSC in blood and marrow in mice, dogs, and humans. In mice, CFU–spleen (S) were found in the marrow in a concentration of 5–50 per 10^5 NC, with blood concentration of about 1/100 that of the marrow. In the dog, about 10 times as many BFU-E and CFU-GM are present in the marrow compared to blood. In humans, BFU-E in the marrow ranged from 50–200 colonies per 10^5 MNC, marrow CFU-GEMM ranged from 5–10 colonies per 10^5 MNC, and marrow CFU-GM ranged from 20–100 colonies per 10^5 MNC. Blood concentrations were considerably lower, approximating 1/50 to 1/400 as many BFU-E, 1/2 to 1/10 as many CFU-GEMM, and 1/10 to 1/1 as many CFU-GM as were present in marrow. Since then, several additional studies have been published (Table 13–1). Ganser et al. (1985) reported a mean CFU-GEMM concentration of 22 (range 4.5–64.5) per ml blood, with a mean CFU-GM concentration of 116 per ml blood. Torres et al. (1985) reported a mean CFU-GM of 72 per 10^5 marrow NC, or about 18 (14–22) per 10^5 marrow MNC. Lasky and Zanjani (1985) reported a mean of 4.2 (0–12)

TABLE 13–1   CFU Concentrations in Marrow and Blood

	Marrow	*Blood*
	Mean (Range)	*Mean (Range)*
CFU-GEMM		
Per 10^5 MNC	8.5 (4–15)	1.7 (0.49–4.5)
Per ml		24.5 (14–46)
CFU-GM		
Per 10^5 MNC	51.5 (18–126)	4.3 (2.5–9)
Per ml		95 (36–227)
BFU-E		
Per 10^5 MNC	15 (11–20)	8.1 (3.8–12.5)
Per ml		186 (68–333)

CFU-GEMM per 10^5 marrow MNC versus 0.65 (0–1.75) CFU-GEMM per 10^5 PB MNC, a ratio of 6.67. Rowley et al. (1987) found a similar marrow CFU-GEMM concentration of 4.6 per 10^5 MNC, while reporting a mean of 126 CFU-GM per 10^5 MNC. Visani et al. (1988) found 41 CFU-GM per 10^5 marrow MNC (range 4–121), while Socinski et al. (1988) reported mean marrow concentrations of 21 CFU-GM and 11 BFU-E per 10^5 MNC. These latter investigators also found mean (range) concentrations of 36 (0–165) CFU-GM and 68 (0–454) BFU-E per ml blood. These are PB/BM ratios of about 1:20 and 1:5, respectively. To et al. (1987) reported a mean blood concentration in normal humans of 14 (5–52) CFU-GEMM and 227 (100–339) CFU-GM per ml blood. These numbers are roughly equivalent to 4–14 CFU-GM and 0.25–2 CFU-GEMM per 10^5 PB MNC.

Recently, Broxmeyer et al. (1989) have renewed interest in cord/placental blood as a rich source of HSC. In a mean volume of 197 ml, Broxmeyer et al. (1989) collected 2.4×10^9 NC containing 34 CFU-GEMM, 158 CFU-GM, and 80 BFU-E per 10^5 NC, or assuming that 35% of cord NC are MNC, 96 CFU-GEMM, 452 CFU-GM, and 227 BFU-E per 10^5 umbilical cord MNC. The total CFU-GM collected were $38.0 \pm 35.3 \times 10^5$, corresponding to about 5×10^4 CFU-GM per kg in the average adult. Theoretically, this number of CFU-GM would be adequate for transplantation. However, these figures are the mean of only three samples, and the large standard deviation reveals the variability inherent in this sort of calculation. On a larger sample of 85 cord blood samples, only $2.34 \pm 0.49 \times 10^5$ CFU-GM in a mean of 56 ml were obtained. These samples were prepared for CFU assay by Ficoll-Hypaque separation, which led to a large (50–90%) loss of MNC and CFUs.

The different measures of cell dosage have been compared in Table 13–2, using as a starting point the empirical/clinical doses of 2 or 3×10^8 NC per kg, and the assumption of a marrow MNC proportion of 20–30%. It

TABLE 13–2 Stem Cell Transplantation: Dosage Considerations

	Nucleated Cells	Mononuclear Cells	CFU-GEMM	CFU-GM
Minimum				
Marrow	2×10^8/kg	0.4×10^8/kg[1]	1.6×10^3/kg[2]	7×10^3/kg[2]
Blood	–	$2–3 \times 10^8$/kg[3]	1.6×10^3/kg	7×10^3/kg
Conservative				
Marrow	3×10^8/kg	0.9×10^8/kg[1]	1.4×10^4/kg[2]	11×10^4/kg[2]
Blood	–	$5–7 \times 10^8$/kg[4]	1.4×10^4/kg	11×10^4/kg

[1] Minimum assumes marrow MNC/NC of 0.2; conservative assumes a ratio of 0.3
[2] Calculated from Table 13–1 using the lower (for minimum value) and upper (for conservative value) range of mean marrow concentration.
[3] Estimated from Table 13–1 using the mean marrow: PB CFU-GEMM ratio of 5 (8.5–1.7).
[4] Estimated using the marrow MNC minimum: conservative ratio of 0.4:0.9.

should be pointed out that most marrow MNC preparations in these studies were obtained using density gradient preparations. Typically, cells lighter than 1.077 g/ml are collected for culture. Thus, marrow MNC collected by density gradient centrifugation may be more or less enriched in certain types of cells than whole marrow MNC isolated using other techniques and counted by standard Wright-Giemsa stains.

As mentioned above, density gradient separation led to very different results compared with unfractioned cells of cord blood. The use of centrifugal counterflow elutriation, and flow cytometry, monoclonal antibodies, and cell sorting have provided insights into the size, appearance, and surface antigens of HSC. Using elutriation, Lasky and Zanjani (1985) reported that blood CFU-GEMM, CFU-GM, and BFU-E coseparated by size, calculated at about 8.7 μm by flow rate and 10.4 μm by eyepiece micrometer. Marrow CFU-GEMM and BFU-E had sizes of 8.6 μm by flow rate and 10.1 μm by micrometer. Although PB CFU-GM were equivalent in size to PB CFU-GEMM and BFU-E, marrow CFU-GM and, especially, CFU–erythroid (E), were larger than the more primitive HSC, with CFU-GM at a flow rate and micrometer size of 9.0 μm and 11.0 μm, respectively, and CFU-E at 9.6 μm and 12.2 μm, respectively. Percoll gradient sedimentation revealed mean (SE) specific gravities of 1.063 (0.001) for all PB HSC and for marrow cells, 1.0641 (0.0017) for BFU-E, 1.0666 (0.002) for CFU-GEMM and CFU-GM, and 1.0762 (0.0041) for CFU-E.

Civin et al. (1984) reported that a mouse monoclonal antibody, raised against the KG-1a human myeloid leukemia cell line, and called anti-My10, recognized a 115-kDa protein present on the surface of immature myeloid cells. Other epitopes of this antigen were recognized by antibody 12.8 (Andrews et al. 1986), BI.3C5 (Trindle et al. 1985), and ICH3 (Watt et al. 1987). The 115-kDa antigen has been categorized as a cluster of differentiation no. 34 (CD34). Other antigens, especially HLA-DR and CD33, are found on HSC as well. Using monoclonal antibodies to CD34, DR, and CD33, HSC have been characterized by morphologic and flow cytometric features. Lu et al. (1987) found that >98% of CFU-GEMM, BFU-E, and CFU-GM coexpressed CD34 and DR. The cloning efficiency of CD34^{+++} DR$^+$ cells was 47%, while that of CD34$^-$ DR$^-$ cells was 0.01%. Civin et al. (1987) found that about 1% of human BM cells were CD34$^+$, and showed that these cells were agranular and slightly larger than lymphocytes. Evidence from a number of laboratories (Andrews et al. 1989; Brandt et al. 1988; Siena et al. 1989) reveals that earlier HSC resemble small, agranular lymphocytes and carry the CD34$^+$ CD33$^-$ DR$^-$ phenotypes, while more committed HSC have a blastlike or large, granular lymphocytic appearance and carry CD34$^+$ CD33$^+$ and CD34$^+$ DR$^+$ phenotypes. Future studies may utilize the techniques of flow cytometry and cell sorting to define the optimum type and number of HSC for transplantation.

Much empirical data attempting to correlate time to hematopoietic recovery with HSC dose have been published. Hematopoietic recovery is usually defined as time to total peripheral blood leucocytes or neutrophils above 1000 or 500 per μl and time to self-sustaining platelet count above 20,000 or 50,000 per μl. In mice undergoing syngeneic transplantation, Jones et al. (1987) showed that the number of CFU-GM or CFU-S was directly related to the speed of recovery, with a minimum number of 1–2 CFU-S being necessary to ensure survival of lethally irradiated mice. These authors also showed a maximum value effect in which CFU-S above a certain number did not speed in recovery. Using other animal models, many earlier investigators reported both minimum and maximum value effects (Debelak-Fehir and Epstein 1975; Nothdurft et al. 1978; Calvo et al. 1976). In humans, this has been more difficult to show, although the preponderance of evidence reveals similar relationships. Several studies in humans have shown no relationship between numbers of CFU-C or CFU-GM infused and hematopoietic recovery (Hartmann et al. 1985; Torres et al. 1985; Atkinson et al. 1985; Visani et al. 1988). In these studies, patients received high numbers of NC and CFU-GM, with nearly all patients receiving more than 3×10^4 CFU-GM per kg, and mean or median values approached or exceeded 10×10^4 CFU-GM per kg. In the study by Hartmann et al. (1985), one patient received only 0.5×10^4 CFU-GM per kg, and this patient required 2.5 and greater than 4 times the median time to recover to 500 neutrophils per μl and 50,000 platelets per μl. Spitzer et al. (1980) correlated ($r = -0.8$) the number of CFU-C infused and hematopoietic recovery. In the patients who received 0.47 to 6.3×10^4 CFU-C per kg, median time to recovery of 500 neutrophils per μl was 22 days, while patients receiving 0 to 0.08×10^4 CFU-C per kg required a median of 58 days. Interestingly, in an early phase I trial utilizing 4-hydroperoxycyclophosphamide (4-HC) as a tumor-purging agent, Kaiser et al. (1985) documented sustained hematopoietic recovery in patients who received autografts with no detectable CFU-C. A trend toward increasing recovery times with increasing concentration of 4-HC was noted, however. Subsequently, the same group reported that residual (post-purging) CFU-GM content was useful in predicting hematologic reconstitution (Rowley et al. 1987). In this study, 4-HC purging led to failure to culture any CFU-GEMM or BFU-E in 67 and 56% of grafts, respectively. Although CFU-GM were grown from all purged grafts, the post-treatment recovery averaged only 8% (an average of only 7.6×10^3 per kg, with the wide range of 5×10^1 to 5.7×10^4 per kg). The log CFU-GM per kg was linearly correlated with time to total blood leucocytes >1000 per μl, granulocytes ≥ 500 per μl, reticulocytes $\geq 2\%$, and time to last platelet transfusion. The r values ranged from -0.60 to -0.80. In a case report, Gluckman et al. (1989) reported that successful transplantation of a 24-month-old infant with Fanconi's anemia using cord blood from his HLA-identical sister. A total of only 0.39×10^5 CFU-GEMM, 1.56×10^5 CFU-GM (methylcellulose), and 3.6×10^5 BFU-E were given. Although the weight of the recipient was not given, this approximates 1 to 2×10^4 CFU-GM per kg. Not surprisingly, engraftment was delayed, with 36 days to granulocytes above 500 per μl

and 50 days to reach a self-sustaining platelet count above 50,000 per μl. However, at 9 months of age the patient was reported well with signs of complete engraftment.

Thus, it would appear that in humans, as in other species, the number of CFUs infused is related to the likelihood of engraftment and to the speed of full hematopoietic reconstitution. Although a simple assay for the toti-potential HSC is not available, CFU assays are useful quality control meas-ures, and can be utilized in two ways: (1) to determine the effects of different processing and storage methods; and (2) to predict (within broad limits) successful engraftment. In particular, the theoretically derived "minimum" and "conservative" doses of CFU-GM (Table 13–2) match reasonably well with empirical results, which suggest a minimum of 1×10^4 CFU-GM per kg and perhaps 5–10 times that many to achieve the shortest period until hematopoietic recovery. It must again be pointed out that some investigators claim far fewer CFU-GM as a minimum dose, e.g., Gorin (1986), who has established a lower threshold for autologous BMT of only 10^3 CFU-GM per kg. The empiric evidence indicates that the risk of delayed engraftment increases sharply at these levels, however. Empirically, very little, if any, increase in recovery rate is seen at doses above 10×10^4 CFU-GM.

All the above referenced studies were performed using bone marrow-derived CFUs. A more recent aspect of the controversy surrounding time to hematopoietic recovery and dose of cells or CFUs infused is related to the use of peripheral blood HSC.

Theoretically, marrow would be a superior source of transplantable HSC, in that the entire spectrum of HSC are present, from totipotential to more committed progenitors, and the addition of marrow supporting cells and stroma might enhance the function of the transplanted HSC. The po-tential advantages of PB HSC are unrelated to the nature of HSC, at least with the current state of knowledge. Instead, these advantages include the possibility that (1) PB is less contaminated with tumor than marrow, of importance in autografting; (2) PB HSC harvesting is a safer procedure than bone marrow harvesting; and (3) PB HSC harvesting offers autografting to patients whose marrow is too damaged by prior therapy to be an effective source of HSC.

The empiric use of PB HSC has produced data similar to that for mar-row, i.e., time to hematopoietic recovery correlates within wide limits to number of PB cells and/or CFUs infused. Reiffers et al. (1987) reviewed 46 cases of PB HSC autografting from six centers in France. Hematopoietic recovery could not be related to patient characteristics, underlying diseases or prior therapy, or number of nucleated cells given. However, CFU-GMs infused had some prognostic value with respect to time to granulocyte re-covery. Patients who received fewer than 12×10^4 CFU-GM per kg reached 500 granulocytes per μl in a median of 19 days versus 12 days for patients receiving more than 12×10^4 CFU-GM per kg. This was significant at $p < 0.025$. Platelet recovery to 50,000 per μl within 30 days separated patients in two groups. Those who recovered within 30 days received a median of 35 × 10⁴ CFU-GM per kg, versus only 2.9 × 10⁴ CFU-GM per kg for patients recovering after 30 days. The wide range, however, prevented these data from reaching statistical significance. From these combined 46 cases, the authors suggested that 15 × 10⁴ CFU-GM per kg are sufficient to ensure hematopoietic recovery after autografting with PB HSC.

Stiff et al. (1987a) compared PB HSC to BM in 12 patients who underwent intensive chemotherapy and autografting for small cell carcinoma of the lung. PB HSC were collected by apheresis during the recovery phase after chemotherapy. An average of 11 × 10⁴ CFU-C per kg from PB versus 7 × 10⁴ CFU-C per kg from BM were given. Neutrophil recovery was equivalent in the two groups, but platelet recovery was significantly delayed in the group receiving PB HSC. Kessinger et al. (1988) found an equivalent speed of hematologic recovery in their patients receiving either PB HSC or autologous BM. Median times to recovery were 22 days for neutrophils >500 per µl and 23 days for a platelet count >20,000 per µl. Relatively high median doses of PB MNC (8.4 × 10⁸ per kg) and CFU-GM (8 × 10⁴ per kg) were given. Late graft failures were not seen. Comparable data have been reported by Williams et al. (1988), who infused a median of 3.8 × 10¹⁰ PB cells (about 4–5 × 10⁸ MNC per kg), and obtained a median time to neutrophil recovery to 500 per µl of 14.5 days and platelet recovery to 50,000 per µl of 49 days. Hurd et al. (1988) reported that autografting with PB was superior to BM in their patients with poor prognosis Hodgkin's disease. Time to recovery of total blood leucocytes to 1000 per µl was 18 days with PB and 27 days with BM; time to independence from platelet support was equivalent (27 versus 25 days). However, patients received a median of 6.8 × 10⁸ PB cells per kg versus only 1.1 × 10⁸ BM cells per kg.

Ventura et al. (1988) and Korbling and Martin (1988) discerned trends towards increasing times to recovery with lower doses of PB CFUs. These latter investigators transplanted four patients using PB HSC collected in the recovery phase after chemotherapy. Two patients received high doses of CFU-GM (6 and 21 × 10⁴ per kg), and recovered rapidly, reaching 500 neutrophils per µl on Days 12 and 10, and self-sustaining platelet counts above 50,000 per µl on Day 10. Two patients who received 1 and 1.6 × 10⁴ CFU-GM per kg had comparable neutrophil recovery (16 and 14 days), but markedly (>100 days) delayed platelet recovery. In this report, the authors summarized 29 patients from eight institutions who received autografts using PB HSC. An increased risk of delayed recovery, especially platelet recovery, was noted when fewer than 2.5 × 10⁴ per kg CFU-GM were infused. Juttner et al. (1988, 1989) are emphatic on this point, reporting that 30–50 × 10⁴ CFU-GM per kg are required to ensure rapid hematologic reconstitution and avoid late graft failure. In their hands, using PB HSC collected during early recovery from induction chemotherapy in patients with acute nonlymphocytic leukemia, hematologic recovery after autografting with PB HSC was found superior to that obtained using autologous BM, and to historical controls receiving allogeneic BM. As for BM HSC, a good match exists between the theoretical and empirical numbers of PB HSC (measured either as MNC per kilogram or CFUs per kilogram) needed to ensure prompt hematologic recovery after autografting.

Very few studies have been published using both BM and PB. Gianni et al. (1989) are proponents of combined sources of HSC, reporting both faster hematopoietic recovery and less need for transfusion support in patients who received autografts with both BM and PB. Much confirmation of this approach is necessary, however, since measures of successful engraftment, such as time to recovery of peripheral blood cell counts, and presence or absence of late graft failure, vary widely from study to study. Differing patient populations and underlying disease states; type and intensity of prior chemotherapy; and complicating factors related to transplantation, especially sepsis and the development of autoimmune cytopenias, are all important variables in the speed and measurement of hematologic recovery, and control and standardization of these factors are often difficult or impossible.

Another important, unanswered question is whether PB HSC are equivalent to BM HSC. In mice, qualitative differences between BM and PB HSC have been discerned, including differing sensitivities to X-irradiation and in the percentage of cells in S phase. These differences narrow when chemical stimulants to mobilization of HSC into the blood are applied (Bell et al. 1986). In humans, the empiric evidence would indicate that few, if any, differences can be discerned between the ability of PB or BM HSC to speed leucocyte recovery. However, several reports of delayed platelet recovery after PB transplantation raise the question of the number and type of megakaryocytic precursors in PB. As pointed out by Stiff et al. (1987a), delayed platelet recovery may be the result of fewer CFU-Mega in PB versus BM. Ganser et al. (1985), using a monoclonal antibody to megakaryocyte antigen GP IIIa to stain CFU-GEMM from PB and BM, found that 87% of BM CFU-GEMM contained megakaryocytic cells versus only 14% for PB CFU-GEMM. Using an antibody to granulocytes as a control, granulocytic cells were identified in an equivalent 95% of CFU-GEMM from either PB or BM. Such qualitative differences may explain the empiric finding that CFU-GEMM are not better than CFU-GM in predicting hematologic recovery (To et al. 1987; Rowley et al. 1987). In an attempt to discern qualitative differences, and to help elucidate the relative roles of PB versus BM HSC in combined transplantation, Siena et al. (1989) studied CD34+ cells from PB and BM using colony-forming assays and surface antigen marker studies. In vitro growth using cultures plated at different cell concentrations was equivalent, with similar growth patterns and cloning efficiencies. Using flow cytometry, PB and BM CD34+ cells were found to possess analogous antigenic patterns and light-scattering properties.

An added complication in the above studies is the timing and nature of PB HSC collection. This topic will be discussed in more detail in a subsequent section; however, it is pertinent here to point out that many stimuli have been reported to increase PB HSC concentration. Specifically, the BM suppression and PB cytopenias induced by high-dose chemotherapy are consistently followed by a rebound in the number of CFU-C, CFU-GM, CFU-GEMM, and BFU-E in the PB (Richman et al. 1976; Abrams et al. 1981; Stiff et al. 1983a).

CFU concentrations are increased 2- to 10-fold or more after chemotherapy, and this effect can be induced as well by adminstration of the hematopoietic growth factor granulocyte/macrophage–colony-stimulating factor (GM-CSF) (Socinski et al. 1988; Siena et al. 1989). Many investigators carefully time PB HSC collections to coincide with hematologic recovery after chemotherapy in order to maximize CFU recovery, while others collect after recovery at a time of a hematologic "steady state." Potentially, qualitative as well as quantitative differences in the mix of HSC in the PB could be induced by these stimuli, especially after administration of hematopoietic growth factors (Migliaccio et al. 1988a and 1988b). Very little data are available to make a scientific judgement about these potential differences. In the study by Ganser et al. (1985) of the megakaryocytic content of PB versus BM CFU-GEMM, normal individuals were studied without chemical stimulation. Siena et al. (1989) studied PB CD34$^+$ cells after high-dose cyclophosphamide with or without supplemental GM-CSF, and as discussed above, found no differences in in vitro growth or immunophenotypes.

Clearly, the very important questions raised by these studies present many opportunities for future research. The greatest challenge lies in improving time to platelet recovery, since GM-CSF and other growth factors, and our greater ability to quantify and characterize neutrophil precursors in PB or BM, will most likely lead to increasingly rapid leucocyte recovery after transplantation.

13.1.2 Bone Marrow Collection and Processing

The large scale *collection* of bone marrow cells and stroma by multiple aspirations has changed very little over the past 20 years. Early investigators determined optimum collection sites, usually the posterior iliac crests, and if necessary, the anterior iliac crests and sternum, and developed intraoperative techniques to break up larger particles while preserving cellular integrity. Thomas and Storb (1970) have described this process in detail. A volume of 600–2500 ml or more can be obtained, representing 10–20% of the blood volume and about 2–5% of the estimated 10^{10} nucleated bone marrow cells per kg (Lowder and Herzig 1990). The typical BM harvest is about 700–1000 ml, with larger volumes being obtained when two transplants (usually autologous) or extensive in vitro manipulation (e.g., tumor purging) are anticipated. Preservative-free heparin, at 5–20 U per ml, is added as an anticoagulant, and some centers use the erythrocyte anticoagulant/preservative solutions acid-citrate-dextrose (ACD) or citratephosphate-dextrose (CPD) in an attempt to prevent cell clumping when extensive processing is performed (Braine et al. 1982; Gilmore et al. 1982). After the addition of 100 to 300 ml of tissue culture medium (e.g., TC199 or RPMI) per 1000 ml of marrow, the marrow is filtered through fine stainless steel mesh to eliminate fat and bone spicules and to break up large particles. It is then placed into plastic blood bags and removed to the bedside for infusion or to the laboratory for processing. The typical BM harvest contains 200–250 ml of erythrocytes, and autologous erythrocytes obtained during processing are routinely washed and returned to the marrow donor.

Buckner et al. (1984) have detailed the effects of BM donation on the donor. Reporting on 1270 harvests in 1160 donors, with donors as young as 6½-months of age, they concluded that BM harvest was generally safe provided stringent criteria were applied. Both general and spinal anesthesia were given, and 81 and 99% of donors left the hospital within 2 and 3 days of the procedure, respectively. No fatalities were ascribed to marrow donation in this series, although a single fatality was reported as part of a larger study of 3290 marrow donations (Bortin and Buckner 1983). Lifethreatening complications occurred in 0.27% of harvests.

Bone marrow *processing* has grown increasingly complex over time. It remains the preference of many centers to manipulate BM as little as possible to avoid HSC loss or damage, and with allogeneic transplants, the harvested BM is often infused intravenously into the recipient within hours of collection. Processing of allogeneic grafts is generally undertaken either to remove ABO/Rh blood group incompatible erythrocytes or plasma, or to remove or kill contaminating lymphocytes to attempt an amelioration of the graft versus host reaction. ABO/Rh blood group incompatibility is clearly not a contraindication to BMT, as many centers have shown (Buckner et al. 1978; Braine et al. 1982; Blacklock et al. 1982; Dinsmore et al. 1983; Warkentin et al. 1985). However, serious acute and/or delayed hemolytic transfusion reactions have been reported subsequent to incompatible BM infusion, due to extensive contamination of the graft with peripheral blood (Buckner et al. 1978; Braine et al. 1982; Warkentin et al. 1983). Initial attempts to avoid hemolysis concentrated on lowering recipient anti-A or anti-B titers by intensive plasma exchange (10 liters per day for 2 or 3 days), followed by deliberate infusion of incompatible erythrocytes to neutralize blood group antibodies prior to marrow infusion (Buckner et al. 1978; Berkman et al. 1978; Hershko et al. 1980). Subsequently, BM processing to remove incompatible erythrocytes or plasma was utilized by several centers, with acceptable clinical results (Dinsmore et al. 1983; Warkentin et al. 1985; Braine et al. 1982; Blacklock et al. 1982). The simplest technique involves the use of 6% hydroxyethyl starch (HES). Using a BM/HES ratio of 8:1 and a sedimentation period of 90 to 180 min, Reich and colleagues reduced erythrocyte contamination to a mean of 23 ml while retaining about 77%

of nucleated cells (Reich et al. 1980; Dinsmore et al. 1983). Warkentin et al. (1985) used 6% HES at a 7:1 ratio in a double sedimentation procedure that reduced mean marrow volume by 75% and mean erythrocyte contamination to a very low 2 ml, while retaining 86% NC and 98% CFU-C. We obtained equally good results in a comparative study of these two methods (Ciavarella and Ahmed 1987).

Presently, most centers utilize one of several available automated cell processors and separators for this purpose (Ciavarella 1989). These cell processors have the advantage of rapid and reproducible erythrocyte and plasma depletion, and acceptable results (final erythrocyte volume <20 ml) have been reported using the COBE 2991 (COBE Laboratories, Denver, CO) (Gilmore et al. 1982; 1983; Jin et al. 1987; Law et al. 1988), the COBE 2997 (Foradji et al. 1988), the COBE Spectra (Lyding et al. 1990), the Haemonetics 30 (Haemonetics Corp., Braintree, MA) (Braine et al. 1982), the Haemonetics V50 (Law et al. 1988; Raijmakers et al. 1986), and the Fenwal CS-3000 (Baxter Healthcase Corp., Fenwal Division, Deerfield, IL) (Carter et al. 1989). For some applications, such as T lymphocyte depletion or tumor cell depletion, automated density gradient separation using Ficoll-Hyapaque, Ficoll-Metrizoate, or Percoll can be performed.

Using the COBE 2991 without Ficoll, Gilmore et al. (1983) reduced marrow volume about 80%, with final volumes in the range of 50–200 ml. Nucleated cell recovery ranged from 70–90%. MNC recoveries were not reported. Several investigators have reported processing results using the COBE 2991 with density gradient separation. Gilmore et al. (1982) recovered an average of 25% of BM cells, containing 109% of the original CFU-GM, in a volume of only 65 ml. Erythrocytes were reduced by 99%. These excellent results, however, have not been reported by subsequent investigators. Jin et al. (1987) reduced erythrocyte contamination to a median of 2.6 ml in 30 marrow processings, but reported median losses of NC, MNC, and CFU-GM of 63, 53, and 52%, respectively. English et al. (1989) reported recovery of 69% of marrow MNC (8.9×10^9 MNC), while reducing erythrocyte contamination to 6 ml. These investigators commented on their inability to match the results of Gilmore et al. using their original protocol, which, according to English et al., led to MNC recovery of only 10%.

The Haemonetics 30 has been utilized with and without density gradient separation procedures. Braine et al. (1982) eliminated 94% of erythrocytes, while retaining 75% (3×10^9) MNC and 57% CFU-GM without density gradient separation. Humblet et al. (1988) first recovered a buffy coat containing 19% of the original marrow volume, 10% of the erythrocytes, 54% of the NC, and 71% of the CFU-GM. This buffy coat was further processed by introducing a Percoll gradient (density 1.079 g per ml). This resulted in an average final product containing only 1% of the original erythrocytes and 10% of the original volume (101 ml). Only 16% of NC were recovered, but this included 100% of buffy coat CFU-GM. Raijmakers et al. (1986) utilized the Haemonetics V50 with Percoll gradients to obtain a product containing

2% of original erythrocytes, 17% of original NC, but 67% of original CFU-GM. Using a two-stage procedure involving an initial automated buffy coat concentrate followed by Percoll separation, similar results were obtained, although CFU-GM loss was greater.

Using the COBE 2997 without Ficoll or Percoll, Foradji et al. (1988) recovered 23% of initial marrow NC, which included 80% of initial MNC, in 15% of the original volume. CFU-GM recovery was reported at 83%. Erythrocytes were reduced by 98% and platelets by 60%. These investigators subjected this processed marrow to a T-lymphocyte depletion step without Ficoll separation. Similar results have been reported by Ritchey et al. (1983) and Sniecinski et al. (1985). Lyding et al. (1990) obtained 75% MNC recovery in 20% of the original volume using a COBE Spectra. CFU-GM and BFU-E recovery were reported at >100% of the original marrow. Less data is available using the Fenwal CS-3000 (Lasky et al. 1986a; Carter et al. 1989). Carter reported a Ficoll-Hyapaque method with average recoveries of 74% MNC and 1% erythrocytes in a final volume of 205 ml. CFU data were not reported.

In summary, several automated or semi-automated systems have reproducibly recovered 70% or more of BM MNC, while reducing volume to 200 ml or less. Volume reduction and erythrocyte depletion can be further reduced using density gradient separation methods, but these methods are relatively time consuming and often result in further loss of MNC and/or CFU-GM. The choice of method will depend upon the needs and preferences of the individual center, especially their familiarity with certain equipment, and whether there is a need to obtain the smallest volume and purest MNC preparation possible. The trade-off that must be faced is relative product purity versus MNC or CFU recovery. At present, acceptable but not ideal results can be obtaining using any of these methods.

13.1.3 Peripheral Blood Stem Cell Collection and Processing

The collection and processing of PB MNC in anticipation of transplantation has been growing dramatically. The number of circulating HSC has been discussed previously, along with current controversies in the use of PB MNC, especially qualitative differences between PB and BM MNC, and the utility or necessity of stimulating donors to increase the number and/or type of HSC in circulation. Zander and Cockerill (1987) have reviewed what little is known about the regulation of circulating HSC. In animals, especially mice and dogs, rapidly interchangeable pools of HSC exist in extravascular sites (presumably the marrow) and in the circulating blood. A mean half-life of autologous PB CFU-GM in the blood of dogs was found to be on the order of 10 min, with a calculated turnover of 9×10^5 CFU-GM per kg per day. Even continuous 24-hr leucopheresis was unable to exhaust circulating CFU-GM. The regulation of HSC between the marrow and the blood is not well understood, although a "marrow-blood" barrier has been proposed (Tavassoli 1979). A number of factors have been identified that can alter the number of PB HSC, and several of these have importance in PB HSC harvesting for study or transplantation (McCarthy and Goldman 1984). In animals, long chain polymers such as polyvinyl sulfonic acid and high molecular weight dextran can increase CFU-GM two- to fourfold. In normal humans, exercise, corticosteroids, endotoxin, and dextran cause similar increases.

Most importantly, chemotherapeutic drugs that suppress blood cell production in the marrow are followed in animals and humans by a 10- to 20-fold expansion in circulating HSC during the recovery period (Richman et al. 1976; Abrams et al. 1981; Stiff et al. 1983a; To et al. 1987; Socinski et al. 1988). Unfortunately, the precise timing of the maximal rise in PB HSC after chemotherapy cannot easily be discerned in a given patient, and since CFU assays take a minimum of 7 days, other surrogate measures indicating maximal blood concentration of HSC have been sought. To et al. (1987) have reported on the utility of following the platelet count post-chemotherapy.

In this study, six patients with acute non-lymphocytic leukemia were given standard induction chemotherapy, and serial CFU-GM and CFU-GEMM were obtained from PB on Days 12 through 21 after the end of chemotherapy. CFU-GM and CFU-GEMM rose in unison, peaking at 20 and 12 times the level seen in normal subjects. The best predictor of a marked increase in blood CFU-GM or CFU-GEMM concentration was the period of time beginning with recovery of platelet count to 100,000 per μl (Figure 13–1). Other investigators have not found this correlation, among them Stiff et al. (1983a), who could not relate the CFU-C peak after chemotherapy to the timing of platelet or granulocyte recovery. This is not surprising, in that platelet recovery is highly variable in different patients, especially when differing underlying diseases and previous chemotherapies are compared. Stiff et al. (1983a) demonstrated instead a correlation between PB monocytosis and peak CFU-C levels in their patients with small-cell carcinoma of the lung. Peak CFU-C levels were most reliably seen when monocytes rose to 30% or more of blood leucocytes (Figure 13–2). Other investigators simply begin apheresis upon leucocyte recovery (usually 10–25 days), which will vary depending upon type of therapy and extent of pretreatment. A potentially improved method of determining the optimum time to begin PB HSC harvesting is to follow the appearance of CD34+ cells into circulation. Siena et al. (1989) studied the appearance of CD34+ cells in the blood of four patients treated with high-dose cyclophosphamide with or without recombinant GM-CSF. CD34+ cells were undetectable for 10 days after treatment, then rose to levels as high as 140 per μl (without GM-CSF) and 350 per μl (with GM-CSF) between Days 15 and 20 (without GM-CSF) and Days 11 and 22 (with GM-CSF). The use of flow cytometry to guide PB HSC collections appears very promising, although a better un-

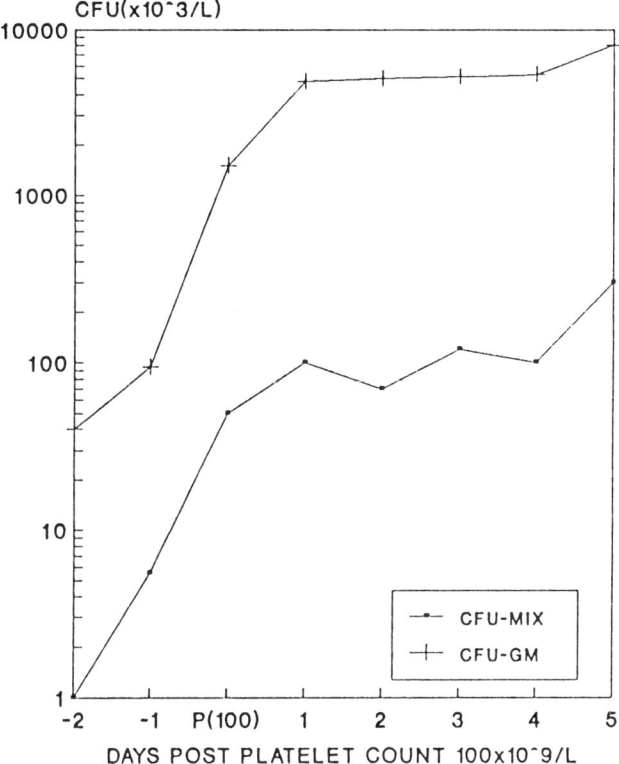

FIGURE 13–1 Correlation of peripheral blood CFU levels with postinduction chemotherapy in patients with acute nonlymphocytic leukemia (ANLL). Adapted with permission from To et al. (1987).

derstanding of the number and mix of CD34$^+$ cells needed for transplantation is necessary.

Other investigators have also documented the effect of hematopoietic growth factors, especially GM-CSF, on the number of PB HSC. Socinski et al. (1988) studied changes in PB CFU-GM and BFU-E in patients with sarcomas following a course of GM-CSF (phase I), chemotherapy plus GM-CSF (phase II), or chemotherapy alone (phase III). In phase I, total leucocytes rose a mean of 4.3 times, while MNC rose 9.2 times. Bone marrow CFU-GM and BFU-E did not change, while 9 of 12 patients had significant increases in PB CFU-GM (median 18-fold increase). PB BFU-E also increased in phase I patients (median eightfold increase). In phase II patients PB CFU-GM and BFU-E were undetectable immediately after chemotherapy, and they rose to peak values by Day 16 after chemotherapy. These peak values were greater in phase II than in phase I, i.e., GM-CSF and chemotherapy were synergistic. In addition, patients who received both

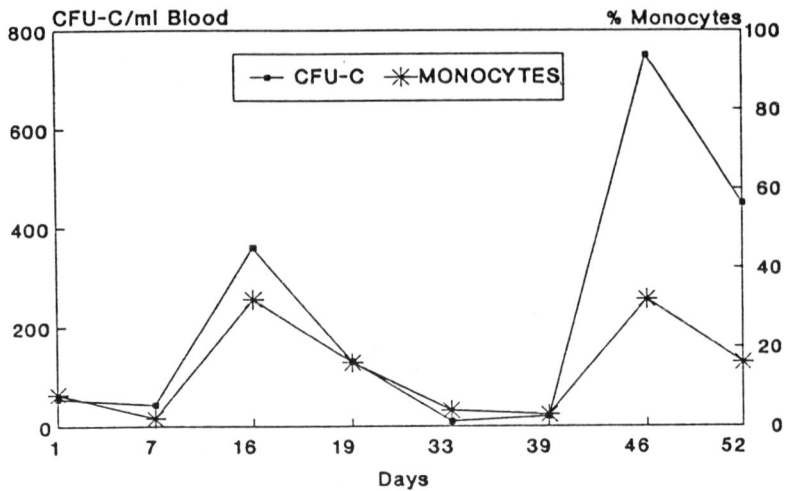

FIGURE 13–2 Correlation of peripheral blood CFU levels with peripheral blood monocytosis during recovery after chemotherapy in patients with small-cell lung carcinoma. Adapted with permission from Stiff et al. (1983a), *Transfusion* 23, 500-503.

stimulants (phase II) achieved peak CFU-GM and BFU-E levels earlier than patients who received chemotherapy alone (phase III). These results are similar to those of Siena et al. (1989), as discussed above.

Thus, it is clear that the use of these techniques will permit the collection of a given number of CFUs in fewer apheresis procedures than in an unstimulated patient. All available cell processors can adequately harvest PB MNC for transplantation. The most important predictors of MNC yield using any machine are the donor's preprocedure MNC count and the volume of blood processed (Hester et al. 1983; Stevenson et al. 1986). Using the Haemonetics V50, a discontinuous flow instrument, Kessinger and colleagues compared two procedures: a standard lymphopheresis protocol and a lymphosurge protocol, which utilizes counterflow elutriation at the time of MNC collection (Schouten et al. 1990). The standard procedure allowed the collection of more MNC (mean 0.76×10^8 per kg) than the surge procedure (mean 0.61×10^8 per kg), albeit at the expense of greater erythrocyte contamination (66 ml versus 33 ml). However, CFU-GM and BFU-E collections per procedure were not different. These investigators subsequently combined these two techniques to produce a modified surge protocol; a standard lymphocytapheresis is followed by reintroduction of the total buffy coat into the Latham bowl and a single surge is performed. In our hands this "Nebraska" surge resulted in a purer product in a smaller volume (97% MNC in 262 ml postsurge versus 59% MNC in 437 ml presurge). However, we averaged only 3.1×10^9 MNC per procedure (Wuest et al. 1989). The highest yields using this machine have been obtained (Kessinger et al. 1987; Burgstaler and Pineda 1990) when collection is extended for 4 hr. There is controversy over whether prolonged collection times diminish the efficiency of PB CFU collection. Keller et al. (1987), using a Haemonetics V50, found that the number of CFU-GEMM collected per pass diminished from the first to the seventh pass. Others, however, have not found a fall off in the efficiency of collection over time (Lasky et al. 1987; Stiff et al. 1983a).

In general, higher yields per unit time are obtained with continuous flow cell processors. Using the Fenwal CS-3000, Lasky et al. (1982) collected 3.5×10^9 MNC containing 300–1180 CFU-GEMM per ml of apheresis product. Collection times ranged from 1–2 hr. These investigators have adapted this method to the collection of PB HSC from children as well (Lasky et al. 1990). Sniecinski et al. (1989) processed 9.5 liters of PB using the CS-3000 and the standard granulocyte separation chamber. They obtained MNC counts of $9.85 \pm 4.6 \times 10^9$ cells per procedure, with $97 \pm 4\%$ MNC purity, in a volume of about 200 ml. We have obtained similar yields, albeit at the expense of 15 to 20% granulocyte contamination. Using the COBE Spectra, a continuous flow device, Padley et al. (1990) processed 10 liters of blood in a mean of 145 min. They obtained 5.7×10^9 MNC per procedure, with a 28% granulocyte contamination. In 23 procedures using the COBE Spectra, we obtained a mean of 7.1×10^9 MNC, with 18% granulocyte and 3.3 ml erythrocyte contamination in a final average volume of 162 ml (D. Ciavarella, unpublished results).

Thus, the typical PB MNC collection procedure harvests 0.4–1.0×10^8 MNC per kg. This corresponds to approximately 0.1–0.9×10^4 CFU-GM per kg in an adult patient. In order to reach 5×10^8 MNC per kg, or 1–5×10^4 CFU-GM per kg, approximately 5–12 procedures may be needed. In this context, it can be seen that if collection procedures can be performed after chemotherapy and/or GM-CSF, sufficient CFU-GM or CFU-GEMM may be collected in two to five procedures. However, as previously discussed, much remains to be learned about the relative utility of PB HSC collected during the recovery phase after chemotherapy or after GM-CSF. In addition, the optimum use of available cell processors remains to be determined. All these machines use a combination of centrifugation and optical sensing devices (which measure size) to trigger collection of the cell population or interest. These machines have an inherent trade-off, i.e., purity versus yield. It is as yet unclear whether MNC fractions containing cells of a wider density range, e.g., from granulocytes to small lymphocytes, are preferable to preparation containing lighter cells only. The data available on the size and density of CFUs and CD34+ cells indicate that these cells are represented both by small, agranular lymphocytes and by larger, blastlike MNC. The complete range of their size and density are not known, however, nor is it understood which cells are optimum in the transplant setting. Thus, clearly, unless similar machines using similar collection parameters are used,

MNC collections from different centers are not likely to be uniform, and may in fact represent considerably different cell populations. When these variables are added to other variables, such as potential qualitative differences between PB HSC harvested before or after GM-CSF or chemotherapy, clear and simple statements as to the best manner in which to collect PB HSC are not possible. Although much progress in understanding these issues has been made, PB HSC transplantation remains very much an empiric science at present.

13.1.4 Cell-Specific Manipulation

In addition to removal of erythrocytes and buffy coat concentration, it may be necessary or desirable to further manipulate BM or PB in vitro. At present, cell-specific manipulation techniques are utilized for three purposes: (1) to remove T lymphocytes, or T-lymphocyte subsets, from allogeneic grafts to influence the graft-versus-host reaction; (2) to remove residual tumor cells prior to autografting (tumor "purging"); and (3) to obtain purified HSC (CD34$^+$ enriched populations). This latter technique is often referred to as "positive purging," to distinguish it from tumor (negative) purging. It should be pointed out that BM or PB manipulation for any of these three purposes is still experimental. The theoretical appeal of removing T lymphocytes and residual tumor (either by negative or positive purging) is very strong, and many different methods to achieve the stated goals have been devised and tried. The true clinical role of cell-specific manipulation, however, remains very unclear, especially regarding tumor cell purging.

13.1.5 T-Lymphocyte Depletion

The T lymphocyte has been clearly identified as the predominant cause of the graft-versus-host reaction in animals and humans. Graft-versus-host disease (GVHD) is a major complication of allogeneic transplantation, occurring in some 30–70% of patients receiving HLA-identical grafts, and has been estimated to contribute to the death of 10% or more of these patients (O'Reilly 1983; Herve et al. 1987). The clinical features, which involve the lymphoid, hematopoietic, gastrointestinal, and dermatological systems, have been well described (Hansen et al. 1981; O'Reilly 1983). The risk of GVHD rises as the degree of HLA mismatch between donor and recipient increases (Beatty et al. 1985). The now well-publicized difficulties in obtaining HLA-identical sibling matches for patients who would benefit from allogeneic BM transplantation have led to differing strategies to increase the likelihood of finding suitable donors for these patients.

These strategies include the use of HLA-nonidentical related donors and HLA-identical unrelated donors. T-lymphocyte depletion would be especially important in these latter situations. Available techniques include monoclonal antibodies (MoAb) and complement, MoAb bound to cellular toxins such as ricin, soybean lectin agglutinin, and centrifugal counterflow elutriation.

These techniques are capable of 2–3 log reductions in BM T-cells. However, not unexpectedly, technical problems are considerable. The efficacy of MoAb clearly depends upon the specificity chosen, and most often, several MoAb specificities are used. Target membrane antigen loss or modulation are documented problems, as is the source, purity, and toxicity of the complement source. Human complement is not efficacious in this ex vivo setting, and the best results have been reported with rabbit serum (Bell 1989). Usually, Ficoll-Hyapaque separated BM cells are incubated with a "cocktail" of MoAb, often anti-CD2, -CD3, -CD5, and/or -CD7, followed by an additional incubation step with rabbit serum. Using these techniques, Herve et al. (1987) achieved 94 ± 4% T-cell depletion, infusing a mean (± SD) of $0.99 \pm 0.65 \times 10^6$ T cells per kg into patients. Mitsuyasu et al. (1986) infused fewer than 1×10^5 T cells per kg after MoAb and complement. Gilmore et al. (1986), using MoAb MBG6 and RFT8 and rabbit serum, removed 95% to > 99% of T cells, with a final patient dose of $1–5 \times 10^5$ T cells per kg. Martin et al. (1985), using eight MoAb and complement, also removed > 99% of T cells. Using soybean agglutinin and E rosetting, Kernan et al. (1986) demonstrated a 3–4 log depletion of T cells, and could reduce the inoculum of donor T cells to $< 10^5$ T cells per kg in most cases.

Counterflow centrifugal elutriation has several potential advantages over MoAb for T-cell depletion (de Witte et al. 1986; Noga et al. 1987). These include obviation of the need for Ficoll-Hyapaque separation, which is time consuming and can result in greater HSC loss. In addition, it has the potential to standardize the number of T cells given. De Witte et al. (1986) achieved a mean T-cell depletion of 98%, infusing a mean of 0.63×10^6 T cells per kg. CFU-GM levels were well maintained, ranging from $2.8–31.3 \times 10^4$ per kg. Noga et al. (1987) reported a standardized protocol that delivered 1×10^6 T cells per kg to all patients. We have reproducibly achieved a similar level of depletion (M. Heil, T. Ahmed, and D. Ciavarella, unpublished data).

Thus, multiple techniques are available for T-cell depletion. Clinical results have, in general, been promising regarding a diminution in the incidence and severity of GVHD. However, two serious drawbacks to the widespread use of T-cell depletion have been found. These are an increased incidence of both graft failure or rejection and/or leukemic relapse in these patients. Graft failure or rejection may be expected when either inadequate recipient immunosuppression or too extensive donor T-lymphocyte depletion alters the immunologic balance in favor of recipient (host) T cells (Gale and Reisner 1986). An increased incidence of leukemic relapse may be a result of lack of a graft-versus-leukemia effect (GvLeu) in T-cell depleted patients. The inverse relationship between GVHD and leukemic relapse was first reported by Weiden et al. (1979, 1981). This effect is presumed due to cytolysis of leukemic cells by donor T cells, although it is unclear whether the same effector cells operate in GVHD and GvLeu.

Martin et al. (1985) reported a diminished 15% incidence (3/20) of GVHD, but a 35% incidence (7/20) of irreversible graft failure after 2–3 log T-cell depletions. The authors ascribed some graft failure to inadequate recipient immunosuppression as well. Mitsuyasu et al. (1986), Herve et al. (1987), de Witte et al. (1986), Kernan et al. (1986), Gilmore et al. (1986), and Noga et al. (1987) all reported a markedly decreased incidence of GVHD in their T-cell depleted allografts, with the incidence of GVHD ≤ grade I ranging from 60–97%. The cumulative results suggest that the incidence of GVHD ≥ grade II falls when total T cells infused are 1×10^5 or below, although the correlation is neither precise nor statistically significant. An important uncontrolled variable in making such an association is the method of T-cell depletion, specifically, the subsets of T cells removed or infused. The precise lymphocyte subsets involved in GVHD are unknown and this issue remains a very important line of research.

Leukemic relapse was reported as "relatively high" (13/24) in the series by Herve et al. (1987). Mitsuyasu et al. (1986) reported relapse in 7/20 T-cell depleted versus 2/20 unmanipulated allografts. This was not statistically significant. Much more attention has been given to the incidence of graft failure in these series. Kernan et al. (1986), Gilmore et al. (1986), de Witte et al. (1986), Noga et al. (1987), and Herve et al. (1987) all reported sustained engraftment in all evaluable patients. Once again, several patient variables, including primary disease, relative risk category, and incidence of cyto-megalovirus disease or other viral disorders, will affect the likelihood of leukemic relapse or graft failure in these patients.

13.1.6 Tumor Cell Purging

An enormous amount of literature has been devoted in recent years to the subject of removal of residual tumor cells in autografts. A variety of reviews and symposium proceedings are available, and only the main features will be discussed here (Gee and Boyle 1988; Gee and Gross 1987; Macintyre 1986; Gross et al. 1990; Bell 1989). The theoretical utility of tumor cell purging derives from a consideration of the tumor load that exists in patients with malignancies. Maximal BM tumor loads would, of course, be expected in hematopoietic malignancies. It has been estimated that a patient with acute leukemia presents with a tumor load of 10^{12} cells, and may be left with as many as 10^{10} malignant cells after induction therapy. If we assume that all residual 10^{10} leukemic cells are present in the BM, then the typical harvest of 2–5% of the marrow pool may contain in excess of 10^8 leukemic cells. This figure can be corroborated by simple calculations which assume that methods to detect residual disease are not sensitive below 1/100 to 1/1000 cells (Bell 1989). A BM harvest of $1–2 \times 10^{10}$ cells may thus contain 1×10^7 to 2×10^8 undetected leukemic cells. In this "worst case" scenario, a 7–8 log kill is necessary to assure a tumor-free graft. Whether absolute purging is necessary to prevent reseeding by tumor autografts is unclear, and there may exist a minimum number, perhaps 10^3 or more, of cells that must be infused to permit tumor regrowth.

A larger question with respect to tumor relapse after autologous grafting is the extent of endogenous tumor eradication by the pretransplant chemo-radiotherapy. Until that question can be adequately answered, the role of tumor cell purging will remain very unclear. A recent survey by the American Medical Association's Diagnostic and Therapeutic Technology Assessment (DATTA) Panel (1990) on autologous bone marrow transplantation found no consensus on the safety or effectiveness of purging techniques. No technique was rated as established by a majority of panelists, and the great majority of panelists believed the safety and effectiveness of purging required further investigation. Nevertheless, the theoretical advantages of tumor cell purging continue to make it an active research area. The most experience has been gained using antitumor antibodies or chemotherapy, although a wide variety of techniques have been reported (Table 13–3).

The in vitro effectiveness of purging techniques has been tested in a variety of model systems using malignant cell lines. Bast et al. (1985) were able to eliminate up to 5 logs of a Burkitt's lymphoma cell line inoculated in vitro into human BM. They utilized MoAb and complement, finding that repeated treatments were superior to a single treatment and that mul-

TABLE 13–3   Tumor Cell Purging Techniques

Immunological[1]
   Monoclonal antibodies with complement
   Monoclonal antibodies with toxins
   Monoclonal antibodies bound to magnetic microspheres

Pharmacological
   4-Hydroperoxycyclophosphamide (4HC)[1]
   4-Sulfoethylthiocyclophosphamide (mafosamide; Asta-Z-7557)[1]
   VP 16-213
   Alkyl-lysophospholipids
   Glucocorticoids
   Verapamil
   I-B-D arabinofuranosylcytosine (Ara-C)
   Deoxycytidine
   6-Hydroxydopamine

Other
   Photoradiation
   Centrifugal counterflow elutriation
   Lymphokine-activated killer cells (LAK cells)

[1] Most frequently used alone or in combination.

tiple MoAb were better than a single MoAb. CFUs were not lost in these treatments. Very similar results were obtained by Strong et al. (1985), who utilized MoAbs bound to the ribosomal inhibitor ricin to effect cell kill. T-cell lines from patients with acute lymphocytic leukemia were used. A cell kill of >4 logs was achieved using a mixture of four different MoAb, a combination that was superior to a single MoAb-ricin alone. CFU-GEMM were inhibited by 20%, while CFU-GM and BFU-E were inhibited by 30 to 40%. Treleaven et al. (1984) utilized six MoAb and a system of magnetic microspheres made of polymerized styrene divinylbenzene. Sheep anti-mouse IgG was bound to the microspheres, and after incubation of the MoAb with BM, the microspheres were incubated with the BM for 2 hr at 4°C. Using a model that inoculated 5×10^6 malignant neuroblasts (using a cell line) into 5×10^8 NC, they obtained a consistent 3 log cell kill. Others using immunomagnetic techniques have obtained similar or better results (Reading et al. 1987).

The two pharmacologic agents most often utilized are derivatives of cyclophosphamide, 4-hydroperoxycyclophosphamide (4HC), and mafos-amide. In a series of studies, optimum conditions for 4HC purging have been determined and applied to animal models and humans (Sharkis et al. 1990; Yeager et al. 1986; Rowley et al. 1989). Three to 4 log reductions in tumor cells in model systems have been described. A 4.7 log kill has been reported using mafosamide (Uckun et al. 1985). These latter investigators also reported that mafosamide plus an immunotoxin (MoAb bound to poke-weed antiviral protein) was superior to either alone.

The clinical efficacy of these and other purging techniques remains un-clear and controversial (DATTA Panel 1990). No randomized trials have been published, and a multitude of uncontrolled variables make evaluation of current data difficult. It would appear that purging techniques are avail-able to achieve a 5–6 log tumor cell kill, which approaches the degree nec-essary to obtain tumor-free (or nearly so) autografts. Therefore, the proper role of tumor cell purging will undoubtedly become clearer in the next few years.

Lastly, the identification of the HSC antigen CD34 has led to the de-velopment of positive purge methods, designed to isolate HSC for either study or transplantation. Since CD34 may be present on cells from patients with acute leukemia and squamous cell cancer of the lung, these patients are not candidates for positive purging. Significantly, however, CD34 has not been detected on lymphoma, myeloma, breast or small-cell lung car-cinoma, or neuroblastoma. Bensinger et al. (1990) have utilized anti-CD34 and an avidin-biotin system to obtain purified CD34$^+$ cells. A mouse IgM CD34 antibody (antibody 12-8) was incubated with BM samples (either contaminated with CD34-lymphoma cell lines or from patients with neuro-blastoma or breast carcinoma). A biotin conjugate of goat antimouse IgM antiserum was then added and the BM-anti-CD34-biotin conjugate mixture was passaged down an immunoadsorption column containing avidin-linked polyacrylamide gel. Adherent CD34$^+$ cells were recovered by agitation. A >3 log reduction of contaminating lymphoma cells was seen. In patient BM samples, an average of 1.2% of original NC were recovered, which contained a mean 86 times more CFU-GM than unprocessed marrow. A similar enrichment of CFUs was noted after passage of PB MNC through this system. In four evaluable patients transplanted with CD34$^+$ cells isolated in this fashion, granulocyte recovery to >500 per μl was seen 21 to 45 days posttransplant. These patients received 50–260 $\times$ 10^6 selected cells, of which 62–92% were CD34$^+$.

At present, commercial systems are available or under development which will utilize membrane-bound MoAb to attempt selective cell collection from either BM or PB.

13.2 HEMATOPOIETIC STEM CELL STORAGE

Optimization of storage conditions for the preservation of HSC would be desirable for many reasons. The growing utility of BM using unrelated donors often leads to marrow harvesting in a city distant to the recipient, and therefore, travel problems can complicate the transplant. In the autologous setting, where both liquid and frozen storage have been utilized, the utility of cell-specific manipulation could be increased if better storage methods (liquid and frozen) for HSC were available.

It should be pointed out that the studies discussed in this section address only the short-term preservation of HSC, rather than their culture. Preservation in this context implies maintenance of the HSC "status quo," without any attempt to induce differentiation, stimulate growth or metabolic activity, or change the relative ratios of the different HSC in a preparation.

Perhaps this approach, which has been satisfactorily applied to storage of nonnucleated blood cells, is too narrow for nucleated cells, particularly given the variety of HSC. In mice, CFU-S can be preserved, even increased, in short-term (4 day) culture by using growth factors (Migliaccio and Visser 1986). However, the ratios of the heterogeneous subtypes of CFU-S were altered by these culture conditions, and the desirability of this sort of modulation of human HSC prior to transplantation has not been studied. Nevertheless, although the present focus of the clinical HSC processing laboratory necessarily remains the preservation of HSC, rather than their culture, future processing laboratories will need to apply some of the lessons of tissue culture to modulate and regulate HSC during and after in vitro manipulation procedures.

13.2.1 Liquid (Nonfrozen) Storage
Published data on the viability, CFU activity, and transplantation potential of liquid stored marrow cells allow few incontrovertable conclusions. In addition to the inherent variability of the CFU assays, storage conditions

have been poorly standardized and potentially important variables, such as the concentrations of calcium, glucose, and hydrogen ion, have largely not been studied.

Wells and Cline (1976) studied 1- to 2-ml aliquots of marrow cells processed in four different ways by comparing CFC-C activity over a 3-day period. Marrow nucleated cells stored in the presence of erythrocytes preserved greater CFU-C activity than cells separated from erythrocytes before storage. Both 25 and 37°C storage were better than 4°C storage, with the best results showing 90% recovery of CFC-C activity at 24 hr and 48% at 72 hr.

Mangalik et al. (1979) studied CFC-C activity in stored marrow containing 14 ml of CPD and 555 units of heparin per 100 ml of marrow. Preservation at 10°C was slightly better than 4°C storage, and both were clearly better than 20°C storage. CFU-C activity was detected at levels as high as 58 and 43% of baseline at 4 and 7 days, respectively. Kohsaki et al. (1981) demonstrated only 30% recovery of baseline CFU-C after 3 days of 4°C storage as unprocessed marrow. Superior results were obtained, however, using alpha medium with 40% fetal calf serum (FCS), 12.3% CPD, and 25 mM HEPES buffer, with retention of 80-90% activity at 3 days and 40% activity at 7 days. In this study, 4°C was superior to 14°C or 24°C, and 40% FCS was superior to 20% FCS and alpha medium alone. Kohsaki et al. commented on the possible importance of the physiologic pH maintained with the FCS, CPD, and HEPES combination. Delforge et al. (1983) reported the best CFU-C recovery data ever published in their 10-day study of marrow stored at 4°C with only 10 U per ml heparin and 10% TC199 medium. A mean recovery of CFU-C of 98% after 4 days was reported, with one experiment finding no loss of CFU-C activity at 9 days of storage. Nearly diametric results were published by Millar and Smith (1984), who reported a linear reduction in CFU-GM activity with time at 4°C. Only 15% activity was recovered at 72 hr, although storage conditions were not discussed, and raw data, including the number of experiments, were not revealed. Hartmann et al. (1985) reported no fall off in CFU-GM when marrow was stored unprocessed for 24 hr. Unfortunately, longer periods were not studied.

Two more recent studies have produced comparable results. Takahaski and Singer (1985) studied CFU-C activity at 4°C under two conditions of storage: the first utilizing unprocessed marrow and the second utilizing a marrow buffy coat in alpha medium with 20% FCS. At 4 days, mean CFU-C preservation was 48% in unprocessed and 18% in processed marrow. These results were comparable to findings by Kohsaki et al. (1981) with 20% FCS stored marrow, but better than their results using unprocessed marrow. In an original and interesting experiment, these authors took subcultures of the 4°C stored marrow at various intervals and used them to establish long-term cultures. The cumulative number of colonies harvested from long-term cultures averaged 67% of Day 0 control at both 4 and 7 days of subculture. These results indicated that 4°C storage for up to 7 days may conserve totipotential HSC and nonhematopoietic elements, and in the opinion of Takahaski and Singer, justified a clinical trial of BMT using marrow stored for 7 days at 4°C. Finally, Lasky et al. (1986b) found that 4°C storage without agitation was preferable to higher temperatures in the preservation of nucleated cells, CFU-GEMM, CFU-GM, and BFU-E. The best results were CFU-GEMM AND CFU-GM recoveries of 70–80% at Day 3, although Day 7 recoveries were <10%.

In summary, CFU activity appears stable for at least 24 hr in liquid storage, and declines to 50–80% of baseline activity after 3–4 days. The preponderance of evidence points to 4°C as the best storage temperature, and the most consistent results have been reported when storing unprocessed marrow. The excellent results reported by Kohsaki et al. (1981), however, using a buffy coat preparation in alpha medium with 40% FCS, CPD, and HEPES hold promise for storage of concentrated HSC preparations. Erythrocytes and serum appear in some studies to have relative protective effects on HSC during storage. Lasky et al. (1986b) have suggested a role for the erythrocyte as oxygen scavenger, a plausible theory in light of the reported deleterious effects of oxygen radicals on progenitor cells in culture (Meagher et al. 1988).

13.2.2 Frozen Storage

The general principles of the cryobiology of cells and tissues have been known for at least 25 years. Unfortunately, a complete understanding of the science of cryopreservation has not been achieved, and, like HSC transplantation in toto, empiricism dictates many procedures and protocols.

Two general mechanisms are believed important in the susceptibility of cells to freezing injury (Meryman 1971, 1989). The first is cellular damage caused by the formation of intracellular (IC) ice. During cooling ice forms first in the extracellular (EC) solution (Pegg 1976), due both to its greater proximity to the cooling apparatus and to the known "insulating" effects of the cell membrane. The temperature at which EC ice forms is inversely proportional to the number and size of crystal-promoting "nuclei," e.g., proteins, in solution. At temperatures of −40°C or so, spontaneous ice formation can occur, since random aggregations of water at this temperature are sufficient to initiate ice formation. At temperatures of −115 to −120°C, the solution viscosity has risen high enough to prevent the arrangement of water into crystalline ice, and a noncrystalline "ice" or glass is formed. This is the temperature of glass transformation (Tg), or vitrification. In general, this EC ice is not directly harmful to cells in solution. At slower cooling rates, IC water exits the cell as the EC water concentration falls during ice formation, and eventually, IC water reaches a concentration at which IC ice cannot form. At more rapid cooling rates, however, IC nucleation occurs, and much evidence documents the detrimental effects of IC ice on cell viability on rewarming.

Unfortunately, the second mechanism of cell injury, which is less well understood than IC ice formation, operates when freezing rates are too low. The cellular dehydration that occurs during slow freezing, and which prevents IC ice injury, is associated with cellular damage on thawing. The precise cause of the second mechanism, which has been called "solute or solution effects" or dehydration injury, is not known. Early investigators believed it was caused by high concentrations of NaCl (both EC and IC) caused by EC ice formation and IC water loss. However, Meryman et al. (1977), studying different species, related cellular injury to the degree of cellular volume reduction, not to NaCl concentration or osmolarity. Upon warming, the dehydrated cell takes up water and swells, and the proximate cause of cell death is the inability of the cell to withstand the resultant hypotonic stress. The mechanism of cellular damage caused by dehydration, i.e., how this process leads to increased sensitivity to osmotic stress, is unclear, although workers in this field have presented data that support the theory of cell membrane damage (perhaps an alteration in the lipid composition or a structural change in the membrane itself) during the volume reduction (dehydration) phase (Meryman 1971; Pegg 1976).

Thus, freezing must be rapid enough to prevent dehydration injury but slow enough to prevent IC ice formation. In mammalian systems, a cooling rate to balance these forces has not been found, since dehydration injury occurs before IC water concentration has fallen sufficiently to prevent IC ice formation. For this reason, chemical additives—cryoprotectants—are needed to alter the balance between these two mechanisms of injury.

Cryoprotectants are classically of two types: penetrating or intracellular cryoprotectants, which pass through the cell membrane, and nonpenetrating or extracellular cryoprotectants, which do not enter the cell. Dimethyl sulfoxide (DMSO) and glycerol are the prototypical IC cryoprotectants, while examples of EC cryoprotectants are sucrose, hydroxyethyl starch (HES), and polyvinylpyrrolidine (PVP). A variety of other compounds provide protection against freezing injury, but do not meet the criteria for nontoxicity that a clinically useful cryoprotectant must possess, at least at the concentrations required.

IC cryoprotectants protect against both types of freezing injury. DMSO penetrates the cell rapidly (within 3 min at 5°C, and within 30 sec at 22°C), reducing on a colligative basis the amount of IC ice formed at any temperature, and increasing solution viscosity, thereby decreasing the flux of water and the rate of EC ice formation, and increasing the Tg (Meryman 1971, 1989).

EC cryoprotectants alone do not prevent dehydration injury; in fact, by raising EC osmolality they may accelerate it. However, when used with IC cryoprotectants, they increase solution viscosity, thus decreasing the flux of water and raising the Tg. Thus, slower cooling rates can be utilized, since the time to dehydration injury is prolonged. They may also act by "stabilizing" the cell membrane on rewarming.

In practice, both DMSO and glycerol have been reported effective in cryopreserving BM cells (Lewis et al. 1966, 1967; O'Grady and Lewis 1972). Although O'Grady and Lewis (1972) reported better results with glycerol than DMSO, others have reported the opposite (Ragab et al. 1977), and DMSO, at a final concentration of 10%, is the more accepted IC cryoprotectant (Gorin 1986). Pegg (1976) found that the combination of an IC (glycerol) and EC (PVP) cryoprotectant was superior to the IC agent alone, but most workers utilize DMSO alone. However, several studies have shown that DMSO with 10 or 20% plasma or serum (either bovine or human) is superior to DMSO alone, raising the possibility that serum or plasma may be functioning as an EC cryoprotectant (Lewis et al. 1966; Ragab et al. 1977; Grilli et al. 1980). Serum or plasma is thus routinely used in most centers. Early experiments related cell recovery to freezing rates, with optimum conditions reported at a freezing rate of -1 to $-3°C$ per min until -30 or $-40°C$, followed by long-term storage in liquid nitrogen ($-196°C$) (Rowe 1966; Rowe and Rinfret 1962; Lewis et al. 1966, 1967). In addition, minimizing the time spent in the "freezing plateau," the temperature increase seen as supercooled solution freezes (heat of fusion) was also judged important (Rowe 1966; Lewis et al. 1967). Thus, controlled rate freezers, which temporarily decrease chamber temperature during the phase transition to shorten the freezing plateau, became standard (Ragab et al. 1977; Wells et al. 1979). Cell concentrations are usually within the range of 1×10^6 to 2×10^7 per ml, with Wells et al. (1979) reporting poor results at concentrations above 15×10^7 ml. Thawing protocols involve rapid warming in water baths (37–42°C), followed by immediate infusion into the patient. Attempts at washing out excess DMSO prior to infusion often leads to cell clumping and cell loss (Wells et al. 1979).

In recent years, alternatives to this standard approach (i.e., 10% DMSO plus 10% serum or plasma, frozen at $-1°C$ per min to $-40°C$, followed by storage in liquid nitrogen) have been reported. Niskanen and Pirsch (1983) studied freezing rates as rapid as $-10°C$ per min using 10% DMSO and 10% FCS. They obtained better CFU-C recovery between -3 and $-7°C$ per min. Berthier et al. (1983) reported a two-step method using 10% DMSO and 20% calf serum. After a 10-min equilibration on crushed ice, 1-ml vials containing 20×10^6 BM cells and cryoprotectant were plunged into mineral baths ranging from -29 to $-43°C$ for 20 min, followed by liquid nitrogen storage. Good (85 to 92%) recovery of CFU-GM was seen at bath temperatures near the temperature of spontaneous ice formation (-37.5 to $-43.3°C$). The investigators commented on the ease and simplicity of this method.

Stiff et al. (1983b) reported a successful cryopreservation protocol for human BM cells that violated many established procedures. Using experience gained in freezing granulocytes, a cryoprotectant mixture containing 5% DMSO, 6% HES, and 4% albumin (final concentrations) was tried. A programmed freezer was not used; instead the cells were simply placed in a mechanical freezer set at $-80°C$. Freezing rates of -2 to $-4°C$ per min were obtained, with a prolonged (10–14 min) freezing plateau. Storage at $-170°C$ or at $-80°C$ for up to 16 months resulted in CFU recoveries of 100%. Importantly, post-thaw cellular clumping, usually caused by nucleoproteins released from damaged granulocytes, was not observed.

Scheiwe et al. (1981) reported an interesting variation on the thawing procedure. They have developed a "revitalization" treatment post-thaw, delivering an amino acid solution by atomizer to the thawed marrow to avoid agitation and minimize osmotic stress. The improvement in MNC recovery over standard thaw procedures was minimal (90% versus 82%).

The optimum length of storage of HSC using these protocols has not yet been determined. Most studies demonstrated some fall off in CFU recovery with prolonged storage (1 year or more), but CFU activity has been demonstrated after up to 121 months of storage (O'Grady and Lewis 1972; Heal and Brightman 1977; Rybka et al. 1980). At present, long-term storage at temperatures below Tg (-130 to $-140°C$) is most appropriate, since metabolic activity ceases at this temperature range. We have reinfused marrow stored at $-80°C$ for 2 years using Stiff's technique, with good cell recovery and successful engraftment (Ahmed et al. 1991). However, long-term storage at this temperature remains experimental at present.

Largely equivalent clinical results have been reported by many investigators using either type of freezing protocol (Visani et al. 1988; Hartmann et al. 1985; Kessinger et al. 1988; Gulati et al. 1988; Stiff et al. 1987a, 1987b). There is both experimental (Fabian et al. 1982) and clinical evidence to suggest that megakaryocytic precursors are more susceptible than other HSC to freezing damage. Future studies will need to concentrate on optimal conditions to preserve megakaryocytic HSC, and on improved methods of assessing freezing injury. Boswell et al. (1983) have used long-term cell cultures for this purpose, and their data suggest that different progenitors possess different cryobiologic properties. Perhaps several different techniques should be used on a single HSC preparation so that all cell lines may be optimally preserved.

13.3 SUMMARY AND CONCLUSIONS

The techniques to collect, process, and store HSC in anticipation of transplantation are now widely available. Important unresolved issues revolve around the as yet imperfect identification and classification of totipotential progenitors. However, much progress has been and will continue to be made despite this limitation. Research priorities of present and future stem cell processing laboratories should include:

1. Optimization of liquid (nonfrozen) storage techniques. This will permit more complex cell-specific manipulations, such as T-lymphocyte subset

selection, isolation of CD34+ populations, treatment in vitro with growth factors, gene transfer experiments, and long-range transport of HSC, to be performed while preserving HSC integrity.

2. A better understanding of the regulation and kinetics of peripheral blood and umbilical cord HSC, to allow optimum collection procedures that do not require marrow harvesting.

3. An intensive study into the optimum conditions of collection, processing, and storage of megakaryocytic progenitors to decrease the long platelet-transfusion dependency of the myeloablated patient.

4. A search for a simple in vitro correlate of engraftment potential of a stem cell preparation.

This will greatly improve the quality control functions of the laboratory as well as contribute to better patient selection for transplantation.

REFERENCES

Abrams, R.A., Johnston-Early, A., Kramer, C., et al. (1981) *Cancer Res.* 41, 35–41.

Ahmed, J., Wuest, D., Ciavarella, D., et al. (1991) *Acta Haematologica* (in press).

Andrews, R.G., Singer, J.W., and Bernstein, I.D. (1986) *Blood* 67, 842–845.

Andrews, R.G., Singer, J.W., and Bernstein, I.D. (1989) *J. Exp. Med.* 169, 1721-1725.

Atkinson, K., Norrie, S., Chan, P., Downs, K., and Biggs, J. (1985) *Br. J. Haematol.* 60, 245–251.

Bast, R.C., DeFabritiis, P., Lipton, J., et al. (1985) *Cancer Res.* 45, 499–503.

Beatty, P.G., Clift, R.A., and Mickelson, E.M. (1985) *N. Engl. J. Med.* 313, 765–771.

Bell, A.J. (1989) *Transfusion Sci.* 10, 39–50.

Bell, A.J., Hamblin, T.J., and Oscier, D.G. (1986) *Bone Marrow Transplantation* 1, 103–110.

Bensinger, W.I., Berenson, R.J., Andrews, R.G., et al. (1990) *J. Clin. Apheresis* 5, 74–76.

Berkman, E.M., Caplan, S., and Kim, C.S. (1978) *Transfusion* 18, 504–508.

Berthier, R., Kaufmann, A., Schweitzer, A., Thevenon, D., and Hollard, D. (1983) *Cryobiology* 20, 637–643.

Blacklock, H.A., Prentice, H.G., Evans, J.P.M., et al. (1982) *Lancet* 2, 1061–1064.

Bortin, M.M., and Buckner, C.D. (1983) *Exp. Hematol.* 11, 916–919.

Boswell, H.S., Niskanen, E.O., Coppda, M.A., and Quesenberry, P.J. (1983) *Exp. Hematol.* 11, 315–323.

Braine, H.G., Sensenbrenner, L.L., Wright, S.K., et al. (1982) *Blood* 60, 420–425.

Brandt, J., Baird, N., Lu, L., Srour, E., and Hoffman, R. (1988) *J. Clin. Invest.* 82, 1017–1021.

Broxmeyer, H.E., Douglas, G.W., Hangoc, G., et al. (1989) *Proc. Natl. Acad. Sci. USA* 86, 3828–3832.

Buckner, C.D., Clift, R.A., Sanders, J.E., et al. (1978) *Transplantation* 26, 233–238.

Buckner, C.D., Clift, R.A., Sanders, J.E., et al. (1984) *Blood* 64, 630–634.

Burgstaler, E.A., and Pineda, A.A. (1990) *J. Clin. Apheresis* 5, 156 (abstr.).

Calvo, W., Fliedner, T.M., Herbst, E., Hugl, E., and Bruch, C. (1976) *Blood* 47, 593–598.

Carter, C.S. Goetzman, H., Yu, M., Wilson, W., and Read, E.J. (1989) *Transfusion* 29 (suppl), 65 (abstr.).

Ciavarella, D. (1989) *Transfusion Science* 10, 165–184.

Ciavarella, D., and Ahmed, T. (1987) *Cancer Invest.* 5, 541–544.

Civin, C.I., Strauss, L.C., Brovall, C., et al. (1984) *J. Immunol.* 133, 157–165.

Civin, C.I., Banquerigo, M.L., Strauss, L.C., and Loken, M.R. (1987) *Exp. Hematol.* 15, 10–17.

DATTA (Diagnostic and Therapeutic Technology Assessment) Panel on Autologous Bone Marrow Transplantation (1990) *J. Am. Med. Assoc.* 263, 881–887.

Debelak-Fehir, K.M., and Epstein, R.B. (1975) *Transplantation* 20, 63–69.

Delforge, A., Ronge-Collard, E., Stryckmans, P., Spiro, T., and Malarone, M.A. (1983) *Br. J. Haematol.* 53, 49–54.

De Witte, T., Hoogenhout, J., de Pauw, B., et al. (1986) *Blood* 67, 1302–1308.

Dinsmore, R.E., Reich, L.M., Kapoor, N., et al. (1983) *Br. J. Haematol.* 54, 441–449.

English, D., Lamberson, R., Graves, V., et al. (1989) *Transfusion* 29, 12–16.

Fabian, I., Dover, D., Wells, J.R., and Cline, M.J. (1982) *Exp. Hematol.* 10, 119–122.

Fauser, A.A., and Messner, H.A. (1979) *Blood* 53, 1023–1027.

Foradji, A., Andreu, G., Pillier-Loriette, C., et al. (1988) *Vox Sang.* 55, 133–138.

Gabrilove, J.L. (1989) *Semin. Hematol.* 26 (suppl 2), 1–4.

Gale, R.P., and Reisner, Y. (1986) *Lancet* 2, 1468–1470.

Ganser, A., Elstner, E., and Hoelzer, D. (1985) *Br. J. Haematol.* 59, 627–633.

Gee, A.P., and Boyle, M.D.P. (1988) *J. Natl. Cancer Inst.* 80, 154–159.

Gee, A.P., and Gross, S. (eds.) (1987) in *Bone Marrow Transplantation* (suppl. 2), pp. 1–150, Macmillan, London.

Gianni, A.M., Bregni, M. Siena, S., et al. (1989) *Hematol. Oncol.* 7, 139–146.

Gilmore, M.J.M.L., Prentice, H.G., Blacklock, H.A., et al. (1982) *Br. J. Haematol.* 50, 619–626.

Gilmore, M.J.M.L., Prentice, H.G., Corringham, R.E., Blacklock, H.A., and Hoffbrand, A.V. (1983) *Vox. Sang.* 45, 294–302.

Gilmore, M.J.M.L., Patterson, J., Ivory, K., et al. (1986) *Brit. J. Haematol.* 64, 69–75.

Gluckman, E., Broxmeyer, H.E., Anerbach, A.D., et al. (1989) *N. Engl. J. Med.* 321, 1174–1178.

Gorin, N.C. (1986) *Clin. Haematol.* 15, 19–47.

Grilli, G., Porcellini, A., and Lucarelli, G. (1980) *Cryobiology* 17, 516–520.

Groopman, J.E., Molina, J.-M., and Scadden, D.T. (1989) *New England J. Med.* 321, 1449–1459.

Gross, S., Gee, A., and Worthington-White, D.A. (eds.) (1990) in *Progress in Clinical and Biological Research*, vol. 333, Wiley-Liss, New York.

Gulati, S.C., Shank, B., Black, P., et al. (1988) *J. Clin. Oncol.* 6, 1303–1313.

Hansen, J.A., Woodruff, J.M., and Good, R.A. (1981) in *Immunodermatology*, vol. 7 (Safai, B., and Good, R.A., eds.), pp. 229–257, Plenum Publishing Corp., New York.

Hartmann, O., Beaujean, F., Bayet, S., et al. (1985) *Eur. J. Cancer Clin. Oncol.* 21, 53–60.

Heal, J.M., and Brightman, A. (1987) *Transfusion* 27, 19–22.

Hershko, C., Gale, R.P., Ho, W., and Fitchen, J.W. (1980) *Br. J. Haematol.* 44, 65–73.

Herve, P., Cahn, J.Y., Flesch, M., et al. (1987) *Blood* 69, 388–393.

Hester, J.P., Kellogg, R.M., and Freireich, E.J. (1983) *J. Clin. Apheresis* 1, 197–201.

Humblet, Y., Le febvre, P., Jacques, J.L., et al. (1988) *Bone Marrow Transplantation* 3, 63–67.

Hurd, D., Lasky, L., Haake, R., et al. (1988) *Proc. Asco* 7, 177 (abstr.).

Jin, N-R., Hill, R., Segal, G., et al. (1987) *Exp. Hematol.* 15, 93–98.

Jones, R.J., Sharkis, S.J., Celano, P., et al. (1987) *Blood* 70, 1186–1192.

Juttner, C.A., To, L.B., Ho, J.Q.K., et al. (1988) *Transplant Proc.* 20, 40–43.

Juttner, C.A., To, L.B., Haylock, D.N., et al. (1989) *Transplant Proc.* 21, 2929–2931.

Kaiser, H., Stuart, R.K., Brookmeyer, R., et al. (1985) *Blood* 65, 1504–1510.

Keller, D.J., Alter, R., Landmark, J.D., Kessinger, A., and Weisenburger, D.D. (1987) *Blood* 70, 321a (abstr.).

Kernan, N.A., Collins, N.H., Juliano, L., et al. (1986) *Blood* 68, 770–773.

Kessinger, A., Landmark, J.D., Smith, D.M., et al. (1987) *Blood* 70, 321a (abstr.).

Kessinger, A., Armitage, J.O., Landmark, J.D., Smith, D.M., and Weisenburger, D.D. (1988) *Blood* 71, 723–727.

Kohsaki, M., Yanes, B., Ungerleider, J.S., and Murphy, M.J. (1981) *Stem Cells* 1, 111–123.

Korbling, M., and Martin, H. (1988) *Plasma Ther. Transfus. Technol.* 9, 119–132.

Lasky, L.C., and Zanjani, E.D. (1985) *Exp. Hematol.* 13, 680–684.

Lasky, L.C., Ash, R.C., Kersey, J.H., Zanjani, E.D., and McCullough, J. (1982) *Blood* 59, 822–827.

Lasky, L.C., Smith, J., Fautsch, S., Kenyon, P., and McCullough, J. (1986a) *Plasma Ther. Transfus. Technol.* 7, 424.

Lasky, L.C., McCullough, J., and Zanjani, E.D. (1986b) *Transfusion* 26, 331–334.

Lasky, L.C., Smith, J.A., McCullough, J., and Zanjani, E.D. (1987) *Transfusion* 27, 276–278.

Lasky, L.C., Smith, J.A., and Bostrom, B. (1990) *Transfusion* 29 (suppl.), 59S (abstr.).

Law, P., Dooley, D.C., Alsop, P., et al. (1988) *Transfusion* 28, 145–150.

Lewis, J.P., Passovoy, M., and Trobaugh, F.E. (1966) *Cryobiology* 3, 47–52.

Lewis, J.P., Passovoy, M., Conti, S.A., McFate, P.A., and Trobaugh, F.E. (1967) *Transfusion* 7, 17–32.

Lowder, J.N., and Herzig, R.H. (1990) in *Modern Transfusion Therapy* (Dutcher, J.P., ed.), pp. 230–250, CRC Press, Boca Raton, FL.

Lu, L., Walker, D., Broxmeyer, H.E., et al. (1987) *J. Immunol.* 139, 1823–1829.

Lyding, J., Zander, A., Rachelle, M., et al. (1990) *J. Clin. Apheresis* 5, 156 (abstr.).

Macintyre, E.A. (1986) *Clin. Haematol.* 15, 249–267.

Mangalik, A., Robinson, W.A., Drebing, C., Hartmann, D., and Joshi, J.H. (1979) *Exp. Hematol.* 7 (suppl 5), 76–94.

Martin, P.J., Hansen, J.A., Buckner, C.D., et al. (1985) *Blood* 66, 664–672.

McCarthy, D.M., and Goldman, J.M. (1984) *CRC Critical Rev. in Clin. Lab Sci.* 20, 1–24.

Meagher, R.C., Salvado, A.J., and Wright, D.G. (1988) *Blood* 72, 273–281.

Meryman, H.T. (1971) *Cryobiology* 8, 173–183.

Meryman, H.T. (1989) in *Clinical Practice of Blood Transfusion*, 2nd ed. (L.D. Petz, and S. Swisher, eds.), pp. 297–314, Churchhill-Livingston, New York.

Meryman, H.T., Williams, R.J., and Douglas, M.S.J. (1977) *Cryobiology* 14, 287–302.

Migliaccio, A.R., and Visser, J.W.M. (1986) *Exp. Hematol.* 14, 1043–1049.

Migliaccio, A.R., Migliaccio, G., and Adamson, J.W. (1988a) *Blood* 72, 1387–1392.

Migliaccio, G., Migliaccio, A.R., and Visser, J.W.M. (1988b) *Blood* 72, 944–951.

Millar, J.L., and Smith, I.E. (1984) in *Autologous Bone Marrow Transplantation and Solid Tumors* (McVie, J.G., Dalesio, O., and Smith, I.E., eds.), pp. 9–12, Raven Press, New York.

Mitsuyasu, R.T., Champlin, R.E., Gale, R.P., et al. (1986) *Annals Internal Med.* 105, 20–26.

Niskanen, E., and Pirsch, G. (1983) *Cryobiology* 20, 401–406.

Noga, S.J., Donnenberg, A.D., and Santos, G.W. (1987) *Bone Marrow Transplantation* 2, 18–22.

Northdurft, W., Fliedner, T.M., Calvo, W., et al. (1978) *Scand. J. Haematol.* 21, 115–122.

O'Grady, L.F., and Lewis, J.P. (1972) *Transfusion* 12, 312–316.

O'Reilly, R.J. (1983) *Blood* 62, 941–964.

Padley, D.J., Wieland, M., Randels, M.J., et al. (1990) *J. Clin. Apheresis* 5, 176 (abstr.)

Pegg, D.E. (1976) *J. Clin. Pathol.* 29, 271–285.

Ragab, A.H., Gilkerson, E., and Myers, M. (1977) *Cryobiology* 14, 125–134.

Raijmakers, R., De Witte, T., Koekman, E., Wessels, J., and Haanen, C. (1986) *Vox Sang.* 50, 146–150.

Reading, C.L., Thomas, M.W., Hickey, C.M., et al. (1987) *Leuk. Res.* 11, 1067–1077.

Reich, L.M., Self, S.Z., and Mayer, K. (1980) *Transfusion* 20, 640 (abstr.).

Reiffers, J., Castaique, S., Tilly, H., et al. (1987) *Plasma Ther. Transfus. Technol.* 8, 360–362.

Richman, C.M., Weiner, R.S., and Yankee, R.A. (1976) *Blood* 47, 1031–1039.

Ritchey, B., Henry, S., Forman, S.J., et al. (1983) *Transplantation* 35, 638–639.

Rowe, A.W. (1966) *Cryobiology* 3, 12–18.

Rowe, A.W., and Rinfret, A.P. (1962) *Blood* 20, 636–637.

Rowley, S.D., Zvehlsdorf, M., Braine, H.G., et al. (1987) *Blood* 70, 271–275.

Rowley, S.D., Jones, R.J., Piantadosi, S., et al. (1989) *Blood* 74, 501–506.

Rybka, W.B., Mittermeyer, K., Singer, J.W., Buckner, C.D., and Thomas, E.D. (1980) *Cryobiology* 17, 424–428.

Scheiwe, M.W., Pusztai-Markos, Z.S., Essers, U., et al. (1981) *Cryobiology* 18, 344–356.

Schouten, H.C., Kessinger, A., Smith, D.M., et al. (1990) *J. Clin. Apheresis* 5, 140–144.

Sharkis, S.J., Santos, G.W., and Colvin, M. (1990) *Blood* 55, 521–523.

Siena, S., Bregai, M., Brando, B., et al. (1989) *Blood* 74, 1905–1914.

Sniecinski, I., Henry, S., Ritchey, B., Branch, D.R., and Blume, K.G. (1985) *J. Clin. Apheresis* 2, 231–234.

Sniecinski, I., Alfonso, G., Giunta, K., et al. (1989) *Transfusion* 29 (suppl.), 45S (abstr.).

Socinski, M.A., Elias, A., Schnipper, L., et al. (1988) *Lancet* 1, 1194–1198.

Spitzer, G., Verma, D.S., Fisher, R., et al. (1980) *Blood* 55, 317–323.

Stevenson, H.C., Stevenson, G.W., Leitman, S.F., et al. (1986) *Plasma Ther. Transfus. Technol.* 7, 365–371.

Stiff, P.J., Murgo, A.J., Wittes, R.E., DeRisi, M.F., and Clarkson, B.D. (1983a) *Transfusion* 23, 500–503.

Stiff, P.J., Murgo, A.J., Zaroulis, C.G., DeRisi, M.F., and Clarkson, B.D. (1983b) *Cryobiology* 20, 17–24.

Stiff, P.J., Koester, A.R., and Engleton, L.E. (1987a) *Transplantation* 44, 585–588.

Stiff, P.J., Koester, A.R., Weidner, M.K., Dvorak, K., and Fisher, R.I. (1987b) *Blood* 70, 974–978.

Storb, R., Prentice, R.L., and Thomas, E.D. (1977) *N. Engl. J. Med.* 296, 61–66.

Strong, R.C., Uckun, F., Youle, R.J., Kersey, J.H., and Vallera, D. (1985) *Blood* 66, 627–635.

Takahaski, M., and Singer, J.W. (1985) *Exp. Hematol.* 13, 691–695.

Tavassoli, M. (1979) *Br. J. Haematol.* 41, 297–302.

Thomas, E.D., and Storb, R. (1970) *Blood* 36, 507–515.

Tindle, R.W., Nichols, R.A.B., Chan, L., et al. (1985) *Leuk. Res.* 9, 1–9.

To, L.B., Dyson, P.G., and Juttner, C.A. (1986) *Lancet* 1, 404–405.

To, L.B., Dyson, P.G., Branford, A.L., et al. (1987) *Exp. Hematol.* 15, 351–354.

Torres, A., Alonso, M.C., Gomez-Villagran, J.L., et al. (1985) *Blut* 50, 89–94.

Treleaven, J.G., Ugelstad, J., Philip, T., et al. (1984) *Lancet* 1, 70–73.

Uckun, F.M., Ramakrishnan, S., and Houston, L.L. (1985) *Cancer Res.* 45, 69–75.

Ventura, G.J., Swan, F., Cabanillas, F.F., and Hester, J.P. (1988) *Proc. Asco* 7, 244.

Visani, G., Dinota, A., Verlicchi, F., et al. (1988) *Bone Marrow Transpl.* 3, 599–605.

Warkentin, P.I., Yomtovian, R., Hurd, D., et al. (1983) *Vox Sang.* 45, 40–47.

Warkentin, P.I., Hilden, J.M., Kersey, J.H., Ramsay, N.K.C., and McCullough, J. (1985) *Vox Sang.* 48, 89–104.

Watt, S.M., Karhi, K., Gatter, K., et al. (1987) *Leukemia* 1, 417–426.

Weiden, P.L., Flournoy, N., Thomas, E.D., et al. (1979) *N. Engl. J. Med.* 1068–1073.

Weiden, P.L., Sullivan, K.M., Flournoy, N., Storb, R., and Thomas, E.D. (1981) *N. Engl. J. Med.* 304, 1529–1535.

Wells, J.R., and Cline, M.J. (1976) *Transplantation* 22, 568–571.

Wells, J.R., Sullivan, A., and Cline, M.J. (1979) *Cryobiology* 16, 201–210.

Williams, W.J. (1983) in *Hematology*, 3rd ed. (Williams, W.J., Beutler, E., Ersler, A.J., and Lichtman, M.A., eds.), p. 29, McGraw-Hill, New York.

Williams, S.F., Bitran, J.D., Richards, J.R., et al. (1988) *Proc. AACR* 29, 181 (abstr.).

Wuest, D., Ciavarella, D., and Ahmed, T. (1989) *J. Clin. Apheresis* 5, 55 (abstr.).

Yeager, A.M., Kaizer, H., Santos, G.W., et al. (1986) *New England J. Med.* 315, 141–147.

Zander, A.R., and Cockerill, K.J. (1987) *J. Clin. Apheresis* 3, 191–201.

Erythropoietin: Its Role in the Regulation of Erythropoiesis and as a Therapeutic in Humans

John W. Adamson

A factor that affects red blood cell production was recognized over a century ago when Bert (1882) associated the hypoxia of altitude with secondary erythrocytosis. Subsequently, Carnot and Deflandre (1906) described an activity in anemic rabbit plasma that stimulated red blood cell production in normal rabbits. We now recognize that factor as a glycoprotein hormone, erythropoietin (Epo), which is the first recombinant hematopoietic growth factor to have found a place—likely forever—in the treatment of human disease. This review summarizes the molecular, cellular, and clinical biology of Epo.

14.1 THE ERYTHROPOIETIN (EPO) GENE

The first reports of the cloning and expression of the human Epo gene appeared in 1985 (Jacobs et al. 1985; Lin et al. 1985) and were confirmed subsequently with localization of the gene to chromosome 7 (Powell et al. 1986). The gene encoding the mature protein is comprised of four exons and three introns. There is also a sequence that codes for a secretory leader piece 27 amino acids in length. The structural gene for human Epo exists as a single copy in the human genome and has been mapped to the long arm of chromosome 7 in the region of 7q21 (Law et al. 1986).

Although the coding sequence predicts a polypeptide of 166 amino acids, the mature protein is made up of 165 amino acids with a molecular weight of 18,400 Da (Figure 14–1). The biologically active hormone exists as a monomer with two internal disulfide bonds linking cysteines at positions 7 and 161, and positions 29 and 33. The molecular weight of the biologically active hormone is approximately 34,000 Da. The higher weight is due to glycosylation at three Asn-linked sites at positions 24, 38, and 83, and a single O-linked glycosylation site at the serine at position 126. Glycosylation of the molecule is required for its survival in vivo. If the carbohydrates are removed by treatment of Epo with glycosidases, biological activity in vitro is retained. Site-directed mutagenesis, which replaces the carbohydrate-linked Asn sites, also results in a molecule with comparable biological activity and protein as determined by radioimmunoassay (RIA). Detailed in vitro studies of completely nonglycosylated Epo are difficult, however, because of the problems encountered with aggregation of the molecule. The carbohydrate constructions of the recombinant human Epo (rHuEpo) and that of native Epo appear to be virtually identical, and both native and

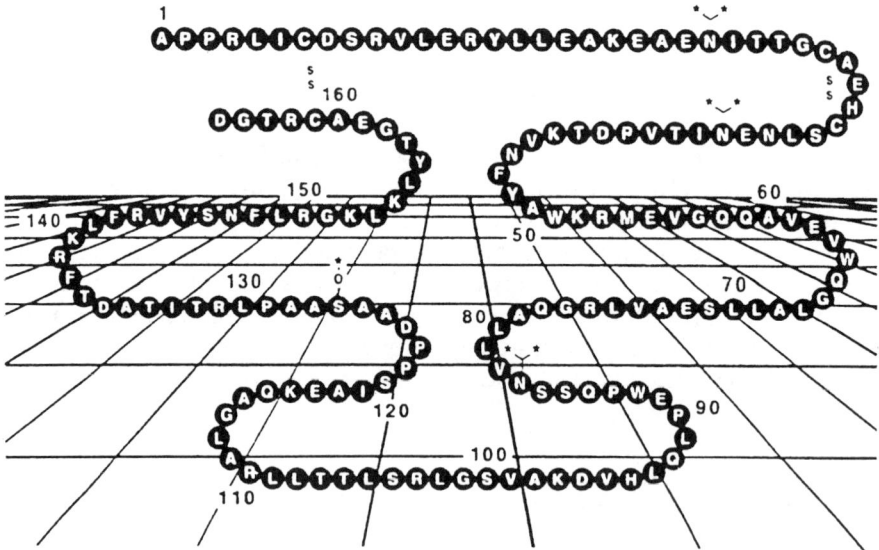

FIGURE 14–1 A schematic of the human erythropoietin molecule. Recombinant erythropoietin has the identical 165–amino acid sequence as does native erythropoietin and its molecular weight is 30,400 Da. There are two internal disulfide bonds that stabilize the structure. The three sites of N-linked glycosylation are shown by paired stars; a single serine-linked glycosylation site is shown at position 126.

recombinant molecules function equally well in vivo and in vitro (Egrie et al. 1986).

Other than the role of the carbohydrate in promoting the in vivo survival of Epo, little is known of the structure/function relationships of the polypeptide. Sytokwski and Donohue (1987) have suggested that amino acids 99–129 may be important in the binding of Epo to its receptor. These workers produced a series of peptides spanning a large portion of the Epo molecule. Although these peptides themselves did not have Epo-like activity and did not block the in vitro action of Epo, antibodies prepared to those peptides spanning the 99–129 region interfered with the in vitro action of Epo.

The cDNAs for mouse (McDonald et al. 1986) and monkey Epo (Lin et al. 1986) have also been cloned and sequenced. There is a high degree of homology among all three species and, overall, >90% homology between human and monkey Epo. In addition, the first introns of the human and mouse Epo genes have a high degree of homology, while homology is not maintained in other introns. Such retained homology suggests an important regulatory function for the first intron, but the precise significance of this is unknown.

14.2 REGULATION OF EPO PRODUCTION

Physiological studies that began nearly 30 years ago clearly implicated the kidney as the important site of Epo production in mammals (Krantz and Jacobson 1970). Animals subjected to hypoxia produce high levels of Epo that can be measured in the plasma by bioassay or RIA. Nephrectomized animals do not produce measurable levels of Epo, while animals subjected to ureteral ligation and subsequent hypoxia have an unimpaired Epo response.

The availability of probes for Epo mRNA has allowed further dissection of its cellular sources. Schuster et al. (1987) reported that Epo mRNA could not be detected in preparations of whole cellular RNA from normal rat kidneys but, when the rats were exposed to severe hypoxia, Epo-specific mRNA appeared within minutes and peaked within a few hours. The appearance of mRNA was followed shortly by the appearance of Epo bioactivity in the plasma of the animals. With cessation of hypoxia, mRNA levels fell and bioactivity declined. These same workers prepared tubular cells and glomeruli by differential sedimentation from rat kidneys. Only mRNA prepared from isolates rich in tubular cells contained Epo-specific transcripts. The preparations of glomeruli had no detectable Epo mRNA. Lacombe et al. (1988) reported similar findings using the technique of in situ hybridization. Rather than the tubular cells themselves, peritubular interstitial cells (presumably endothelial cells) were implicated as the source of Epo. Koury et al. (1988) have confirmed these observations using similar techniques.

In addition, they have found that not all peritubular interstitial cells produce Epo—only a small subset. With the onset of hypoxia, the amount of Epo mRNA per cell does not increase. Rather, it is the number of cells producing Epo that increases in a manner precisely equivalent to the total increase in Epo mRNA (Koury et al. 1989). Furthermore, most of the Epo-producing cells are located in the cortex of the kidney.

Recently, insight into the oxygen-sensing mechanism has been provided by Goldberg et al. (1988). Their work with Epo-producing (and hypoxia responsive) hepatoma cell lines led to the model of a heme-containing protein that bound oxygen and resulted in the regulation of Epo production. Paralysis of this oxygen-binding protein blocked Epo production in the face of hypoxia. This protein remains to be more fully characterized as does the important issue of whether the Epo-producing cells in the kidney are also the cells that monitor oxygen availability.

14.3 PLASMA EPO

The mature protein circulates in the plasma at a concentration of 15–25 mU/ml as determined by current sensitive RIAs (Garcia et al. 1982). Under normal circumstances, the activity determined by RIA parallels that determined by bioassay. Epo production increases with hypoxia or anemia. In a variety of carefully studied human conditions, an inverse log/linear relationship between plasma or urinary Epo levels and hemoglobin concentration was found. Thus, with established anemia and a hemoglobin of 5–8 g/dl, plasma Epo levels of 400–1000 mU/ml would be anticipated. In addition, there is a diurnal variation in plasma and urinary Epo levels, suggesting that Epo production varies during different parts of the day, being highest in the afternoon and evening, and lowest in the overnight hours (Adamson et al. 1966). This was first shown for normal individuals and found to be exaggerated in anemic individuals. The physiological mechanisms underlying the diurnal variation are unknown.

Preliminary studies of rHuEpo pharmacokinetics in patients with chronic renal failure (CRF) indicate a half-life on the average of 6–7 hr (range 5–11) with a single exponential of decay (Egrie et al. 1988; Cotes et al. 1989). We have carried out a study of the plasma half-life of native human Epo in one patient with CRF, who was subsequently treated with rHuEpo. The half-life of the native human Epo was 6.7 hr, while that of rHuEpo was 7.2 hr (Adamson, J.W., Egrie, J.C., and Eschbach, J.W., unpublished observations). Thus, this single study indicates that there are likely to be no significant differences between the half-life of the native and recombinant molecule in humans.

14.4 INTERACTIONS OF EPO WITH ITS TARGET CELL

14.4.1 Studies In Vitro

The cellular anatomy of erythropoiesis has been established by studies of various populations of progenitor cells that can be detected only through their growth in culture and their relative dependence on specific growth factors, including Epo. The earliest recognizable erythroid progenitor in culture is the erythroid burst-forming cell (BFU-E). This progenitor has the capacity to form large hemoglobinized colonies that contain as many as 10^4 cells. Epo is required for the hemoglobinization of the cells in these colonies, but the early divisions of the progenitor are dependent upon other growth factors referred to as having erythroid burst-promoting activity (BPA). In the absence of BPA, despite the presence of Epo, large erythroid bursts will not form (Yen et al. 1985; Migliaccio et al. 1988). Growth factors that have now been isolated and their genes cloned, and which have BPA, include granulocyte/macrophage colony-stimulating factor (GM-CSF) (Kaushansky et al. 1986) and interleukin 3 (IL-3) (Yang et al. 1987).

BFU-E, as they mature and differentiate, give rise to erythroid colony-forming cells (CFU-E), which can be assayed independently in vitro. These progenitor cells have an absolute requirement for Epo in order to form colonies in culture and do not appear to respond in a meaningful way to other growth factors. Recent studies by Koury and Bondurant (1990) propose a model of Epo action that involves the reversal of programmed cell death (apoptosis). Thus, progression along the differentiation/maturation pathway is initially governed by early acting growth factors and, in the absence of Epo, later progenitors and erythroblasts fail to survive. In this model, Epo is a survival factor. Experiments with other growth factors will be useful to see if this pattern of cell growth regulation is a more general phenomenon in hematopoiesis.

14.4.2 In Vivo Studies

Studies in experimental animals and humans suggest that Epo acts primarily on pregenitors intermediate between BFU-E and CFU-E (Adamson et al. 1978; Kimura et al. 1986). This conclusion is supported by several lines of evidence. First, if mice are hypertransfused to produce a state of polycythemia, the numbers and cell cycle kinetics of marrow BFU-E are unchanged, while the numbers of CFU-E decline to approximately 20–25% of normal (Adamson et al. 1978). Second, if rats are made chronically anemic (average hematocrit 15%) through sustained iron deficiency, and then the numbers and cell cycle kinetics of the various progenitors are determined, only the CFU-E compartment is affected (Kimura et al. 1976). No alterations in the numbers or cell cycle kinetics of BFU-E were found while a three- to fourfold increase in total CFU-E was seen (Figure 14–2). From such studies, the concept has emerged that Epo acts during the terminal differ-

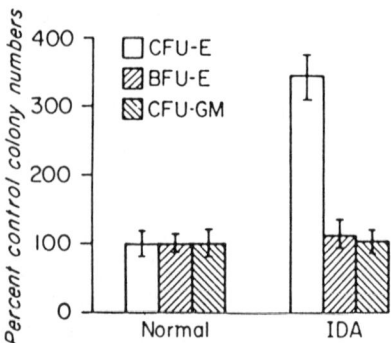

FIGURE 14-2 The total body numbers (marrow and spleen) of BFU-E, CFU-E, and CFU-GM in the chronically anemic (hematocrit 15) rat. Only the CFU-E population was increased and no alteration in cell cycle kinetics was seen in BFU-E or CFU-GM. IDA, iron deficiency anemia. Data from Kimura et al. (1986).

entiation and maturation steps of erythropoiesis, and Epo is required for the maintenance of CFU-E numbers (see Section 14.4.1).

Although Epo is thought generally to be a lineage-specific factor, several workers have shown that Epo also stimulates acetylcholinesterase generation by murine megakaryocytes (Ishibashi et al. 1987), or enhances megakaryocyte colony formation (McLeod et al. 1976). Thus, at least in vitro, Epo appears to be able to induce or accelerate megakaryocyte growth and cytoplasmic maturation.

14.4.3   Studies of the Epo Receptor

Epo interacts with its target cells through a plasma membrane receptor. The availability of recombinant human Epo has triggered a series of early studies of receptor biology. Sawyer et al. (1987), using spleen cells from Friend virus (anemia strain)–infected mice, demonstrated binding of radioiodinated Epo to cells. Using an enriched progenitor cell population, they reported an average of 300 high affinity receptors per cell. Initial studies suggested two classes of receptors having different affinities. This has not been confirmed by other workers, however, and most investigators have reported a single class of receptors as determined by Scatchard analysis.

Problematic in the analysis of these data is the fact that heavily radioiodinated Epo is not biologically active. Nevertheless, displacement analyses by Broudy et al. (1988), as well as binding studies carried out with metabolically labeled recombinant human Epo, confirm the results obtained with the radioiodinated hormone.

If target cells are exposed to radioiodinated Epo, and their membranes solubilized and subjected to polyacrylamide gel electrophoresis, two bands of radioactivity are found (Sawyer et al. 1987). If the molecular weight of Epo is subtracted, the results indicate a receptor of at least two subunits having molecular weights of 85,000 and 100,000.

Broudy et al. (1988) studied a number of murine and human cell lines for Epo-binding properties. They found that a Rauscher virus-transformed erythroleukemia cell line, reported to be responsive to Epo, had an average of 1700 receptors per cell. A single class of receptor was found with an affinity of 440 pM. In contrast, the human erythroleukemia cell line OCIM1 had an average of 3000 receptors per cell with an affinity of 280 pM. The receptor on these cells had at least two subunits of 105,000 and 95,000 molecular weights. Of interest, receptor numbers could be modulated by exposure of cells either to the tumor promoter TPA or to dimethylsulfoxide (DMSO). TPA-stimulated cells down-regulated receptor numbers to 350 per cell, while the cell line GM-979, exposed for 96 hr to DMSO, increased receptor numbers from 1600 to 8400 per cell.

Detailed studies by Sawada et al. (1990) have demonstrated the progressive increase in numbers of Epo receptors and progenitor cell responses from BFU-E to CFU-E. As CFU-E mature through to the erythroblast stage, Epo receptor numbers decline again (Sawada et al. 1987).

D'Andrea et al. (1989) reported the successful cloning of the murine Epo receptor. It is a single polypeptide with a typical membrane-spanning region. Of interest, the receptor expressed on transfected cells displayed two binding constants—one of high affinity (30 pM) and one of low affinity (210 pM). This development should aid substantially in our understanding of the mechanism by which the Epo signal is transduced and how Epo affects nuclear function. Studies of growth factor–dependent cell lines, including Epo-dependent lines, will also be helpful in providing information about the elements of transmembrane signaling and nuclear activation that may be common to more than one growth factor. With labeled rHuEpo now readily available, as well as probes for the human Epo receptor gene, it will soon be possible to study a variety of human disorders, such as recessively expressed familial erythrocytosis (Adamson 1975), to determine if abnormalities in Epo receptor function underlie the clinical phenotype.

14.5 EPO AS A THERAPEUTIC TREATMENT

14.5.1 Chronic Renal Failure

The classic disorder known to be associated with decreased Epo production is CRF. A hypoproliferative anemia is an almost invariable consequence of this condition (Adamson et al. 1968). Most investigators believe that inadequate Epo production is the prime cause of the anemia.

Because of the unique relationship between renal function and Epo production, patients with CRF were the logical choice for Epo treatment when the recombinant hormone became available. Clinical trials with rHuEpo were initiated in 1985 in Seattle (Eschbach et al. 1987) and in 1986 in London (Winearls et al. 1986), and were targeted to anemic patients on hemodialysis. Multicenter trials in the USA (Eschbach et al. 1989a), Western Europe (Bommer et al. 1988a), and Japan (Akizawa et al. 1988) have been conducted. The results support the concept that Epo deficiency is the major mechanism responsible for the anemia.

In the initial U.S. trial, rHuEpo was administered as an intravenous bolus injection three times a week immediately following dialysis. Erythropoiesis was monitored by measurements of the reticulocyte count, quantitative ferrokinetics, transfusion requirements, and hematocrit (Eschbach et al. 1987). The lowest rHuEpo doses used, 1.5 and 5 U/kg body weight, were ineffective. Initial responses were seen at 15 U/kg per dose and, in some patients, this dose resulted in a cessation of transfusion requirements and a partial correction of the anemia (Figure 14–3). At all doses >15 U/ kg, effective responses were observed and the anemia was corrected. As shown in Figure 14–4, the rate of correction of the anemia was correlated directly with the initial dose of rHuEpo.

At the highest doses employed, increases of as much as 10 hematocrit points were observed within 2 to 3 weeks after therapy was started.

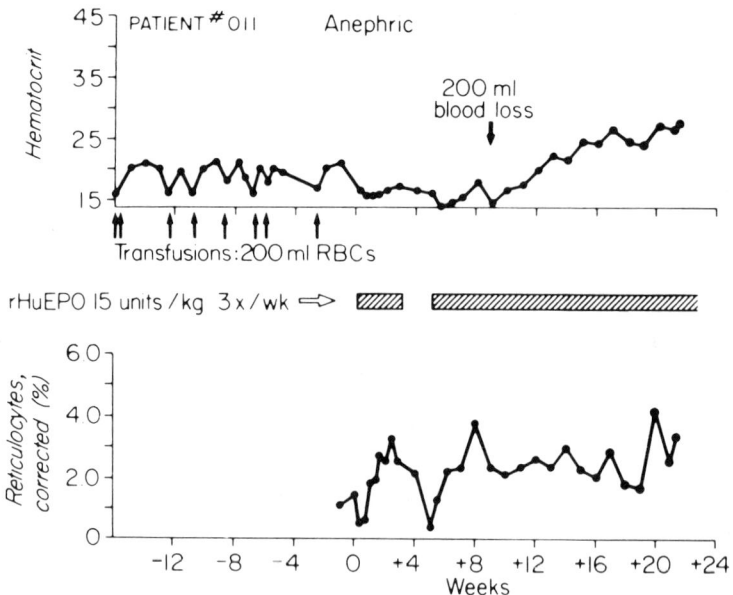

FIGURE 14–3 Response to rHuEpo, 15 U/kg, given intravenously three times weekly to a patient on hemodialysis. Transfusion requirements ceased after the onset of therapy. The patient's hematocrit subsequently peaked at 25 with this dose of rHuEpo (top). The reticulocyte count, corrected for the anemia, rose with therapy and returned to baseline levels when therapy was interrupted for 2 weeks (bottom). Reproduced with permission from Adamson and Eschbach (1989).

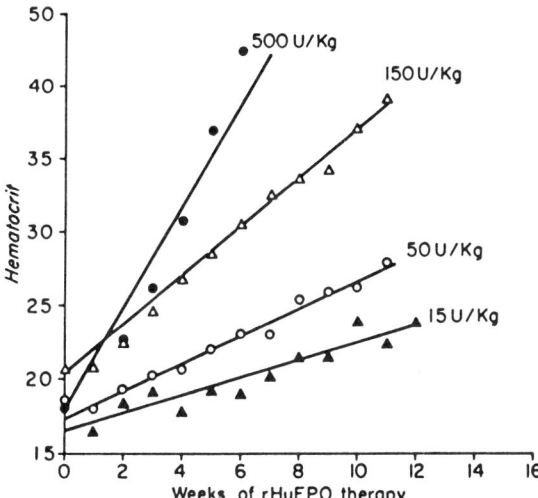

FIGURE 14-4 The hematocrit response to various doses of rHuEpo in hemodialysis patients. Four or five patients comprise each treatment group and the data represent the mean weekly hematocrits. Reproduced with permission from Eschbach et al. (1987).

Anemic hemodialysis patients are now being treated throughout the world. Although the dosage protocols have varied, the findings of the initial studies have been confirmed with a nearly 100% response rate. Patients who have entered maintenance therapy have continued to respond to rHuEpo without evidence of resistance and, at this writing, there have been no reports of patients who have developed antibodies to rHuEpo (Eschbach et al. 1989a). Thus, rHuEpo is effective and well tolerated.

We have also reported on 17 patients with the anemia (hematocrit < 30) of progressive renal failure not yet requiring dialysis (Eschbach et al. 1989b). These patients have responded to rHuEpo in a manner similar to hemodialysis patients. While most of the information available has been obtained with the intravenous administration of the drug, results in pre-dialysis patients (Eschbach et al. 1989b) and from several dialysis centers (Bommer et al. 1988b; Granolleras et al. 1989) indicate that subcutaneous dosing with rHuEpo may be more effective, although the precise correlation with intravenous dosing and details of pharmacokinetics have not been reported.

14.5.2   Use of rHuEpo in Promoting Blood Donations for Autologous Use

Because of the concerns for the safety of blood used in transfusions, a multicenter clinical trial was carried out to determine whether rHuEpo enhanced the ability of individuals to donate blood for self-use. The results of those studies have been published recently (Goodnough et al. 1989). The dose of rHuEpo was 600 U/kg twice weekly, and the study was placebo-controlled and double-blinded. The mean number of units collected from the group receiving rHuEpo was 5.4 ± 0.2 versus 4.1 ± 0.2 (SE) for the placebo-treated group. When the average hematocrit of the units drawn was taken into consideration, rHuEpo allowed the collection of 41% more red blood cells than was possible otherwise. Furthermore, the mean hematocrit of the two patient groups was significantly different at the end of the trial (38.6 versus 35.2 for the rHuEpo-treated and placebo-control groups, respectively). Thus, rHuEpo has potential to increase the number of surgical procedures that can be carried out with the sole use of autologous blood. It is difficult to know how widespread the use of rHuEpo will become in this setting, however, and, in the first report, there was no reduction in the use of homologous blood. As familiarity with the drug increases and strategies for optimizing iron delivery for erythropoiesis improve, it is hoped that adjunctive therapy with rHuEpo in this setting would become increasingly important.

14.5.3 Treatment of Other Chronic Anemias

Considerable interest exists in using rHuEpo in patients with anemia associated with chronic inflammatory or infectious diseases or malignancies. There is a feature of inflammation that might blunt the response to rHuEpo, however. Specifically, chronic inflammatory diseases of many kinds are associated with a reduced release of iron from storage sites and a reduced percent of transferrin saturation. This alteration in internal iron metabolism is referred to as reticuloendothelial iron blockade and is one of the major criteria for the diagnosis of the anemia of chronic disease (Douglas and Adamson 1975). Because this relative iron deficiency has been so refractory to simple manipulations such as oral or parenteral iron supplementation, it might be anticipated that rHuEpo will be less effective in this setting than in otherwise normal individuals.

Rheumatoid arthritis (RA) is a disorder characterized by this kind of anemia. Although the number of rHuEpo-treated RA patients reported to date is small, the drug has been shown to be effective in this group. The initial report included two patients whose hematocrits were normalized with doses of rHuEpo ranging from 150–200 U/kg given intravenously three times weekly (Means et al. 1989). The degree of reticulocytosis was not as great as in patients with CRF. Furthermore, it was not clear that joint symptoms improved in these patients or that quality of life was enhanced as a result of rHuEpo treatment. As more data accumulate in patients with active inflammatory disease, a pattern of relative rHuEpo resistance may emerge, and it is possible that the doses required to maintain target hematocrit in these patients will be higher than the doses required in uncomplicated CRF patients. More experience with rHuEpo in this group of patients will be necessary before firm conclusions can be drawn, but it is encouraging that at least the anemia of one form of chronic disease appears responsive to rHuEpo therapy.

Anemia is also common in patients with acquired immunodeficiency syndrome (AIDS). The anemia is frequently much more severe in those individuals receiving zidovudine (AZT) therapy. Some patients develop red cell aplasia and become transfusion-dependent. In a preliminary report, rHuEpo was shown to decrease by over 50% the number of red cell transfusions in selected groups of AIDS patients treated wtih AZT (Fischl et al. 1990). Some patients not only became transfusion-independent but also normalized their hemoglobin and hematocrit. Those patients whose plasma Epo levels were <500 mU/ml prior to rHuEpo therapy had a greater likelihood of responding to the Epo than those whose endogenous Epo levels were >500 mU/ml. The improvement in hematocrit by rHuEpo in these patients was associated with an improved quality of life. The fact that rHuEpo will reduce the dependence of this group on red blood cell transfusions will save community resources in the form of blood and blood products.

14.5.4   Use of rHuEpo in Other Conditions

Table 14-1 summarizes the known and proposed indications for the use of rHuEpo as a therapeutic treatment. Clearly, patients with anemia associated with CRF or progressive renal failure are the prime candidates for rHuEpo treatment. One would also predict that some patients with myelodysplastic syndromes and, perhaps, chronic anemia associated with malignancy, would respond to rHuEpo with an improvement in hemoglobin and hematocrit, and a reduction in transfusion requirements. Elimination of transfusion dependency would be an important end-point for rHuEpo therapy in these patients, but it is uncertain whether quality of life will be improved demonstrably. It is also uncertain if rHuEpo will be effective in patients with

TABLE 14-1   Defined and Potential Clinical Uses for rHuEpo

1. Anemia of renal failure (both dialysis and predialysis patients)
2. Autologous blood donation
3. Recovery from chemotherapy or radiation therapy
4. Anemia of prematurity
5. Chronic anemias
   a. Inflammation (e.g., rheumatoid arthritis)
   b. Infection (e.g., AIDS with or without AZT therapy)
   c. Neoplasia
   d. Aplastic anemia
   e. Myelodysplastic syndromes
   f. Hemoglobinopathies (e.g., sickle-cell disease)

forms of marrow failure such as aplastic anemia or pure red cell aplasia. Nevertheless, the availability of rHuEpo promises to be a major therapeutic advance for nephrologists, hematologists, and medical oncologists.

14.6 SUMMARY

The application of recombinant DNA technology to the field of hematology has contributed greatly to our understanding of Epo gene structure and regulation, cellular expression and regulation of hormone production, pharmacokinetics, receptor biology, and ultimately, the value of this hormone as a therapeutic treatment. Areas that will undoubtedly prove fruitful for future research include the mechanisms by which hypoxia influences gene expression, structure/function relationships of the Epo molecule, mechanisms of transmembrane signaling and nuclear activation, and the application of rHuEpo in the treatment of other anemias. Epo is but one example of the contribution that modern biology has made to the understanding of hematopoietic regulation and to the availability of these regulators for the treatment of human disease.

REFERENCES

Adamson, J.W. (1975) *Semin. Hematol.* 12, 383–396.
Adamson, J.W., and Eschbach, J.W. (1989) *Quart. J. Med.* 73, 1093–1101.
Adamson, J.W., Alexanian, R., Martinez, C., and Finch, C.A. (1966) *Blood* 28, 354–364.
Adamson, J.W., Eschbach, J.W., and Finch, C.A. (1968) *Am. J. Med.* 44, 725–733.
Adamson, J.W., Torok-Storb, B., and Lin. N. (1978) *Blood Cells* 4, 89–103.
Akizawa, T., Koshikawa, S., Takaku, F., et al. (1988) *Int. J. Artif. Organs* 11, 343–350.
Bert, P. (1882) *Compt. Rend. Acad. Sci.* 94, 805–809.
Bommer, J., Kugel, M., Schoeppe, W., et al. (1988a) *Contrib. Nephrol.* 66, 85–93.
Bommer, J., Ritz, E., Weinreich, T., Bommer, G., and Ziegler, T. (1988b) *Lancet* 2, 406.
Broudy, V.C., Lin, N., Egrie, J., et al. (1988) *Proc. Nat. Acad. Sci. USA* 17, 6513–6517.
Carnot, P., and Deflandre, C. (1906) *Compt. Rend. Acad. Sci.* 143, 432–435.
Cotes, P.M., Pippard, M.J., Reid, C.D.L., et al. (1989) *Quart J. Med.* 10, 113–137.
D'Andrea, A.D., Lodish, H.F., and Wong, G.G. (1989) *Cell* 57, 277–285.
Douglas, S.W., and Adamson, J.W. (1975) *Blood* 45, 55–65.
Egrie, J.C., Strickland, T.W., Lane, J., et al. (1986) *Immunobiology* 172, 213–224.
Egrie, J.C., Eschbach, J.W., McGuire, T., and Adamson, J.W. (1988) *Kid. Inter.* 33, 262A.
Eschbach, J.W., Egrie, J.C., Downing, M.R., Browne, J.K., and Adamson, J.W. (1987) *N. Engl. J. Med.* 316, 73–78.

Eschbach, J.W., Abdulhadi, M.D., Browne, J.K., et al. (1989a) *Ann. Int. Med.* 111, 992–1000.

Eschbach, J.W., Kelly, M.R., Haley, N.R., Abels, R.I., and Adamson, J.W. (1989b) *N. Engl. J.Med.* 321, 158–163.

Fischl, M., Galpin, J.E., Levine, J.D., et al. (1990) *N. Engl. J. Med.* 322, 1488–1493.

Garcia, J.F., Ebbe, S.N., Hollander, L., et al. (1982) *J. Lab. Clin. Med.* 99, 624–631.

Goldberg, M.A., Glass, G.A., Cunningham, J.M., and H.F. Bunn. (1988) *Proc. Natl. Acad. Sci. USA* 84, 7972–7976.

Goodnough, L.T., Rudnick, S., Price, T.H., et al. (1989) *N. Engl. J. Med.* 321, 1163–1168.

Granolleras, C., Branger, B., Beau, M.C., et al. (1989) *Contr. Nephrol.* 76, 143–148.

Ishibashi, T., Koziol, J.A., and Burstein, S.A. (1987) *J. Clin. Invest.* 79, 286–289.

Jacobs, K., Shoemaker, C., Rudersdorf, R., et al. (1985) *Nature* 313, 806–810.

Kaushansky, K., O'Hara, P.J., Berkner, K., et al. (1986) *Proc. Nat. Acad. Sci. USA* 83, 3101–3105.

Kimura, H., Finch, C.A., and Adamson, J.W. (1986) *J. Cell. Physiol.* 126, 298–306.

Koury, M.J., and Bondurant, M.C. (1990) *Science* 248, 378–381.

Koury, S.T., Bondurant, M.C., and Koury, M.J. (1988) *Blood* 71, 524–527.

Koury, S.T., Koury, M.J., Bondurant, M.C., Caro, J., and Graber, S.E. (1989) *Blood* 74, 645–651.

Krantz, S.B., and Jacobson, L.O. (1970) *Erythropoietin and the Regulation of Erythropoiesis,* University of Chicago Press, Chicago.

Lacombe, C., Da Silva, J.-L., Bruneval, P., et al. (1988) *J. Clin. Invest.* 81, 620–623.

Law, M.L., Cai, G.-Y., Lin F.-K., et al. (1986) *Proc. Nat. Acad. Sci. USA* 83, 6920–6924.

Lin, F.-K., Suggs, S., Lin, C.H., et al. (1985) *Proc. Nat. Acad. Sci USA* 92, 7580–7585.

Lin, F.-K., Lin, C.H., Lai, P.H., et al. (1986) *Gene* 44, 201–209.

McDonald, J.D., Lin, F.-K., and Goldwasser, E. (1986) *Molec. Cell Biol.* 6, 842–848.

McLeod, D.L., Shreeve, M.M., and Axelrad, A.A. (1976) *Nature* 261, 492–494.

Means, R.T., Olsen, N.J., Krantz, S.B., et al. (1989) *Arthritis Rheum.* 32, 638–642.

Migliaccio, G., Migliaccio, A.R., and Adamson, J.W. (1988) *Blood* 72, 1387–1392.

Powell, J.S., Berkner, K.L., Lebo, R.V., and Adamson, J.W. (1986) *Proc. Natl. Acad. Sci. USA* 83, 6465–6469.

Sawada, K., Krantz, S.B., Kans, J.S., et al. (1987) *J. Clin. Invest.* 80, 357–366.

Sawada, K., Krantz, S.B., Dai, C.-H., et al. (1990) *J. Cell. Physiol.* 142, 219–230.

Sawyer, S.T., Krantz, S.B., and Luna, J. (1987) *Proc. Nat. Acad. Sci. USA* 84, 3690–3694.

Schuster, S.J., Wilson, J.H., Erslev, A.J., and Caro, J. (1987) *Blood* 70, 316–318.

Sytowski, A.J., and Donohue, K.A. (1987) *J. Biol. Chem.* 262, 1161–1165.

Winearls, C.G., Oliver, D.O., Pippard, M.J., et al. (1986) *Lancet* 2, 1175–1178.

Yang, Y.-C., Ciarletta, A.B., Temple, P.A., et al. (1986) *Cell* 47, 3–12.

Yen, Y.P., Zabala, P., Doney, K., et al. (1985) *J. Lab. Clin. Med.* 106, 384–392.

Hematopoietic Colony-Stimulating Factors

Kenneth Kaushansky
Christopher B. Brown
Stephen Petersdorf

The hematopoietic colony-stimulating factors (CSFs) are a group of acidic glycoproteins required for the survival, proliferation, and differentiation of hematopoietic progenitor cells in semisolid culture, and are involved in the functional activation of mature leukocytes at sites of inflammation. In addition, several CSFs have recently been shown to activate nonhematopoietic tissues including endothelium and the placenta. At recent count, at least 12 biotechnology groups have an active research and development program in one or another of the CSFs. This intense interest stems from the recent demonstration that CSFs act clinically to speed hematopoietic recovery from marrow suppressive therapy, induce the recovery of bone marrow function in several naturally occurring states of marrow failure, and stimulate the functional activation of mature monocytes and neutrophils (Groopman et al. 1989). It is envisioned that the use of these potent biological-response modifiers will significantly alter the morbidity and mortality from natural and treatment-related states of marrow failure, and may potentially play a role in a number of infectious diseases.

The development of a semisolid technique for the culture of blood cells in the mid-1960s (Pluznick and Sachs 1965; Bradley and Metcalf 1966) led to the realization that hematopoietic cells are absolutely dependent on a source of proteins that promote their survival and growth. As the semisolid nature of the culture system resulted in the clonal expansion of single-progenitor cells, the presence of isolated colonies containing one or more of the recognizable blood cell types led to a model of hematopoiesis that still persists (Figure 15–1).

Initially, hemopoietic cultures were supported by cellular feeder layers. Eventually, it was recognized that medium conditioned by a number of cell types could support the developing blood-cell colonies. The protein(s) present in these conditioned culture media were, thus, termed CSFs. Individual CSFs were originally defined by the cell type in the developing hematopoietic

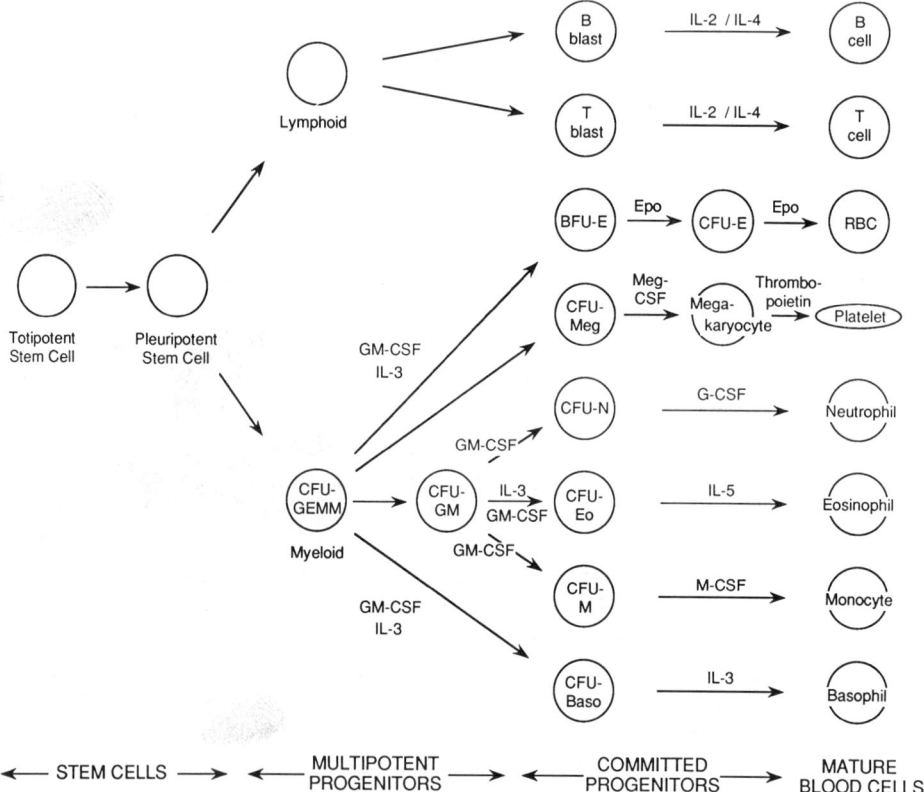

FIGURE 15–1 A model of hematopoiesis. The earliest progenitor cells are thought to undergo a series of cell divisions in which they gradually lose the capacity for pleuripotent differentiation, and eventually develop into progenitors committed to a single-cell lineage. CFU, colony-forming unit; BFU-E, burst-forming unit; Eo, eosinophil; N, neutrophil; E, erythroid; Meg, megakaryocyte; M, macrophage; Baso, basophil; GM, granulocyte-macrophage.

colony. Hence, granulocyte-macrophage (GM)–CSF was defined as a protein that supported the development of colonies composed of granulocytes, macrophages, or a combination of both cell types. In this way, G-CSF, M-CSF, and multi-CSF (also known as interleukin-3, IL-3; capable of stimulating the formation of colonies containing not only myeloid cells [granulocytes and macrophages] but also erythroid and megakaryocytic cells) were also recognized. Subsequently, all four of these distinct activities were biochemically isolated and molecularly cloned from human (Kawasaki et al. 1985; Nagata et al. 1986a; Wong et al. 1985; Yang et al. 1986) and murine (Rajavashisth et al. 1987; Tsuchiya et al. 1986; Gough et al. 1984; Fung et al. 1984) sources.

With the availability of recombinant protein, the CSFs were extensively characterized. GM-CSF, G-CSF, and multi-CSF are biochemically similar. They are all small (M_r 15–35 kDa), acidic (pK 4.5–5.8), secreted, single-chain molecules that contain one or two intrachain disulfide bonds required for biological activity. In contrast, M-CSF is a homodimeric molecule, which contains up to nine disulfide bonds, is relatively large (M_r 45–70 kDa) and exists in both membrane-bound and secreted forms.

The availability of large amounts of purified recombinant CSFs has also allowed their more extensive functional characterization, both in vitro and in vivo. Several characteristics not previously recognized due to a lack of purified CSF from natural sources have now been described. Although originally defined as leading to the formation of only monocyte and granulocytic colonies, GM-CSF has been shown to stimulate at least five cell lineages—neutrophilic, eosinophilic, monocytic, erythroid, and megakaryocytic—to proliferate and differentiate. In addition, M-CSF has been shown to have significant effects on placental tissue (Bartocci et al. 1986), G-CSF and GM-CSF act on endothelial cells (Bussolino et al. 1989), and G-CSF acts together with IL-3 on multipotential hematopoietic progenitor cells (Ikebuchi et al. 1988).

A number of the CSFs act in concert with other cytokines to stimulate the development of a multitude of progenitor cells. The development of serum-free culture systems has led to the realization that multiple CSFs may be required for the complete development of a fully mature leukocyte from its hematopoietic progenitor (Migliaccio et al. 1989). For example, although multi-CSF or GM-CSF can act to stimulate the development of a multitude of cell types in serum-containing cultures, colonies fail to develop in serum-depleted cultures. Addition of G-CSF will allow the development of neutrophil-containing colonies, addition of M-CSF will allow the development of monocyte-containing colonies, addition of erythropoietin will allow development of erythroid colonies, and addition of eosinophil differentiation factor (IL-5) will allow the development of eosinophil-containing colonies. In this way, multi-CSF and GM-CSF are thought to be "early-acting" factors, and G-CSF, M-CSF, erythropoietin, and IL-5 are "late-acting" factors. These findings have expanded the original model of hematopoiesis such that specific differentiation pathways are now recognized to be influenced by more than a single hematopoietic growth factor (Figure 15–1).

The administration of recombinant CSFs to normal and marrow-suppressed animals, and specific immunoassays have led to a greater understanding of the role of the CSFs in hematopoiesis. In vivo administration of CSFs, alone or in combination, results in expansion of the pool of hematopoietic progenitor cells, an increase in the cell cycle status of these cells, an expansion of marrow cellularity, and in most cases, an increase in the numbers and functional state of peripheral blood leukocytes. In addition, using specific assays, M-CSF, G-CSF, and GM-CSF can be detected in the sera of mice infected with the intracellular pathogen *Listeria monocytogenes* (Cheers et al. 1988). Administration of the inflammatory mediators tumor-necrosis-factor, interleukin-1, or lymphotoxin results in the production of M-CSF and GM-CSF in the spleens and lungs of mice (Kaushansky et al. 1988a). In these ways, the CSFs are thought to be critical in the host response to infectious, inflammatory, and immunologic stimuli.

15.1 BIOTECHNOLOGY OF CSFs

Recombinant CSFs have been produced in bacteria, yeast, filamentous fungi, and in a large number of mammalian cell types. Table 15–1 lists some sources of the recombinant CSFs presently undergoing clinical trials. With the exception of M-CSF, all of the CSFs can be produced in any of the conventional vehicles for the production of foreign protein, and no one

TABLE 15–1   Sources of Production of the CSFs for Use in Preclinical and Clinical Trials

Recombinant Protein	Producer[1]	Source[2]
M-CSF	Cetus, Inc.	M, B
	Genetics Institute	M
G-CSF	AMGen, Inc.	M, B
	Chugai Pharmaceuticals	B
GM-CSF	Sandoz Pharmaceuticals	M
	Immunex, Inc.	Y
	Schering-Plough	B
	Hoechst/Behringwerke	B
	ZymoGenetics, Inc.	M, F
Multi-CSF	Genetics Institute	M, B
	Immunex/Behringwerke	M, Y, B
	DNAX	B

[1] The list of manufacturers is not necessarily complete.
[2] M = mammalian cells, F = filamentous fungi, Y = yeast, B = bacteria.

production source has been proven superior in terms of bioavailability or biological effectiveness. With the exception of the initial distribution phase, the presence of carbohydrate has not been shown to vastly alter the pharmacokinetics of the various forms of the CSFs. However, a recent report suggests that the source of production may influence the immunogenicity of the resultant recombinant protein. Recombinant human GM-CSF produced in yeast and administered to a group of patients undergoing combination chemotherapy for malignancy was shown to evoke a specific immunologic response in 4 of 13 patients (Gribben et al. 1990). The absence of O-linked carbohydrate, combined with the altered N-linked carbohydrate structure characteristic of yeast-derived protein was felt to be responsible for the immunologic responses of these patients. In only one of the patients was the antibody response considered significant, manifest as failure to maintain adequate serum levels of the drug and failure to mount a hematopoietic response. This study, although preliminary, should presage the careful examination of recombinant proteins in humans.

15.2 MULTI-CSF

Multi-CSF, or IL-3, was originally characterized as a cytokine that stimulated production of the lymphocyte enzyme 20-alpha-steroid dehydrogenase (Ihle et al. 1981). Simultaneously, several other groups described activities including WEHI-3 growth factor, P-cell stimulating factor, mast cell growth factor, and histamine-producing cell-stimulating factor, which ultimately proved to be multi-CSF (Ihle et al. 1983). As its name implies, multi-CSF stimulates a wide range of progenitor cells to produce hematopoietic colonies in vitro and stimulates a number of mature blood cell types to become functionally activated. In the presence of growth factors that stimulate the terminal differentiation of the developing cells, multi-CSF acts during the G_1 phase of the cell cycle (Kelvin et al. 1986) to stimulate erythroid, eosinophilic, basophilic, monocytic, and megakaryocytic progenitors to proliferate and differentiate in semisolid culture (Valent et al. 1989a; Lopez et al. 1989; Messner et al. 1987; Bot et al. 1989; Metcalf et al. 1989; Chen and Clark 1986). In addition, multi-CSF promotes the maturation of mature megakaryocytes (Burstein 1986), the functional activation of eosinophils (Rothenberg et al. 1988) and the proliferation of mature monocytes and macrophages (Whetton et al. 1986).

Like other CSFs, multi-CSF acts synergistically with other hematopoietic growth factors, both in vitro (Sieff and Ekern 1988) and in vivo (Broxmeyer et al. 1987a). In combination with either GM-CSF or M-CSF, the concentrations of multi-CSF required to increase hematopoietic progenitor cell cycling and to increase the number of progenitor cells in the marrow and spleens of animals was reduced 10- to over 400-fold (Broxmeyer et al. 1987b). Unlike the other CSFs, multi-CSF is capable of supporting the self-renewal of pleuripotent stem cells, at least those which give rise to murine spleen colonies and human blast-cell colonies (Spivak et al. 1985; Leary et al. 1987; Kobayashi et al. 1989). Thus, multi-CSF acts on the earliest hematopoietic cells yet identified. These findings support the hypothesis that multi-CSF is the primary growth factor for pleuripotential progenitors.

The regulation of multi-CSF production has been intensely studied during the past few years. A number of investigators have determined that the primary source of multi-CSF is a subpopulation of stimulated T-lymphocytes (Niemeyer et al. 1989; Wimperis et al. 1989; Kaushansky et al. 1989a). However, recent studies utilizing the polymerase chain reaction from other cell types and Northern blot analysis using large amounts of poly-A-enriched RNA have demonstrated the presence of multi-CSF-specific transcripts in monocytes and mast cells (Ernst et al. 1989; Wodnar-Filipowicz et al. 1989). The physiologic significance of these alternate sources of multi-CSF is, however, uncertain.

The gene for human multi-CSF is located on the long arm of chromosome 5 (LeBeau et al. 1987), in very close proximity to the gene for GM-CSF (Barlow et al. 1987). This, and other evidence, point to a common ancestral gene for these two growth factors. The genetic elements required for the highly inducible and tissue-specific expression characteristic of multi-CSF are now under study. The 5'-flanking region of the human and murine multi-CSF genes contain a number of homologies with other lymphokine genes. Specifically, a decanucleotide sequence termed CK-1 has been found in the genes encoding multi-CSF, IL-2, the IL-2 receptor, leukemia inhibitory factor, G-CSF, and GM-CSF (Stanely et al. 1985; Kaushansky 1987; Shannon et al. 1988). An octanucleotide sequence termed CK-2 is also found in the multi-CSF and GM-CSF genes (Shannon et al. 1988). These sequences have been proposed to function to enhance GM-CSF expression, but their role in the regulation of multi-CSF is uncertain (Shoemaker et al. 1990; Mathey-Prevot et al. 1990). Also present in the 5'-flanking region of the multi-CSF gene is a heptanucleotide activator protein-1 (AP-1) site, a sequence that binds the transcription factors encoded by the cellular proto-oncogenes *c-jun* and *c-fos,* and the related genes *jun-B, jun-D,* and *fra-1* (Turner and Tjian 1989). This site, located approximatly 300 bp upstream of the site of multi-CSF transcription initiation, has recently been shown to functionally upregulate multi-CSF gene expression in human T-lymphocyte cell lines (Shoemaker et al. 1990; Mathey-Prevot et al. 1990), as well as IL-2 in murine T-cells (Muegge et al. 1989). Finally, a second important enhancer region recognized in the multi-CSF gene is located approximately 150 bp upstream of the transcription initiation site, and appears to bind a novel transcriptional regulator (Shoemaker et al. 1990).

Most of what is known of the structure-function relationships of multi-CSF comes from the analysis of the murine growth factor, a molecule that shares 25% sequence identity with the human cytokine. The cDNA for murine multi-CSF encodes a polypeptide predicted to contain a 27-amino acid secretory leader and a 139-amino acid secreted polypeptide (Fung et al. 1984). The human cDNA encodes a 19-residue leader and a 133-residue mature polypeptide (Yang et al. 1986). Like GM-CSF, murine multi-CSF contains two disulfide bonds, located between Cys_{16} and Cys_{79} (no. 1 = mature N-terminal Ala) and between Cys_{78} and Cys_{139} (Clark-Lewis et al. 1988a). In contrast, human multi-CSF contains only the first pair of these Cys residues. In keeping with these findings, studies utilizing a series of substitution mutants of murine multi-CSF revealed that only the first of these two disulfide bonds is critical for its biological function (Clark-Lewis et al. 1988b).

Murine multi-CSF contains a large amount of N-linked carbohydrate, present on four Asn residues. Human multi-CSF is predicted to contain N-linked carbohydrate on two Asn residues. In contrast to the other CSFs, O-linked carbohydrate modification has not been detected on murine multi-CSF (Ziltener et al. 1988). But, like other molecules in which the question has been addressed (e.g., interferon gamma and the IL-2 receptor), murine multi-CSF displays microheterogeneity attributable to varying patterns of carbohydrate modification (Ziltener et al. 1988). This degree of size and charge heterogeneity, characteristic of the CSFs in general, is dependent on the cells or cell lines used for production of the growth factor.

Multi-CSF acts on hematopoietic progenitor cells and mature leukocytes by binding to specific cell-surface receptors. Receptors for multi-CSF have been identified upon cells of multiple-hematopoietic lineages, including erythroid precursors, and eosinophils, monocytes, and mast cells (Nicola and Metcalf 1986; Park et al. 1986). Most investigators have reported the presence of a low number (approximately 100) of IL-3 receptors on normal marrow cells that display a single high-affinity binding constant (K_m approximately 100 pM). However, other studies have suggested the presence of an additional class of low-affinity (K_m approximately 10 nM) receptors, primarily upon monocytes (Nicola and Metcalf 1986; Park et al. 1989).

Using chemical cross-linking agents, several groups have identified two or three cell-surface polypeptides that bind to ^{125}I-multi-CSF. Studies in murine cells have identified 65- to 70-kDa and 140-kDa polypeptides, although the smaller polypeptide may represent a proteolytic breakdown product generated during processing of the cell membranes (Isfort et al. 1988a). A number of groups have demonstrated a role for protein phosphorylation during signal transduction mediated by multi-CSF binding (Kanakura et al. 1989). This observation correlates with the ability of other tyrosine-kinase receptors to abrogate the dependence on multi-CSF when transfected into multi-CSF-dependent cell lines (Isfort et al. 1988b). In keeping with these findings, several groups have shown that the larger IL-3 binding polypeptide undergoes rapid phosphorylation on tyrosine in response to ligand binding (Koyasu et al. 1987; Sorenson et al. 1989).

Although binding of multi-CSF to its receptor was initially thought to be specific, recent studies have led investigators to reexamine these conclusions. Using human monocytes, leukemic cell lines, and fresh acute mye-logenous leukemia (AML) cells that contain binding sites for both growth factors, several groups have reported that multi-CSF and GM-CSF have been able to partially to completely cross-compete for binding to the het-erologous receptor (Gesner et al. 1989; Kannourakis et al. 1989; Park et al. 1989; Budel et al. 1990). Additional evidence that may help to explain these findings is based upon cross-linking studies using human AML blast cells and monocytes. Using human ^{125}I-multi-CSF, cross-linked bands of 175, 105, and 85 kDa have been detected. Of considerable interest, ^{125}I-GM-CSF is also cross-linked to three polypeptides, one or more of the bands in common with those cross-linked to ^{125}I-IL-3, suggesting that the IL-3 and GM-CSF receptors might share a common subunit (Kannourakis et al. 1989; Budel et al. 1990). In addition, binding of ligand to either receptor results in the rapid phosphorylation of several membrane and cytoplasmic proteins. One group has reported that a cytoplasmic phosphoprotein, p93, is phos-phorylated in response to binding of either multi-CSF or GM-CSF (Kan-akura et al. 1989). The significance of these findings is uncertain, although the similarity of response of cells that bear receptors to both multi-CSF and GM-CSF (Elliot et al. 1989) may be explained on the basis of a shared signal-transduction pathway.

Recently, a multi-CSF-binding protein was cloned from a mast cell line that displays a low-affinity binding site for the growth factor. When trans-fected into fibroblasts that are devoid of multi-CSF receptors, the cDNA encodes a polypeptide with a 22-amino acid secretory leader sequence, a 417-amino acid extracellular domain, a 26-residue transmembrane domain, and a 413-amino acid cytoplasmic domain. The transfected cDNA led to the production of a surface-membrane protein that bound multi-CSF with an affinity equal to that of the cell line from which the cDNA library was produced. In addition, the affinity of this multi-CSF binding protein was similar to that of the low-affinity receptor found on normal monocytes and cross-linking studies of these transfected cells revealed bands characteristic of those found on normal monocytes.

A number of other cytokine receptors have recently been cloned and a superfamily of such proteins is now recognized. Included in this family are the receptors for IL-2, erythropoietin, GM-CSF, IL-4, IL-6, and IL-7. Thus far, the low-affinity binding protein for all of these receptors have been identified. Further, in the case of IL-2 and IL-6, a second binding protein has been identified that, when complexed to the low-affinity binding poly-peptide, results in a dimeric receptor of high-binding affinity. It is thus anticipated that a second subunit of the multi-CSF receptor will be iden-tified.

The role of multi-CSF in hematopoiesis in vivo is less clear than for the other CSFs. Using long-term marrow culture as a model of the marrow microenvironment, several groups have been able to demonstrate the pres-ence of M-CSF, G-CSF, and GM-CSF in the stromal cells that support these cultures (Nemunaitis et al. 1989; Simmons et al. 1990). In contrast, multi-CSF has yet to be identified in these cultures. In addition, multi-CSF has not been detected in the serum of infected animals (Cheers et al. 1988), or in bone marrow cells or cells at the site of inflammation (Kaushansky et al. 1988a). Despite these findings, a case can be made for a role for multi-CSF in allergic reactions through its effects on basophils, eosinophils, and plasma histamine levels (Valent et al. 1989b). In addition, multi-CSF is the only CSF, yet identified, capable of supporting the self-renewal and proliferation of pleuripotent hematopoietic progenitor cells. The high specific activity of multi-CSF coupled with the low abundance of hematopoietic progenitor cells likely accounts for the failure to detect multi-CSF at sites of hematopoiesis.

The presence of receptors for multi-CSF on most cases of fresh leukemic cells (Budel et al. 1989), the correlation between the presence of these receptors and a proliferative response to exogenous growth factor (Vallenga et al. 1987), and the production of the growth factor from fresh leukemic cells in some patients (Demetri and Griffen 1989) suggest that multi-CSF may be involved in the development of myeloid leukemia. In this regard, the continuous administration of recombinant protein through the transplantation of multi-CSF-producing cells into irradiated syngeneic mice led to a hematopoietic myeloproliferative syndrome (Chang et al. 1989). In addition, the recent demonstration that intracellular multi-CSF can act as an autocrine growth stimulator (Dunbar et al. 1989) provides circumstantial evidence that multi-CSF may be involved in the multi-step pathogenesis of leukemia, and has led to a substantial level of current research.

Finally, despite the many questions about the physiological role of the protein, it is clear that the pharmacologic administration of recombinant multi-CSF leads to a significant hematopoietic response. This is particularly true when administered in combination with other CSFs and cytokines. In preparation for preclinical trials, a number of studies of the pharmacokinetics of both recombinant murine and human multi-CSF have been performed. Both native and recombinant E. coli murine multi-CSF were found to have a circulatory half-life of 3–5 min (Metcalf and Nicola 1988). Most of the administered protein was eliminated in the liver and kidneys, but labeled multi-CSF was also detectable in hematopoietic organs. Similar results were obtained with human multi-CSF administered to primates (Donahue et al. 1988).

In the initial trials of murine multi-CSF, minimal effects on the peripheral blood cell counts were noted with the infusion of moderately high doses of recombinant protein. A significant effect on the numbers of hematopoietic progenitor cells was noted, however, and in keeping with the in vitro effects on progenitor cells of multiple lineages, this expansion was as marked for erythroid progenitors as for myeloid progenitor cells (Kindler et al. 1986). In studies that utilized recombinant human multi-CSF administered to rhesus monkeys, peripheral leukocytosis has been more significant.

In several studies, peripheral blood basophil and eosinophil counts have been noted to rise substantially, and peripheral neutrophil counts have increased to a lesser extent (Donahue et al. 1988). In a number of models of chemotherapy-induced and natural states of marrow failure, human multi-CSF has been shown to improve blood cell production (Gillio et al. 1989; Ganser et al. 1989). For example, in two series of patients with spontaneous marrow failure, a high proportion of patients have shown a significant hematopoietic response with multiple lineage improvement (Kurzrock et al. 1989; Ganser et al. 1989). The effects on hematopoietic progenitor cell numbers in primates and humans has been similar to those in mice; they induce a rise in the numbers of all types of marrow cells (Donahue et al. 1988) and in the levels of circulating progenitor cells (Monroy et al. 1989).

The expansion of the pool of marrow and circulating progenitor cells would be predicted to prime the hematopoietic system to respond to the later-acting growth factors. Recent preclinical studies in primates have confirmed these predictions. Studies combining human multi-CSF and GM-CSF have resulted in substantial stimulation of hematopoiesis, with increases in circulating progenitor cells and mature blood cells being far greater than that produced by either agent alone (Donahue et al. 1988; Gesner et al. 1989; Mayer et al. 1989). These effects to increase circulating progenitor cells have also suggested a role for multi-CSF in providing adequate numbers of stem cells for transplantation, a role previously restricted to bone marrow harvesting.

Taken together, the current trend is for the use of multi-CSF to expand the number of progenitor cells that will respond to a later-acting CSF. Thus, it is envisioned that multi-CSF will assume a clinical role of hematopoietic primer, its use restricted primarily in combination with other agents. Its role in providing increasing numbers of circulating stem cells for transplantation is provocative, and the recent demonstration of the feasibility of using peripheral blood stem cells (To and Juttner 1987) should provide an additional clinical role for this multipotential hemopoietin in the coming years.

15.3 GM-CSF

GM-CSF is a glycoprotein with an approximate molecular weight of 25 kDa. Although originally named for its action on hematopoietic precursors of the myelomonocytic lineage, GM-CSF has now been shown to influence non-myeloid precursors and to functionally activate mature myelomonocytic cells. GM-CSF is produced and secreted by several cell types, including T-lymphocytes (Kaushansky et al. 1989a), monocytes (Thorens et al. 1987), fibroblasts (Kaushansky et al. 1988b), endothelial cells (Broudy et al. 1986), and bone marrow stromal cells (Lovhaug et al. 1986). Although not detectable in cultures of unstimulated cells, GM-CSF production and secretion occurs in response to both humoral and physical stimuli. The involvement in the stimulation of hematopoiesis or in the inflammatory response is common to all of the various stimulators of GM-CSF gene expression.

GM-CSF mRNA or secreted, immunologically or biologically detectable protein is produced by T-cells in response to culture in the presence of IL-1 (Bagby et al. 1981) and IL-2 (Kelso et al. 1986), by fibroblasts after stimulation with IL-1 (Kaushansky et al. 1988b), by endothelial cells stimulated with IL-1 (Bagby ct al. 1986; Sieff et al. 1987), or tumor-necrosis factor (TNF) (Broudy et al. 1986), by marrow stromal cells growing in long-term culture under the influence of IL-1 and bacterial lipopolysaccharide (LPS) (Rennick et al. 1987), and by macrophages stimulated by LPS (Thorens et al. 1987). Of interest, fibronectin (which binds to specific receptors on macrophage membranes and can act as a surface for macrophage attachment and migration at sites of tissue injury) and the process of phagocytosis both induce GM-CSF gene expression in macrophages (Thorens et al. 1987). Gamma interferon and, not surprisingly, the powerful anti-inflammatory agent dexamethasone have been shown to inhibit the LPS-induced increase in GM-CSF mRNA by macrophages (Thorens et al. 1987).

GM-CSF is capable of influencing both hematopoietic progenitor cells and the function of mature blood cells. Although originally defined as supporting the development of colonies containing only granulocytes and macrophages, recombinant GM-CSF, acting in the G_1 phase of the cell cycle (Pluznik et al. 1984), induces the formation of granulocyte, macrophage, granulocyte/macrophage, eosinophil, megakaryocytic, and, in the presence of erythropoietin, erythroid and mixed erythroid-nonerythroid colonies (Sieff et al. 1985; Metcalf et al. 1986; Tomonaga et al. 1986: Kaushansky et al. 1986). Experiments done in serum-free conditions suggest that GM-CSF acts in concert with other cytokines, such as G-CSF, M-CSF, and erythropoietin, to exert its full spectrum of activities (Bot et al. 1989; Migliaccio et al. 1989).

GM-CSF exerts a variety of effects on mature blood cells that assist, in an orderly fashion, to localize neutrophils and monocytes at sites of tissue injury or invasion, enhance recognition and phagocytosis of foreign material, and to augment the production of oxidative metabolites. This occurs, at least in part, in several ways. After a brief exposure to GM-CSF, neutrophils display a threefold increase in receptors for the bacterial chemoattractant fMet-Leu-Phe (Weisbart et al. 1986), and enhanced chemotaxis in response to this formyl peptide. In addition, mediated by an increase in cell surface adherence molecules such as the integrins and the complement breakdown product $C3b_i$, neutrophils become adherent to vascular endothelium (Ross and Lambris 1982; Anderson and Looney 1986; Zimmerman and McIntyre 1988). This is the critical step in the migration of phagocytes into the inflammatory focus. With longer exposure, GM-CSF acts as a migration-inhibition factor (Gasson et al. 1984), localizing the phagocytes at the site of the inflammatory process. GM-CSF also enhances phagocytosis of bacteria (Fleischmann et al. 1986) and yeast (Metcalf et al. 1986), and increases the expression of F_c receptors for immunoglobulin G (IgG) (Perussia et al. 1987) and IgA (Weisbart et al. 1988) on neutrophils, facilitating antibody-dependent cellular cytotoxicity. Finally, while not directly increasing oxidative metabolism in neutrophils, additional exposure to GM-CSF primes neutrophils to respond to endogenous ($C5_a$ and Leukotriene B_4) and exogenous (fMet-Leu-Phe and zymosans) inflammatory mediators to augment superoxide anion production (Weisbart et al. 1985, 1987).

GM-CSF likely carries out these activities by acting in a highly localized manner. GM-CSF is only very rarely found circulating in biologically relevant concentrations. However, using sensitive immunoassays, GM-CSF can be detected at sites of ongoing inflammation, such as in the synovial fluid of patients with acute rheumatoid arthritis (Xu et al. 1989). In addition, GM-CSF has been found adherent to glycosaminoglycans in the stroma of long-term marrow cultures (Gordon et al. 1987), apparently poised to act on hematopoietic progenitors, again in a highly localized manner. Taken together, these data are consistent with the hypothesis that GM-CSF plays an important role in host-defense mechanisms. The cytokine is produced in response to known mediators of the inflammatory reaction by those cells most likely to be present at sites of tissue injury and pathogenic invasion. It acts to initiate and amplify the defense mechanisms of mature cells of the immune system. GM-CSF can also act in the marrow to stimulate the production and recruitment of additional phagocytes into host defenses.

Recent data have demonstrated that GM-CSF influences a number of cell types and tissues not considered part of the hematopoietic system, including the placenta (Wegman et al. 1989), endothelial cells (Bussolino et al. 1989), and osteoblasts. The role played by GM-CSF in these nonhematopoietic tissues is unclear, but the boundaries of activity of this cytokine remain to be fully defined.

The cloning of the gene and cDNAs for human and murine GM-CSF has led to a better understanding of the control of GM-CSF gene expression (Gough et al. 1984; Wong et al. 1985; Stanley et al. 1985; Miyatake et al. 1985; Kaushansky et al. 1986). The GM-CSF gene exists in single copy on the long arm of human chromosome 5 (LeBeau et al. 1986) and on murine chromosome 11 (Barlow et al. 1987), respectively. The human gene is clustered with several other cytokine genes, including IL-3, IL-4, IL-5, M-CSF, platelet-derived growth factor (PDGF), and epidermal growth factor (EGF), as well as that for the M-CSF receptor, encoded by the *c-fms* proto-oncogene and the PDGF receptor.

The organization of the human and murine GM-CSF genes are similar (Miyatake et al. 1985). Four exons separated by three introns span 2.5 kbp of the chromosome. Except for a 9-bp deletion in the first exon of the murine gene, the exons of the murine and human genes are exactly the same length. Overall, the two genes share 70% sequence homology. However, the most highly conserved region is located in the 5'-flanking region of the human and murine GM-CSF genes. This finding suggests the presence of sequence elements involved in the transcriptional regulation of the cytokine.

Using a series of promoter-reporter gene constructions, several groups have shown that the first 63 bp 5' of the cap site are all that are necessary for fully inducible GM-CSF gene expression (Nimer et al. 1988; Kaushansky 1989). Located between 49 and 63 bp upstream of the cap site are repeating CATT(T/A) sequences. This region has been shown to have upregulatory function on reporter gene expression in both endothelial cells (Kaushansky 1989) and lymphocytes (Nimer et al. 1988), and was shown to bind a nuclear protein or protein complex by DNAase footprinting (Nimer et al. 1988), lending additional evidence to its possible regulatory role. However, an octanucleotide sequence present in the murine and human GM-CSF genes, as well as the human and murine IL-3 genes, a decanucleotide sequence present in a number of inflammatory response genes, including GM-CSF, IL-3, G-CSF, and the IL-2 receptor, and a negative regulatory element shown by functional analysis to lie between 179 and 193 bp upstream of the cap site may also play a role in the regulation of GM-CSF gene expression (Shannon et al. 1988; Nimer et al. 1988; Kaushansky 1989).

Posttranscriptional mechanisms also play an important role in control of GM-CSF production. The 3'-untranslated region (UTR) of GM-CSF mRNA contains a region rich in adenine and uracil nucleotides (AU rich) also found in the 3'-UTR of a number of other cytokines and proto-oncogenes. The presence of this region has been shown to impart a remarkable degree of instability upon the mRNA in which it resides (Shaw and Kamen 1986), and the resultant short half-life of the transcript might allow regulation through a mechanism involving stabilization of the GM-CSF message. The site at which this regulation is exerted and its mechanism is a topic of much recent interest (Bagby et al. 1988).

The cloning of the cDNAs for murine and human GM-CSF have also allowed investigation into the structure/function relationships of the proteins. The murine and human cDNAs encode polypeptides of 141 and 144 amino acids, respectively, which share 54% amino acid sequence identity. Each cDNA encodes a 17 amino acid leader sequence that is cleaved to form mature polypeptides of 124 and 127 amino acids. Despite this high degree of sequence homology, a conserved pattern of disulfide bonding, and similar physicochemical parameters (e.g., helical content) there is no cross-species activity. Human and murine GM-CSF contains two sites of N-linked and three or four sites of O-linked carbohydrate modification. Variable glycosylation gives GM-CSF its characteristic size heterogeneity. Thus far, however, no functional role in the production, secretion, or in vivo specific activity of GM-CSF has been determined for the carbohydrate modification characteristic of the molecule (Kaushansky et al. 1987; Kaushansky 1990).

Studies using deletion and insertion mutants (Shanafelt and Kastelein 1989; Gough et al. 1987), synthetic polypeptides (Clark-Lewis et al. 1988b), interspecies hybrids (Kaushansky et al. 1989c), and monoclonal-antibody-epitope mapping (Brown et al. 1990) have identified regions critical to the functional activity of GM-CSF. Portions of both the amino and carboxyl regions of the protein seem to be involved in the receptor binding and biological activity of human GM-CSF. Recent physical evidence suggests that these two regions, separated by at least 40 residues in the linear sequence of the protein, are brought into close juxtaposition in the tertiary structure of human GM-CSF (Brown et al. 1990).

Using a number of computer algorithms and physical measurements it can be predicted that GM-CSF is composed of several alpha helixes (Wingfield et al. 1988). Parry and coworkers have recently suggested that GM-CSF is arranged in a four-helix bundle, a structure similar to that recently determined by X-ray crystallography for IL-2 (Parry et al. 1988). The first of the helixes is amphiphilic, and contains on its hydrophilic surface many of the residues shown to be functionally important for receptor-binding and biological activity (Kaushansky et al. 1989b). A long loop between the third and fourth helixes contain the carboxyl region of the protein identified as part of the active site of the molecule. Computer modeling has suggested that these two regions (helix and loop) are also in close juxtaposition (Kaushansky et al. 1990).

The initial step in the action of the CSFs is their binding to specific cell-surface receptors. Specific high-affinity GM-CSF receptors exist on hematopoietic cells responsive to the growth factor (Walker and Burgess 1985; Gasson et al. 1986; Park et al. 1986; DiPersio et al. 1988). In addition, GM-CSF receptors have been identified on endothelial cells (Bussolino et al. 1989), placental membranes (Gearing et al. 1989), small-cell lung carcinoma cell lines, and on SV40 transformed simian COS cells (Cocita Bladwin et al. 1989). The role of the receptors in these latter settings is unclear, however, the recent demonstration of a biological effect of GM-CSF on cultured endothelium (Bussolino et al. 1989) may provide insights into a previously unrecognized role for GM-CSF in vascular biology.

When cross-linking techniques have been used to characterize the receptor on a variety of cells an 84-kDa protein is consistently identified (DiPersio et al. 1988; Kannourakis et al. 1989; Gearing et al. 1989; Gesner et al. 1989; Cocita Baldwin et al. 1989). A smaller, molecular weight binding protein has also been demonstrated on some cells by some, but not all, investigators (DiPersio et al. 1988; Cocita Baldwin et al. 1989; Kannourakis et al. 1989). Receptor density on receptor-bearing cells is quite low. On normal human myeloid cells, the receptor number increases with maturation (DiPersio et al. 1988). However, mature neutrophils and eosinophils only contain about 1000 receptors per cell while immature human myeloid cell lines contain only a few hundred receptors per cell (DiPersio et al. 1988).

The binding characteristics of the GM-CSF receptor are a source of controversy. The existence of a high-affinity receptor on human cells is widely accepted. This site binds with a K_d of <100 pM and can mediate all observed functions of GM-CSF in human hematopoietic cells (Gasson et al. 1986). A second, low-affinity class of receptors has been observed by some authors, but not by others using the same cells (Elliot et al. 1989; DiPersio et al. 1988; Gasson et al. 1986; Gearing et al. 1989). When detected, this low-affinity binding site usually exists on cells that also express high-affinity receptors. In contrast, placental membranes and COS cells have only a low-affinity binding site (Gearing et al. 1989; Cocita Baldwin et al. 1989). The explanation for this apparent variability in expression of high- and low-affinity receptors is unclear.

Studies of the specificity of the GM-CSF receptor have indicated a unique relationship between GM-CSF and IL-3. In binding studies using human cells, where the possibility of receptor internalization was carefully excluded, GM-CSF and IL-3 partially to completely cross-compete for binding to each others' receptors on cells that express both IL-3 and GM-CSF receptors (Gesner et al. 1989; Kannourakis et al. 1989; Park et al. 1989; Budel et al. 1990; Elliot et al. 1989; Lopez et al. 1989). There is no correlation between the expression of one or two GM-CSF binding site types and the ability of IL-3 to compete for receptor occupancy. No other cytokines have been found to date that demonstrate such cross competition.

Using a human placental cDNA library expressed in COS cells, a cDNA clone was recently obtained that binds ^{125}I-GM-CSF with an affinity characteristic of intact placental membranes (low-affinity binding site [Gearing et al. 1989]). The clone contains a 1.2-kb open-reading frame that is predicted to encode a 22-amino acid leader peptide and a mature protein of 378 amino acids (of which 320 reside in an extracellular domain). The discrepancy between the predicted M_r of 43 kDa and the observed 84 kDa found in cross-linking studies of a number of cell types may be ascribed to the presence of 11 potential sites of glycosylation in the extracellular domain. The short 54-amino acid intracellular domain does not contain sequence homology to a tyrosine-kinase catalytic domain. The structure of the GM-CSF receptor is most homologous to the human IL-6 receptor but is part of a family of receptors that also includes the erythropoietin receptor, the IL-2 receptor beta chain, the IL-3 receptor, the IL-4 receptor, the IL-7 receptor, and the prolactin receptor.

Evidence that this cloned product represents at least a portion of a GM-CSF receptor complex rests on several observations. Northern blot analysis demonstrates that GM-CSF receptor-bearing cells, in contrast to receptor-negative cells, express mRNA coding for the cloned message. In addition, ^{125}I GM-CSF can be cross-linked to an 84-kDa protein in receptor-negative cells transfected with the GM-CSF receptor cDNA. Taken together, these data suggest that the recently reported cDNA clone does encode one subunit of the GM-CSF receptor, but based on the arrangement of the IL-2 receptor, it is very likely that a second subunit will be described that converts this cDNA product into a high-affinity, functional receptor.

Based on the biological profile of GM-CSF and promising preclinical animal studies, recombinant human GM-CSF has been studied in a number of clinical trials involving patients with a variety of states of marrow failure (reviewed in Groopman et al. 1989; Mitsuyasu and Golde 1989). In most situations, the drug has proven highly efficacious.

In the setting of autologous bone marrow transplantation for solid tumors or lymphoid malignancies, the administration of GM-CSF led to a significant reduction in the length of the period of leukopenia and in the incidence of infectious complications. In addition, the drug reduced the number of platelet transfusions and shortened the period of hospitalization. GM-CSF administered to a group of patients undergoing conventional chemotherapy for sarcoma raised the nadir leucocyte counts, and shortened the period of pancytopenia and the interval between cycles of chemotherapy. Similar effects on leukocyte counts have been reported in other studies. The role of GM-CSF in aplastic anemia and in myelodysplastic syndromes (MDS) is less clear. Improvements in cell counts of treated patients have been modest and restricted to the granulocyte lineage. In addition, patients with high initial myeloblast counts with MDS have had an apparent acceleration of their disease course. Given that the defect in these disorders is at the level of the hematopoietic stem cell, it is unlikely that GM-CSF alone would be useful in the treatment of these disorders.

15.4 G-CSF

G-CSF, as its name implies, primarily affects cells of the granulocytic series. Murine G-CSF is a 22- to 25-kDa glycoprotein initially characterized from serum and from lung-conditioned medium from mice injected with endotoxin (Burgess and Metcalf 1980). Human G-CSF is a 19-kDa protein first isolated from placental-conditioned medium and the bladder carcinoma cell line 5637 (Welte et al. 1985). In contradistinction to GM-CSF, and perhaps due to the higher degree of homology among the sequences of different species, G-CSF activity crosses a wide spectrum of species boundaries (Zsebo et al. 1986).

The cellular sources of G-CSF include stimulated monocytes (Vallenga et al. 1988), endothelial cells (Zsebo et al. 1988), marrow stomal cells (Fibbe et al. 1988), mature neutrophils (Lindemann et al. 1987), and fibroblasts (Kaushansky et al. 1988b). As with GM-CSF, the production of G-CSF is controlled by other cytokines and inflammatory mediators (Fibbe et al. 1988; Hermann et al. 1986; Lindemann et al. 1987; Munker et al. 1986; Rambaldi et al. 1987; Vallenga et al. 1988; Zsebo et al. 1988).

In contrast to IL-3 and GM-CSF, and in common with M-CSF, G-CSF acts only upon more committed hematopoietic progenitor cells, stimulating the formation of CFU-G almost exclusively (Platzer et al. 1985; Souza et al. 1986; Welte et al. 1985; Zsebo et al. 1986). And like the other CSFs, G-CSF also activates mature neutrophils. For example, G-CSF induces the expression of IgA F_c receptors (Weisbart et al. 1988) and enhances the binding of the bacterial chemoattractant fMet-Leu-Phe to the surface of neutrophils. And like GM-CSF, G-CSF promotes neutrophil chemotaxis (Bonilla et al. 1988) and augments the antibody-dependent cellular cytotoxicity of neutrophils (Avalos et al. 1987; Platzer et al. 1986).

The cloning of the genes and cDNAs of human and murine G-CSF has led to a better understanding of the regulation and mechanisms of action of G-CSF (Souza et al. 1986; Tsuchiya et al. 1986, 1987; Nagata et al. 1986a and 1986b). The human gene has been localized to chromosome 17, proximal to the breakpoint of the chromosomal translocation (15:17) in acute promyelocytic leukemia (LeBeau et al. 1987; Simmers et al. 1987). There are five exons and four introns in the human and murine genes. The second intron in the human gene contains an alternative splice site accounting for the finding of two cDNAs in a squamous carcinoma cell line encoding polypeptides that differ from each other by three amino acids (Nagata et al. 1986b). The biological activity of the longer polypeptide is approximately 100-fold less than the shorter protein (Nagata et al. 1986b). However, analogous alternative splicing has not been identified in the murine system, making the biological significance of the longer form of G-CSF unclear.

The shorter transcript of human G-CSF encodes a polypeptide of 204 amino acids with a 30-amino acid leader peptide. The mature 174-amino acid peptide has no sites of N-linked glycosylation but has O-linked sugar moieties that terminate in sialic acid at Thr_{133}. Human G-CSF has two disulfide bonds between Cys_{36} and Cys_{42}, and between Cys_{64} and Cys_{74} (Nicola et al. 1983). In addition, human and murine G-CSF each contain a nondisulfide-linked cysteine residue. G-CSF has been predicted to assume a four-helix bundle structure similar to that described for IL-2 and GM-CSF (Parry et al. 1988).

As with the other CSFs, G-CSF initiates its activity by binding to a specific cell-surface receptor. The receptor is present in low numbers ($<$ 1000) on normal neutrophils and promyelocytes (Budel et al. 1989; Begley et al. 1988; Uzumaki et al. 1988), and can be down-modulated by inflammatory mediators such as fMet-Leu-Phe, LPS, and GM-CSF (Nicola et al. 1986). In addition, G-CSF receptors are present on the human myeloid leukemic cell line HL60 and have been found consistently on fresh human leukemic blast cells, in numbers and with binding characteristics similar to normal cells (Evans et al. 1990; Begley et al. 1987; Budel et al. 1989). Most leukemic blasts bearing G-CSF receptors will proliferate in response to the cytokine, but the presence of receptors is not completely predictive of response (Begley et al. 1987; Budel et al. 1989).

The G-CSF receptor displays a single high-affinity binding site with a dissociation constant of approximately 400 pM (ranging from 60–900 pM in various reports [Nicola and Metcalf 1984; Uzumaki et al. 1988]). Cross-linking studies suggest that both the murine and human G-CSF receptor is a 150-kDa protein (Nicola and Peterson 1986; Uzumaki et al. 1988), and of the cytokines, only murine G-CSF can compete for receptor binding of labeled human G-CSF (Begley et al. 1987; Nicola and Metcalf 1984). G-CSF signal transduction mechanisms have not yet been fully elucidated. Rapid phosphorylation of a 75-kDa protein has been demonstrated after incubation of receptor-bearing cells with G-CSF, suggesting tyrosine-kinase activity (Evans et al. 1990). However studies using murine NFS-60 cells have suggested the involvement of GTP binding proteins and the adenylate cyclase system (Matsuda et al. 1989).

In the past few years, G-CSF, like the other hematopoietic growth factors, has been utilized in clinical trials of patients with a variety of malignant and nonmalignant states of marrow insufficiency (reviewed in Groopman et al. 1989; Gabrilove and Jakubowski 1989). G-CSF has proven to be as efficacious as GM-CSF in most clinical situations and has not been associated with any serious side effects.

G-CSF was administered to patients undergoing chemotherapy for urogenital cancer in one study and to patients with squamous cell cancer of the lung in a second study. In comparison to those patients not receiving the drug, G-CSF administration was associated with a shortened period of chemotherapy-induced neutropenia, a decrease in febrile days, and a decrease in antibiotic requirement. A higher percentage of patients were able to receive scheduled chemotherapy on time, and the patients noted a decrease in the frequency and severity of mucositis. A separate group found that beneficial responses could be obtained whether the G-CSF was administered intravenously or subcutaneously.

The cytokine has also been used to treat the neutropenia associated with hairy cell leukemia and bone marrow transplant recipients with encouraging results, including significant increases in circulating neutrophil counts and resolution of ongoing infectious complications. Finally, although the responses to G-CSF therapy have been comparable to those obtained with GM-CSF in most clinical settings, superior clinical results have been obtained with G-CSF in the treatment of patients with myelodysplastic syndromes, particularly patients with high, initial blast cell counts.

15.5   M-CSF

M-CSF (CSF-1) is a hematopoietic growth factor required for the proliferation, differentiation, and survival of cells of the monocyte-macrophage lineage (Das and Stanley 1982). The growth factor is produced by several cell types, and in addition to its role in blood cell production, the growth factor appears to be involved in the activation of mature macrophages (Stanley et al. 1983), placental development (Pollard et al. 1987), and possibly, in neoplasia. Early investigation of M-CSF employed growth factor purified from murine L-cell fibroblasts (Stanley and Heard 1977) or purified growth factor from human urine (Stanley et al. 1975). More recently, cDNA clones for M-CSF have been obtained from a human pancreatic carcinoma cell

(Kawasaki et al. 1985) and trophoblastic cell lines (Wong et al. 1987). The availability of both human and murine recombinant M-CSF has led to an increased understanding of the mechanism of action of this pleotropic cytokine.

Both native human M-CSF and murine M-CSF are heavily glycosylated homodimeric proteins linked by multiple disulfide bonds. The molecular weight of the protein ranges from 45 to 70 kDa, the heterogeneity is secondary to variable degrees of glycosylation (Das and Stanley 1982). The nonglycosylated subunits of both species have a molecular weight of 14.5 kDa. As several human and murine cDNAs of M-CSF predict a subunit of 26 kDa, substantial posttranslational processing occurs (Ralph et al. 1986). Nonglycosylated M-CSF has full biological activity, but reduction of disulfide bonds renders the growth factor inactive.

CSFs from both species promotes differentiation and proliferation of monocyte-macrophage progenitors in the bone marrow. While there is significant sequence homology between the two proteins, murine M-CSF does not function on human macrophage precursors, whereas the human growth factor stimulates both murine and human macrophages, and monocytes (Ralph et al. 1986; Wong et al. 1987). The ability of M-CSF to stimulate bone marrow progenitors in these assays is increased by low concentrations of GM-CSF (Caracciolo et al. 1987; Broxmeyer et al. 1987a). It has been suggested that GM-CSF enhances the action of M-CSF by stimulating M-CSF receptor expression in otherwise uncommitted progenitor cells.

In addition to its effect on the commitment of progenitor cells to the monocyte-macrophage line, M-CSF is also required for the survival of circulating monocytes and tissue macrophages. Sensitive immunoassays have demonstrated the continual presence of M-CSF in the serum of normal animals (Hanamura et al. 1988). Furthermore, M-CSF stimulates numerous immune effector functions of mature monocytes and macrophages. The immune functions enhanced by M-CSF include increased phagocytic activity and microbial killing such as enhanced intracellular killing of yeast such as *Candida* species (Karbassi et al. 1987), improved resistance to viral infection (Lee and Warren 1987), tumoricidal activity (Ralph and Nakoinz 1987), improved antibody-mediated cytotoxicity (Garnick et al. 1989), and enhanced monocyte production (Broxmeyer et al. 1987b). While many of these functions are the direct result of M-CSF activity, some of these actions may be secondary to the increased viability of the macrophages so that these cells can produce other inflammatory cytokines. Macrophages produce IL-1 (Moore et al. 1980), interferon, and TNF (Warren and Ralph 1986) in response to M-CSF. Furthermore, M-CSF induces production of oxygen metabolites (Wing et al. 1985), plasminogen activator (Lin et al. 1979), and prostaglandin E (Kurland et al. 1979) by macrophages in the inflammatory response.

Most cellular sources that produce M-CSF contain transcripts of variable size. While some of the heterogeneity of M-CSF mRNA is secondary to different sites of polyadenylation, most of the variability is secondary to alternate splicing at one or both of two sites. The prominent species of transcripts are 4.3 and 1.8 kbp in size (Kawasaki et al. 1985; Wong et al. 1987). Several cDNA clones for M-CSF have been extensively studied and have provided insights into the heterogeneity of the size of M-CSF. Two frequently isolated clones are 4 kbp (Wong et al. 1987) and 1.6 kbp (Kawasaki et al. 1987) in size, predicting primary translation products of 554 and 256 amino acids, respectively. These two clones differ primarily by the presence of an 894-bp insert in the open-reading frame of the cDNA. Comparison of the two cDNA clones to the genomic sequence of M-CSF reveals that these two clones arise from alternate splicing at the sixth exon, with an 894-bp insert, as well as at the polyadenylation site in the ninth exon (Figure 15–2) (Ladner et al. 1987; Wong et al. 1987). The two clones encode polypeptides, which are identical from the 32-amino acid secretory leader sequence through the 149-amino terminal residues of the mature growth factor. Following residue 149 (Gln), the larger clone includes an additional 298-amino acid domain. Following position 447, the sequence again matches that of the smaller clone following residue 149. Both clones have a short 23-amino acid hydrophobic transmembrane region just upstream of the carboxyl-terminus (Rettenmier and Roussel 1988).

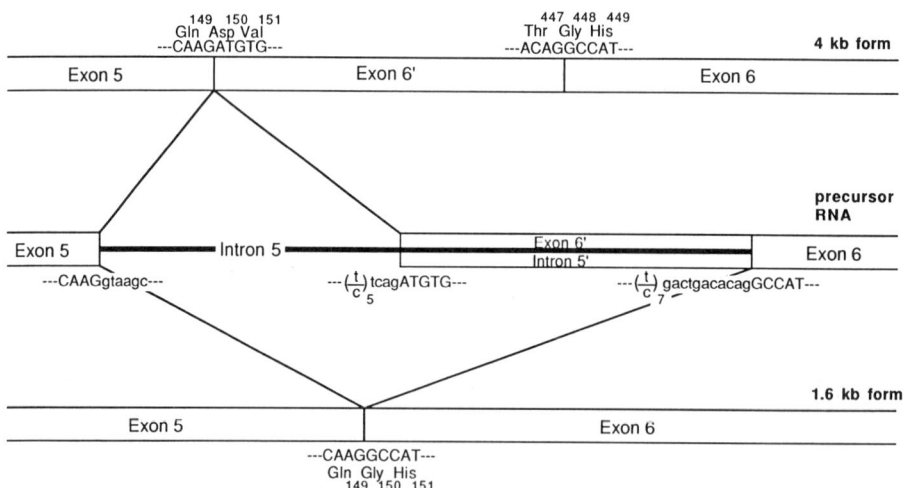

FIGURE 15–2 The site of alternative splicing from the precursor RNA in the sixth exon to produce the 4- and 1.6-kbp forms. The splice donor and splice acceptor sites are shown. The M-CSF gene is composed of 10 exons, with the mature protein being encoded in the second through eighth exons. The sixth exon is variable in size, depending on which splice-acceptor site is used. The gene is approximately 20 kbp in size (Ladner et al. 1987).

Finally, a third cDNA clone obtained from the pancreatic cancer cell line, Mia-PaCa-2, has been described and encodes a 438-amino acid polypeptide (Ceretti et al. 1988). This third clone is again produced by alternate splicing in the sixth exon producing a different sequence in the membrane bound precursor region of the protein.

The 554-, 438-, and 256-amino acid precursors all undergo proteolytic processing. The 256-amino acid form of M-CSF is processed to a final subunit that is approximately 160 amino acids in size. Interestingly, the proteolytic processing of the 554- and 438-amino acid precursors yield identical 223-amino acid subunits. This indicates that processing of the membrane-bound precursor probably occurs upstream of residue Thr-363, since this residue is the last amino acid common to both cDNAs upstream of the transmembrane region (Ceretti et al. 1988).

The biologic significance of the various forms of M-CSF are now beginning to be better understood. Due to the presence of the 23-amino acid hydrophobic domain, all of the forms of M-CSF are thought to first exist as a membrane-bound protein. While the mature polypeptides derived from the 256- and 554-amino acid products have equal biologic activity, they differ in their carbohydrate composition, and their membrane-bound precursors are processed by alternative mechanisms. The 256-amino acid precursor is maintained as a membrane-bound dimeric glycoprotein throughout its intracellular transport. It is expressed at the cell surface in a stable manner where two identical subunits of about 160 amino acids are slowly released (Rettenmier et al. 1987). Consequently, this smaller precursor is found primarily as a membrane-bound form. By contrast, the larger 554-amino acid precursor is rapidly cleaved within the cell and its homodimeric product is efficiently secreted, resulting in little or no detectable plasma-membrane-bound intermediate (Rettenmier and Roussel 1988).

Despite the differences in kinetics of production and release from cells that produce these two forms of the growth factor, both recombinant molecules are equally active in assays of progenitor cell proliferation. Since the amino terminal 149 amino acids are identical in the two forms, it appears that this amino terminal region is entirely responsible for the biologic activity of the growth factor (Wong et al. 1987). The importance of this region in biologic activity is further emphasized when comparing the amino acid sequence homology between murine and human M-CSF. The greatest degree of homology occurs at the amino-terminal portion of the protein. The 23 amino acids of the secretory leader and the first 227 amino acids of the mature protein are 82 and 87% conserved, respectivly. Between residues 228 and 339, the homology clearly deteriorates with only 49% conservation (Ladner et al. 1987). However, the presence of two distinct methods of processing M-CSF precursors suggests that there may be different functions for the soluble and plasma-membrane-bound form of the growth factor.

M-CSF is produced by several different types of cells. Constitutive expression can be detected in a number of cell culture systems and additional protein can be induced by a variety of stimulatory signals. Fibroblasts and bone marrow stromal cells produce M-CSF constitutively and may be responsible for the steady-state levels of the growth factor in the serum of normal individuals (Rajavashisth et al. 1987; Hunt et al. 1987). In addition, M-CSF expression is inducible in monocytes, macrophages, lymphocytes, endothelial cells, and uterine-glandular epithelial cells. Monocytes and macrophages synthesize M-CSF mRNA following exposure to phorbol esters (Horiguchi et al. 1986) and stimulation by inflammatory cytokines such as interferon-gamma (Rambaldi et al. 1987), GM-CSF (Horiguchi et al. 1987), and TNF-alpha (Oster et al. 1987). The fact that the same cells that are stimulated by the growth factor also produce the same protein suggests that activated mononuclear phagocytes may enhance their activity and proliferation at the site of inflammation through a positive feedback mechanism. In addition to fibroblasts, stromal cells, and monocytes, M-CSF is produced by lymphocytes in response to lectin (Wong et al. 1987), and endothelial cells produce M-CSF in response to IL-1 or TNF (Seelentag et al. 1987).

M-CSF also appears to play a role in placental development (Pollard et al. 1987; Arceci et al. 1989). In the mouse, a 10,000-fold elevation of mouse uterine M-CSF is observed during pregnancy as demonstrated by radioimmunoassay (Bartocci et al. 1986). The levels of M-CSF increase during pregnancy and are greatest at parturition. The M-CSF is produced by the uterine-glandular epithelium in response to the synergistic action of progesterone and estradiol (Pollard et al. 1987). The production of M-CSF by uterine epithelial cells thus appears to play an important role in placental development.

M-CSF exerts its action by binding to a single class of high-affinity receptors on the surface of mononuclear phagocytes (Guilbert and Stanley 1986). The cell-surface receptor, demonstrated by binding studies with radiolabeled M-CSF is the *c-fms* proto-oncogene product (Sherr et al. 1985). The receptor is a transmembrane glycoprotein with tyrosine-kinase activity (Yeung et al. 1987). When M-CSF binds to the receptor, the tyrosine residues on the receptor autophosphorylate along with a number of other cellular proteins. With the exception of the 185-kDa M-CSF receptor, a 133- and a 260-kDa protein, all of the resultant phosphoproteins (at least 15 have been identified) are cytoplasmic (Sengupta et al. 1988). While these proteins have not yet been characterized, it appears that phosphorylation of their tyrosine residues must play a role in M-CSF signal transduction.

The binding of M-CSF to its receptor initiates numerous rapid cellular and biochemical responses, including membrane ruffling and vacuolization (Tushinski et al. 1982), alteration of the Na^+/H^+ exchange leading to immediate cytoplasmic alkalinization (Vairo and Hamilton 1988), increased hexose transport (Hamilton et al. 1988), and expression of numerous early response genes including *c-myc* and *c-fos* (Orlofsky and Stanley 1987). Expression of these early genes by themselves is not sufficient for proliferation of monocytes and macrophages as M-CSF is required throughout the entire G_1 phase of the cell cycle. After these early events, there is a decrease in protein turnover within 2 hr and stimulation of DNA synthesis within 8 to 12 hr (Tushinski and Stanley 1985).

M-CSF receptors are detectable in low numbers on immature bone marrow precursors. However, in the presence of IL-1 or IL-3, an increase in the number of precursors that respond to M-CSF are noted, suggesting that IL-1 or IL-3 induce expression of the M-CSF receptor (Chen and Clark 1986). Furthermore, it has been suggested that the effect of IL-1 is indirect and that IL-6 is the mediator of M-CSF-receptor expression (Ikebuchi et al. 1987). In addition to cytokine-induced upregulation, M-CSF receptor levels increase with monocyte maturation, and high levels of receptors are present on mature, nondividing macrophages and monocytes. In this setting, continued M-CSF stimulation is thought to be essential for survival.

There is considerable interest in the role of M-CSF and its receptor in the neoplastic process. The genes encoding M-CSF and its receptor are on the long arm of chromosome 5, the same region as the genes for GM-CSF (Huebner et al. 1985), IL-3 (LeBeau et al. 1987), and the PDGF receptor. The GM-CSF and IL-3 genes are clustered near the centromeric end of the gene (Yang et al. 1988), whereas the M-CSF and PDGF receptor genes are located near the telomeric end of the chromosome (Roberts et al. 1988). Deletions of the chromosome, 5q, are seen in patients with myelodysplasia or acute nonlymphocytic leukemia, thus chromosomal aberrations of the genes for M-CSF and its receptor may play a role in the development of these disorders. In addition, the evidence that the M-CSF receptor is the normal cellular counterpart of the oncogene of a transforming feline retrovirus (*v-fms*) raises the possibility that activation mutations within the *c-fms* gene might contribute to hematopoietic malignancies involving receptor-positive cells of the monocytic series. Considerable effort is currently underway to determine whether mutations of the *c-fms* gene are detectable in leukemic blasts.

The production of recombinant human M-CSF (rh M-CSF) has permitted extensive evaluation of the in vitro and in vivo effects of the growth factor. While there is a relative paucity of human clinical trials using rh M-CSF, there are currently a number of completed studies evaluating the role of the growth factor in hematologic reconstitution following chemotherapy, enhancement of immunologic function against infection in bone marrow transplantation, and as a possible antitumor agent. This latter aspect of M-CSF has drawn particular interest since recent studies have established that rh M-CSF in concert with murine IgG 3 antibodies is able to activate monocytes for tumoricidal activity against neuroblastoma, melanoma, and colon carcinoma (Garnick et al. 1989).

The in vivo effects of M-CSF have been examined extensively in animal models. When the growth factor is given to cyclophosphamide-treated mice, the numbers of CFU-M and CFU-GM, as well as splenic CFU-GM, CFU-GEMM, and BFU-E, are increased (Broxmeyer et al. 1987a, 1987b). However, as M-CSF does not stimulate erythroid precursors in vitro, these findings suggest that at least some of the in vivo effects of M-CSF may be mediated indirectly by accessory cells.

Several studies have shown that M-CSF administered to mice produces a monocytosis, neutrophilia, and lymphopenia. Ulich et al. (1990) recently examined the hematologic response when a single dose of rh M-CSF was given to Lewis rats. A relatively high dose of 200–400 mg/kg was required to develop a monocytosis. The administration of M-CSF produces an initial peripheral monocytopenia that may be secondary to sequestration or margination of the cells. A similar response occurs shortly after the in vivo administration of GM-CSF. The monocytopenia is followed by a monocytosis and an accompanying increase in marrow monoblasts. The monocytosis peaks at 28–30 hr. Dexamethasone administered 1 min prior to M-CSF administration partially inhibited the monocytosis. The M-CSF administration also causes mild neutrophilia and lymphopenia that develops from 2–16 hr after growth factor administration. Since IL-1 and TNF both induce neutrophilia and lymphopenia, and M-CSF is known to induce expression of these cytokines, it has been suggested that the neutrophilia and lymphopenia is an indirect effect of M-CSF (Ulich et al. 1990).

Preclinical toxicology studies of M-CSF have been performed in rats and cynomolgus monkeys (Stoudemire et al. 1989). In rats receiving M-CSF for 14 days at doses of 100, 500, and 1,000 µg/kg body weight, an apparently reversible thrombocytopenia was the only hematologic toxicity noted. In monkeys, M-CSF produced marked monocytosis, as well as a dose-related reversible thrombocytopenia and a reversible decrease in hemoglobin concentration.

With these studies as a background, therapeutic applications in humans for M-CSF are now being explored. Trials in Japan with partially purified human-urinary M-CSF led to an increased neutrophil count in children with chronic neutropenia (Komiyama et al. 1988). This result suggested that M-CSF may be useful in preventing myelosuppression. Based on its in vitro activities, other applications being currently explored include its usefulness as an agent to treat malignancy, as well as its role in the treatment of chronic bacterial and fungal infections. While M-CSF has been shown useful in providing resistance to viral infections in vitro, M-CSF may be contraindicated in HIV infections as this growth factor along with GM-CSF enhances viral replication in mononuclear phagocytes (Gendelman et al. 1988).

15.6 SUMMARY

In summary, hematopoietic growth factors have been discovered, biochemically characterized, cloned, produced by recombinant DNA technology, and put into clinical use in a period of 25 years. We are approaching a greater understanding of the cellular anatomy and molecular mechanisms that regulate production of the CSFs, the ways in which the CSFs interact with their cell surface receptors and trigger their biological effects, the nature of these receptors themselves and their mechanisms of signal transduction, and the effects of the CSFs in vitro and in vivo on hematopoietic progenitor cells and mature leukocytes. However, many questions remain. What is the mechanism that couples growth-factor binding to the triggering of cellular proliferation? How do multi-CSF and GM-CSF cross-compete at the level of the cell-surface receptor, and yet show no primary amino acid sequence homology? What are the mechanisms that regulate the tissue expression profile of multi-CSF compared to the genetically similar growth factor GM-CSF? And, what are the optimal dosages, schedules of administration, and combinations of CSFs optimal for each of several conditions of marrow failure? These are but a few of the questions that continue to occupy much current research interest.

REFERENCES

Anderson, C.L., and Looney, R.J. (1986) *Immunol. Today* 7, 264.

Arceci, R.J., Shanahan F., Stanley, E.R., et al. (1989) *Proc. Natl. Acad. Sci.* 86, 8818–8822.

Avalos, B., Hedzat, C., Baldwin, G.C., et al. (1987) *Blood* 70 (suppl. 1), 165a.

Bagby, G.C., Rigas, V.D., Bennett, R.M., et al. (1981) *J. Clin. Invest.* 68, 56–63.

Bagby, G.C., Shaw, G., Brown, M.A., et al. (1988) *Blood* 72 (suppl. 1), 109a.

Bagby, G.C., Dinarello, C.A., Wallace, P., et al. (1986) *J. Clin. Invest.* 78, 1316–1323.

Barlow, D.P., Bucan, M., Lehrach, H., Hogan, B.L.M., and Gough, N.M. (1987) *EMBO J.* 6, 617–623.

Bartocci, A., Pollard, J.W., and Stanley, E.R. (1986) *J. Exp. Med.* 164, 956–961.

Begley, C.G., Metcalf, D., and Nicola, N.A. (1987) *Leukemia* 1, 1–8.

Begley, C.G., Metcalf, D., and Nicola, N.A. (1988) *Exp. Hematol.* 16, 71–79.

Bonilla, M.A., Gillo, A.P., Ruggerio, M., et al. (1988) *Blood* 72 (suppl. 1), 349.

Bot, F.J., van Eijk, L., Schipper, P., and Lowenberg, B. (1989) *Blood* 73, 1157–1160.

Bradley, T.R., and Metcalf, D. (1966) *Aust. J. Exp. Biol. Med. Sci.* 44, 287–299.

Broudy, V., Kaushansky, K., Segal, G.M., Harlan, J., and Adamson, J.W. (1986) *Proc. Natl. Acad. Sci. USA* 83, 7467.

Brown, C.B., Hart, C.E., Curtis, D.M., Bailey, M.C., and Kaushansky, K. (1990) *J. Immunol.* 144, 2184–2189.

Broxmeyer, H.E., Williams, D.E., Hangoc, G., et al. (1987a) *Proc. Natl. Acad. Sci. USA* 84, 3871–3875.

Broxmeyer, H.E., Williams, D.E., Cooper, S., et al. (1987b) *J. Clin. Invest.* 79, 721–730.

Budel, L.M., Touw, I.P., Delwel, R., Clark, S.C., and Lowenberg, B. (1989) *Blood* 74, 565–571.

Budel, L.M., Elbaz, O., Hoogerbrugge, H., et al. (1990) *Blood* 75, 1439–1445.

Burgess, A., and Metcalf, D. (1980) *Int. J. Cancer* 56, 809.

Burstein, S.A. (1986) *Blood Cells* 11, 469–479.

Bussolino, G., Wang, J.M., Defilippi, P., et al. (1989) *Nature* 337, 471.

Caracciolo, D., Shirsat N., and Wong, G.G. (1987) *J. Exp. Med.* 166, 1851–1856.

Ceretti, D.P., Wignall, J., Anderson, D., et al. (1988) *Molec. Immunol.* 25, 761–770.

Chang, J.M., Metcalf, D., Lang, R.A., Gonda, T.J., and Johnson, G.R. (1989) *Blood* 73, 1487–1497.

Cheers, C., Haigh, A.M., Kelso, A., et al. (1988) *Infection Immunity* 56, 247–251.

Chen, B.D.-M., and Clark, C.R. (1986) *J. Immunol.* 137, 563–570.

Clark-Lewis, I., Aebersold, R., Ziltener, H., et al. (1986) *Science* 231, 134–139.

Clark-Lewis, I., Hood, L.E., and Kent, S.B.H. (1988a) *Proc. Natl. Acad. Sci. USA* 85, 7897–7901.

Clark-Lewis, I., Lopez, A.F., To, L.B., et al. (1988b) *J. Immunol.* 141, 881.

Cocita Baldwin, G., Gasson, J.C., Kaufman, S.E., et al. (1989) *Blood* 73, 1033–1037.

Das, S.K., and Stanley, E.R. (1982) *J. Biol. Chem.* 257, 13679–13684.

Demetri, G.D., and Griffen, J.D. (1989) *Hematology/Oncology Clinics North America* 3, 535–551.

DiPersio, J., Billing, P., Kaufman, S., et al. (1988) *J. Biol. Chem.* 263, 1834–1841.

Donahue, R.E., Seehra, J., Metzger, M., et al. (1988) *Science* 241, 1820–1823.

Dunbar, C.E., Browder, T.M., Abrams, J.S., and Nienhuis, A.W. (1989) *Science* 245, 1493–1496.

Elliot, M.J., Vadas, M.A., Eglington, J.M., et al. (1989) *Blood* 74, 2349–2359.

Ernst, T.J., Ritchie, A.R., Stopak, K.S., and Griffin, J.D. (1989) *Blood* 74 (suppl. 1), 116a.

Evans, J.P.M., Mire-Sluis, A.R., Hoffbrand, A.V., and Wickremasinghe, R.G. (1990) *Blood* 75, 88–95.

Fibbe, W.E., van Damme, J., Billiau, A., et al. (1988) *Blood* 71, 430–435.

Fleischmann, J., Golde, D.W., Weisbart, R.H., et al. (1986) *Blood* 68, 708.

Fung, M.C., Hapel, A.J., Ymer, S., et al. (1984) *Nature* 307, 233–236.

Gabrilove, J.L., and Jakubowski, A. (1989) *Hematology/Oncology Clinics North America* 3, 427–440.

Ganser, A., Lindemann, A., Seipelt, G., et al. (1989) *Blood* 74 (suppl. 1), 50a.

Garnick, M., and O'Reilly, R. (1989) *Hematology/Oncology Clinics North America* 3, 497–509.

Garnick, M., Stoudemire J., Donahue, R.E., et al. (1989) *Proc. ASCO* 8, 184.

Gasson, J.C., Weisbart, R.H., Kaufman, S.E., et al. (1984) *Nature* 226, 1339–1342.

Gasson, J.C., Kaufman, S.E., Weisbart, R.H., Tomonaga, M., and Golde, D.W. (1986) *Proc. Natl. Acad. Sci. USA* 83, 669–673.

Gearing, D.P., King, J.A., Gough, N.M., and Nicola, N.A. (1989) *EMBO J.* 8, 3667–3676.

Gendelman, H., Orenstein, J.M., Martin, M.A., et al. (1988) *J. Exp. Med.* 167, 1428–1435.

Gesner, T., Mufson, R.A., Turner, K.J., and Clark, S.C. (1989) *Blood* 74, 2652–2656.

Gillio, A.P., Laver, J., Abboud, M., et al. (1989) *Blood* 74 (suppl. 1), 117a.

Gordon, M.Y., Riley, G.P., Watt, S.M., and Greaves, M.F. (1987) *Nature* 326, 403–405.

Gough, N.M., Gough, J., Metcalf, D., et al. (1984) *Nature* 309, 763–767.

Gough, N.M., Grail, D., Gearing, D.P., and Metcalf, D. (1987) *Eur. J. Biochem.* 169, 353.

Gribben, J.G., Devereux, S., Thomas, N.S.B., et al. (1990) *Lancet* 335, 434–437.

Groopman, J.E., Molina, J.-M., and Scadden, D.T. (1989) *N. Engl. J. Med.* 321, 1449–1459.

Guilbert, L.J., and Stanley, E.R. (1986) *J. Biol. Chem.* 261, 4024–4032.

Hamilton, J.A., Vairo, G., Lingelbach, S.R., et al. (1988) *J. Cell Physiol.* 134, 405–409.

Hanamura, T., Motoyoshi K., Yoshida, K., et al. (1988) *Blood* 72, 886–892.

Horiguchi, J., Warren, M.K., and Ralph, P. (1986) *Biochem. Biophys. Res. Commun.* 141, 924–927.

Horiguchi, J., Warren, M.K., and Kufe, D. (1987) *Blood* 69, 1259–1261.

Huebner, K., Isobe, M., Croce, C.M., et al. (1985) *Science* 230, 1282–1285.

Hunt, P., Robertson, D., Weiss, D., et al. (1987) *Cell* 48, 997–1007.

Ihle, J.N., Pepersack, L., and Rebar, L. (1981) *J. Immunol.* 126, 2184–2189.

Ihle, J.N., Keller, J., Oroszlan, S., et al. (1983) *J. Immunol.* 131, 282–287.

Ikebuchi, K., Clark, S.C., Ihle, J.N., Souza, L.M., and Ogawa, M. (1988) *Proc. Natl. Acad. Sci. USA* 85, 3445–3449.

Ikebuchi, K., Wong, G.G., Clark, S.C., et al. (1987) *Proc. Natl. Acad. Sci. USA* 84, 9035–9039.

Isfort, R.J., Stevens, D., May, W.S., and Ihle, J.N. (1988a) *Proc. Natl. Acad. Sci. USA* 85, 7982–7986.

Isfort, R., Huhn, R.D., Frackelton, A.R., and Ihle J.N. (1988b) *J. Biol. Chem.* 263, 19203–19209.

Kanakura, Y., Drucker, B., Furukawa, Y., et al. (1989) *Blood* 74 (suppl. 1), 194a.

Kannourakis, G., Jubinsky, P., and Sieff, C.A. (1989) *Blood* 74 (suppl. 1), 153a.

Karbassi, A., Becker, J.M., Foster, J.S., et al. (1987) *J. Immunol.* 139, 417–421.

Kaushansky, K. (1987) *Blood Cells* 13, 3–13.

Kaushansky, K. (1989) *J. Immunol.* 143, 2525–2529.

Kaushansky, K. (1990) in *Hematopoiesis: UCLA Symposia on Molecular and Cellular Biology* vol. 120, 20–26 Feb. 1989, Tamarron, CO, pp. 1–10, Wiley-Liss, New York.

Kaushansky, K., O'Hara, P.J., Berkner, K., et al. (1986) *Proc. Natl. Acad. Sci. USA* 83, 3101–3105.

Kaushansky, K., O'Hara, P.J., Hart, C.E., Forstrom, J.W., and Hagen, F.S. (1987) *Biochemistry* 26, 4861–4867.

Kaushansky, K., Broudy, V.C., Harlan, J.M., and Adamson, J.W. (1988a) *J. Immunol.* 141, 3410–3415.

Kaushansky, K., Lin, N., and Adamson, J.W. (1988b) *J. Clin. Invest.* 81, 92.

Kaushansky, K., Miller, J.E., Morris, D.R., Wilson, C.B., and Hammond, W.P. (1989a) *Cell. Immunol.* 122, 62–70.

Kaushansky, K., Shoemaker, S.G., Alfaro, S.A., and Brown, C.B. (1989c) *Proc. Natl. Acad. Sci. USA* 86, 1213–1217.

Kaushansky, K., Brown, C.B., and O'Hara, P.J. (1990) *Int. J. Cell Cloning* 8 (suppl. 1), 26–34.

Kawasaki, E.S., Ladner, M.B., Wang, A.M., et al. (1985) *Science* 230, 291–296.

Kelvin, D.J., Chance, S., Shreeve, M., et al. (1986) *J. Cell. Physiol.* 127, 403–409.

Kindler, V., Thorens, B., de Kossodo, S., et al. (1986) *Proc. Natl. Acad. Sci. USA* 83, 1001–1005.

Kobayashi, M., Van Leeuwen, B.H., Elsbury, S., et al. (1989) *Blood* 73, 1836–1841.

Komiyama, A., Ishiguro, A., Kubo, T., et al. (1988) *Blood* 71, 41–46.

Koyasu, S., Tojo, A., Miyajima, A., et al. (1987) *EMBO J.* 6, 3979–3984.

Kurland, J.I., Pelus, L.M., Ralph, P., et al. (1979) *Proc. Natl. Acad. Sci. USA* 76, 2326–2330.

Kurzrock, R., Talpaz, M., Salewski, E., and Gutterman, J.U. (1989) *Blood* 74 (suppl. 1), 154a.

Ladner, M.B., Martin, G.A., Noble, J.A., et al. (1987) *EMBO J.* 6, 2693–2698.

Leary, A.G., Yang, Y.-C., Clark, S.C., et al. (1987) *Blood* 70, 1343–1348.

LeBeau, M.M., Westbrook, C.A., and Diaz, M.O. (1986) *Science* 231, 984–987.

LeBeau, M.M., Epstein, N.D., O'Brien, S.J., et al. (1987) *Proc. Natl. Acad. Sci. USA* 84, 5913–5917.

LeBeau, M.M., Lemons, R.S., Ccarrino, J.J., et al. (1987) *Leukemia* 1, 795–799.

Lee, M.T., and Warren, M.K. (1987) *J. Immunol.* 138, 3019–3030.

Lindemann, A., Oster, W., Riedel, D., et al. (1987) *Blood* 70 (suppl. 1), 223.

Lopez, A.F., To, L.B., Yang, Y-C., et al. (1987) *Proc. Natl. Acad. Sci. USA* 84, 2761–2765.

Lopez, A.F., Eglinton, J.M., Gillis, D., et al. (1989) *Proc. Natl. Acad. Sci. USA* 86, 7022–7026.

Lovhaug, D., Pelus, L.M., Nordlie, E.M., Boyum, A., and Moore, M.A.S. (1986) *Exp. Hematol.* 14, 1037–1042.

Mathey-Prevot, B., Andrews, N.C., Murphy, H.S., et al. (1990) *Proc. Natl. Acad. Sci. USA* 87, 5046–5050.

Matsuda, S., Shirafuji, N., and Asano, S. (1989) *Blood* 74, 2343–2348.

Mayer, P., Valent, P., Liehl, E., and Bettelheim, P. (1989) *Blood* 74 (suppl. 1), 124a.

Messner, H.A., Yamasaki, K., Jamal, N., et al. (1987) *Proc. Natl. Acad. Sci. USA* 84, 6765–6769.

Metcalf, D., and Nicola, N.A. (1988) *Proc. Natl. Acad. Sci. USA* 85, 3160–3164.

Metcalf, D., Begley, C.G., Johnson, G.R., et al. (1986) *Blood* 67, 37–45.

Migliaccio, G., Migliaccio, A.R., Kaushansky, K., and Adamson, J.W. (1989) *Exp. Hematol.* 17, 110–115.

Mitsuyasu, R.T., and Golde, D.W. (1989) *Hematology/Oncology Clinics of North America* 3, 411–425.

Miyatake, S., Otsuka, T., Lee, F., and Arai, K. (1985) *EMBO J.* 4, 2561–2568.

Monroy, R.L., Davis, T.A., Donahue, R.E., and Macvittie, T.J. (1989) *Blood* 74 (suppl. 1), 75a.

Moore, R.N., Oppenheim, J.J., Farrar, J.J., et al. (1980) J. Immunol. 150, 231–235.

Muegge, K., Williams, T.M., Kant, J., et al. (1989) *Science* 246, 249–251.

Munker, R., Gasson, J.C., Ogawa, M., et al. (1986) *Nature* 323, 79–82.

Nagata, S., Tsuchiya, M., Asano, S., et al. (1986a) *Nature* 319, 415–418.

Nagata, S., Tsuchiya, M., Asano, S., et al. (1986b) *EMBO J.* 5, 575–581.

Nemunaitis, J., Andrews, D.F., Crittenden, C., Kaushansky, K., and Singer, J.W. (1989) *J. Clin. Invest.* 83, 593–601.

Nicola, N.A., and Metcalf, D. (1984) *Proc. Natl. Acad. Sci. USA* 81, 3765–3769.

Nicola, N.A., and Metcalf, D. (1986) *J. Cell. Physiol.* 128, 180–188.

Nicola, N.A., and Peterson, L. (1986) *J. Biol. Chem.* 261, 12384–12389.

Nicola, N., Metcalf, D., Matsumoto, M., et al. (1983) *J. Biol. Chem.* 258, 9017–9023.

Nicola, N.A., Vadas, M.A., and Lopez, A.F. (1986) *J. Cell. Phys.* 128, 501–509.

Niemeyer, C.M., Sieff, C.A., Mathey-Prevot, B., et al. (1989) *Blood* 73, 945–951.

Nimer, S.D., Morita, E.A., Martis, M.J., Wachsman, W., and Gasson, J.C. (1988) *Molec. Cell. Biol.* 8, 1979–1984.

Orlofsky, A., and Stanley, E.R. (1987) *EMBO J.* 6, 2947–2952.

Oster, W., Lindemann, A., Horn, S., et al. (1987) *Blood* 70, 1700–1703.

Park, L.S., Friend, D., Gillis, S., and Urdal, D.L. (1986) *J. Biol. Chem.* 261, 4177–4183.

Park, L.S., Friend, D., Price, V., et al. (1989) *J. Biol. Chem.* 264, 5420–5427.

Parry, D.A.D., Minasian, E., and Leach, S.J. (1988) *Molec. Recognition* 1, 107–110.

Perussia, B., Kobayashi, M., Rossi, M., et al. (1987) *J. Immunol.* 138, 765.

Platzer, E., Welte, K., Gabrilove, J., et al. (1985) *J. Exp. Med.* 162, 1788–1801.

Platzer, E., Oez, S., Welte, K., et al. (1986) *Immunobiology* 172, 185–193.

Pluznick, D.H., and Sachs, L. (1965) *J. Cell Comp. Physiol.* 66, 319–324.

Pluznik, D.V., Cunningham, R.E., and Noguchi, P.D. (1984) *Proc. Natl. Acad. Sci. USA* 81, 7451–7455.

Pollard, J.W., Bartocci, A., Arceci, R., et al. (1987) *Nature* 330, 484–486.

Rajavashisth, T.B., Eng, R., Shadduck, R.K., et al. (1987) *Proc. Natl. Acad. Sci. USA* 84, 1157–1161.

Ralph, P., and Nakoinz, I. (1987) *Cell Immunol.* 105, 270–274.

Ralph, P., Warren, M.K., Ladner, M.B., et al. (1986) *Cold Spring Harbor Symposium on Quantitative Biology* vol. LI, Cold Spring Harbor, NY, pp. 679–683, Cold Spring Harbor Press.

Rambaldi, A., Young, D.C., and Griffin, J.D. (1987) *Blood* 69, 1409.

Rennick, D., Yang, G., Gemmell, L., and Lee, F. (1987) *Blood* 69, 682–691.

Rettenmier, C.W., Roussel, M.F., Ashmun, R.A., et al. (1987) *Mol. Cell Biol.* 7, 2378–2387.

Rettenmier, C.W., and Roussel, M.F. (1988) *Mol. Cell Biol.* 8, 5026–5034.

Roberts, W.M., Look, A.T., Roussel, M.F., et al. (1988) *Cell* 55, 655–661.

Ross, G.D., and Lambris, J.D. (1982) *J. Exp. Med.* 155, 96.

Rothenberg, M.E., Owen, W.F., Silberstein, D.S., et al. (1988) *J. Clin. Invest.* 81, 1986–1992.

Seelentag, W.K., Mermod, J.J., Montesano, R., et al. (1987) *EMBO J.* 6, 2261–2265.

Sengupta, A., Liu, W-K., Yeung, Y.G., et al. (1988) *Proc. Natl. Acad. Sci. USA* 85, 8062–8066.

Shanafelt, A.B., and Kastelein, R.A. (1989) *Proc. Natl. Acad. Sci. USA* 86, 4872–4876.

Shannon, M.F., Gamble, J.R., and Vadas, M. (1988) *Proc. Natl. Acad. Sci. USA* 85, 674–678.

Shaw, G., and Kamen, R. (1986) *Cell* 46, 659–667.

Sherr, C.J., Rettenmier, C.W., Sacca, R., et al. (1985) *Cell* 41, 665–676.

Shoemaker, S., Hromas, R., and Kaushansky, K. (1990) *Proc. Natl. Acad. Sci. USA* 87, 9650–9654.

Sieff, C.A., and Ekern, S. (1988) *Blood* 74 (suppl. 1), 134a.

Sieff, C.A., Emerson, S.G., Donahue, R.E., and Nathan, D.G. (1985) *Science* 230, 1171–1173.

Sieff, C.A., Tsai, S., and Faller, D.V. (1987) *J. Clin. Invest.* 79, 48–51.

Simmers, T.N., Webber, L.M., Shannon, M.F., et al. (1987) *Blood* 70, 330.

Simmons, P.J., Kaushansky, K., and Torok-Storb, B. (1990) *Proc. Natl. Acad. Sci. USA* 87, 1386–1390.

Sorenson, P., Mui, A.L.-F., and Krystal, G. (1989) *J. Biol. Chem.* 264, 19253–19258.

Souza, L.M., Boone, T.C., Gabrilove, J., et al. (1986) *Science* 232, 61–65.

Spivak, J.L., Smith, R.R.L., and Ihle, J.N. (1985) *J. Clin. Invest.* 76, 1613–1621.

Stanley, E.R., and Heard, P.M. (1977) *J. Biol. Chem.* 252, 4305–4312.

Stanley, E.R., Hansen, G., Woodcock, J., et al. (1975) *Fed. Proc.* 34, 2272–2278.

Stanley, E.R., Guilbert, L.J., Tushinski, R.J., et al. (1983) *J. Cell Biochem.* 21, 151–159.

Stanley, E., Metcalf, D., Sobieszczuk, P., Gough, N.M., and Dunn, A.R. (1985) *EMBO J.* 4, 2569–2573.

Stoudemire, J.B., Metzger, M., Timony, G., et al. (1989) *Proc. AACR* 30, 538.

Thorens, B., Mermod, J.J., and Vassali, P. (1987) *Cell* 48, 671.

To, L.B., and Juttner, C.A. (1987) *Br. J. Hematol.* 66, 285–288.

Tomonaga, M., Golde, D.W., and Gasson, J.C. (1986) *Blood* 67, 31–36.

Tsuchiya, M., Asano, S., Kaziro, Y., and Nagata, S. (1986) *Proc. Natl. Acad. Sci. USA* 83, 7633–7637.

Turner, R., and Tjian, R. (1989) *Science* 243, 1689–1694.

Tushinski, R.J., and Stanley, E.R. (1985) *J. Cell Physiol.* 122, 221–228.

Tushinski, R.J., Oliver, I.T., Guilbert, L.J., et al. (1982) *Cell* 28, 71–81.

Ulich, T.R., del Castillo, J., Watson, C.R., et al. (1990) *Blood* 75, 846–850.

Uzumaki, H., Okabe, T., Sasaki, N., et al. (1988) *Biochem. Biophys. Res. Commun.* 156, 1026–1032.

Vairo, G., and Hamilton, J.A. (1988) *J. Cell Physiol.* 134, 13–19.

Valent, P., Schmidt, G., Bessemer, J., et al. (1989a) *Blood* 73, 1763–1769.

Valent, P., Volc-Platzer, B., Mayer, P., and Bettelheim, P. (1989b) *Blood* 74 (suppl. 1), 232a.

Vallenga, E., Ostapovicz, D., O'Rourke, B., and Griffin, J.D. (1987) *Leukemia* 1, 584–589.

Vallenga, E., Rambaldi, A., Ernst, T.J., et al. (1988) *Blood* 71, 1529.

Walker, F., and Burgess, A.W. (1985) *EMBO J.* 4, 933–939.

Warren, M.K., and Ralph, P. (1986) *J. Immunol.* 137, 2281–2285.

Wegman, T.G., Athanassakis, I., Guilbert, L., et al. (1989) *Transpl. Proc.* 21, 566–568.

Weisbart, R.H., Golde, D.W., and Gasson, D.W. (1986) *J. Immunol.* 137, 3584.

Weisbart, R.H., Golde, D.W., Clark, S.C., et al. (1985) *Nature* 314, 361.

Weisbart, R.H., Kwan, L., Golde, D.W., et al. (1987) *Blood* 69, 18.

Weisbart, R.H., Kacena, A., Schuh, A., et al. (1988) *Nature* 332, 647.

Welte, K., Platzer, E., Lu, L., et al. (1985) *Proc. Natl. Acad. Sci. USA* 82, 1526–1530.

Whetton, A.D., Monk, P.N., Consalvey, S.D., and Downes, C.P. (1986) *EMBO J.* 5, 3281–3286.

Wimperis, J.Z., Niemeyer, C.M., Sieff, C.A., et al. (1989) *Blood* 74, 1525–1530.

Wing, E.J., Ampel, N.M., Waheed, A., et al. (1985) *J. Immunol.* 135, 2052–2056.

Wingfield, P., Graber, P., Moonen, P., Craig, S., and Pain, R.H. (1988) *Eur. J. Biochem.* 173, 65–72.

Wodnar-Filipowicz, A., Heusser, C.H., and Moroni, C. (1989) *Nature* 339, 150–152.

Wong, G.G., Witek, J.S., Temple, P.A., et al. (1985) *Science* 228, 810–815.

Wong, G.G., Temple, P.A., Leary, A.C., et al. (1987) *Science* 235, 1504–1509.

Xu, W.D., Firestein, G.S., Taetle, R., Kaushansky, K., and Zvaifler, N.J. (1989) *J. Clin. Invest.* 83, 876–882.

Yang, Y-C., Ciarletta, A.B., Temple, P.A., et al. (1986) *Cell* 47, 3–10.

Yang, Y-C., Kovacic, S., Kriz, R., et al. (1988) *Blood* 71, 958–961.

Yeung, Y.G., Jubinsky, P.T., Sengupta, A., et al. (1987) *Proc. Natl. Acad. Sci. USA* 84, 1268–1271.

Ziltener, H.J., de St. Groth, B.F., Leslie, K.B., and Schrader, J.W. (1988) *J. Biol. Chem.* 263, 14511–14517.

Zimmerman, G.A., and McIntyre, T.M. (1988) *J. Clin. Invest.* 81, 531.

Zsebo, K.M., Cohen, A.M., Murdock, D.C., et al. (1986) *Immunobiology* 172, 175–184.

Zsebo, K.M., Yuschenkoff, N.V., Shiffer, S., et al. (1988) *Blood* 71, 99–103.

Long-Term Bone Marrow Cell Cultures

August J. Salvado

The biotechnology for the growth of hematopoietic progenitors in tissue culture has been available to the scientific community for nearly a quarter of a century. In its earlier years, this technology centered around clonal-assay systems capable of growing committed progenitor cells with lineage specificity and little capacity for self-replication (see below). Later modifications to clonal-assay methods allowed the growth of multipotent progenitors with greater potential for self-renewal. Altogether, clonal methods have contributed enormously to our current understanding of the cell biology of normal hematopoiesis. They have been indispensable tools in shaping our concept of the physiologic hierarchy that distinguishes a morphologically anonymous realm of hematopoietic progenitors (Figure 16–1). Moreover, they helped identify several of the growth factors described in previous chapters and constructed the foundation for understanding their roles in the regulation of each of the hematopoietic lineages. Finally, they were instrumental in the development of recombinant proteins for many of these factors.

The opinions expressed herein are the private views of the author and should not be construed as official or as reflecting the views of the U.S. Department of the Army or the Department of Defense.

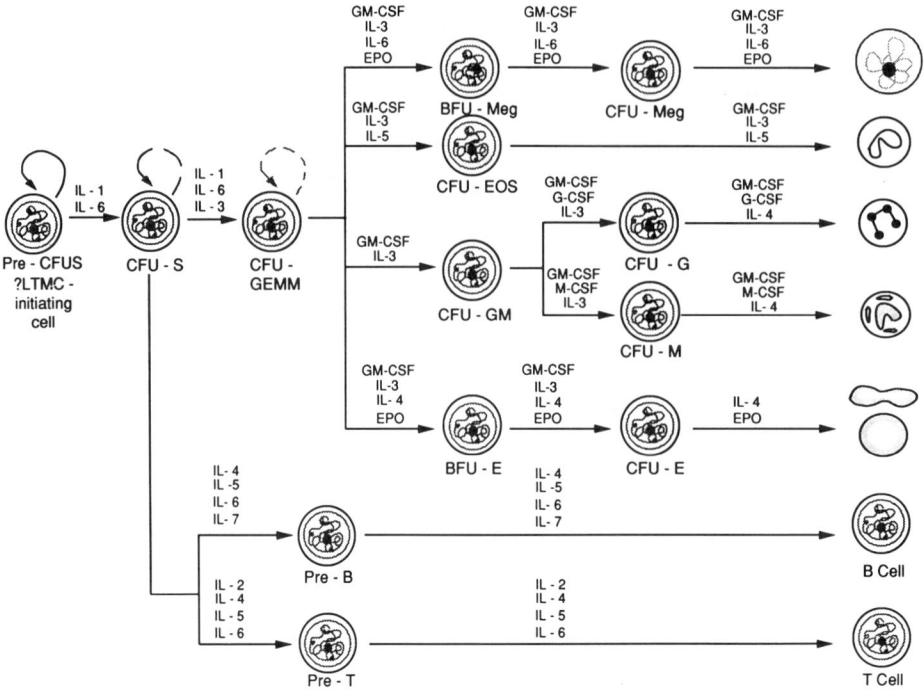

FIGURE 16-1 Schematic hierarchy of hematopoietic progenitors identified through a combination of spleen-colony assays (mice), in vitro clonal assays (mice and humans), and long-term bone marrow cultures (mice and humans). Included are growth factors with reported regulatory/modulatory effects on one or more progenitor cell type(s).

These assets notwithstanding, the limitations of the clonal systems and the cells maintained within these systems invited exploration of an in vitro culture system that would allow a closer approximation of the hematopoietic microenvironment and maintenance of cells with a greater capacity for self-renewal. Concurrent with the development of the various clonal assays, the close relationship between hematopoietic cells and marrow "stromal" cells, and the necessity for a competent microenvironment for normal hematopoiesis became increasingly clear (Knospe et al. 1968; Friedenstein et al. 1974). It was not unreasonable, therefore, to suspect that prolonged hematopoiesis in vitro might, in some way, require the participation of these microenvironmental cells. Chang and Anderson showed in 1971 that murine granulopoiesis in vitro lasted several days in the presence of bone marrow-derived adherent "nurse" cells. Later, in England, Dexter et al. (1973) were studying leukemogenesis and also observed prolonged granulopoiesis in cultures that developed an adherent layer of cells. Dexter et al. methodically pursued their observation and were able to demonstrate, by 1976, extended survival of hematopoietic progenitors closely allied to a previously established adherent layer of mixed-cell types derived from syngeneic marrow (Dexter et al. 1976). The success of Dexter et al. with murine cells led other investigators to introduce various modifications to the technique and adapt it to a number of species from hampsters to lower primates and humans (Eastment et al. 1982; Moore et al. 1979; Gartner and Kaplan 1980; Hocking and Golde 1980).

Progress in the development of liquid culture systems presented several intriguing possibilities. First, it allowed ready access to all cells, adherent and nonadherent, for harvest and further investigation. The technique, therefore, permitted the study of both hematopoietic progenitor cells and nonhematopoietic cells forming the microenvironment needed to sustain hematopoiesis. Pertinent interactions between the progenitors and the stromal cells might now be examined in vitro. Next, cultures that produce primitive hematopoietic progenitors for prolonged periods in the absence of added growth factors could also be useful in studying the intrinsic production of these factors and their interactions with hematopoietic cells. Finally, the potential to sustain (and perhaps expand) a population of primitive hematopoietic cells brings exciting possibilities for clinical utility. The remainder of this chapter will examine, then, some of the accomplishments realized in each of these areas with long-term bone marrow cultures during the past 16 years.

16.1 EVOLVING HIERARCHY OF HEMATOPOIETIC PROGENITORS

16.1.1 Clonal Assays

A discussion of clonal assays is mandatory to any examination of long-term marrow cultures since clonal systems have been used extensively to decipher both the basic physiology and perturbations applied to long-term methods.

The first clonal assay was, in fact, an in vivo assay in mice described by Till and McCulloch (1961) (Table 16–1). These investigators showed that bone marrow cells from donor animals would form macroscopic spleen colonies when injected intravenously into syngeneic-irradiated hosts. The pluripotentiality of the progenitors was demonstrated when cells from separate spleen colonies contained different hematopoietic lineages, some colonies included multiple lineages. Additionally, cells from individual colonies could produce secondary colonies in a new host and those derivative colonies would often display the same lineage heterogeneity. The clonality and common origin of these spleen colonies, termed CFU-S, was confirmed in later studies (Becker et al. 1963). More recent investigations indicate that CFU-S counted after 11 days in the host define a pluripotent stem cell with greater capacity for self-renewal than cells that form spleen colonies at earlier time points (Magli et al. 1982).

TABLE 16–1 Clonal Assay Systems in Current Use

Assay	Lineage Restriction	Progeny	Reference
CFU-S	Multipotent	Erythroid, myeloid megakaryocytic	Till and McCulloch (1961)
Blast cell	Multipotent	Erythroid, myeloid megakaryocytic	Leary and Ogawa (1987)
CFU-GEMM	Multipotent	Erythroid, myeloid megakaryocytic	Fauser and Messner (1978)
BFU-E	Unipotent	Erythroid	Axelrad et al. (1973)
CFU-E	Unipotent	Erythroid	Stephenson et al. (1971)
CFU-GM	Bipotent	Granulocyte, macrophage	Bradley and Metcalf (1966); Ichikawa et al. (1966)
CFU-EOS	Unipotent	Eosinophil	Metcalf et al. (1974)
CFU-MEG	Unipotent	Megakaryocytes	Nakeff and Daniels-McQueen (1976); Williams et al. (1978)
CFU-T	Unipotent	T-Lymphocytes	Claesson et al. (1977)
CFU-B	Unipotent	B-Lymphocytes	Metcalf et al. (1975); Kincade et al. (1978)

In vitro clonal assays growing lineage-restricted progenitors were later developed and quickly became the cornerstone in the process of ordering the physiologic layers of hematopoietic progenitor cells in the bone marrow. The general principal underlying each of these assays is the immobilization of the target cells as a single-cell suspension in a semisolid matrix, such as agar, methylcellulose, or clotted plasma. In the presence of appropriate growth factors, a subset of cells within the original population will divide and differentiate to form colonies of morphologically recognizable progeny. The parent cells can then be inferentially identified and quantitated 2 days to 3 weeks after plating through the number and character of the derived colonies.

The first of such assays, in 1966, recognized a cell that was capable of differentiating into colonies containing granulocytes, macrophages, or a mix of the two cell types (CFU-GM). Soon afterward, similar systems were developed to look at progenitors for erythroid cells (BFU-E, CFU-E), as well as megakaryocytes (CFU-MEG), eosinophils (CFU-EOS), and both T and B lymphocytes (CFU-T, CFU-B). More recently, clonal techniques have been utilized to culture multipotent progenitors capable of differentiating into more than one lineage (CFU-GEMM). A second clonal method for multipotent progenitors, the blast-colony assay, quantitates a cell with both multiple potency and a higher capacity for self-replication since many of the progeny are capable of growing secondary colonies on replating experiments (Leary and Ogawa 1987). For more complete reviews on hemato-poietic progenitors, the reader is referred to the reviews by Quesenberry and Levitt (1979), Dexter et al. (1988), and Mazur and Cohen (1989).

16.1.2 Hematopoietic Progenitors in Long-Term Cultures

Most of the clonal progenitors currently defined have been found in the adherent and nonadherent cell fractions of long-term bone marrow cultures. Several observations can be made regarding the proliferation and self re-newal of these progenitors in vitro.

First, evidence indicates that there is a difference in proliferative and self-renewal capacity among cells found in the adherent and nonadherent fractions of the culture. Mauch et al. (1980) found that murine CFU-S derived from the adherent cell layer had greater self-renewal potential than those isolated from the nonadherent fraction. Using in vitro clonal assays, Coulombel et al. (1983a) described that the adherent layers of human long-term marrow cultures consistently contained the majority of the BFU-E and CFU-GM by the third to fourth week of culture. Also, the progenitor-derived colonies from the adherent layer were substantially larger than colonies from nonadherent progenitors.

Another pertinent observation reflecting upon the maintenance of self-renewing progenitors in long-term cultures is the duration of hematopoiesis. At present this parameter is finite for all described variations of the method. Additionally, marked species variability is a hallmark of the systems now available. With some strains of mice, clonigenic cells are recovered from long-term marrow cultures for periods up to 1 year (Sakakeeny and Green-berger 1982) and CFU-S are detectable for over 20 weeks (Greenberger et al. 1983). Human long-term cultures, on the other hand, are essentially devoid of lineage-restricted progenitors by 8–20 weeks (Gartner and Kaplan 1980; Meagher et al. 1988; Moore and Sheridan 1979) and there is no anal-ogous assay available for a cell type equivalent to murine CFU-S. In mice, therefore, the transient maintenance of progenitors with some potency for self renewal was readily documented through spleen colony and in vitro clonal assays, showing that the cumulative sum of these clonigenic progen-itors recovered from the system usually exceeded the initial input. In hu-mans, this observation was more difficult to establish. Investigators have now been able to show, however, with careful quantitation, that the net output of lineage-restricted clonigenic progenitors from the cultures does indeed exceed the basal values (Meagher et al. 1988).

Most recently, studies in both mice and humans suggest the presence of a very primitive cell in long-term marrow cultures that precedes cloni-genic progenitor cells. Spooncer et al. (1985) demonstrated that an enriched population of primitive (Day 12) CFU-S approaching 100% purity could generate foci of hematopoiesis with a typical "cobblestone" appearance in murine long-term cultures, but were unable to establish long-term hematopoiesis or generate secondary CFU-S. One possibility suggested for this observation is that the cells from unseparated marrow that are needed to establish long-term hematopoiesis in vitro and generate secondary CFU-S may come from a different population. Ploemacher and Brons (1989) were then able to physically separate different populations of primitive murine progenitors based upon counterflow elutriation, light scatter properties, and fluorescent staining with rhodamine-123, a supravital dye taken up in mitochondrial membranes (Figure 16–2). Those cells with the greatest rhodamine uptake, Rh-bright, were efficient in forming 12-day spleen colonies in primary hosts, but did not repopulate host marrow with progeny that would produce 12-day spleen colonies in secondary hosts. On the other hand, cells with the least avid uptake of rhodamine, Rh-dull, could not form 12-day spleen colonies efficiently in primary hosts, but did form marrow progeny that were quite efficient in secondary spleen-colony formation. A clear difference in radioprotective ability was also described with far fewer cells

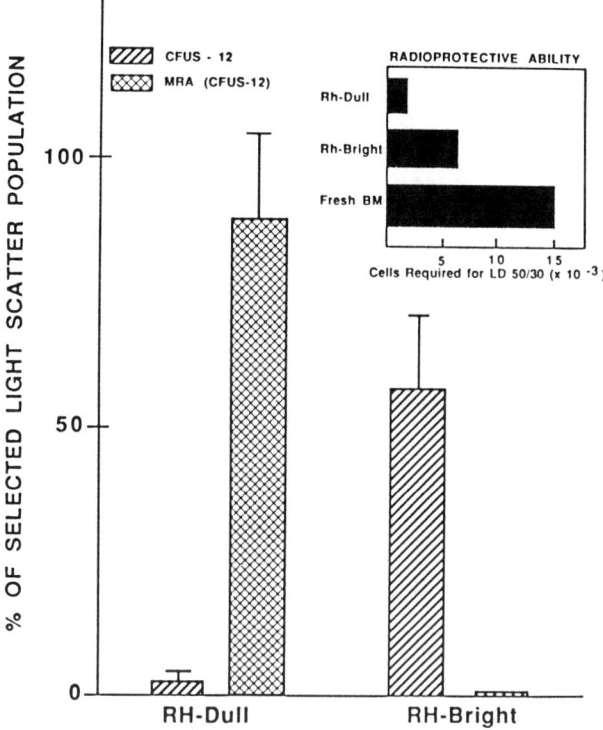

FIGURE 16–2 Ability of elutriated murine bone marrow cells gated on light scatter and sorted on rhodamine-123 uptake to form spleen colonies in vivo (diagonal bars); repopulate bone marrow with cells that form spleen colonies in secondary hosts (crosshatched bars); and salvage 50% of a group of lethally irradiated hosts for 30 days (insert-solid bars). Data from Ploemacher and Brons (1989).

from the Rh-dull fraction being required to salvage 50% of a group of lethally irradiated syngeneic hosts for at least 30 days. Later work from the same laboratory showed that Rh-bright cells formed many foci of cobblestone areas by Day 7 when seeded upon irradiated stromas in long-term cultures, whereas Rh-dull cells formed most cobblestone areas at a later time (Day 28) (Ploemacher et al. 1989). Limiting-dilution assays in long-term cultures suggested a frequency of approximately one per 10,000 marrow cells for these Rh-dull, nonclonigenic "pre-CFU-S."

At about the same time, Sutherland et al. (1989) were able to demonstrate a cell with similar characteristics in humans (Table 16–2). Using density separation at 1.068 g/ml as an initial step, marrow cells were further separated by fluorescence activated cell sorting employing low forward and perpendicular light scatter, very high CD34 reactivity, and negative to low HLA-DR reactivity as the discriminants. Sorted cells possessing these features were also shown to be poorly capable of generating primary clonigenic progenitors in vitro, but able to generate progeny in long-term cultures after 5 weeks that could then form colonies in clonigenic assays. Estimation of the marrow repopulating ability of this group of progenitors will be more difficult to establish in humans. One possible yardstick for estimation of this property may be the correlation between long-term culture repopulating cells and marrow engraftment. Engraftment of human bone marrow has not been shown to correlate well with the results of in vitro clonal assays. It remains to be seen, however, whether determination of long-term culture initiating cells correlates better with the rescue capability of transplanted bone marrow cells.

16.2 HEMATOPOIETIC MICROENVIRONMENT

16.2.1 Accessory Cells and Growth Factors

The continuous production of primitive hematopoietic progenitors in long-term marrow cultures remains an elusive goal with significant clinical implications. However, the transient survival of these cells and establishment

TABLE 16–2   Characteristics of Long-Term Culture (LTC) Initiating Cells

	LTC-Initiating Cells	Clonigenic Cells
Density	<1.068 g/ml	>1.068 g/ml
Forward light scatter	Low–very low	High
My 10 positivity	Highest 2%	Highest 5%
HLA-DR positivity	Negative to ±	High
Percent clonigenic cells in population with column parameters	3.5% (±1.2%)	30% (±2%)
Percent LTMC-initiating cells population with column parameters	8.9% (±3.7%)	1.7% (±0.1%)

Data from Sutherland et al. (1989).

of limited hematopoiesis in the presence of an appropriate microenvironment in long-term cultures lends strength to the concept that extended function of this complex system relative to clonal assays is dependent upon the application and balance of regulatory and/or modulatory controls external to the progenitors. The requirement for these controls is only partially satisfied at present by the close contact of progenitors in the system with nonhematopoietic stromal cells.

The adherent stromal layer is, in fact, composed of a heterogeneous mix of cells that includes macrophages, fibroblasts, adiposites, and "reticular" and endothelial cells (Dexter 1979, 1982; Dexter et al. 1985; Zuckerman and Wicha 1983; Singer et al. 1984). Hocking and Golde (1980) have shown that T lymphocytes also persist for a number of weeks in human long-term cultures. While mature lymphocytes are rapidly lost from murine long-term cultures (Dorshkind and Phillips 1983), progenitor cells that can give rise to lymphocytes persist for several weeks (Fulop and Phillips 1989). As seen in Figure 16-3, many of these cell types are capable of producing factors that can influence hematopoietic progenitors.

The mechanism(s) of the translocation of those factors and their presentation to the progenitor cells is an area of active investigation. However, several possibilities might be considered (Figure 16–4). One would be a paracrine mechanism whereby minute quantities of soluble factors produced by accessory cells are functional over a limited distance. The imposition of range limitations on the efficacy of these factors could be a simple function of rapid dilution as they diffuse through the medium or, alternatively, rapid binding of the factors to extracellular matrix proteins resulting in high localized concentrations of the factors (see below). Aside from soluble factor production, however, the close juxtaposition of progenitors and accessory cells in long-term cultures would easily allow for regulatory interactions through cell-cell contact with the membranes of certain accessory cells directly presenting regulatory molecules to the responding progenitor cells.

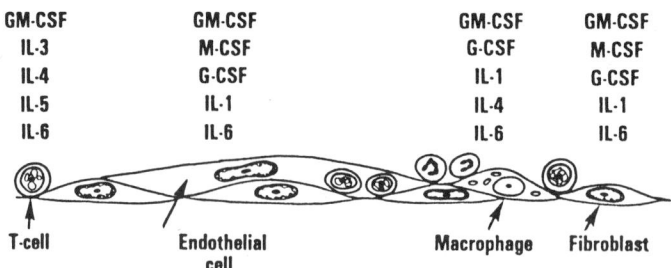

FIGURE 16-3 Hematopoietic growth factors produced by stromal cells in the microenvironment of long-term marrow cultures. Factor production was documented by bioassay of conditioned media from cell lines or mRNA analysis of the cells.

A B C

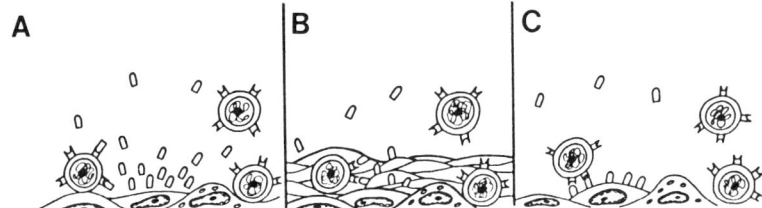

FIGURE 16-4 Mechanisms for localization of factors that stimulate hematopoietic progenitor growth in long-term marrow cultures. (A) Effective local stimulatory concentrations rapidly lost through diffusion of growth factor in medium. (B) High local concentrations of factors maintained near producing cell through rapid binding to extracellular matrix proteins. (C) Membrane-associated molecules directly presented by stromal cell to hematopoietic progenitor.

Initial interests in the production of growth factors by stromal cells focused on factors known at that time to stimulate multipotent or lineage restricted colonies in vitro, the traditional colony stimulating factors (CSFs). Earlier studies were able to clearly document the production of macrophage CSF in murine cultures (Shadduck et al. 1983). Later the remaining CSFs (GM-CSF, Meg-CSF, and G-CSF) were found as well (Heard et al. 1982; Gualtieri et al. 1984; Fibbe et al. 1988; Rennick et al. 1987).

Since the early to mid-1980s, it has become quite clear that cytokines grouped in the family of interleukins are also able to influence hematopoietic progenitors (Table 16-3). As recombinant proteins for most of these factors rapidly became available, investigators discovered that there is significant overlap in the ability of both the interleukins and the CSFs to affect the family of hematopoietic progenitor cells (Figure 16-4). Stated differently, any single protein in this group often has physiologic effects on the proliferation in vitro of multiple members in the hierarchy of increasingly restricted progenitors. Moreover, multiple proteins often act in concert to additively or synergistically enhance colony growth of a single type of progenitor. For more extensive reviews on growth factors the reader is referred to Chapter 15 in this volume, and to Herrmann and Mertelsmann (1989).

Although several of the interleukins (IL-4, IL-5, IL-7) seem to represent or act in conjunction with terminal growth factors to affect lineage restricted progenitors, others (IL-1, IL-3, IL-6) appear to act on more primitive hematopoietic cells (see Figure 16-1 and Table 16-3). IL-1 not only causes the release of CSFs and IL-6 from accessory cells in long-term marrow cultures, but has also been implicated in the direct activation or priming of the predominantly quiescent primitive progenitor cell population. While IL-6 effects several restricted progenitor cells, this factor too has recently been suggested to affect more primitive progenitors, perhaps by shortening the G_0 state in these cells (Ikebuchi et al. 1987). IL-3, like GM-CSF, acts

TABLE 16–3   **Hematopoietic Progenitor Interactions Attributed to Interleukins**

Protein	Interactions
IL-1	May directly affect DNA cycle in primitive hematopoietic proteins
	Stimulates production of GM-CSF, G-CSF, M-CSF, and IL-6 stromal cells
	Stimulates IL-3 and GM-CSF production in T cells
IL-3	Acts as a multipoietin directly stimulating colony formation of multiponent progenitors
	Continues to act synergistically with several terminal growth factors to increase colony formation in vitro of lineage-restricted progenitors
IL-4	Stimulates lymphocyte progenitors
	Synergizes with erythropoietin to stimulate late erythroid progenitors
	Synergizes with GM-CSF to stimulate myeloid progenitors
IL-5	Stimulates lymphocyte progenitors
	Stimulates eosinophil progenitors along with IL-3 and GM-CSF
IL-6	Acts directly upon primitive multipotent progenitors; may shorten G_0 phase of cell cycle; synergizes with IL-3 to enhance multipotent colony growth
	Acts with IL-3 to augment CFU-MEG growth and terminal differentiation
	Acts with IL-4 and erythropoietin to augment erythroid colony growth

directly as a multipoietin, stimulating colony formation of multipotent and early lineage-restricted progenitor cell types.

Recent studies document that bone marrow stromal cells can generate several of the interleukin proteins including IL-1, IL-4, IL-6, and IL-7 (Fibbe et al. 1988; Gimble et al. 1989; King et al. 1988; Nemunaitis et al. 1989). Interestingly, the production of IL-3 has been difficult to demonstrate so far in these cultures (Kodama et al. 1986; Eliason et al. 1988).

The complex protein interactions regulating hematopoiesis in long-term cultures may, in fact, extend beyond those noted above. Cashman et al. (1990) recently showed that activation of quiescent primitive hematopoietic cells in long-term cultures could be elicited with addition of other proteins known to be derived from mesenchymal cell types found in these cultures. Most of these proteins, including platelet derived growth factor (PDGF), IL-2, and T-cell growth factor (TGF)-alpha had a positive effect upon the cycling of early hematopoietic cells. TGF-beta, on the other hand, was noted to be strongly inhibitory to the cycling effect produced by these other proteins. The inhibitory effect of TGF-beta on hematopoiesis in vitro has been noted by others as well. Hayashi et al. (1989) have shown that lymphopoiesis and myelopoiesis are differentially sensitive to this effect in long-term marrow cultures.

This potent inhibitory influence of the stromal environment coupled with unregulated activation of quiescent progenitors in the *absence* of contact with adherent stromal cells (Eaves et al. 1986) led Eaves and Eaves (1988) to propose a model in which dominant stromal inhibitory influences for hematopoiesis are reversibly influenced through the production and local action of stimulatory molecules. The data are also consistent with earlier speculation by others for the existence of "stromal cell domains" in which cells capable of producing positively acting molecules may influence progenitor cell production (Trentin 1972; Dexter et al. 1984).

16.2.2 Extracellular Matrix

Aside from its cellular components, the adherent layer of long-term marrow cultures also contains a skeletal framework of extracellular proteins present on cell surfaces as well as in the intercellular spaces. These proteins, formed by the stromal cells, include several types of collagen (I, III, IV, V), laminin, fibronectin, and proteoglycans (Bentley 1982; Keating and Singer 1983; Zuckerman and Wicha 1983). The proteoglycans are macromolecules consisting of a protein core surrounded by repeating carbohydrate chains (glycosaminoglycans, GAGs) composed primarily of chondroitin sulfate, heparin sulfate, dermatan sulfate, and hyaluronic acid. These proteoglycans are felt to play a significant role in morphogenesis and investigators have recently devoted attention to their potential influence on hematopoiesis in long-term marrow cultures.

Spooncer et al. (1983) showed that β-D-xyloside, which increases freechain GAG synthesis but decreases overall proteoglycan levels, results in greater production of CFU-S from murine cultures. While the mechanism for this observation remains uncertain, heightened interest in proteoglycan effects in this system has now focused on two promising areas.

One of those areas is the binding of hematopoietic growth factors to GAGs. Gordon et al. (1987) were able to document the binding of GM-CSF to stromal layers that had been dehydrated and previously salt extracted. When conditioned medium containing GM-CSF from a bladder-carcinoma cell line were incubated with these treated stromal layers, <5% of the protein was removed. The deleted protein, however, contained 40–60% of the GM-CSF activity. Similar observations were made when conditioned medium or rGM-CSF was incubated in plates coated with bone marrow GAGs. In the latter case, the colony-forming activity could then be recovered by salt extraction of the plates after GM-CSF binding. Furthermore, the binding of GM-CSF was found to be tissue-specific and could not be duplicated with GAGs extracted from fetal liver tissue.

Roberts et al. (1988) then extended these observations to show that recombinant GM-CSF and IL-3 both bound to a commercially prepared extracellular matrix preparation in a way that allowed presentation of the active portions of each molecule. Through selective digestion of separate types of GAGs, this binding was specifically associated with heparin sulfate. Further experiments using intercellular-nonmembrane-associated matrix proteins suggested that growth factors are primarily bound by matrix proteins *associated* with stromal cell membranes. These data would be consistent, then, with a model in which high local concentrations of stimulatory molecules are achieved, and perhaps stabilized, through binding to extracellular matrix proteins at or near the surface of the stromal cells producing the factors.

Though most attention has focused on the production of proteoglycans by stromal cells rather than hematopoietic progenitor cells, Minguell and Tavassoli (1989) have now shown that FDCP-1, a murine factor-dependent hematopoietic cell line, is also capable of proteoglycan production. Unlike stromal cells that produce a variety of proteoglycans with different GAG components, FDCP-1 cells were found to produce a unique membrane-associated chondroitin sulfate proteoglycan. It is not difficult to envision a role for membrane associated proteoglycan on the surface of hematopoietic progenitors in the regulation of cellular relationships in the hematopoietic microenvironment through interactions with other extracellular matrix proteins such as fibronectin or collagen. Alternatively, these investigators theorize a possible role for this molecule in "homing" of progenitor cells to hematopoietic tissues.

16.3 CLINICAL APPLICATIONS—CURRENT AND FUTURE

Though the picture is by no means complete, it should be quite clear from the observations noted above that long-term marrow cultures have contributed substantially to our understanding of both hematopoietic progenitor cells and the extracellular proteins, growth factors, and accessory cells that constitute a microenvironment suitable for hematopoiesis to occur. Those same observations, however, along with some others lead to the exciting prospect that this methodology might well have clinical utility in specific areas.

16.3.1 Autologous Bone Marrow Transplantation

One such area is that of autologous bone marrow transplantation. As already noted, data from several laboratories support the concept that long-term marrow cultures permit the survival and at least limited proliferation of very primitive preclonigenic hematopoietic progenitors. In mice, these cells produce progeny that can repopulate and salvage irradiated-secondary hosts (Ploemacher et al. 1989). In humans, they give rise to clonigenic progeny after several weeks in culture (Sutherland et al. 1989). In addition, Winton and Colenda (1987) have demonstrated that 4-hydroperoxycyclophosphamide, an agent useful in purging malignant cells from bone marrow prior to autologous transplantation, will eliminate clonigenic cells from marrow used to initiate long-term cultures while selectively sparing quiescent nonclonigenic precursors that are then capable of repopulating the adherent layers in the long-term cultures.

Another interesting and somewhat surprising observation came in the early 1980s when investigators initiated human long-term cultures with marrows from patients with chronic myelocytic leukemia in an effort to better clarify abnormalities in the behavior of the neoplastic cell population. After several weeks in culture, the number of Ph[1]-negative clonigenic precursors often rose from barely detectable initial levels to a significant proportion of the remaining population of clonigenic cells (Coulombel et al. 1983b). The results were surprising because standard chemotherapeutic agents used in this disease are clearly unable to decrease the detectable Ph[1] population and even aggressive chemotherapeutic regimens have had only limited success in this respect (Clarkson 1985). Subsequent studies on additional patients with chronic and acute myelocytic leukemias revealed this pattern of behavior in long-term cultures to be relatively common in these diseases (Eaves et al. 1985).

The survival of primitive cells in long-term culture, the sparing of these same cells by chemotherapeutic agents effective in purging malignant cells from marrow, and the apparent selective advantage of normal progenitors in the culture over abnormal cells in some cases of malignant hematopoietic disease all suggest a potential use of the system in preparing marrow for autologous transplant. Indeed, early results have now been reported on nine patients from three separate groups in which long-term marrow cultures have been used for this purpose (Chang et al. 1986; Turhan et al. 1988; Stoppa et al. 1989). All reported patients receiving such transplants to date have had either CML or AML as an underlying problem. The data, though very preliminary, are quite promising and warrant close scrutiny for longer follow up.

16.3.2 Replacement Therapy in Bone Marrow Failure

Another apparent use for a culture system in which self-renewing cells are capable of surviving would be expansion of those critical cells for replacement therapy in marrow failure states such as aplastic anemia. Until now, however, significant expansion of the progenitor cell population in vitro, for human cells at least, has remained an elusive goal. More recent data on growth factors such as IL-1, IL-3, and IL-6, which seem to have significant effects on proliferation of early hematopoietic progenitors, may open new avenues of exploration in this area. In this regard, Okano et al. (1989) reported a significant expansion of murine CFU-S in 6-day liquid-suspension cultures supplemented with both IL-3 and IL-6. Recovery values exceeded input values fivefold, and the ability of cells cultured with or without factors to rescue lethally irradiated syngeneic mice was 89% versus 22%, respectively. Whether analogous expansion of primitive human progenitors can be demonstrated, and additional benefit derived from extended culture of these cells in the presence of an adherent stroma and exogenous factors are open questions of significant importance and interest.

16.3.3 Gene Therapy

As the list of newly revealed DNA codes for critical physiologic proteins grows daily, so does the interest in the use of gene therapy as a potential approach to certain genetic diseases. Marrow cells are often a logical target for such therapy, being easily obtainable and having numerous in vitro systems, including long-term cultures available for their manipulation. Currently, amphotrophic retroviral vectors are available that can mediate gene transfer into mammalian cells but are themselves unable to replicate within those cells after integration into the genome (Miller and Buttimore 1986). Successful gene transfer into hematopoietic progenitors using these vectors has already been accomplished by several groups (Keller et al. 1985; Dick et al. 1985; Eglitis et al. 1985). Among the problems encountered, however, have been low rates of infectivity and transient or unstable levels of expression of the inserted genes with time.

One possibility for these problems would be instability of the vector DNA within the target cell. Alternatively, loss of expression of the inserted gene could result from ineffective integration into the more primitive progenitors with greater capacity for self-renewal. Since studies have shown that DNA replication is required for effective integration of murine leukemia viral genomic material into host cells (Richter et al. 1984), it is assumed that similar principles would apply in targeting selected genes within viral vectors to early hematopoietic progenitors. Unfortunately, the vast majority of the target cells of interest are likely to be in a "rest" or G_0 state (Becker et al. 1965; Lajtha 1979).

Recent efforts by a number of investigators, then, have focused on improving the efficiency of targeting genes to more primitive progenitors by manipulating the DNA synthetic cycle of these cells prior to or during long-term culture. At least two laboratories have now reported increased efficiency of retroviral gene transfer into human and canine hematopoietic progenitor cells maintained in long-term marrow cultures (Schuening et al. 1989; Hughes et al. 1989).

It is well known that progenitors in long-term cultures are triggered into DNA synthesis immediately following a feeding cycle (Dexter et al. 1977). Schuening et al. (1989) took advantage of this observation by infecting canine marrow cells immediately prior to their addition to long-term culture and supplying them with several "boosts" of virus-laden supernate with repetitive feeds over a 6-day period of culture. The investigators noted approximately fourfold increases over uncultured infected cells in expression of two different genes in the CFU-GM produced by these cells.

Hughes et al. (1989), on the other hand, infected human marrow cells by coculture with viral producer cells in the presence or absence of exogenous IL-1β and GM-CSF. After the 72-h "infection" phase of culture, they added the cells to long-term cultures. Following the progeny of the hematopoietic progenitors for 6 weeks, they showed a six- to eightfold increase in gene insertion when growth factors were present during the initial infection culture with stable expression of the gene during the entire long-term culture period. These data are in agreement with those of Bodine et al. (1989), who showed an increase in gene insertion into murine hematopoietic cells infected in the presence of both virus and exogenous growth factors that affect primitive progenitors (IL-3 and IL-6). Gene expression in transfected cells was detectable in vivo for up to 12 months.

Though gene therapy remains poised on the horizon of medical science and continued refinement of the methodology will undoubtedly occur for some time, long-term cultures in some combination with progenitor preselection and growth factor manipulation are quite likely to play a significant role in the progression of this sophisticated therapy from the bench to the bedside.

REFERENCES

Axelrad, A.A., McLeod, O.L., Shreeve, M.M., and Heath, D.S. (1973) in *Proceedings of the 2nd International Workshop on Hematopoiesis in Culture* (Robinson, W.A., ed.), pp. 226–234, Grune and Stratton, New York.

Becker, A.J., McCulloch, E.A., and Till, J.E. (1963) *Nature* 197, 452–454.

Becker, A.J., McCulloch, E.A., Siminovich, L., and Till, J.E. (1965) *Blood* 26, 296–308.

Bentley, S.A. (1982) *Br. J. Haematol.* 50, 491–497.

Bodine, D.M., Karlsson, S., and Nienhuis, A.W. (1989) *Proc. Natl. Acad. Sci. USA* 86, 8897–8901.

Bradley, T.R., and Metcalf, D. (1966) *Aust. J. Exp. Biol. Med. Sci.* 44, 287–300.

Cashman, J.D., Eaves, A.C., Raines, E.W., Ross, R., and Eaves, C.J. (1990) *Blood* 75, 96–101.

Chang, V.T., and Anderson, R.N. (1971) *J. Reticuloendothel. Soc.* 9, 568–579.

Chang, J., Morgenstern, G., Deakin, D., et al. (1986) *Lancet* 8, 294–295.

Claesson, M.H., Rodger, M.B., Johnson, G.R., Whittingham, S., and Metcalf, D. (1977) *Clin. Exp. Immunol.* 28, 526–534.

Clarkson, B. (1985) *J. Clin. Oncol.* 3, 135–139.

Coulombel, L., Eaves, A.C., and Eaves, C.J. (1983a) *Blood* 62, 291–297.

Coulombel, L., Kalousek, D.K., Eaves, C.J., Gupta, C.M., and Eaves, A.C. (1983b) *N. Engl. J. Med.* 308, 1493–1498.

Dexter, T.M. (1979) *Clin. Haematol.* 8, 453–468.

Dexter, T.M. (1982) *J. Cell Physiol.* 1 (suppl), 87–94.

Dexter, T.M., Allen, T.D., Lajtha, L.G., Schofield, R., and Lord, B.I. (1973) *J. Cell. Physiol.* 83, 461–474.

Dexter, T.M., Allen, T.D., and Lajtha, L.G. (1976) *J. Cell Physiol.* 91, 335–344.

Dexter, T.M., Wright, E.G., Krisza, F., and Lajtha, L.G. (1977) *Biomedicine* 27, 344–349.

Dexter, T.M., Simmons, P., Purnell, R.A., Spooncer, E., and Schofield, R. (1984) in *Aplastic Anemia: Stem Cell Biology and Advances in Treatment* (Young, N.S., Levine, A.S., and Humphries, R.K., eds.), pp. 13–33, Alan R. Liss, New York.

Dexter, T.M., Spooncer, E., Simmons, P., and Allen, T.D. (1985) in *Long-Term Marrow Culture* (Wright, D.G., and Greenberger, J.S., eds.), pp. 57–96, Alan R. Liss, New York.

Dexter, T.M., Ponting, I.L., Roberts, R.A., and Spooncer, E. (1988) *Soc. Gen. Physiol. Ser.* 43, 25–38.

Dick, J.E., Magli, M.C., Huszar, D., Phillips, R.A., and Bernstein, A. (1985) *Cell* 42, 71–79.

Dorshkind, K., and Phillips, R.A. (1983) *J. Immunol.* 131, 2240–2245.

Eastment, C.E., Denholm, E., Katsnelson, I., Arnold, E., and T'so P. (1982) *Blood* 60, 130–135.

Eaves, A.C., and Eaves, C.J. (1988) *Blood Cells* 14, 355–368.

Eaves, C., Coulombel, L., Dube, I., et al. (1985) in *Hematopoietic Stem Cell Physiology* (Cronkite, E.P., Dainiak, N., McCaffrey, R.P., Palek, J., and Quesenberry, P.J., eds.), pp. 403–413, Alan R. Liss, New York.

Eaves, A.C., Cashman, J.D., and Gaboury, L.A. (1986) *Proc. Natl. Acad. Sci. USA* 83, 5306–5310.

Eglitis, M.A., Kantoff, P., Gilboa, E., and Anderson, W.F. (1985) *Science* 230, 1395–1398.

Eliason, J.F., Thorens, B., Kindler, V., and Vasalli, P. (1988) *Exp. Hematol.* 16, 307–312.

Fauser, A.A., and Messner, H.A. (1978) *Blood* 52, 1243–1248.

Fibbe, W.E., van Damme, J., Billian, A., et al. (1988) *Blood* 71, 430–435.

Friedenstein, A.J., Chailakhyan, R.K., Latsinik, N.V., Panasyuk, A.F., and Keiliss-Borok, I.V. (1974) *Transplantation* 17, 331–340.

Fulop, G.M., and Phillips, R.A. (1989) *Blood* 74, 1537–1544.

Gartner, S., and Kaplan, H. (1980) *Proc. Natl. Acad. Sci. USA* 77, 4756–4759.

Gimble, J.M., Pietrangeli, C., Henley, M.A., et al. (1989) *Blood* 74, 303–311.

Gordon, M.Y., Riley, G.P., Watt, S.M., and Greaves, M.F. (1987) *Nature* 326, 403–405.

Greenberger, J.S., Hoffman, N., Lieberman, M., Botnick, L., and Sakakeeny, M.A. (1983) *J. Natl. Cancer Inst.* 70, 323–331.

Gualtieri, R.J., Shadduck, R.K., and Quesenberry, P.J. (1984) *Blood* 64, 516–525.

Hayashi, S., Gimble, J.M., Henley, A., Ellingsworth, L.R., and Kincade, P.W. (1989) *Blood* 74, 1711–1717.

Heard, J.M., Fichelson, T., and Varet, B. (1982) *Blood* 59, 761–767.

Herrmann, F., and Mertelsmann, R. (1989) *Blut* 58, 117–128.

Hocking, W.G., and Golde, D.W. (1980) *Blood* 56, 118–124.

Hughes, P.F.D., Eaves, C.J., Hogge, D.E., and Humphries, R.K. (1989) *Blood* 74, 1915–1922.

Ichikawa, Y., Pluznik, D.H., and Sachs, L. (1966) *Proc. Natl. Acad. Sci. USA* 56, 488–495.

Ikebuchi, K., Wong, G.D., Clark, S.C., et al. (1987) *Proc. Natl. Acad. Sci. USA* 84, 9035–9039.

Keating, A., and Singer, J.W. (1983) *Exp. Hematol.* 11 (suppl. 14), 144.

Keller, G., Paige, C., Gilboa, E., and Wagner, E.F. (1985) *Nature* 318, 149–154.

Kincade, P.W., Paige, C.J., Parkhouse, R.M.E., and Lee, G. (1978) *J. Immunol.* 120, 1289–1294.

King, A.G., Wierda, D., and Landreth, K.S. (1988) *J. Immunol.* 141, 2016–2026.

Knospe, W.H., Blom, J., and Crosby, W.H. (1968) *Blood* 31, 400–405.

Kodama, H., Hagiwara, H., Sudo, H., et al. (1986) *J. Cell Physiol.* 129, 20–26.

Lajtha, L.G. (1979) *Differentiation* 14, 23–34.

Leary, A.G., and Ogawa, M. (1987) *Blood* 69, 953–956.

Magli, M.C., Iscove, N.N., and Odartchenko, N. (1982) *Nature* 295, 527–529.

Mauch, P., Greenberger, J.S., Botnick, L., Hannon, E., and Hellman, S. (1980) *Proc. Natl. Acad. Sci. USA* 74, 3879–3882.

Mazur, E.M., and Cohen, J.L. (1989) *Clin. Pharmacol. Ther.* 46, 250–256.

Meagher, R.C., Salvado, A.J., and Wright, D.G. (1988) *Blood* 72, 273–281.

Metcalf, D., Parker, J., Chester, H.M., and Kincade, P.W. (1974) *J. Cell Physiol.* 84, 275–290.

Metcalf, D., Nossal, G.J.V., Warner, N.L., et al. (1975) *J. Exp. Med.* 142, 1534–1549.

Miller, A.D., and Buttimore, C. (1986) *Mol. Cell Biol.* 6, 2895–2902.

Minguell, J.J., and Tavassoli, M. (1989) *Blood* 73, 1821–1827.

Moore, M.A.S., and Sheridan, A.P.C. (1979) *Blood Cells* 5, 297–311.

Moore, M.A.S., Sheridan, A.P.C., Allen, T.D., and Dexter, T.M. (1979) *Blood* 54, 775–793.

Nakeff, A., and Daniels-McQueens, S. (1976) *Proc. Soc. Exp. Biol. Med.* 151, 587–590.

Nemunaitis, J., Andrews, F.D., Mochizuki, Y., Lilly, M.B., and Singer, J.W. (1989) *Blood* 74, 1929–1935.

Okano, A., Suzuki, C., Takatsuki, F., et al. (1989) *Transplantation* 48, 495–498.

Ploemacher, R.E., and Brons, R.H.C. (1989) *Exp. Hematol.* 17, 263–266.

Ploemacher, R.E., van der Sluijs, J.P., Voerman, J.S.A., and Brons, N.H.C. (1989) *Blood* 74, 2755–2763.

Quesenberry, P., and Levitt, L. (1979) *N. Engl. J. Med.* 301, 755–760, 819–823, 868–872.

Rennick, D., Yang, G., Gemmell, L., and Lee, F. (1987) *Blood* 69, 682–691.

Richter, A., Ozer, H.L., Des Groseillers, L., and Jolicoeur, P. (1984) *Mol. Cell Biol.* 4, 151–159.

Roberts, R., Gallagher, J., Spooncer, E., et al. (1988) *Nature* 332, 376–378.

Rozensjan, L.A., Shoham, D., and Kalechman, I. (1975) *Immunology* 29, 1041–1055.

Sakakeeny, M.A., and Greenberger, J.S. (1982) *J. Natl. Cancer Inst.* 68, 305–317.

Schuening, F.G., Storb, R., Stead, R.B., et al. (1989) *Blood* 74, 152–155.

Shadduck, R.K., Waheed, A., Greenberger, J.S., and Dexter, T.M. (1983) *J. Cell Physiol.* 114, 88–92.

Singer, J.W., Keating, A., and Cattner, J.M. (1984) *Leuk. Res.* 8, 535–545.

Spooncer, E., Gallagher, J.T., Kriza, F., and Dexter, T.M. (1983) *J. Cell Biol.* 96, 510–514.

Spooncer, E., Lord, B.I., and Dexter, T.M. (1985) *Nature* 316, 62–64.

Stephenson, J.R., Axelrad, A.A., McLeod, D.L., and Schreeve, M.M. (1971) *Proc. Natl. Acad. Sci. USA* 68, 1542–1546.

Stoppa, A.M., Maraninchi, D., Lafage, M., et al. (1989) *Br. J. Haematol.* 72, 519–523.

Sutherland, H.J., Eaves, C.J., Eaves, A.C., Drogowska, W., and Lansdorp, P.M. (1989) *Blood* 74, 1563–1570.

Till, J.E., and McCulloch, E.A. (1961) *Radiat. Res.* 14, 213–222.

Trentin, J.J. (1972) in *Regulation of Hematopoiesis* (Gordon, A.S., ed.), pp. 161–186, Meredith, New York.

Turhan, A.G., Eaves, C.J., Humphries, R.K., et al. (1988) *Blood* 72 (suppl.), 184a.

Williams, N., Jackson, H., Sheridan, A.P.C., et al. (1978) *Blood* 51, 245–255.

Winton, E.F., and Colenda, K.W. (1987) *Exp. Hematol.* 15, 710–714.

Zuckerman, K.S., and Wicha, M.S. (1983) *Blood* 61, 540–547.

PART
IV

Blood-Borne Viral Diseases

Inactivation of Viruses Found with Plasma Proteins

Bernard Horowitz

Blood derivatives prepared from plasma pools have never been safer. Safety arises from improvements in donor recruiting and donor blood screening and the development of procedures that remove or inactivate viruses. It is the intent of this chapter to review the advances which have led to this conclusion and to estimate the remaining viral risk. Earlier reviews from this laboratory (Prince et al. 1987) and from others (Menache and Aronson 1985; Gomperts 1986; Burnouf et al. 1987; Mannucci and Colombo 1988) should also be consulted for additional detail.

17.1 VIRAL RISK FROM SINGLE UNITS OF BLOOD

Individually collected units of blood harbor viruses in proportion to their prevalence in the donor population. This is especially true for viruses causing chronic disease since individuals with acute infections refrain from donating blood or are excluded at the time of the physical examination. Viral risk of donated units is reduced further through the application of sensitive

Appreciation is extended to Marcia Franklin-Henry for her dedication in helping prepare this manuscript. Work supported in part by Grant no. HL41221 from NHLBI.

screening tests for virus presence. As a consequence of improved methods of donor selection and blood screening, the risk of a single blood unit transmitting any of the three principal viruses of concern in a transfusion setting, hepatitis B virus (HBV), non-A, non-B hepatitis virus (NANBHV), and human immunodeficiency virus (HIV), has been reduced substantially and is now estimated at 0.05% (Aach et al. 1981), 1% (Aach et al. 1981; Cossart et al. 1982), and 0.0005% (Schorr et al. 1985; Bove 1987; Peterman et al. 1987; Ward et al. 1988), respectively. Based on these estimates and from the number of units of blood transfused, the risk to the recipient can be calculated from simple probability theory. A patient receiving 20 units of blood (or blood components such as red cell or platelet concentrates) would have a risk of exposure to these three viruses of 1%, 18%, and 0.01%, respectively. The principal causative agent of NANBHV, termed hepatitis C virus (HCV), has now been identified (Vanderpoel 1990; Weiner et al. 1990); the use of a blood screening test for HCV antibody should reduce HCV transmission by 80% or more.

17.2 VIRAL RISK FROM PLASMA PROTEIN FRACTIONS

17.2.1 Fresh Frozen Plasma and Cryoprecipitate

Fresh frozen plasma and cryoprecipitate, each of which is currently prepared by blood banks and remains the product of single or a limited number (e.g., eight) of donors, have a risk of virus presence similar to the units of blood from which they were derived. Elimination of virus risk from these components awaits the development and implementation of virus sterilization technology compatible with these complex mixtures. The leading candidates for this application are the use of solvent/detergent technology (Horowitz et al. 1985a) and passage through filters with a pore size that is smaller than viruses (Hamamoto et al. 1989).

17.2.2 Immune Serum Globulin and Albumin

Technology developed by Edwin Cohn and his associates in the 1940s made possible the preparation of human serum albumin, immune serum globulin, and other protein fractions, thus deriving a whole family of products from plasma and expanding its utility (Cohn et al. 1946). The plasma fractionation industry has exploited these techniques fully. As a consequence, plasma from twenty thousand or more donors is pooled prior to fractionation into separate products. A probability calculation similar to that described above indicates that such a plasma pool is virtually certain to contain HBV, HCV, and HIV. The safety of products prepared from this pool depends on the initial viral load, for which estimated values are given in Table 17–1 and four principal factors: immune neutralization with antibody from one unit neutralizing virus contained in another, removal of virus through purifi-

TABLE 17-1 Estimated Values of Initial Viral Load in Plasma Pools

	HBV	HCV[1]	HIV
Units pooled	24,000	24,000	24,000
Prevalence in donated blood	0.05%	1%	0.0005%
Infected units in pool	12	240	0.12
Virus titer in infected unit (ID_{50}/ml)	10^2	$10^{2.5}$	10^2
Viral load in pool	3×10^5	1.9×10^7	3×10^3

[1] Before screening for HCV antibody.

cation, serendipitous inactivation occurring during processing or storage, and virus sterilization.

Albumin and intramuscular immune globulins prepared from such plasma pools have admirable records of virus safety (Murray and Ratner 1953; Finlayson 1979; Gerety and Aronson 1982; Iwarson et al. 1985). Albumin safety is so widely recognized that it is frequently used as a formulation aid in other products. The relative contribution of each of the four factors affecting viruses in the pool is shown in Table 17-2. While little is known about the fractionation properties of HCV, a minimum removal and inactivation of 10^{10} ID_{50} has been reported for HBV and HIV in the preparation of albumin and immune serum globulin. To estimate the probability of safety, the initial viral load can be divided by the reported extent of removal and inactivation and the number of vials produced from a single lot. For HBV and HIV, the extent of virus removal and inactivation during the preparation of immune serum globulin and albumin is many orders of magnitude above the estimated viral load, thus providing a wide safety margin (Table 17-3). This corresponds with the clinical observations of safety. A similar calculation for HCV must await the development of a sufficiently sensitive antigen assay or a quantitative assessment of HCV

TABLE 17-2 Effect on ID_{50} of Virus Removal and Inactivation[1]

	Immune Globulin			Albumin		
Mechanism	HBV	HCV	HIV	HBV	HCV	HIV
Immune neutralization	5	unk[2]	unk	5	unk	unk
Purification	5	unk	4.1	1.3–2.6	unk	≥3
Serendipitous inactivation	unk	unk	11[3]	unk	unk	3
Virus sterilization	0	0	0	4	≥4	≥6
Total	10	unk	15.1	10.3–11.6	≥4	≥12

[1] Values shown are expressed as the $\log_{10}$ decline in ID_{50}.

[2] unk, unknown. Screening tests that eliminate antibody-positive donors may reduce potential benefit.

[3] Some studies indicate substantially less (Prince et al. 1986a).

TABLE 17–3   Calculated Probability of Virus Safety[1]

	Number of Vials per ID_{50}[2]		
Product	HBV	HCV	HIV
Immune Globulin	8×10^7	ND	5×10^{15}
Albumin	3×10^9	ND	3×10^{12}

[1] Calculated assuming a 24,000-unit plasma pool (5,000 liters), 2 vials of albumin/l, and 0.5 vials of immune globulin/l.

[2] ID_{50}, the viral dose that will infect 50% of recipients; ND, not determined.

mRNA, perhaps through a modification of the polymerase chain reaction (PCR) procedure.

Immune neutralization, removal, and serendipitous inactivation of HCV during fractionation must be substantial (e.g., $\geq 10^{10}$ removal/inactivation), given the clinical safety of intramuscular immune globulin solutions. However, it should be noted that intravenous immune globulin preparations, also prepared by cold ethanol techniques, occasionally have transmitted NANBHV (Horowitz and Piet 1986). This suggests that serendipitous inactivation is an important variable in the preparation of immune globulin solutions and that a smaller margin of safety occurs with HCV than with HBV and HIV. Safety should be achievable through the use of validated procedures for virus sterilization, and several have been described (Stephan and Dichtelmuller 1983; Edwards et al. 1987; Uemura et al. 1989).

17.2.3   Coagulation Factor Concentrates

Antihemophilic factor concentrate (AHF) and prothrombin complex concentrates, as traditionally manufactured, transmitted HBV, HCV, and HIV. The prevalence of markers of infection from these viruses among adult hemophiliacs varies from 50–100%. Since typical virucidal procedures applicable to viral vaccines would modify and inactivate the coagulation factor of interest and might also have adverse immunological consequences in repeatedly infused recipients, new methods of virus inactivation had to be developed that were compatible with protein structure and function. The best characterized methods can be grouped into four classes: 1) wet heat or pasteurization, in which proteins are heated in solution; 2) dry heat, in which protein solutions are lyophilized prior to heating; 3) cold sterilization or chemical approaches, in which chemicals known to modify viruses are added to protein solutions; and 4) methods of irradiation, e.g., with ultraviolet (UV) light or gamma rays, in which proteins in solution or following lyophilization are irradiated. In addition, other modern procedures can result in substantial virus removal. Laboratory, animal, and clinical data for the effects of each of the four approaches on virucidal activity have been reviewed recently (see Introduction); thus, only the most important observations, coupled with new information, will be presented here.

17.2.3.1 Wet Heat. While some proteins can be heated for prolonged periods in solution at 60°C, most, including coagulation factors, require the addition of a stabilizer. Stabilization of protein can occur without stabilization of virus when specific ligands that interact only with binding sites on the protein are used. Examples of specific stabilization include the addition of fatty acid to albumin; 2,3-DPG to hemoglobin; and substrate to enzyme. If no specific stabilizers have been identified, nonspecific stabilizers such as high concentrations (0.5-3 M) of simple salts, sugars, and/or amino acids can substitute. Although this approach also stabilizes virus (Horowitz et al. 1985b), a favorable balance between protein recovery and virus kill can be achieved, depending on the conditions utilized. HBV was not adequately inactivated on heating a prothrombin complex concentrate at 60°C for 10 hours in the presence of 3 M potassium citrate (Menache and Aronson 1985). Nonetheless, coagulation factor concentrates from Behringwerke (Marburg, Germany), heated at 60°C for 10 hours in the presence of 2.75 M glycine and 50% sucrose, combined with purification, have a greatly reduced incidence of virus transmission (Table 17–4). AHF recovery after heating is approximately 60%.

17.2.3.2 Dry Heat. Lyophilized factor VIII concentrates retain $\geq 90\%$ of activity after heating at 60°C for 72 hours. This method offers the advantage

TABLE 17–4 Clinical Viral Safety of Coagulation Factor Concentrates Treated by Wet Heat[1]

| Units Infused | Clinical Results (Infected/Total No. Patients Treated) | | | Reference |
	HBV	HCV	HIV	
200,000			0/18	Mosseler et al. 1985
251,000	0/10	0/26	0/26	Schimpf et al. 1987a
311,000	2/x			Brackmann and Egli 1988
15.9 × 10⁶			0/155	Schimpf et al. 1989
1.89 × 10⁶	0/31	2/31		Heimburger et al. 1987
129,000	0/13	0/13	0/13	Auerswald et al. 1989
75,000	0/14	0/25	0/25	Mannucci et al. 1989
18.8 × 10⁶	2/(68 + x)	2/95	0/237	

[1] Treatment conditions: 2.75 M glycine, 50% sucrose, 60°C, 10 hours. x is some unknown number of patients.

that it can be performed in final vials, eliminating the possibility of recontamination following virus sterilization. However, numerous studies, presented in summary form in Table 17–5, demonstrate that while HIV transmission can be reduced by heating lyophilizates at 60 or 68°C for up to 72 hours, HCV continues to be transmitted at high frequency, even when heating is performed on a slurry suspended in organic solvent. The failure to achieve clinical safety is mirrored by the relatively low efficacy of these processes in inactivating HIV, HBV, and/or HCV in laboratory or animal studies. It seems likely that the heterogeneity of the lyophilized cake with respect to both water and salt contributes to treatment failures.

To address the above failure, one manufacturer, Immuno (Vienna, Austria), heats lyophilized product in bulk in the presence of a defined water vapor atmosphere at 60°C for 10 hours (AHF) and also at 80°C for 1 hour (other products). The results indicated the occasional transmission of HBV, but no transmission of NANBHV or HIV (Table 17–6) under these conditions. Another manufacturer, the British Blood Products Laboratory (Elstree, England), heats the lyophilized product at 80°C for 72 hours. Results indicated no transmission of HBV, NANBHV, or HIV (Table 17–7). However, the purification scheme employed in the manufacture of AHF had to be altered to withstand this heat cycle (Winkelman et al. 1989).

17.2.3.3 Chemical Approaches. Pioneering studies on the sterilization of human plasma (Lo Grippo and Hartman 1958; Lo Grippo et al. 1964) have led to the routine use of β-propiolactone (BPL) and UV light by Biotest (Frankfurt, Germany). However, this approach has not been widely adopted since BPL alkylates protein and is a carcinogen, though there is convincing

TABLE 17–5 Viral Safety of Coagulation Factor Lyophilizates Treated by Dry Heat

Treatment Procedure	Validation (Log$_{10}$ Kill)[1]			Clinical Safety (Infected/Total)		
	HBV	HCV	HIV	HBV	HCV	HIV
60°C, 30 hours	<3.5	>3.5	2	0/2	2/2	2/90
60°C, 72 hours	<2.5	Endog	~1	0/12	15/19	0/24
68°C, 72 hours	ND	>3.4	>2.8	NA	1/6	0/6
Heated at 60°C, 20 hours, + heptane	>2.7–4.0	2.7–<3.5	>3.2	0/18	8/37	0/37
68°C, 72 hours, + chloroform	ND	ND	ND	0/3	3/3	ND
				0/35	29/67	2/157

[1] Endog, an unknown quantity of virus was present endogenously; ND, not determined; NA, not applicable.

TABLE 17–6   Clinical Viral Safety of Coagulation Factor Lyophilizates Heated with Increased Water Vapor[1]

| Product[2] | Units Infused | Clinical Results (Infected/Total No. Patients Treated) | | | Reference |
		HBV	NANBHV	HIV	
AHF	180,000	4/14	0/28	0/28[3]	Mannucci et al. 1988
AHF	80,000	0/14	0/22	0/16	Immuno (Vienna, Austria) 1988
AHF	700,000			0/60	Schimpf et al. 1987b
PCC	23,000	0/16	0/17	0/13	Preiss et al. 1989
Factor VII, PCC	12,000	0/2	0/3	0/3[4]	Schimpf 1987
PCC, FEIBA	100,000			0/21	Schimpf 1988
	1,095,000	4/46	0/70	0/110	

[1] AHF heated at 60°C for 10 hours at 190 mbar water vapor. Other products additionally heated at 80°C for 1 hour at 1375 mbar water vapor.

[2] AHF, antihemophilic factor concentrate; PCC, prothrombin complex concentrate; Factor VII, Factor VII concentrate; FEIBA, Factor VIII inhibitor bypassing activity.

[3] Included in Schimpf et al. 1987.

[4] Included in Schimpf et al. 1989.

TABLE 17–7   Clinical Viral Safety of Coagulation Factor Lyophilizates Heated at 80°C for 72 Hours

| Units Infused | Clinical Results (Infected/Total No. Patients Treated) | | | Reference |
	HBV	NANBHV	HIV	
104,320	0/16	0/32	0/32	Colvin et al. 1988

evidence that unreacted BPL hydrolyzes completely in plasma to noncarcinogenic β-hydroxypropionic acid. While recovery of factor IX on treatment is acceptable (approximately 50%), the recovery of factor VIII is not. The process has been validated extensively (Prince et al. 1983), and published results on a modest number of patients indicate the absence of transmission of HBV, NANBHV, and HIV (Table 17–8). Recent unpublished reports from several clinical centers indicate the transmissions of HIV attributable to a single lot of prothrombin complex concentrate. Assuming adherence to sterilization protocols, it seems most likely that the treated product came in contact with a virus source, perhaps untreated plasma.

An improved chemical method utilizes the organic solvent, tri-(*n*-butyl)phosphate (TNBP), typically together with a detergent, to disrupt the envelope of lipid-enveloped viruses (Horowitz et al. 1985a; Prince et al.

TABLE 17-8 Clinical Viral Safety of Coagulation Factor Concentrates Treated with BPL and UV Light

Units Infused	Clinical Results (Infected/Total No. Patients Treated)			Reference
	HBV	NANBHV	HIV	
2,500	0/5	0/5		Heinrich et al. 1982
700,000	0/6	0/6	0/6	Heinrich et al. 1987
702,500	0/11	0/11	0/6	

1986b). HBV, all forms of transfusion-transmitted NANBHV (including HCV), HIV, all retroviruses, and cytomegalovirus (CMV), and other herpes-class viruses are all enveloped. Nonenveloped viruses are unaffected. Because the chemical action is directed against lipid only, the recovery of protein, including both factor VIII and factor IX, exceeds 95% (Edwards et al. 1987). Published results indicate the lack of transmission of HBV, NANBHV, and HIV by solvent/detergent-treated concentrates (Table 17–9).

17.2.3.4 Irradiation. Ultraviolet (UV) irradiation was one of the first approaches explored in order to reduce the viral infectivity of plasma (Oliphant and Hollaender 1946; Blanchard et al. 1948). Used alone, this method was inadequate (Murray et al. 1955). A more recent exploration of UV treatment (Kallenbach et al. 1989) led to the conclusion that $\geq 10^5$ ID_{50} of some but not all viruses would be inactivated under conditions where factor VIII recovery was 85%. The use of gamma irradiation (Horowitz et al. 1985b) and photoactive compounds (Singer et al. 1988) have been explored in the laboratory, but thus far they have not been used in the treatment of a commercially available product.

17.2.3.5 Purification. Purification can result in substantial removal of virus, thus contributing to overall safety as discussed above for both immune serum globulin and albumin prepared by methods of precipitation. Virus removal with chromatographic procedures can be even more efficient. For example, we have shown that affinity chromatography of fibronectin on gelatin-Sepharose reduces the level of hepatitis B surface antigen (HBsAg), intentionally added to the sample just prior to chromatography, by $\geq 10^{10}$. Piszkiewicz et al. (1989) have shown that the purification of AHF by immune affinity chromatography eliminates 10^4 ID_{50} of HIV. On theoretical grounds, it seems unlikely that such procedures, by themselves, can achieve total

TABLE 17-9 **Clinical Viral Safety of Coagulation Factor Concentrates Treated by Solvent/Detergent**

Product	Units Infused	Clinical Results (Infected/ Total No. Patients Treated)			Reference
		HBV	NANBHV	HIV	
AHF	145,000	NA	0/17	0/18	Horowitz et al. 1988 Horowitz and Valinsky 1990
AHF	234,000	NA	0/18	0/18	Gazengel et al. 1989 Guerois et al. 1990
Factor IX	169,600	NA	0/3	0/3	Verroust et al. 1990
AHF & Factor IX	ND	NA	0/19	ND	Guerois et al. 1990[2]
AHF	1,008,000	NA	NA	0/9	Brackmann 1987
PCC	943,200	0/15	0/18	0/18	Gonzaga 1988 Gonzaga, personal communication
AHF	2,418,480	0/6	0/8	0/68	Gonzaga, personal communication
AHF	1,371,600	NA	0/23	0/40	Panicucci et al. 1990
AHF	1,632,000	NA	NA	0/24	Mariani 1990
AHF	ND	NA	0/38	0/55	Gomperts et al. 1991
AHF	541,000	NA	NA	0/18	Perret et al. 1989
PCC	265,000	NA	NA	0/8	Perret et al. 1989
	8,727,880	0/21	0/144	0/279	

[1] ND, not determined; NA, not applicable.
[2] Abstract cites 0/40; however, it includes patients reported in above two studies.

safety; rather safety can be enhanced when coupled with methods of inactivation.

Attempts to remove virus by binding it to affinity or hydrophobic chromatographic resins have failed to provide safe product, despite encouraging laboratory evidence (Einarsson et al. 1981). Failure may result from the differences between laboratory and manufacturing settings, including such factors as column size and extent of employee training.

The filtration removal method could be very useful since proteins should remain unaffected. Recent evidence indicates that $\geq 10^{4.5}$ ID_{50} of HIV can be removed by filtration (Hamamoto et al. 1989), but it remains to be seen whether adequate reproducibility can be achieved in routine manufacture of these filters.

17.2.3.6 Combined Approaches.
Use of two or more methods of removal and inactivation operating independently can enhance the probability of

viral safety. Biotest utilizes both BPL and UV irradiation in the manufacture of its factor IX concentrate; Baxter utilizes TNBP/detergent together with immune affinity purification; Immuno Corp. and Alpha Therapeutics (Los Angeles, CA) explored the use of dry heat with the lyophilized cake suspended as a slurry in organic solvent. One advantage of combining methods is that validation studies can be conducted separately on each procedure, thereby yielding data on a higher level of kill/removal than would be possible for any one method, in part due to limitations of the titer of available virus stocks. For example, the highest dose of HCV available for use in a sterilization procedure is about 10^5 ID_{50}, making it impossible to directly measure a kill higher than 10^5. However, if a 10^5 kill is shown for that procedure and a 10^5 kill is also reported for an independent procedure, a total kill/removal for the two procedures of 10^{10} would be validated. Validation studies of this nature would provide enhanced confidence to manufacturer, regulatory agency, and user alike.

17.3 SUMMARY AND CONCLUSION

Plasma protein solutions such as albumin and intramuscular immune globulin have long histories of viral safety. Coagulation factor concentrates as traditionally manufactured frequently transmitted HBV, HCV, and HIV. Indeed, it is probable that every vial of concentrate contained infectious HCV. Modern coagulation factor concentrates have a greatly improved

TABLE 17–10 Viral Safety of Concentrates Treated by Various Procedures[1]

Treatment Procedure	Validation (Log_{10} Kill)			Clinical Safety (Infected/Total)		
	HBV	HCV	HIV	HBV	HCV	HIV
Heated at 80°C, 72 hours, as lyophilizate	ND	ND	ND	0/16	0/32	0/32
Heated at 60°C, 72 hours, as lyophilizate + vapor	ND	ND	>6.0	4/46	0/70	0/110
Heated at 60°C, 10 hours, in solution, + sucrose, glycine	>5.6	>5.5	>5.0	2/68	2/95	0/237
Heated at 60°C, for 30 hours as lyophilizate, + purification	3.0^2	$>5.5^2$	>6.0	NA	0/19	0/19
TNBP/detergent	>6.0	>5.0	>10.0	0/10	0/45	0/115
TNBP/detergent, + purification	$>8.0^2$	$>7.0^2$	>14.0	NA	0/29	0/45
BPL + UV	6.9	>4.5	>6.0	0/11	0/11	0/6
				6/151	2/301	0/564

[1] ND, not determined; NA, not applicable.
[2] Estimated values.

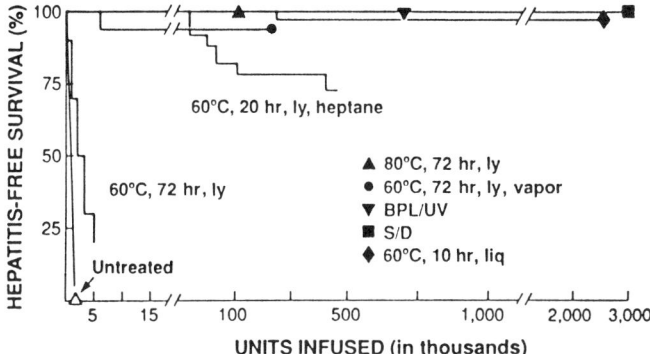

FIGURE 17-1 Hepatitis safety of coagulation factor concentrates. The reported incidence of hepatitis transmission of untreated AHF concentrates and of coagulation factor concentrates treated by the methods indicated is provided as a function of the number of units transfused. Abbreviations: ly, lyophilized; S/D, solvent/detergent; liq, liquid.

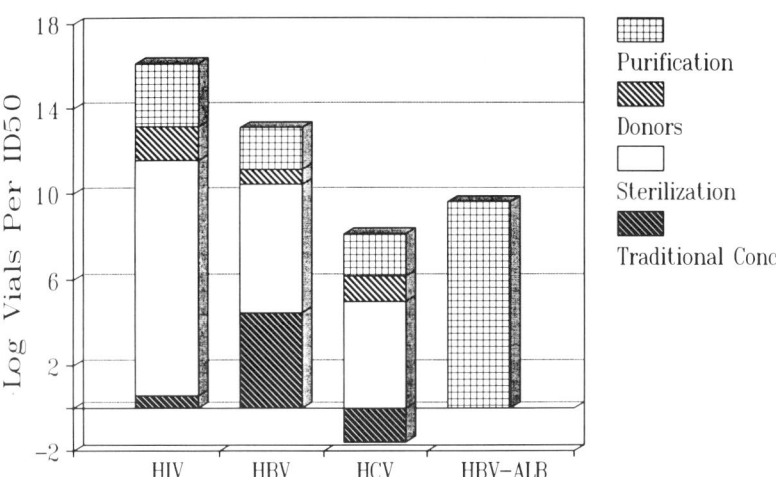

FIGURE 17-2 Calculated probability of viral safety for AHF concentrates and albumin. The probability of viral safety, expressed as the number of vials calculated to contain 1 ID_{50}, was calculated for a solvent/detergent-treated AHF concentrate for HIV, HBV, and HCV, and also for albumin with regard to HBV. The calculations were based on the expected initial virus load in a 5,000-liter plasma pool; estimates of immune neutralization, virus removal, serendipitous inactivation, and demonstrated virus kill on sterilization; and an anticipated process recovery of 750 vials of AHF and 10,000 vials of albumin.

safety record arising, principally, from the implementation of virucidal procedures. It is interesting to note that the same methods that failed to substantially reduce NANBHV transmission in clinical studies are those that were found to inactivate $<10^4$ ID_{50} of HIV, HBV, and/or HCV in preclinical studies (Table 17–5). Implementation even of these methods nearly eliminated the transmission of HIV by coagulation factor concentrates.

A summary of the results of the most successful procedures is given in Table 17–10. The results show 0/564 patients had evidence of HIV transmission, 6/151 patients had evidence of HBV transmission, and 2/301 patients had evidence of HCV transmission. As compared with those procedures described in Table 17–5, the greater kill of HIV, HBV, and NANBHV demonstrated preclinically, and the improved clinical results, are most notable. The data, examined in terms of units transfused, are presented in Figure 17–1. Since the average adult hemophiliac in the United States receives 80,000 units of clotting factor per year, the best of the concentrates show safety over the equivalent of at least 10-human-years of treatment.

Are the best of today's coagulation factor concentrates safe from the transmission of HBV, NANBHV (including HCV), and HIV. Given the limited number of patients eligible for clinical studies, and the length, difficulty, and expense of such studies, the best answer comes from a knowledge of the initial virus load coupled with information regarding virus removal, serendipitous inactivation, and intentional sterilization. A recently completed analysis of these factors (Horowitz 1990) indicates that the best of the modern coagulation factor concentrates are likely to be as safe as albumin (Figure 17–2).

REFERENCES

Aach, R.D., Szmuness, W., Mosley, J.W., et al. (1981) *N. Engl. J. Med.* 304, 989–994.

Auerswald, G., Mittler, U., and Popp, M. (1989) *Thromb. Haemost.* 62, 455 (abstr.).

Blanchard, M.C., Stokes, J., Jr., Hampil, B., Wade, G., and Spizizen, J. (1948) *J. Am. Med. Assoc.* 138, 341–343.

Bove, J.R. (1987) *N. Engl. J. Med.* 317, 242–245.

Brackmann, H.-H. (1987) in *Symposium zur Hamophiliebehand-lung: Virussicherheit eines Kaltsterilisierten Faktor-VIII-Parparates*, Munster, Universitats-Kinderklinik, Münster, Germany.

Brackmann, H.H., and Egli, H. (1988) *Lancet* 2, 967.

Burnouf, T., Martinache, L., and Goudemand, M. (1987) *Nouv. Rev. Fr. Hematol.* 29, 93–96.

Cohn, E.J., Strong, L.E., Hughes, W.L., Jr., et al. (1946) *J. Am. Chem. Soc.* 68, 459–475.

Colvin, B.T., Rizza, C.R., Hill, F.G.H., et al. (1988) *Lancet* 2, 814–816.

Cossart, Y.E., Kirsch, S., and Ismay, S.L. (1982) *Lancet* 1, 208–213.

Edwards, C.A., Piet, M.P.J., Chin, S., and Horowitz, B. (1987) *Vox Sang.* 52, 53–59.

Einarsson, M., Kaplan, L., Nordenfelt, E., and Miller, E. (1981) *J. Virol. Meth.* 3, 213–228.

Finlayson, J.S. (1979) *Semin. Thromb. Hemost.* 6, 44–74.

Gazengel, C., and the Hemophilia French Study Group (1989) *Thromb. Haemost.* 62, 7455 (abstr.).

Gerety, R.J., and Aronson, D.L. (1982) *Transfusion* 22, 347–351.

Gomperts, E.D. (1986) *Am. J. Hematol.* 23, 295–305.

Gomperts, E.D., Debiasi, R., and the Hemophil study group (1991) in *Excerpta Medica International Congress Series* (Lustier, J.M., ed.), Elsevier Science Publishers, Amsterdam, (in press).

Gonzaga, A.L., Bonecker, C., Pecego, M.M.N., and Cotias, P.M. (1988) in *International Congress of the World Federation of Hemophilia*, Madrid, p. 257 (abstr.).

Guerois, C., Parquet, A., Gazengel, C., et al. (1990) in *XIX International Congress of the World Federation of Hemophilia*, Washington, DC, p. 32 (abstr.).

Hamamoto, Y., Harada, S., Kobayashi, S., et al. (1989) *Vox Sang.* 56, 230–236.

Heimburger, N., Karges, H.E., and Weidmann, E. (1987) *Develop. Biol. Standard* 67, 303–310.

Heinrich, D., Kotitschke, R., and Berthold, H. (1982) *Thromb. Res.* 28, 75–83.

Heinrich, D., Sugg, U., Brackmann, H.H., Stephan, W., and Lissner, R. (1987) *Develop. Biol. Standard* 67, 311–317.

Horowitz, B. (1990) *Yale J. Biol. Med.* 63, 361–369.

Horowitz, B., and Piet, M.P.J. (1986) *Plasma Ther. Tranfus. Technol.* 7, 503–513.

Horowitz, B., Wiebe, M.E., Lippin, A., and Stryker, M.H. (1985a) *Transfusion* 25, 516–522 (abstr.).

Horowitz, B., Wiebe, M.E., Lippin, A., Vandersande, J., and Stryker, M.H. (1985b) *Transfusion* 25, 523–527.

Horowitz, M.S., Horowitz, B., Rooks, C., and Hilgartner, M.W. (1988) *Lancet* 2, 186–189.

Horowitz, M.S., and Valinsky, J. (1990) in *XIX International Congress of the World Federation of Hemophilia*, Washington, DC, p. 27 (abstr.).

Iwarson, S., Steen, Y., Rybo, G., et al. (1985) *Transfusion* 25, 15–17.

Kallenbach, N.R., Cornelius, P.A., Negus, D., et al. (1989) in *Virus Inactivation in Plasma Products*, (Morgen Thaler, J.-J., ed.), pp. 70–82, Karger, Basel.

Lo Grippo, G., and Hartman, F. (1958) *Bibl. Haematol.* 7, 255–230.

Lo Grippo, G.A., Wolfram, B.R., and Rupe, C.E. (1964) *J. Am. Med. Assoc.* 187, 722–726.

Mannucci, P.M., and Colombo, M. (1988) *Lancet* 2, 782–785.

Mannucci, P.M., Zanetti, A.R., and Colombo, M. (1988) *Br. J. Haemotol.* 68, 427–430.

Mannucci, P.M., Schimpf, K., Brettler, D.B., et al. (1989) *Thromb. Haemost.* 62, 180 (abstr.).

Mariani, G. (1990) *XVII Congresso triennale sui Problemi Clinici e sociali dell' Emofilia,* Italy.

Menache, D., and Aronson, D. (1985) *Prog. Clin. Biol. Res.* 182, 407–423.

Mosseler, J., Schimpf, K., Auerswald, G., et al. (1985) *Lancet* 1, 111.

Murray, R., and Ratner, F. (1953) *Proc. Soc. Exp. Biol. Med.* 83, 554–555.

Murray, R., Oliphant, J.W., Tipp. J., et al. (1955) *J. Am. Med. Assoc.* 157, 8–14.

Oliphant, J.W., and Hollaender, A. (1946) *Pub. Health Rep.* 61, 598–602.

Panicucci, F., Felloni, L., Vannini, F., et al. (1990) in *XIX International Congress of the World Federation of Hemophilia*, Washington, DC, p. 20 (abstr.).

Perret, B.A., Baumgartner, C., and Meili, E. (1989) *Schweiz. Med. Wschr.* 119 (suppl.), 28.

Peterman, T.A., Lui, K.-J., Lawrence, D.N., and Allen, J.R. (1987) *Transfusion* 27, 371–374.

Piszkiewicz, D., Sun, C.-S., and Tondreau, S.C. (1989) *Thromb. Res.* 55, 627–634.

Preiss, D.U., Eberspächer, B., and Rosner, I. (1989) *Thromb. Haemost.* 62, 1462 (abstr.).

Prince, A., Stephan, W., and Brotman, B. (1983) *Rev. Infect. Dis.* 5, 92–107.

Prince, A.M., Piet, M.P.J., and Horowitz, B. (1986a) *N. Engl. J. Med.* 314, 386–387.

Prince, A.M., Horowitz, B., and Brotman, B. (1986b) *Lancet* 1, 706–710.

Prince, A.M., Horowitz, B., Horowitz, M.S., and Zang, E. (1987) *Eur. J. Epidemiol.* 3, 103–118.

Schimpf, K. (1987) in *3. Rundtischgesprach uber Aktuelle Probleme der Substitutionstherapie Hamophiler* (Landbeck, G., and Schimpf, K., eds.), pp. 69–79, Springer-Verlag, Frankfurt.

Schimpf, K. (1988) *Hämophilie-Blatter* 22, 31–36.

Schimpf, K., Mannucci, P.M., Kreutz, W., et al. (1987a) *N. Engl. J. Med.* 316, 918–922

Schimpf, K., Brackmann, H.H., Bock, D., et al. (1987b) *Thromb. Haemost.* 58, 346 (abstr.).

Schimpf, K., Brackmann, H.H., Kreuz, W., et al. (1989) *N. Engl. J. Med.* 321, 1148–1152.

Schorr, J.B., Berkowitz, A., Cumming, P.D, Katz, A.H., and Sandler, S.G. (1985) *N. Engl. J. Med.* 313, 384–385.

Singer, C.R.J., Azim, T., and Sattentau, Q. (1988) *Br. J. Haematol.* 69, 31 (abstr.).

Stephan, W., and Dichtelmuller, H. (1983) *Lancet* 2, 1486.

Uemura, Y., Uriyu, K., Hirao, Y., et al. (1989) *Vox Sang.* 56, 155–161.

Vanderpoel, C.L., Reesink, H.W., Schaasberg, W., et al. (1990) *Lancet* 335, 558–560.

Ward, J.W., Holmberg, S.D., Allen, J.R., et al. (1988) *N. Engl. J. Med.* 318, 473–477.

Weiner, A.J., Kuo, G., Bradley, D.W., et al. (1990) *Lancet* 335, 1–3.

Winkelman, L., Owen, N.E., Evans, D.R., et al. (1989) *Vox Sang.* 57, 97–103.

Inactivation of Viruses Found with Cellular Components

Bernard Horowitz
Jay Valinsky

Despite considerable effort in the 1950s directed toward the identification and evaluation of virus sterilization procedures that would reduce the viral risk of blood transfusions, it was not until the discovery of the hepatitis B virus (HBV) (Blumberg et al. 1969; Prince 1968), together with appropriate serologic reagents to detect its presence, that significant progress in reducing virus transmission was made. Routine testing for hepatitis B surface antigen (HBsAg), and the drive to eliminate paid donors, substantially reduced hepatitis transmission rates (Alter et al. 1972). The discovery of the AIDS virus (HIV-1) in 1983/1984 (Barr-Sinoussi et al. 1983; Popovic et al. 1984) and of HTLV-1 in 1980 (Poiesz et al. 1980), the implementation of surrogate tests for non-A, non-B hepatitis (e.g., alanine aminotransferase (ALT) and antibody to the core antigen of HBV (anti-HBc)) (Menitove 1988) and donor self-exclusion protocols (Mayer and Pindyck 1988), and the recent discovery of hepatitis C virus (HCV) in 1989 (Vanderpoel et al. 1990; Weiner et al. 1990) have all led to marked improvements in the safety of the blood supply.

Appreciation is extended to Marcia Franklin-Henry for her dedication in helping prepare this manuscript. Work supported in part by Grant no. HL41221 from NHLBI.

Nonetheless, it is clear that absolute safety cannot be achieved with serologic testing alone. Viral antigens may be present long before the appearance of detectable levels of antibodies, all screening and confirmatory tests are limited by sensitivity and specificity, and mistakes are made. The current risk of HBV, HIV-1, and HCV transmission from a single unit of blood is estimated at 0.05%, 0.0005%, and 0.05%, respectively (see Chapter 17; Horowitz 1990). However, many recipients of blood transfusions are exposed to 10 or more units of cellular components, whole blood, or plasma, thus multiplying the risk. While the risk of viral infection can be eliminated for some patients by using methods that avoid blood transfusion or that employ autologous blood, the only approach to achieving total safety applicable to all patients receiving blood would appear to be virus sterilization. This point is convincingly illustrated by a review of the record of viral safety of inactivated blood protein products. Prior to virus sterilization, essentially every vial of a coagulation factor concentrate contained infectious virus. With virus sterilization, these products have achieved an admirable safety record and are now safer than the individual units from which they were derived (see Chapter 17).

This chapter reviews current progress in the development of virus sterilization methods applicable to blood cell components. The principal approaches under investigation are listed in Table 18–1. As compared with virus sterilization of protein products, sterilization of cellular products is more difficult. Blood cells are more complex and fragile than proteins, and

TABLE 18–1   Principal Methods for the Inactivation of Viruses in Cellular Blood Components

Virus removal
Immune neutralization
Hydrolyzable chemical agents
  β-Propiolactone
  Methyl amines
  Aryl diol epoxides
  Ozone
  Halogenated oxidizing agents
Photosensitization with:
  Hematoporphyrin derivatives
  Merocyanine
  Phthalocyanine
  Hypericin
  Psoralen
  Sapphyrins
Irradiation
  Ionizing radiation
  Ultraviolet light

virus is present in both cell-free and cell-associated forms. Nonetheless, since neither erythrocytes nor platelets are replicating cells, methods that modify membranes or nucleic acid may prove useful. The challenge, then, is to effectively eliminate virus infectivity without adversely affecting cell circulatory or functional properties, and without generating new structures that would be immunogenic, i.e., neoimmunogens.

18.1 PHYSICAL METHODS FOR VIRUS REMOVAL OR INACTIVATION

Physical techniques for the elimination of viruses from cellular blood components can be divided into two main types: removal by washing, centrifugation, or filtration; and inactivation by irradiation (see Section 18.5). These methods do not require the addition of chemicals or other agents to the blood to effect sterility, and automated instrumentation (e.g., cell washers) and equipment (leukocyte filters) are already in place that could be modified for batch processing of units of blood. However, the equipment and consumables required for these procedures are costly and the extensive processing required may lead to additional problems (e.g., hemolysis, extensive cell loss, platelet aggregation).

As in the cases of chemical, photochemical, or immunological approaches, these physical methods must be efficient, not only in the removal or inactivation of viruses free in the plasma, but also of the cell-associated forms. Titers of free viruses in plasma vary greatly, for example, as few as 50 infectious particles/ml for HIV-1 to $>10^9$ infectious particles for hepatitis B. Titers of cell-associated virus are much harder to ascertain, especially in the case of retroviruses, which can exist in latent proviral forms incorporated in host cell DNA. Estimates for the infection of T-helper ($CD4^+$) cells by HIV-1, for example, are as low as 1 cell in 10,000.

18.1.1 Virus Removal by Washing, Centrifugation, and Filtration Techniques

Removal of viruses by washing blood cells free of plasma is intuitively the simplest method for lowering virus titer in infected units of blood. These techniques are more amenable to red cell concentrates than to more fragile components like platelets. Automated cell washers currently in use in most blood banks are capable of removing residual plasma from red cell concentrates. In a typical run on a commercial cell washer, residual plasma, and presumably the titer of free virus, can be reduced 25–35 fold. For HIV, this may be sufficient to bring the titer below detectable levels, but perhaps not below the level required to eliminate infectivity. Washing is unlikely to prove effective for viruses present at high titer, such as HBV. While additional washing cycles with larger volumes of saline might be expected to improve the efficacy of virus removal, recent evidence presented by the American National Red Cross indicates that this is not the case (L. Friedman, personal communication) and there are a number of drawbacks to the additional processing steps, including hemolysis and reduced recovery of red cells. The extensive processing required by this method may preclude its use with platelet concentrates.

It is expected that differential centrifugation would be even less effective in the removal of leukocyte-associated viruses since it cannot remove enough of the leukocytes, which typically contaminate red cell and platelet products. In centrifugation procedures, $0.1-1 \times 10^9$ leukocytes per unit remain as contaminants of red cell concentrates (Meryman and Hornblower 1986). The same situation holds for platelet concentrates; the number of residual contaminating leukocytes may vary considerably ($1-10 \times 10^8$/unit) depending upon the efficacy of the initial centrifugation steps (Shoendorfer et al. 1983). Newer techniques for automated platelet production, in which platelets are prepared from the buffy coat, may produce more highly leukocyte-depleted ($5-10 \times 10^7$/unit) products (van Marwijk et al. 1990).

Filtration may provide a more efficient alternative for the removal of leukocytes from red cell and platelet concentrates. High-efficiency leukocyte removal from red cell concentrates has been achieved using polyester, cotton wool, and other materials (Sirchia et al. 1987; Riverberi and Menini 1990; van Marwijk-Kooy et al. 1990). Removal of more than 99% of contaminating leukocytes, with only small losses of platelets or red cells, has been reported. Nonetheless, these procedures may leave as many as $0.1-1 \times 10^7$ leukocytes per unit. While this may be very effective in reducing febrile or other tranfusion reactions, filtration would not be expected to remove cell-free virus and might not eliminate high titers of cell-associated virus.

However, there has been some success in the removal of cell-associated virus by leukocyte depletion in the case of cytomegalovirus (CMV), which is always found in a cell-associated form (Schrier et al. 1985). Several multicenter studies (Gilbert et al. 1989; Tegtmeier 1989) indicate that when high-capacity leukocyte filters are employed, the transmission rate of CMV infections to children to CMV-negative mothers is minimal. In one study, for example, none of the 59 CMV-seronegative patients who received transfusions of leukocyte-depleted platelets seroconverted. In this instance, 99.9% of the leukocytes in platelet concentrates were removed by a combination of centrifugation and filtration (de Graan-Hentzen et al. 1989).

In contrast, Rawal et al. (1989) examined the efficacy of several leukocyte filters in the removal of cell-associated HIV-1. Units of blood were inoculated either with lymphocytes obtained from HIV-1-infected cell cultures or from HIV-1-seropositive individuals. Filtration reduced the cell-associated virus "titer" by nearly 6 logs for the cultured lymphocytes and 2 logs for lymphocytes from donors. However, residual cell-associated virus was detected in the filtered products by polymerase chain reaction (PCR) and by lymphocyte coculture techniques.

18.2 IMMUNE NEUTRALIZATION

The immune neutralization of free viruses in whole blood and blood components is a reasonable alternative to the addition of chemical agents. The high specificity of naturally occurring or monoclonal reagents should, in principle, make them good agents for attack on potential pathogens, and perhaps without the induction or display of neo-antigens. Brummelhuis and Over (1989) have demonstrated the efficacy of virus neutralization techniques in the case of HBV transmission in plasma derivatives. However, there are no reports of the use of neutralizing antibodies in transfusions of red cells or platelet components. Little information is available regarding the activity, existence, or efficacy of neutralizing antibodies reactive with HCV in this context either (Habibi and Garretta 1990).

Neutralizing antibodies have been discovered in individuals infected with HIV-1 (Cao et al. 1990). Some of these antibodies have neutralizing activity in vitro (Goudsmit et al. 1988; Palker et al. 1988), but not in vivo, as demonstrated in chimpanzee studies (Prince et al. 1988). In addition, antibodies reactive with particular epitopes may enhance HIV-1 replication rather than neutralizing it (Bernard et al. 1990), suggesting that any attempts to use naturally occurring immunoglobulins as neutralizing reagents could have the opposite effect.

18.3 HYDROLYZABLE CHEMICAL AGENTS

18.3.1 β-Propiolactone

The use of β-propiolactone (BPL) either alone or together with ultraviolet (UV) irradiation has been extensively studied in the viral sterilization of cell-free and cell-containing blood components (for review, see LoGrippo 1959; Prince et al. 1983). BPL has several notable advantages: 1) it inactivates a wide variety of viruses and microorganisms; 2) it hydrolyzes under physiological conditions to a nontoxic form; 3) the structure and function of a variety of plasma proteins and of red cells is reasonably well maintained following treatment; and 4) direct reaction with nucleic acid is possible. On the other hand, covalent reactions with proteins and other macromolecules occur readily, hydrolysis results in acid formation which needs to be controlled, and, in its unreacted form, BPL is a proven carcinogen. One company (Biotest, Frankfurt, Germany), routinely uses BPL together with UV treatment in the preparation of a prothrombin complex concentrate and immune globulin-containing solutions. Red blood cells treated with BPL alone stored satisfactorily for 21 days, but they gave a shortened recovery/half-life on infusion into humans (LoGrippo 1959).

18.3.2 Methyl-bis (β-Chloroethyl) Amine Hydrochloride (MCA)

MCA is a nitrogen mustard shown to hydrolyze to nontoxic forms. Hartman et al. (1949) studied the use of MCA for the treatment of both plasma and whole blood. Treatment with 0.3–0.5 mg/ml was shown to inactivate 10^6 infectious doses (ID_{50}) of vesicular stomatitis virus (VSV). Whole human blood treated with 0.45–0.6 mg/ml MCA had normal storage properties; however, treatment of dog blood with >0.25 mg/ml MCA resulted in marked lysis.

18.3.3 Aryl Diol Epoxides

Aryl diol epoxides are sufficiently hydrophobic to penetrate cell membranes and have been shown to react readily with DNA. Like BPL, aryl diol epoxides are readily hydrolyzed in aqueous solution to nonreactive, nontoxic forms. Aryl diol epoxides are known to form adducts with DNA in vitro and in vivo (Gamper et al. 1980), and several studies have indicated some antiviral effects of these agents (Lockhart et al. 1986; Chang et al. 1981; Bowden et al. 1986).

Williams et al. (1988) showed that treatment with benzopyrene diol epoxide inactivated $\geq 10^6$ tissue culture infectious doses ($TCID_{50}$) of VSV when present in tissue culture media. However, in the presence of 1×10^9 red blood cells/ml, virus kill was reduced by approximately 3 $\log_{10} TCID_{50}$. Treatment did not induce lysis, and osmotic fragility was unaffected.

18.3.4 Ozone

Ozone is a strong oxidant previously shown to inactivate viruses in sewage and water effluents (Biedermann and Katzenelson 1976; Akey and Walton 1985; Harakeh et al. 1985). In uncontrolled studies, ozone treatment in humans has been reported to be effective against herpes simplex virus (Mattassi et al. 1985; Konrad 1985) and hepatitis virus (Konrad 1985). Freeberg and Carpendale (1988) recently reported the inactivation of $\geq 10^4$ $TCID_{50}$ of HIV-1 following the treatment of serum with 4 mg/l ozone. While no data on virucidal activity was provided, Zee (1986) described the treatment of blood and blood components with 1–100 mg/l of ozone.

A preliminary set of experiments showed that exposure of either whole blood or plasma to ozone (50 mg/l) failed to kill VSV, Sindbis virus, or encephalomyocarditis virus (B. Horowitz, unpublished observations).

18.3.5 Halogenated Oxidizing Agents

Halogenated oxidizing agents such as chlorous acid (Rubinstein 1990) and UG40 (A.R. Globus, personal communication) may be useful in inactivating viruses in whole blood and blood components. Thus far, data from this approach have not been published. In general, oxidizing agents might not have the required specificity. Considerable inhibition of viral kill at a given concentration of reagent by plasma and cells would be expected.

18.4 PHOTOSENSITIZATION TECHNIQUES

In 1900, Raab discovered that microorganisms are killed on exposure to sensitizing dyes, light, and oxygen (Raab 1900). Interest in the use of photosensitizing dyes has been sparked by their potential in the treatment of tumors (Nelson et al. 1990; Henderson and Bellinier 1989). Specificity presumably arises from the enhanced uptake and reactivity of these dyes by tumor cells as compared with normal cells. Photolysis of membrane or cell-bound dyes with appropriate excitation wavelengths, in the visible or in the UV, may result in cell and/or virus killing. Damage to nucleic acids, proteins, and carbohydrates has also been demonstrated (Spikes and Straight 1967; Wells 1962). Unsaturated molecules in cell membranes (Lamola et al. 1973) and the guanine ring of nucleic acids (Cadet et al. 1982) are damaged apparently through oxidative mechanisms. While the reaction mechanisms are complex (Foote 1968, 1984; Bensasson and Land 1978), oxidation appears to occur through intermediates that are formed following excitation of the dye into the triplet state; energy is subsequently transferred to oxygen to form reactive singlet species. Biological substrates can react directly with excited dyes, activated forms of molecular oxygen, and with radical species formed from other organic molecules in the environment.

Since photosensitizing dyes differ from one another in their hydrophobicity, absorption spectra, and the half-lives of the activated species, a systematic study of a variety of dyes for the treatment of blood and blood components appears to be warranted. It should be emphasized that photosensitizing dyes whose absorption spectra are significantly red-shifted relative to the absorption spectrum of hemoglobin may be the most useful for the treatment of whole blood or red cell concentrates. The chemical structures of some commonly used photosensitizing dyes are shown in Figures 18-1 through 18-6.

18.4.1 Hematoporphyrin Derivatives

Hematoporphyrin derivatives (HPDs) were among the first photosensitizers to be evaluated as antiviral agents in the treatment of blood cell components (Matthews 1988) because of their prior use in man for treatment of tumors. HPDs are typically mixtures of hematoporphyrins. One of the photochemically active components of these mixtures is shown in Figure 18-1. Specificity in the treatment of blood cell concentrates apparently is the result of the concentration of the dye by viral membranes. Matthews (1988) showed that treatment of herpes simplex virus type 1 (HSV-1) in culture medium

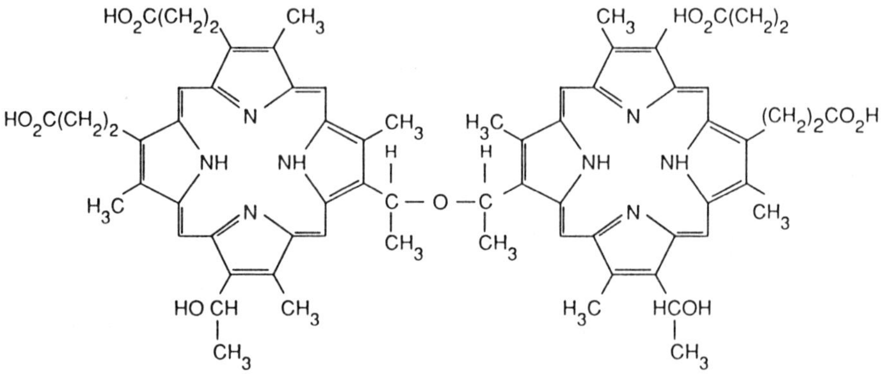

FIGURE 18-1 A hematoporphyrin derivative.

with 2.5 μg/ml of HPD and 5 J/cm^2 of light inactivated $\geq$5.5 log$_{10}$ TCID$_{50}$. However, treatment of whole blood under the same condition or with 20 μg/ml of HPD inactivated only 2.9 log$_{10}$ TCID$_{50}$ of HSV-1. Reduced virus kill probably resulted in part from the overlap in the absorption spectrum of HPD with that of hemoglobin. HIV-1 (3.3 log$_{10}$ TCID$_{50}$) in culture medium was also inactivated at 10 and 20 μg/ml, but not 2.5 μg/ml, of HPD. No significant red cell lysis was detected on treatment of whole blood, and prothrombin time, activated partial thromboplastin time, fibrinogen, serum protein electrophoresis, and white blood cell count and differential were unaffected. A significant drop in platelet count was observed, however.

More recently, Neyndorff et al. (1990) reported that a benzoporphyrin derivative (BDP) inactivated virus to a higher extent than HPD. VSV ($\geq$10^7 TCID$_{50}$) added to whole blood was inactivated with BDP at 2-4 μg/ml and a light intensity of 57.6 J/cm^2. Red cell lysis was 1-2%. Improved virus kill probably resulted from the right-shifted light absorbance of BPD vs. HPD.

18.4.2 Merocyanine 540

Merocyanine 540 (MC540) (Figure 18-2) is representative of a class of dyes which show staining specificities for a variety of cell and tissue types (Easton et al. 1978). Because of its specificity for leukemic cells (Valinsky et al. 1978), MC540 has been used as a bone-marrow-purging agent in Phase I clinical trials for transplants in patients with acute leukemias (Sieber and Krueger 1989). MC540 has been shown to inactivate both cell-free and cell-associated forms of Friend erythroleukemia virus, VSV, HSV-1, cytomegalovirus (CMV), HTLV-1, and HIV-1 (Sieber et al. 1987; Cole et al. 1989; Prodouz 1989; Moroff et al. 1989). For cell-free virus, the viral envelope appears to be the primary target (Sieber et al. 1990). However, MC540 antiviral activity is inhibited substantially at hematocrits above 15% and

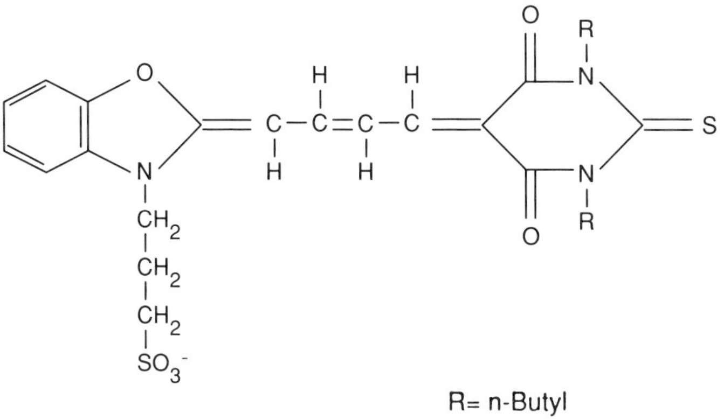

R= n-Butyl

FIGURE 18-2 Merocyanine 540.

in the presence of $\geq 15\%$ plasma (Cole et al. 1989). The treatment of washed platelets with 72 μg/ml MC540 without light suppressed their response to thrombin, while exposure to both MC540 and light (18 J/cm²) induced platelet aggregation (Prodouz 1989). Nonetheless, treatment of whole blood with MC540 (15 μg/ml) and light (70 w/m², 90 min) rendered malarious blood noninfectious (Smith et al. 1990). A recent report (Gulliya et al. 1990) indicated that both antitumor and antiviral activities were retained following photoactivation of MC540 prior to the addition of the dye to the sample, implying that relatively long-lived photoproducts formed and were virucidal.

18.4.3 Phthalocyanines
As compared with HPD or MC540, phthalocyanines (PCs) (Figure 18–3) should have enhanced potential in the treatment of red cell-containing suspensions because of their red-shifted absorption spectra. The absorption maximum of these dyes is approximately 670 nm. Treatment of whole blood or red cell concentrates with aluminum phthalocyanine or its sulfonated forms (2–25 μM, 44–176 J/cm²) resulted in the inactivation of $\geq 10^4$ TCID$_{50}$ of cell-free and cell-associated forms of VSV and HIV-1. Minimal hemolysis was observed ($<2\%$), and red cell osmotic fragility was not increased (Horowitz et al. 1991). It remains to be seen whether treated red cells have normal storage and in vivo properties. In a preliminary report, Singer et al. (1988) noted that treatment of plasma with sulfonated aluminum phthalocyanine resulted in loss of 50% of factor VIII activity.

18.4.4 Hypericin
Hypericin is a multicyclic macromolecule (Figure 18–4) that has been reported to interfere with retrovirus replication (Meruelo et al. 1988; Lavie et al. 1989). Unlike the compounds described above, antiviral activity oc-

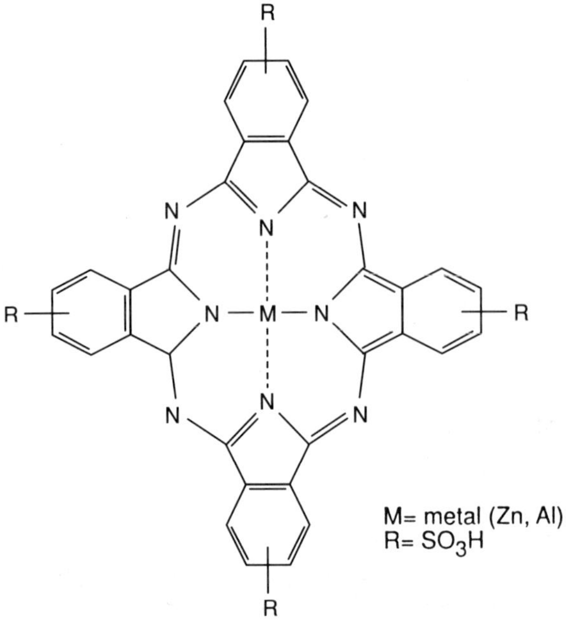

FIGURE 18–3 Phthalocyanine sulfonate.

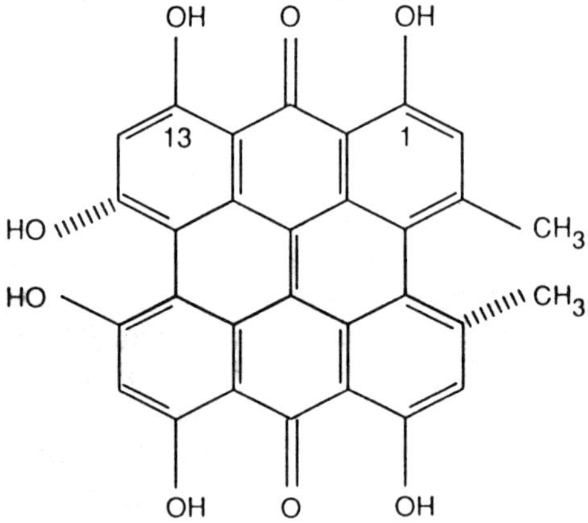

FIGURE 18–4 Hypericin.

curred in the dark, but was potentiated by light (Hudson et al. 1990). The mechanisms responsible for antiviral action in light and dark are probably distinct. In the dark, hypericin is thought to interfere with viral budding. In the presence of light, cell-free mouse CMV was inactivated, possibly in a manner similar to that which occurs with the other photosensitizers described above.

18.4.5 Sapphyrins

Sapphyrins are pentapyrrolic macrocyclic molecules (Figure 18–5) with a strong π–π transition at 680–690 nm. Judy et al. (1991) recently initiated studies designed to elucidate the antiviral properties of two sapphyrins: a more hydrophobic methylated form, and a more hydrophilic, dicarboxylated form. While the dicarboxylated sapphyrin was relatively ineffective in inactivating HSV-1, the methylated form (33 μM, 10 J/cm^2 light) was shown to inactivate ≥ 5 $\log_{10}$ TCID$_{50}$ of HSV-1 added to culture medium. Studies on compatibility with blood and blood components were not reported.

18.4.6 Psoralens

Psoralens (Figure 18–6) are natural furocoumarins found in many foods and have been used therapeutically since antiquity (Fitzpatrick and Pathak 1959). The reactions of psoralens with biologically relevant molecules requires exposure to long-wavelength UV light, and, in the absence of oxygen, the reaction is directed largely toward nucleic acids (Averbeck 1989). Both DNA and RNA viruses can be inactivated with this treatment. Trioxsalen (4,5,8-trimethylpsoralen; TMP) is licensed for use by the U.S. Food and Drug Administration for the treatment of psoriasis. 4'-Aminomethyl-4,5',8-trimethylpsoralen (AMT) has significantly higher water solubility than 8'-methoxypsoralen (8-MOP) (0.8 μg/ml vs. >10,000 μg/ml) (Hanson 1983) and, therefore, it may prove more useful.

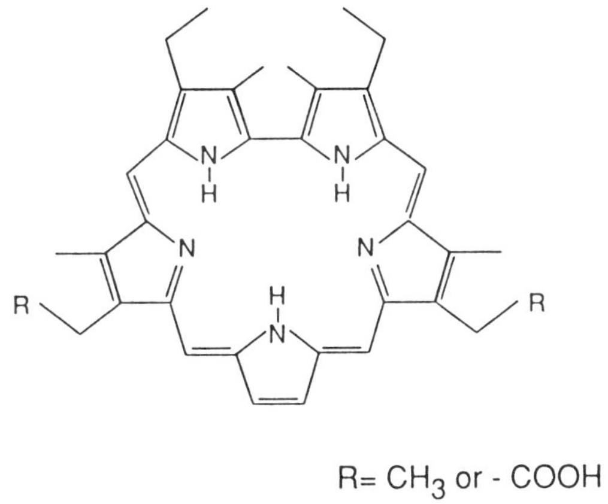

R= CH$_3$ or - COOH

FIGURE 18–5 Sapphyrin.

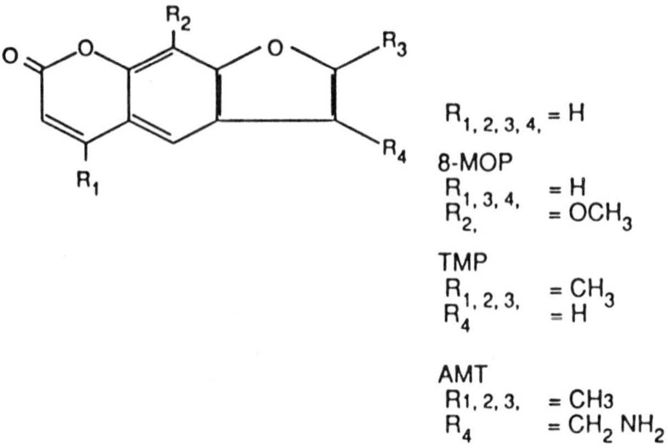

FIGURE 18-6 Psoralen(s). TMP (trioxsalen; 4,5,8-trimethylpsoralen); AMT (aminomethyl-4,5′,8-trimethylpsoralen); and 8-MOP (8′-methoxypsoralen).

Addition of 20 μg/ml AMT and 0.5 μg/ml of TMP every hour for 10 hours to dilute protein solutions, together with UV light exposure, was shown to inactivate both HBV ($\geq 10^{4.5}$ chimpanzee infectious doses (CID$_{50}$)) and NANBHV ($\geq 10^4$ CID$_{50}$) (Alter et al. 1988). Similarly, treatment of dilute protein solution with 0.3 mg/ml 8-MOP and UV for 10 hours was shown to inactivate $\geq 10^{4.5}$ CID$_{50}$ of HBV and $\geq 10^{4.0}$ CID$_{50}$ of NANBHV. However, studies on the inactivation of hepatitis viruses in red cell and platelet concentrates have not been reported. Treatment of a platelet concentrate with 8-MOP (0.3 mg/ml) and UV light (Lin et al. 1989) for up to 6 hours was shown to inactivate ≥ 6.7 log$_{10}$ of *E. coli,* ≥ 6.9 log$_{10}$ of *S. aureus,* ≥ 7.3 log$_{10}$ PFU of phage fd, 2.5 log$_{10}$ of phage R17, and 5.1 log$_{10}$ of feline leukemia virus (FELV). Platelet morphology, yield, and response to the aggregation agent A23187 were comparable to the untreated controls. An attempt to utilize a single addition of 10 μg/ml of AMT was unsuccessful: the presence of as little as 12% plasma substantially inhibited virus kill, and a substantial reduction in platelet aggregation response was noted (Moroff et al. 1989).

18.5 IRRADIATION METHODS

Exposure of cells and viruses to UV light or ionizing radiation results in dose-dependent damage to nucleic acids. In the case of DNA, the efficacy of irradiation depends on the inability of the host cell or virus to repair double- or single-strand breaks by SOS or other repair mechanisms. Following irradiation, it is presumed that: 1) free viruses would be altered such that infectivity would be lost; 2) cell-associated viruses would also be rendered inactive with appropriate doses; and 3) while the biological activities of leukocytes contaminating component preparations might be reduced by doses that inactivate viruses, the functional activities of nonnucleated cells (e.g., red blood cells and platelets) would be maintained. The inactivation of leukocytes might be an additional benefit of this procedure since the production of infectious virus following the activation of latent proviruses may require biologically competent host cells. As an added benefit, in homologous transfusions, graft-versus-host disease might be minimized.

18.5.1 Ionizing Radiation

Irradiation of whole blood and cellular components with gamma radiation (e.g., ^{137}Cs) has been in common practice in blood banks for a number of years. Irradiation is typically performed to inactivate lymphocytes in the various blood components, thereby reducing the incidence of graft-versus-host disease and other immunologic transfusion reactions (Leitman and Holland 1985). In typical irradiation protocols, 1.5–3.0 Gy are delivered over a period of several minutes. These doses do not significantly affect the viability and functional activities of red cells, platelets, and neutrophils (Button et al. 1981).

Doses of ionizing radiation in this range have only limited effects on the viability of many viruses, however. For example, plaque-forming activity of strains of herpes simplex virus is destroyed only in the range of 1–3 kGy of ^{60}Co irradiation (Rosen et al. 1987). In this case, it was suggested that the viral capsid may play a protective role. Kitchen et al. (1989) tested the effects of gamma irradiation on HIV-1 added to plasma or to purified coagulation factors. HIV-1 was inactivated more rapidly (0.164 $TCID_{50}$ dose/ml/kGy) than the factors (e.g., 0.00173 log_{10} units/ml/kGy). Experiments describing HIV-1 inactivation in cellular components were not reported. Bigbee et al. (1989) reported that 25 kRad of X irradiation was required to completely inactivate HIV-1 in samples of human body fluids used in forensic analyses.

18.5.2 Ultraviolet Light

HIV-1 can be inactivated by ultraviolet (UV) light in the absence of photosensitizing agents (e.g., psoralens) or alkylating agents (BPL), but with reduced efficacy. Nakashima et al. (1986) demonstrated that 5000 J/m^2 of UV radiation completely inactivated HIV-1 with regard to proliferation in MT-4 cells in vitro. However, HIV-1 clones could be derived from virus irradiated at 2000 J/m^2. Rasheed et al. (1986) demonstrated that UV-irradiated HIV-1 could induce cytopathic effects in cultured cells but could not induce the formation of proviral DNA from the endogenous RNA template.

In what is one of the few recent studies aimed at evaluation of the efficacy of UV irradiation of viruses in blood products, Prodouz et al. (1987)

used a pulsed XeCl excimer laser to inactivate polio virus in platelet concentrates or plasma. At doses of UV (308 nm) from 10–20 J/cm^2 the virus was inactivated under conditions in which platelet and plasma protein functions were substantially maintained.

While these studies demonstrate the possibility that UV irradiation might be a useful method for virus inactivation under certain restricted conditions, one important caveat should be considered, namely, virus replication or transcriptional activity can be induced following exposure to doses of UV which are suboptimal for virus inactivation (Stein et al. 1989a, 1989b). For example, significant activation (up to 150-fold) of HIV-1 transcriptional activity has been demonstrated in chimeric constructs containing HIV-1 long terminal repeats that had been transfected into lymphoid cell lines or HeLa cells (Valerie et al. 1988).

18.6 SUMMARY AND CONCLUSIONS

Blood routinely released for transfusion by blood banks has never been safer. Improved safety is the direct result of better donor education and selection and implementation of sensitive assays for viral markers. Nonetheless, a small but definable virus risk remains. This risk can become substantial in chronically transfused and immune-suppressed patients. Elimination of all contaminating viruses will undoubtedly require the use of virus removal or chemically based virucidal procedures, provided they are effective against both intracellular and extracellular virus, do not adversely affect the functioning of normal blood cells, and implementation can be cost effective. Based on a viral titer of 10^2-10^3 ID_{50}/ml in a unit of blood that screens as negative, we calculate that a reduction in transmission rate requires a method that can eliminate $\geq 10^6$ ID_{50} of virus.

While some mention has been given to cost savings of the use of virucidal procedures through the elimination of one or more screening tests, we find this a most unlikely scenario. Virus elimination methods will need to be used in addition to, and not instead of, the procedures now in place. Virucidal procedures that are broad-based and that inactivate both lipid-enveloped and protein-coated viruses may obviate the necessity to implement future testing for newly discovered or rare viruses.

What can we conclude concerning progress in this new, but burgeoning field? A brief summary of the attributes of the methods under investigation is provided in Table 18–2.

Methods of virus removal, e.g., washing by centrifugation or elimination of leukocytes by filtration, are the most advanced and raise the fewest questions regarding cell structure and function. Both platelets and red cells could be sterilized using these methods. The removal of leukocytes by filtration has been reported to reduce the risk of CMV, but it seems unlikely that the risk of HBV, HCV, or HIV-1 infections may be significantly reduced, since

TABLE 18–2   General Attributes of the Five Methods of Virus Elimination

	Removal	Immune Neutralization	Hydrolyzable Chemical Agent	Photosensitization	Irradiation
Intra-/extracellular virus	Both	Both	Both	Both	Both
Lipid-/protein-enveloped virus	Both	Both	Both	Lipid only[1]	Both
RBC/platelet concentrate	Both	Both	Both	Both	Both
Reagent removal	No	No	Yes?	Yes	No
Carcinogenic	No	No	Yes	Unknown	No?
Neoimmunogens	No	No	Unknown	Unknown	Unknown
Cost	Moderate to high	Moderate	Modest	Moderate to high	Moderate

[1] Psoralens active against both protein- and lipid-enveloped viruses.

these viruses also exist in cell-free forms in blood. On the other hand, since transmission of HTLV-1 by transfusion is associated with cellular components (Okochi 1987), a response analogous to that found for CMV might be anticipated. Despite advances in automation, implementation remains labor intensive and utilizes costly materials. An additional caveat must be noted. The types of leukocytes that contaminate these preparations varies with the method, and in most cases, they have only been poorly characterized (Bodensteiner 1989). Thus, a method that depletes the vast majority of polymorphonuclear leukocytes but concentrates mononuclear cells (monocytes and lymphocytes) would be unacceptable. Thus, the use of on-line filters to remove leukocytes during platelet or red cell transfusions, though not completely effective in removing contaminating virus, may nonetheless provide an increment of increased safety, which may offset the substantial cost of the procedures.

The removal of virus free in the plasma by neutralizing antibodies would appear to be an attractive alternative procedure. The removal of hepatitis B virus from plasma products using high-titer antibodies has been the primary success of this technique. Studies on neutralizing antibodies reactive with HIV-1 indicate that while some may be active in vitro in preventing infectivity, in vivo neutralization of HIV-1 remains problematic. Nonetheless, the massive effort currently underway to develop an HIV-1 vaccine may lead to reagents with enhanced potential for virus neutralization in blood components.

The potential for the removal/inactivation of cell-associated virus using antibody neutralization techniques has not been fully explored and still represents a fruitful area of investigation. Neutralization techniques of this sort would depend on the recognition of cell-surface antigens specific to virus-infected cells, for example, virus-induced cellular antigens, cell-surface-bound viral antigens, or cellular antigens characteristic of virus-infected cells. For example, elimination of CD4+-positive cells from transfused units of blood might provide an incremental improvement. Clearly, more investigations and the preparation of high-titer, specific, human antibodies (natural or monoclonal) will be required to bring this approach to fruition.

The use of hydrolyzable chemical agents has several notable advantages: they are inexpensive and their use would not be labor- nor material-intensive. However, prior to routine use, there would need to be a fundamental change in society's attitude toward the use of compounds with carcinogenic potential.

Of all the virucidal methods described in this chapter, photoactive compounds have received the greatest attention. While they appear to have sufficient virucidal activity, it has yet to be shown that cells treated with photosensitizing dyes survive and function normally in the circulation. There appear to be several additional disadvantages to the use of dyes: 1) Since they bind to cellular membranes, substantial quantities will be transfused, provoking questions about their general toxicity and phototoxicity.

2) The reaction mechanisms all rely on oxidation. 3) The potential exists for the generation of neo-immunogens and possibly for a shortened circulatory half-life.

The irradiation protocols commonly used to inactivate lymphocytes in transfused blood do not create conditions sufficient for the inactivation of blood-borne viruses like HIV-1 or hepatitis B virus. Irradiation by ionizing radiation (gamma emitters, X-ray) seems possible in principle, but there are few reports detailing the efficacy of such methods on the inactivation of the relevant viruses in cellular components. Doses that inactivate viruses also have consequences for the biological activities of cells of the immune system, but this may not be a major problem when transfusing red cell and platelet concentrates.

Similar arguments can be made about the potential of UV-irradiation methods. The distinct advantage of these methods over techniques that employ photosensitizers or alkylating agents is simply that they do not require the addition of chemical agents. Once again, there is limited data on the inactivation of the relevant viruses in cellular blood products (Capon et al. 1990). A potential disadvantage of UV-irradiation techniques is that, under some conditions, virus replication or at least transcriptional activities may be induced rather than inactivated. Finally, there are significant effects of UV and ionizing irradiation on accessory cell function (Rich et al. 1987; Rouse et al. 1989). Thus, transfusion with components that have been UV irradiated could have consequences for immune modulation, especially in immunocompromised recipients.

Thus we are still at the early stages of discovery in this field. Absolute safety or near-absolute safety of transfused blood cell components is unlikely unless virucidal methods can be applied to cell components. Further investigations will be required in order to determine the usefulness of this approach, its beneficial potential, and the risks associated with its use.

REFERENCES

Akey, D.H., and Walton, T.E. (1985) *Applied Environ. Micro.* 50, 882–886.

Alter, H.J., Holland, P.V., Purcell, R.H., et al. (1972) *Ann. Intern. Med.* 77, 691–699.

Alter, H.J., Creagan, R.P., Morel, P.A., et al. (1988) *Lancet* 2, 1446–1450.

Averbeck, D. (1989) *Photochem. Photobiol.* 50, 859–882.

Barr-Sinoussi, F., Chermann, J.C., Rey, F., et al. (1983) *Science* 220, 868–871.

Bensasson, R., and Land, E.J. (1978) in *Photochemical and Photobiological Reviews,* vol. 3 (Smith, K.C., ed.), pp. 163–191, Plenum Publishing Corp., New York.

Bernard, J., Reveil, B., Naymon, I., et al. (1990) *AIDS Res. Human Retrovir.* 6, 243–249.

Bigbee, P.D., Sarin, P.S., Humphreys, J.C., et al. (1989) *J. Forensic Science* 34, 1303–1310.

Biedermann, N., and Katzenelson, E. (1976) *Water Res.* 10, 629–631.

Blumberg, B.S., Sutnick, A.I., and London, W.T. (1969) *J. Am. Med. Assoc.* 207, 1895–1896.

Bodensteiner, D.C. (1989) *Transfusion* 29, 651–653.

Bowden, G.T., Ossanna, N., and Hurd, E. (1986) *Chemico-Biological Interact.* 58, 333–344.

Brummelhuis, H.G., and Over, J. (1989) *Current Studies Hematol. Blood Trans.* 56, 128–137.

Button, L.N., DeWolf, W.C., Newburger, P.E., et al. (1981) *Transfusion* 21, 419–426.

Cadet, J., Decarroz, C., Wang, S.Y., and Midden, W.R. (1982) *Photochemistry Photobiology* 44 (suppl.), 126s (abstr.).

Cao, Y.Z., Friedman-Kien, A.E., Mirabelle, M., et al. (1990) *J. Acquired Imm. Def. Syndrome USA* 3, 195–199.

Capon, S.M., Sacher, R.A., and Deeg, H.J. (1990) *Transfusion* 30, 678–681.

Chang, G.T., Harvey, R.G., and Weiss, S.B. (1981) *Biochem. Biophys. Res. Comm.* 100, 1337–1346.

Cole, M., Stromberg, R., Friedman, L., et al. (1989) *Transfusion* 29 (suppl.), 42s (abstr.).

de Graan-Hentzen, Y.C.E., Gratama, J.W., Mudde, G.C., et al. (1989) *Transfusion* 29, 757–760.

Easton, T.E., Valinsky, J.E., and Reich, E. (1978) *Cell* 13, 475–486.

Fitzpatrick, T.B., and Pathak, M.A. (1959) *J. Invest. Dermatol.* 32, 229–231.

Foote, C.S. (1968) *Science* 162, 963–970.

Foote, C.S. (1984) in *Porphyrin Localization and Treatment of Tumors* (Dolron, D.R., and Gomer, C.J., eds.), pp. 3–18, Alan R. Liss, Inc., New York.

Freeberg, J.K., and Carpendale, M.T. (1988) *OzoNachrichten* 7, 1 (abstr.).

Gamper, H.B., Straub, K., Calvin, M., and Bartholomew, J.C. (1980) *Proc. Nat. Acad. Sci. USA* 77, 2000–2004.

Gilbert, G.L., Hayes, K., Hudson, I.L., et al. (1989) *Lancet* 1, 1228–1231.

Goudsmit, J., Debouch, C., Meloen, R.H., et al. (1988) *Proc. Nat. Acad. Sci. USA* 85, 4478–4482.

Gulliya, K.S., Pervaiz, S., Chanh, T.C., Newman, J., and Matthews, J.L. (1990) *Photochemistry Photobiology* 51 (suppl.), 65s (abstr.).

Habibi, B., and Garretta, M. (1990) *Lancet* 335, 855–856.

Hanson, C.V. (1983) in *Medical Virology II* (de la Maza, L.M., and Peterson, E.M., eds.), pp. 45–79, Elsevier Science Publishing Co., New York.

Harakeh, M.S., et al. (1985) *Science Engineer.* 6, 235–243.

Hartman, F.W., Mangun, G.H., Feeley, N., and Jackson, E. (1949) *Proc. Soc. Exp. Biol. Med.* 70, 248–254.

Henderson, B.W., and Bellinier, D.A. (1989) *CIBA Foundation Symp.* 146, 112–125.

Horowitz, B. (1990) *Yale J. Biol. Med.* 63, 361–369.

Horowitz, B., Williams, B., Rywkin, S., et al. (1991) *Transfusion* 31, 102–108.

Hudson, J.B., Lopez-Bazzochi, I., and Towers, G.H.N. (1990) *Photochemistry Photobiology* 51 (suppl.), 52s (abstr.).

Judy, M.M., Matthews, J.L., Newman, J.T., et al. (1991) *Photochemistry Photobiology* 53, 101–107.

Kitchen, A.D., Mann, G.F., Harrison, J.F., and Zuckerman, A.J. (1989) *Vox Sang* 56, 223–229.

Konrad, H. (1985) in *Medical Applications of Ozone* (LaRaus, J., ed.), pp. 140–146, International Ozone Association, Norwalk, CT.

Lamola, A.A., Yamane, T., and Trozzolo, A.M. (1973) *Science* 179, 1131–1133.

Lavie, G., Valentine, F., Levin, B., et al. (1989) *Proc. Nat. Acad. Sci. USA* 86, 5963–5967.

Leitman, S.F., and Holland, P.V. (1985) *Transfusion* 25, 293–300.

Lin, L., Wiesehahn, G.P., Morel, P.A., and Corash, L. (1989) *Blood* 74, 517–525.

Lockhart, M.L., Ungers, G.E., Deutsch, J.F., et al. (1986) *Chemico-Biological Interact.* 58, 217–231.

LoGrippo, G.A. (1959) *Ann. NY Acad. Sci.* 83, 578–594.

Mattassi, R., et al. (1985) in *Medical Applications of Ozone* (LaRaus, J., ed.), pp. 134–137, International Ozone Association, Norwalk, CT.

Matthews, J.L. (1988) *Transfusion* 28, 81–83.

Mayer, K., and Pindyck, J. (1988) in *AIDS: Etiology, Diagnosis, Treatment, Prevention,* 2nd ed. (DeVita, V.T., Hellman, S., and Rosenberg, S.A., eds.), pp. 375–384, Lippincott, New York.

Menitove, J.E. (1988) *Transfusion Med. Rev.* 2, 65–75.

Meruelo, D., Lavie, G., and Lavie, D. (1988) *Proc. Nat. Acad. Sci. USA* 85, 5230–5234.

Meryman, H.T., and Hornblower, M. (1986) *Transfusion* 26, 101–106.

Moroff, G., Benade, L.E., Dabay, M., et al. (1989) *Transfusion* 29 (suppl.), 42s (abstr.).

Nakashima, H., Koyangi, Y., Horada, S., and Yamamoto, N. (1986) *J. Invest. Dermatol.* 87, 239–243.

Nelson, J.S., Liaw, L.H., Lahliem, R.A., et al. (1990) *J. Nat. Cancer Inst.* 82, 868–873.

Neyndorff, H.C., Bartel, D.L., Tufaro, F., and Levy, J.F. (1990) *Transfusion* 30, 485–490.

Okochi, K. (1987) in *AIDS: The Safety of Blood and Blood Products* (Petricciani, J.C., Gust, I.D., Hoppe, P.A., and Krijnen, H.W., eds.), p. 231–233, John Wiley and Sons, New York.

Palker, T.J., Clark, M.E., Langlois, A.J., et al. (1988) *Proc. Nat. Acad. Sci. USA* 85, 1932–1936.

Poiesz, B.J., Ruscetti, F., Gazdar, A.F., et al. (1980) *Proc. Nat. Acad. Sci. USA* 77, 7415–7419.

Popovic, M., Sarngadharan, M.G., Read, E., and Gallo, R.C. (1984) *Science* 224, 497–500.

Prince, A.M. (1968) *Proc. Nat. Acad. Sci. USA* 60, 814–821.

Prince, A.M., Stephan, W., and Brotman, B. (1983) *Rev. Infectious Diseases* 5, 92–107.

Prince, A.M., Horowitz, B., Baker, L., et al. (1988) *Proc. Nat. Acad. Sci. USA* 85, 6944–6948.

Prodouz, K.N. (1989) *Transfusion* 29 (suppl.), 42s (abstr.).

Prodouz, K.N., Fratantoni, J.C., Boone, E.J., and Bonner, R.F. (1987) *Blood* 70, 589–592.

Raab, O. (1900). *Zeitschrift Biol.* 39, 524.

Rasheed, S., Gottlieb, A.A., and Garry, R.F. (1986) *Virology* 154, 395–400.

Rawal, B.D., Busch, M.P., Endow, R., et al. (1989) *Transfusion* 29, 460–462.

Rich, E.A., Elmets, C.A., Fujiwara, H., et al. (1987) *Clin. Exp. Immunol.* 70, 116–120.

Riverberi, R., and Menini, C. (1990) *Vox Sang* 58, 188–191.

Rosen, A., Taylor, D.M., and Darai, G. (1987) *Int. J. Radiat. Biol. Related Stud. Phys. Chem. Med.* 52, 795–804.

Rouse, B.T., Hartley, D., and Doherty, P.C. (1989) *Viral Immunol.* 2, 69–78.

Rubinstein, A.I., and Rubinstein, D.B. (1990) *Transfusion* 30 (suppl.), 408a (abstr.).

Schrier, R.D., Nelson, J.A., and Oldstone, B.A. (1985) *Science* 230, 1048–1051.

Shoendorfer, D.W., Hansen, L.E., and Kenny, D.M. (1983) *Transfusion* 23, 182–189.

Sieber, F., and Krueger, G.J. (1989) *Semin. Hematol.* 26, 35–39.

Sieber, F., O'Brien, J.M., Krueger, G.F., et al. (1987) *Photochem. Photobiol.* 46, 707–711.

Sieber, F., O'Brien, J.M., and Gaffney, D.K. (1990) *Photochemistry Photobiology* 51 (suppl.), 96s (abstr.).

Singer, C.R.J., Azim, T., and Sattentau, Q. (1988) *Brit. J. Hematol.* 70 (suppl.), 111 (abstr.).

Sirchia, G., Rebulla, P., Paravicini, A., et al. (1987) *Transfusion* 27, 402–405.

Smith, O.M., Traul, D.L., McOlash, L., and Sieber, F. (1990) *Photochemistry Photobiology* 51 (suppl.), 67s (abstr.).

Spikes, J.D., and Straight, R. (1967) *Ann. Rev. Phys. Chem.* 18, 409–436.

Stein, B., Rahmsdorf, H.J., and Steffan, A. (1989a) *Mol. Cell Biol.* 9, 5169–5181.

Stein, B., Kramer, M., Rahmsdorf, H.J., et al. (1989b) *J. Virol.* 63, 4540–4544.

Tegtmeier, G.E. (1989) *Arch. Pathol. Lab. Med.* 113, 236–245.

Valerie, K., Delers, A., Bruck, C., et al. (1988) *Nature* 333, 78–81.

Valinsky, J.E., Easton, T.E., and Reich, E. (1978) *Cell* 13, 487–499.

Vanderpoel, C.L., Reesink, H.W., Schaasberg, W., et al. (1990) *Lancet* 335, 558–560.

van Marwijk-Kooy, M., van Prooijen, H.C., Borghuis, L., et al. (1990) *Transfusion* 30, 34–38.

Weiner, A.J., Kuo, G., Bradley, D.W., et al. (1990) *Lancet* 335, 1–3.

Wells, C.F. (1962) *J. Chem. Soc.* 1962, 3100–3104.

Williams, B., Horowitz, B., Geacintov, N., and Valinsky, J.E. (1988) *Blood* 72, 287a (abstr.).

Zee, Y.C. (1986) U.S patent no. 4,632,980.